STATISTICS
An Intuitive Approach

2. The percentile rank of the term X.

$$PR_X = \left[\frac{C + \left(\frac{X - L}{i} \right) F}{N} \right] 100 \tag{6.2}$$

E. Standard Deviation of Means of Samples of Size N

$$\sigma_{\bar{X}} = \frac{\sigma}{\sqrt{N}} \tag{8.1}$$

F. Testing Hypotheses Using Normal Curve Areas

1. Value of z for binomial variable.

$$z = \frac{(O \pm .5) - NP}{\sqrt{NP(1 - P)}} \tag{11.1}$$

2. Value of z for sample mean.

$$z = \frac{\bar{X} - \mu}{\frac{\sigma}{\sqrt{N}}} \tag{11.2}$$

G. Estimation

1. Unbiased estimate of population variance from ungrouped data.

$$s^2 = \frac{\Sigma (X - \bar{X})^2}{N - 1} \tag{12.1}$$

$$s^2 = \frac{\Sigma X^2 - \frac{(\Sigma X)^2}{N}}{N - 1} = \frac{N \Sigma X^2 - (\Sigma X)^2}{N(N - 1)} \tag{12.2}$$

2. Unbiased estimate for population variance from grouped data.

$$s^2 = \frac{\Sigma FX^2 - \frac{(\Sigma X)^2}{N}}{N - 1} \tag{12.3}$$

3. 95% confidence interval for the mean.

$$\bar{X} - 1.96 \frac{\sigma}{\sqrt{N}} < \mu < \bar{X} + 1.96 \frac{\sigma}{\sqrt{N}} \tag{12.4}$$

4. 99% confidence interval for the mean.

$$\bar{X} - 2.58 \frac{\sigma}{\sqrt{N}} < \mu < \bar{X} + 2.58 \frac{\sigma}{\sqrt{N}} \tag{12.5}$$

H. The t Distribution

1. Estimate of standard error of mean.

$$s_{\bar{X}} = \frac{s}{\sqrt{N}} \tag{13.2}$$

2. Value of t.

$$t = \frac{\bar{X} - \mu}{s_{\bar{X}}} \tag{13.3}$$

3. Estimate of standard deviation of distribution of differences of means of samples.

$$s_d = \sqrt{\frac{(N_1 - 1)s_1^2 + (N_2 - 1)s_2^2}{N_1 + N_2 - 2}} \sqrt{\frac{1}{N_1} + \frac{1}{N_2}} \tag{13.4}$$

Fourth Edition

STATISTICS An Intuitive Approach

George H. Weinberg
Consulting Psychologist, New York City

John A. Schumaker
Professor of Mathematics, Rockford College

Debra Oltman
Department of Mathematics, Oral Roberts University

Brooks/Cole Publishing Company
Monterey, California

Brooks/Cole Publishing Company
A Division of Wadsworth, Inc.

Printed in the United States of America

10 9 8 7 6 5 4 3 2 1

Library of Congress Cataloging in Publication Data

Weinberg, George H
Statistics: an intuitive approach.

Includes index.
1. Statistics. I. Schumaker, John Abraham, joint author. II. Oltman, Debra, joint author. III. Title.
QA276.12.W4 1981 001.4'22 80-18474
ISBN 0-8185-0426-9

Project Development Editor: *Ray Kingman*
Production Coordinator: *Marilu Uland*
Production Editor: *Celia Hirsch*
Manuscript Editor: *Evelyn Tucker*
Design: *Marsha Cohen*
Illustrations: *Sirius Productions*
Production Service: *Cobb/Dunlop Publisher Services, Inc., New York*

Preface

This book is designed to provide a sound introduction to the various topics usually covered in a first course in statistics. It is well suited for use at either the undergraduate or graduate level. The primary emphasis is on proper insight into concepts; hence, only a minimal background in mathematics is required.

Our sincere intent is to impart to students a clear understanding without subjecting them to the drudgery often associated with such a course. By not posing problems until each new concept is thoroughly developed, we have minimized the ''math anxiety'' often encountered by many students. This approach, along with some carefully placed humor and examples set in numerous practical situations, will allow the student actually to enjoy this introduction to statistics.

Although we are greatly pleased by the success of the first three editions of this book, we feel that substantial improvements have been made during revision. Many of the changes assure that we remain abreast of current trends (increased emphasis on nonparametric methods, for example); others reflect the many excellent suggestions of instructors and students who have used the text. The good qualities of the previous editions, especially readability, have been retained.

One of the most visible changes in the Fourth Edition is the reorganization of topics. The first six chapters provide a general treatment of descriptive statistical methods. Chapters 7 and 8 concentrate on the normal distribution and the central limit theorems. In Chapter 9 is found an expanded discussion of probability, followed in Chapter 10 by an introduction to decision making and risk, one of the unique features of this text. Chapters 11 through 14 present the most common parametric statistical tests. Regression, correlation, and linear prediction are addressed in Chapters 15 through 17, and the final two chapters deal with nonparametric tests. A new feature in the Fourth Edition is the inclusion of four review sections placed at intervals throughout the book.

A number of new topics have been incorporated into this writing. Among them are a review of basic mathematical concepts, the method for finding percentile ranks from grouped data, chi square goodness-of-fit tests, the Mann-Whitney test, and the Wilcoxon signed-ranks test. Other areas have been expanded or improved. Much has been added to the sections on probability and the normal approximation to the binomial distribution. The chapter concerning decision making and risk now includes calculations of type-one and type-two errors for situations involving distributions of sample means. An additional step has been added to the formal hypothesis testing procedure, and there is a new emphasis on interpretation of results. The entire six-step test is presented in many examples.

All exercise sets have been redone. There are two sets of problems at the end of each chapter, beginning with Chapter 2. Each set gives full coverage of the topics in the chapter. The problem sets are parallel; they provide comparable exercises in both sets. Answers for the odd-numbered problems are provided; solutions for the even-numbered problems are available to the instructor. There are enough exercises to serve both as study aids for the student and as diagnostic tools for the instructor. The situations encountered in the problems reflect the wide applicability of statistical methods; this should reinforce the course's value in the mind of the serious student.

For carefully reviewing this and the previous edition for revision purposes, we wish to thank: Henry Africk, New York City Community College; William N. Bailey, Mississippi College; Stephen F. Bollinger, Jackson State University; Charles E. Cliett, East Carolina University; Gary Davis, Southwest Texas State University; Michael Halliwell, California State University (Long Beach); Bud L. Kintz, Western Washington University; Dennis Leitner, Southern Illinois University; Eric S. Lubot, Bergen Community College; Dennis L. Nunes, St. Cloud State University; and Terry L. Watt, Cities Service Oil Company. Each comment and suggestion was examined, and we incorporated many of them into the text. And again we wish to express our gratitude to the many instructors and students who supplied comments on the manuscripts and published versions of the First, Second, and Third Editions.

We are also grateful to Sir Ronald A. Fisher, F.R.S., Cambridge, to Dr. Frank Yates, F.R.S., Rothamsted, and to Messrs. Oliver and Boyd, Edinburgh, for permission to reprint, in abridged form, Tables III and IV from their book *Statistical Tables for Biological, Agricultural, and Medical Research;* to the Institute of Mathematical Statistics for permission to reprint, in abridged form, a table from page 66 of volume 14 of the *Annals of Mathematical Statistics;* and to Professors George W. Snedecor and William G. Cochran and the Iowa State University Press for permission to use the *F* table from their book *Statistical Methods, Sixth Edition*.

Mrs. Oltman would like to thank her husband Bob, her daughter Betsy, and her parents for the patience and interest they exhibited during this writing. There is no substitute for a supportive family. The guidance and help of Ray Kingman and the rest of the Brooks/Cole staff have certainly made this an exciting and rewarding adventure. And finally, a word of thanks to a very special friend, John Block. His interest in this project and many words of encouragement greatly contributed to the final work.

George H. Weinberg
John A. Schumaker
Debra Oltman

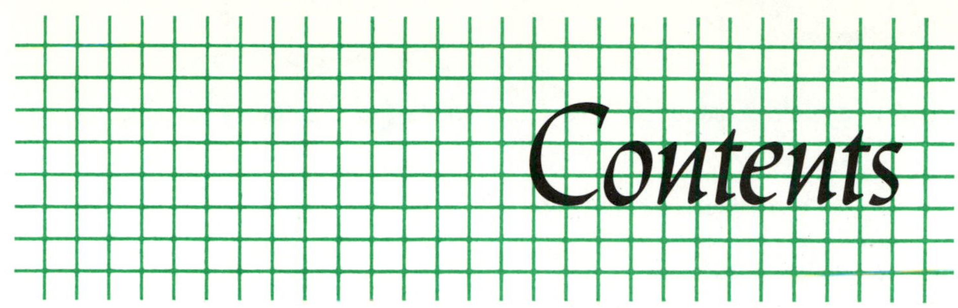

Contents

1 Orientation and Introduction — 1

1.1 Statistics and You — 1
1.2 The History of Statistics — 2
1.3 Descriptive and Sampling Statistics — 4
1.4 The Random Sample and the Stratified Random Sample — 5
1.5 The Abuses of Statistics — 6
1.6 Skepticism and the Counterresistance to Statistics — 7
Problems — 9

2 The Mode, the Median, and the Mean — 11

2.1 Definitions — 11
2.2 Two Properties of the Mean — 14
2.3 How Changing the Terms Affects the Mean — 17
2.4 Some Applications of the Mean, the Median, and the Mode — 20
2.5 Notation — 21
Problems — 22

3 Variability and Two Measures of Variability — 28

3.1 Variability — 28
3.2 The Variance — 30
3.3 The Standard Deviation — 32
3.4 How Changing the Terms in a Distribution Affects the Standard Deviation and the Variance — 34
3.5 Applications of the Variance — 36
3.6 Notation — 36
3.7 Computational Formula for the Variance — 39
Problems — 40

4 Ways to Describe the Position of a Term in a Distribution 46

4.1 Percentile Ranks and Percentiles 46
4.2 z Scores 49
4.3 z Score Averages 52
4.4 No Inherent Relationship between z Scores and Percentile Ranks 53
4.5 The Mean and Standard Deviation of a Set of z Scores 55
4.6 Standard Scores or T Scores 56
4.7 Applications 57
4.8 Notation 57
Problem 58

5 Grouping Data and Drawing Graphs 64

5.1 Discrete and Continuous Variables 64
5.2 Tabulating and Graphing the Values of a Discrete Variable 65
5.3 Rounding Off 67
5.4 Grouping Data–The Discrete Case 69
5.5 Grouping Data–The Continuous Case 72
5.6 The Curve as an Approximate Graph 75
5.7 The Meaning of Area 77
5.8 Data That Account for the Shape of a Graph 78
5.9 Applications 80
Problems 81

6 Computing Various Measures from Grouped Data 89

6.1 Introduction 89
6.2 The Median and Percentiles from a Histogram 89
6.3 Computation of the Median and Other Percentiles 92
6.4 Quartiles and Deciles 94
6.5 Percentile Ranks from Grouped Data 94
6.6 The Cumulative Curve 95
6.7 The Mode for Grouped Data 97
6.8 The Mean for Grouped Data–Direct Method 97
6.9 The Mean for Grouped Data–Coded Method 99
6.10 The Variance and the Standard Deviation for Grouped Data–Direct Method 103
6.11 The Variance and the Standard Deviation for Grouped Data–Coded Method 105
Problems 107

Review I 113

7 The Normal Distribution 122

7.1 Theoretical Distributions 122
7.2 The Normal Distribution and the Normal Curve 122
7.3 Four Properties of the Normal Distribution 124
7.4 Making Inferences about Terms Using the Normal Distribution Table 126
7.5 The Distribution of Shots 132
7.6 Determining Scores from Percentile Ranks 135
7.7 Applications 136
Problems 137

8 Distributions of Sample Sums and Sample Means 144

8.1 The Central Limit Theorem 144
8.2 How Nature Uses Theorem 8.1 146
8.3 The Central Limit Theorem for Sample Means 147
8.4 Properties of the Distribution of Sample Means 149
8.5 Locating a Sample Mean in the Distribution of Sample Means 152
Problems 156

9 Probability 162

9.1 Probability 162
9.2 The Word "Probability" in Everyday Language 164
9.3 Sample Spaces and Events 165
9.4 Probability and the Sampling Experiment 167
9.5 Probability and the Normal Distribution 168
9.6 Probability Statements about Sample Means 171
9.7 The Normal Approximation to the Binomial Distribution 173
Problems 176

10 Decision Making and Risk 181

10.1 Introduction 181
10.2 Hypothesis Testing 183
10.3 The Long-Run Outcomes with a Fair and a Biased Coin 186
10.4 Computing Errors Using Distributions of Sample Means 191
10.5 A Final Word on Risk 194
Problems 195

Review II 200

11 The Hypothesis-Testing Procedure 208

11.1 Introduction 208
11.2 The Null Hypothesis 209
11.3 When to Reject a Null Hypothesis; The Significance Level 210
11.4 Three Hypothesis-Testing Problems 217
11.5 Some New Language 222
11.6 One-Tail Tests 222
Problems 226

12 Estimation 230

12.1 Introduction 230
12.2 Unbiasedness of an Estimate 230
12.3 The Concept of Degrees of Freedom 235
12.4 Estimating the Variance 237
12.5 Computational Formulas for s^2 241
12.6 Interval Estimation 241
12.7 Interval Estimation of the Population Mean 243
Problems 247

13 The t Distribution and t Tests 252

13.1 Introduction 252
13.2 The t Distribution 255
13.3 The t Test for the Difference between Means 259
13.4 Application of the t Test for the Difference between Means 263
13.5 The t Test for Matched Groups 267
Problems 271

14 The F Test and Analysis of Variance 278

14.1 Introduction 278
14.2 The Theoretical Model of the F Distribution 279
14.3 Comparing More Than Two Sample Means 286
14.4 Introduction to Analysis of Variance 287
14.5 The Technique of Analysis of Variance 293
Problems 298

Review III **306**

15 Regression and Prediction 316

15.1 Introduction 316
15.2 Blind Prediction 317
15.3 Predictor and Predicted Variables 320
15.4 Regression 329
Problems 332

16 Correlation 337

16.1 The Concept of Correlation 337
16.2 Correlation Construed as the Relationship between Two Sets of z Scores 346
16.3 The Correlation Coefficient 350
16.4 Computing the Correlation Coefficient 353
16.5 The Test of Significance for a Correlation Coefficient 357
Problems 359

17 Correlation and Linear Regression 364

17.1 Introduction 364
17.2 The Geometric Meaning of the Correlation Coefficient 367
17.3 The Best-Fitting Line in the Scattergram of Original Data 370
17.4 The Best-Fitting Line and Prediction 374
17.5 Correlation and the Accuracy of Linear Prediction 379
Problems 382

18 Chi Square Tests for Independence and "Goodness-of-Fit" 387

18.1 Introduction 387
18.2 Independence 388
18.3 Illustrations of Relatedness 389
18.4 Presentation of Findings in a Contingency Table 391
18.5 Expected Frequencies 392
18.6 The Chi Square Test 394
18.7 Chi Square for More Complex Problems 397
18.8 Chi Square and "Goodness-of-Fit" 401
Problems 404

19 Nonparametric Statistical Tests 409

19.1 Introduction 409
19.2 The Mann-Whitney Test 410
19.3 The Sign Test–"Before" and "After" Data 413
19.4 The Wilcoxon Signed-Ranks Test–Naturally Paired Samples 416
19.5 The Rank Correlation Test–Alternative to Significance of Pearson r 419
19.6 The Runs Test for Randomness 423
Problems 427

Review IV 434

Appendixes A1

Index A30

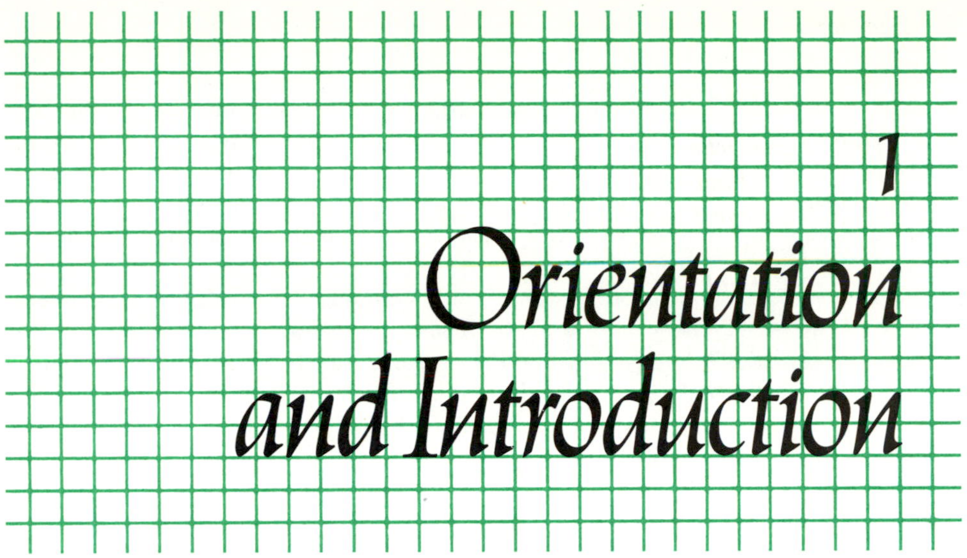

1 Orientation and Introduction

1.1 Statistics and You

You, the reader, are very possibly coming to this book with little or no previous knowledge of the subject. It may be that the very word *statistics* conjures up a confusing mass of data, tables, and charts. You've heard news analysts, politicians, and even sports announcers give statistics as if they were highly informative. There are statistics on the cost of living, on population growth, and on the batting averages of baseball players. In fact, there seem to be statistics on virtually every subject of interest these days.

To those without previous training such arrays of numbers can be discouraging to view. However, some of the original figures or data have already been condensed into more meaningful and useful figures, such as "indexes" of the cost of living and baseball "averages." These numbers are also appropriately called "statistics." In fact, the singular, *statistic*, may properly be used to refer to a specific value of one of these measures. This is a usage that we shall develop more fully later in the text.

The word *statistics* is also frequently used as a short way of saying "statistical methods." **Statistical methods** may be described as methods used in the collection, presentation, analysis, and interpretation of data. An example would be the methods used in going from the data on each separate time "at bat" for a baseball player to the player's final "batting average."

We shall be concerned primarily with developing an understanding of the basic statistical concepts and methods. Most of these methods may be understood in the light of common-sense reasoning, and it is our major purpose to show how this reasoning can be done. In fact, *statistical methods may be described as an application of common-sense reasoning to the analysis of data.*

Statistical methods are essential whenever useful information is to be distilled from large masses of data. You may be embarking on a field of endeavor in which you will need to derive useful results from data that you or others have collected. On the other hand, you may not foresee that you will ever have to do this and, under these circumstances, you may well ask the value of a knowledge of statistical concepts. We would call your attention to two major reasons why you should become statistically literate.

The literature of most professional fields contains the results of experiments conducted by researchers in the field. These published reports often use the language of statistics in such a way that they require statistical literacy on the part of any reader who is to make profitable use of them. Such literacy may be essential to a decision as to whether a newly tested method should be used in the reader's own practice in a particular professional field.

One need not turn to professional literature to find widespread use of statistics. The daily newspapers and the popular magazines often use statistical terminology, especially in presenting "averages" and graphs. In many instances statistics are not only used but also abused, as we shall see in a later section. The alert and conscientious citizen needs to be an *intelligent consumer* of statistics when reading such material.

This need for statistical literacy has been kept constantly in mind in the writing of this book. Our goal has been to present the major concepts in statistical method in a manner that will create the understanding needed for such literacy.

It is, of course, necessary in the study of elementary statistics to use a certain amount of arithmetic. A review is presented in Appendix VIII for the reader who would feel more comfortable beginning a study of statistics after reading a short summary of the basic arithmetic and algebraic operations. Symbolic notation is kept to a minimum in this text and is carefully explained when introduced. This book is in no sense a mathematical treatise. We simply want to help you become statistically literate.

1.2 The History of Statistics

The development of statistics has been like that of language itself. Its origins are ancient. For instance, the use of the mean was well known at the time of Pythagoras, and mention of statistical surveys was made in biblical times. Like language, statistics developed gradually where it was needed. As society became more complex, there developed more demand for accurate summary statements and inferences made in numerical form. The discipline of statistics has in recent years gained momentum both in its mathematical development and through its many applications in new fields.

The modern development of statistics began in the 16th century, when governments of various western European countries became interested in collecting information about their citizens. By the 17th century, surveys that closely resemble our modern census were already being conducted. By that time, insurance companies were beginning to thrive and were already compiling mortality tables to determine their life insurance rates. Different vested interests were slowly turning toward enterprises that necessitated the treatment of data and that demanded the use of statistics.

The mathematical basis of statistics is certainly not the subject matter of this book, but its existence, like the engine of an automobile, is vital. Isaac Newton (1642–1727), whose contribution to the invention of calculus was an outstanding event in mathematics, was perhaps the most necessary figure for the development of modern statistics, although Newton could scarcely have heard of the subject. Other mathematicians whose contributions have been primarily in the field of pure mathematics have done more for the development of statistics indirectly than many of those whose names are associated specifically with the field of statistics itself. Perhaps the two most prominent of these mathematicians were Abraham DeMoivre (1667–1754) and Carl Gauss (1777–1855).

As for the statisticians themselves, Adolph Quetelet (1796–1874), a Belgian, was the first to apply modern methods to collected data. Quetelet is sometimes referred to as the father of modern statistics, less because of his contributions than because of his continued emphasis on the importance of using statistical methods. After studying with the best-known mathematicians of his day, Quetelet established a Central Commission for Statistics, which became a model for similar organizations in other countries.

Unexpected as it sounds, Florence Nightingale (1820–1910) was an ardent proponent of the use of statistics. She argued that the administrator could be successful only if guided by statistical knowledge and that both the legislator and the politician often failed because their statistical knowledge was insufficient.

Two other important contributors to statistics were the Englishmen Sir Francis Galton (1822–1911) and Karl Pearson (1857–1936). Galton, a cousin of Charles Darwin, became deeply interested in the problem of heredity, to which he soon applied statistical tools. Among other things, he developed the use of percentiles. Pearson made many statistical discoveries, and both Galton and Pearson contributed greatly to the development of correlation theory, which we shall consider in detail later.

The most prominent contributor to the field of statistics in the 20th century has been Sir Ronald Fisher (1890–1962). Fisher made numerous contributions during the period from 1912 until his death in 1962, and many of them have had great impact on contemporary statistical procedures.[1]

The 20th century has seen the birth and growth of formal statistics instruction in the United States. At the turn of the century only a handful of statistics courses were being given in all the colleges in the country. In fact, most of these were given in economics departments and by instructors whose focus made them unique among their colleagues. During the first 30 years of this century the emphasis on applications of statistics to problems of psychology slowly increased; but it should be noted that during this period psychology itself had a questionable status and was frequently considered only as a branch of philosophy.

Shortly before World War II the number of applications of statistical methods in

[1]The reader who wants an interesting account of the history of statistics up to 1929 should consult *Studies in the History of Statistical Method*, by Helen M. Walker (Baltimore: Williams & Wilkins Co., 1929).

the social sciences began to increase. The number of surveys of all kinds increased, and the need to interpret data in psychology and education made it necessary for workers to have at least a basic understanding of statistics. One by one, statistics courses were added as requirements in psychology departments and in schools of education. Today the worker in any of the human behavior fields, such as psychology, sociology, or anthropology, is expected at least to have what is called *statistical literacy*. In fact, it is virtually impossible to major in any of these fields without having to take at least one statistics course.

One reason for the rapid growth of statistics in recent years is the increasing ease of processing large masses of data. Modern electronic computers make it possible to analyze in a short time huge collections of data that would have been entirely unmanageable just a few years ago. Such data can now be analyzed and the results made available for public and professional consumption while they are still pertinent to the current situation. If you should ever have to make an extensive statistical survey, you will undoubtedly enlist the aid of modern methods of computation. Familiarity with a desk calculator is useful for less extensive computations.

Because of computers, there is much less need for an emphasis on the computational aspect in the teaching and learning of elementary statistical concepts. The time is much more profitably spent in furthering a real understanding of the ideas involved. On the other hand, practice in working with the various statistical measures is actually helpful in developing this understanding. The problems at the ends of chapters in this book are designed for this purpose.

1.3 Descriptive and Sampling Statistics

A given set of values or measures of the same thing is called a **distribution**. For example, the scores on a particular examination constitute a distribution. Most of the discussion in this book is concerned with various types of distributions and their properties.

Two main approaches comprise the subject matter of statistics. One is the **descriptive statistics** approach; the other is that of **sampling statistics**. The methods of descriptive statistics entail specifying a population of interest and then collecting the measurements of all the members of that population. These original measurements or scores are called **raw data**. Raw data themselves are descriptive, but the science of descriptive statistics deals with methods of deriving from raw data measurements that are more tersely descriptive of the original population. In fact, it is the type of measure once removed from the raw data that is of prime importance to the statistician and research worker. For instance, the *average* IQ of members of an army battalion is obviously much more comprehensible and meaningful than the lengthy list of IQ scores as they were originally obtained. But it almost goes without saying that an understanding of the exact meaning of an average is necessary to interpret an average in any particular case.

Let us emphasize what has been said. The descriptive statistical approach makes use of *all* the data concerning a population, and it entails deriving descriptive statistical measurements from data. The word *population* does not imply vastness;

for instance, the marks of 22 students in a section of an elementary statistics course might constitute a population of interest to us. Where our interest is wholly in this population of 22 cases, the average thus computed is a descriptive statement about a population. We are content that our interpretation of the average is a meaningful statement about the class itself. For instance, we might be able to say, "The average mark was 87, indicating that the class grasped the material." Note that we are making use of all the data comprising the population of interest, and, because our concern does not extend beyond the group we have measured, we must be said to be using descriptive statistics.

The approach of sampling statistics becomes relevant when we have access only to a sample drawn from the population of interest and when we do not have in our possession the raw scores comprising the total population. In the case described, we might have the same data for the 22 students comprising the elementary statistics course, but we might wish to generalize about the level of all students of elementary statistics in the college. We again have 22 raw scores in our possession, but now they comprise a sample drawn from a much larger population. Under this condition one might say, "The average score in this section was 87. Judging from this sample, I conclude that the students in the college are tending to grasp the material." A statement of this kind is quite different from the one cited earlier, for it is a generalization from a sample to a population that has not been fully surveyed.

The science of sampling statistics, or **statistical inference** as it is alternatively called, consists of procedures that tell us how much we may generalize and what kinds of statements we may make. Sampling statistics virtually always applies to problems of research, since our investigation is limited to a small group of subjects; but our desire is to make statements about a vast population of similar subjects, including even those as yet nonexistent.

In sum, the distinction is that the methods of descriptive statistics are those of describing a population given all the relevant data. The methods of sampling statistics are those of making inferences from a sample about a larger population that remains unseen but that yielded the sample. The next six chapters of this book deal exclusively with problems of descriptive statistics; the focus of the last 12 chapters is primarily on sampling problems.

1.4 The Random Sample and the Stratified Random Sample

The ability to do accurate research depends on at least some insight into statistical methods. It is becoming increasingly difficult to separate the domain of the research worker from that of the statistician. In particular, the issue of how to gather a sample is, strictly speaking, one for the research worker; but real insight into the problem of sampling depends on at least some appreciation of how the sample is to be treated and used.

A basic concept is that of **randomness**. Consider an illustration. Suppose we wish to get a fair sample of the opinions of all the students in a college. Obviously, it would not be satisfactory merely to question the first 20 students who come out of a room after a freshman English class or the first 20 students whom we might meet at the entrance to the mathematics building. If we are to generalize from a sample it must be random, and such samples would not be, since every member of the population does not have an equal opportunity of being included in the samples.

We can see, merely by applying common-sense reasoning, that our procedure should be such that no one student's opinion is given a better opportunity than any other's of being included in the sample. Such a sample—namely, **one that is gathered so that each member of the population is an equally likely candidate to be included in it—is called a *random sample.***

There are various techniques for gathering random samples. For instance, in the situation described one might put the name of each of the students on a separate card. Then one might shuffle all the cards in a large container and draw at random the number of cards desired for the sample. The final step would be to consider only those students selected and ask them for their opinions, which would constitute the random sample.

A variation of the technique described is that of collecting a **stratified random sample.** This technique ensures in advance that a sample will contain the same proportions of members of different groups as are in the larger population and thus be representative. For instance, in the situation described one might wish to ensure that the same fraction of juniors appears in the sample as is enrolled in the college. That is, if one-fifth of the students in the college are juniors, then one would randomly select exactly enough juniors to comprise one-fifth of the sample. The same procedure would be followed for each of the other classes and perhaps even for the different subject matter areas, if they were considered relevant to the opinions being surveyed.

The word *stratified* is applied to the method described because it entails actually going to different strata or levels in order to complete the sample. The procedure of stratifying a sample is necessary only where subgroups within the population have meaning for the data being gathered, in which case nonproportionality would make a sample nonrepresentative and misleading. The method of stratifying samples is used for such surveys as election polls, where a stratified sample is crucial since different subpopulations often manifest markedly different voting preferences.

1.5 The Abuses of Statistics

The abuses of statistics are many; insight into basic statistical concepts is the best defense against them. Abuses are frequently found in popular publications as well as in technical journals. Often they are the result of ignorance rather than design. For instance, as we shall see, there is quite a difference among the mean, the median, and the mode, although all three are apt to be described as *averages*. It often occurs, especially during disputes, that different vested interests each use the word "average" but have in mind different measures that were actually computed. For instance, where labor and management have clashed, leaders on each side have cited averages that were computed correctly but that were markedly different. Naturally, each of the vested interests cites the value that is most advantageous to its own case.

One of the most famous abuses of research methods involved a fallacious sampling procedure. The flaw was in the method of selecting cases, rather than in statistically treating the data incorrectly, and it was undoubtedly an actual error and not a purposeful distortion. The error was made in 1936 by the *Literary Digest* magazine during its preelection presidential poll. The *Digest* predicted that Alfred Landon would win easily over Franklin Roosevelt, but in the end Roosevelt carried 46 of the 48 states, many of them by a landslide. The *Digest* had made its error in choosing a sample from 10 million persons originally selected from telephone listings and from the list of its own subscribers. In the depression year of 1936, the persons who could afford magazine subscriptions or telephones did not constitute a random sample of the voters in the United States. The voters in the sample selected were predominantly for Landon, whereas the majority in the larger population was for Roosevelt. Incidentally, the *Literary Digest* rapidly lost status after the election, and, undoubtedly as a partial result of its error, it soon ceased to exist. In subsequent years election pollsters took careful note of how the *Digest* had blundered and developed stratified sampling techniques to ensure representativeness.

Abuses are often found even in cases in which extreme care has been exercised to assure a random sample. For example, data obtained from questionnaires and pollings may often be colored by the wording used in asking questions. In other cases the truthfulness of those responding is suspect. Timing also plays a crucial role in many research situations. Responses to the question "What is your favorite spectator sport?" may differ greatly from October (World Series time) to January (Super Bowl time). Information obtained from volunteer subjects might not reflect the feelings or opinions of the entire population even though technically every member of the population *could* have volunteered.

Obviously we are limited in our discussion of the abuses of statistics, since we have not as yet presented the subject matter of statistics. It is enough to say that nearly any statistical procedure may be done incorrectly, either through ignorance or by design. The reader who wishes to look further into the possible abuses of ordinary statistical methods should consult the little book entitled *How to Lie with Statistics*, by Darrell Huff (New York: W. W. Norton & Co., Inc., 1954).

1.6 Skepticism and the Counterresistance to Statistics

Quite naturally, the enormity of the push of statistics has mobilized what is almost a necessary self-protective attitude on the part of the individual. A new orientation is becoming quite common: "Keep statistics away from me." This is unfortunate. Although distortions are in fact often carried out with the aid of statistics, the fault is that individuals use inappropriate methods and it makes no sense to blame the statistical orientation itself.

There are several common objections to the use of statistics. They are (1) that statistics are untrustworthy since by using statistics one can prove anything, (2) that statistical thinking is unfair because it is antiindividual, and (3) that the use of statistics in connection with human problems is cold and unfeeling.

The distrust of statistics as a vehicle that can "prove" anything is undoubtedly a reaction to many current abuses. To go back, it is apparent that to conduct accurate research is a painstaking task. Designing an experiment and actually gathering data require careful planning. Improper procedures may alter results and make an experiment invalid. Obviously, at least some statistical knowledge is needed to interpret data. Therefore, when an organization that obviously has a vested interest omits the details of an experiment because of "time shortage" and merely reports a finding that favors its own product, one must certainly be curious about, if not suspicious of, the report.

It does appear quite remarkable that each of many organizations with competing products can conduct research and find from obtained data that its own product is superior. Obviously it is possible to "slant" research by varying data-gathering techniques or misinterpreting findings. The average person hears conflicting statistics reflecting obvious vested interests nearly every day and quite understandably comes to distrust statistical reports altogether. Thus one finds great distrust of statistical methods expressed by college students and by others, and no wonder.

The fact is that the science of statistics is an enormously powerful tool to facilitate learning. But, like any powerful instrument, it may be used to increase human effectiveness and happiness or it may be abused through conscious misapplication. No statistical method is in itself right or wrong, but each has its time and place. Large organizations, such as advertisers, sometimes deliberately misapply statistical methods. The best defense against being hoodwinked is education. Typically, the brief statements of findings issued by vested interests cannot be properly interpreted without considerably more information. Where enough details are given, there is great advantage in being able to evaluate them. In any event, where there are abuses, the situation is not that figures lie but that liars are apt to figure. The solution is for the intelligent reader to be able to figure too.

The next objection—that statistical thinking is unfair because there is no acknowledgment of the individual—is a subtle one that is often felt but seldom verbalized. As defined, statistics deals with properties in populations; but the great pride of modern society, particularly in American culture, is in the emphasis on the individual. The impulse is to say, "Don't tell me that the majority in my subculture is voting for the candidate whom I dislike." It is to say, "Don't tell me what the average person does or that my expectancy based on my education is to make 200 dollars a week." Statistics, with its emphasis on the population as opposed to the individual, seems to lead to statements that are subtly abrasive to individual freedom. The feeling behind the old maxim "Never generalize" is that to do so is unfair.

Once again the answer is that a general statement, if true, cannot itself be at fault. The statements that the majority in a community is voting for a particular candidate and that the average person in a subgroup makes 200 dollars a week do not imply that any individual must also vote for that candidate or make 200 dollars a week. A general statement that is true may be highly informative. Difficulty arises only when the listener mistakenly interprets such a general statement as if it had to be true for each individual in the group. In each of the cases cited, the listener's objection would undoubtedly be based on such a misinterpretation. It is the listener's own error, or desire to conform, that leads to the conclusion that one must operate as the majority or be in the majority.

The objection to general statements when they are true typically reflects either a misinterpretation of these statements or a fear of them. One has absolute proof of this where the listener replies by pointing out an "exception." Since the general statement is never an invariable one, there is no such thing as an exception. A listener who cites a contradictory instance is evidencing a mistaken translation of the general statement into an invariable one. The solution is to learn to acknowledge general statements one believes to be true and at the same time to appreciate that a general statement is not an attempt to deprive one of one's freedom. "Yes, I understand that the majority in my community is voting for that candidate, but I am not."

The third objection to statistics is that data themselves are numbers; they are "cold." The feeling is that "pure numbers" can never do justice to the material when human beings are concerned. Can numbers really describe human problems or suffering? The answer is that they cannot and that they are not intended to do so. But enumeration has led to social and medical discovery that has already been relevant in alleviating suffering. Numbers themselves obviously do not tell the whole story of humankind but only the enumerable part of it.

As for the complaint that statistics are "cold," one might consider what one's own position would be in the following situation. Suppose it were shown that a two-minute lie detector test leads to a higher percent of correct convictions and acquittals in murder trials than when live juries are used. In other words, an innocent person would have a better chance of escaping sentence by taking the test than by pleading the case before a jury. In such a circumstance, you might ask yourself whether, as an innocently accused person, you would prefer to be tried by the lie detector test or to plead your case, even though your chance of proving your innocence would be less with the jury. The question is obviously posed to contrast cold statistical findings with human judgment, and obviously the answer is not easy.

Fortunately, however, there would always be alternatives. Were the situation to occur, a careful statistical analysis would show where the lie detector test and the jury disagreed in their verdicts. This analysis might prompt further investigations into where juries make their mistakes so that they might be better educated in those areas. The point is that the "coldness" of statistics should not imply that they be disregarded, because the gathering and interpreting of data are procedures aimed at specific ends. Where one has both integrity and knowledge of correct procedures, the benefits are most likely to be great.

Problems

1.1 Listen for mention of the word *statistics* on radio and TV newscasts. In what context is the word used?

1.2 Look for references to *statistics* in newspapers and magazines. In what context is the word used?

1.3 Explain how baseball "batting averages" are computed. With respect to a player's batting average, discuss how many hits a player may be expected to make in the next 10 times at bat.

1.4 Find newspaper or magazine statements about "average" values. Is any mention made of the type of average being used?

1.5 Comment on a sales manager's remark that no salesperson should have a monthly sales amount below the average for that month.

1.6 Bring in any examples that you think illustrate abuses of statistical method and discuss them in class.

1.7 An advertising claim states, "Product X is preferred by 9 out of 10 men." State why you would or would not accept this as a valid claim to product superiority.

1.8 A television commercial proclaims that Product X is "now 30% better, faster acting." What criticism do you have of this commercial as "statistical proof of product superiority"?

1.9 Bring in the latest survey or poll result from the Sunday newspaper. Is there any mention of the sampling technique used?

1.10 Discuss the success or failure of the predictions made about the outcome of the last presidential election.

1.11 Report on some abuse of statistical method discussed in *How to Lie with Statistics*.

1.12 Find out anything you can about the technique known as "area sampling."

1.13 Describe a method not mentioned in the text that might be useful in obtaining a fair random sample of the students in a college.

1.14 Find references to statistics and statistical data in a text in your major field.

1.15 Obtain more information about Quetelet, Galton, Pearson, Fisher, and Nightingale and their work or interest in statistics.

1.16 Consult the indexes in the library and ascertain the names of five governmental agencies that publish statistics.

1.17 Examine a daily report of New York Stock Exchange transactions. The usual data given for a particular stock are total shares traded, high price, low price, and closing price for the day. Is the closing price an average of the high and low prices? Can you calculate from the given data the dollar value of the shares traded for a particular stock? Why?

1.18 How would you proceed to obtain a sample of one-syllable words used in this textbook? How would you apply the sample results to cover the whole textbook?

1.19 Your college administration undertakes a study of the total hours a student devotes to homework per week. How would you select a suitable sample of students for this study?

1.20 A new type of electric light bulb is developed and produced in small quantities for experimental study. Suggest a suitable sampling procedure for determining the life of these new bulbs.

1.21 A local radio station conducts a one-hour call-in survey to determine whether city residents favor a new trash pick-up proposal. Of the 215 calls received, 198 are opposed. What criticism do you have of this procedure as statistical proof that the public opposes the new plan?

1.22 Which group(s) would be underrepresented in "random" samples taken from (a) telephone directories, (b) volunteers, (c) social registers, (d) compiled lists of credit card customers, (e) welfare rolls, and (f) student directories?

1.23 Explain how public opinion polls may influence the opinions they are trying to measure.

1.24 Find a newspaper article reporting on the results of a study or poll that refers to a "margin of error." What does "margin of error" mean in this context?

1.25 Explain why descriptive methods would be inappropriate for determining the breaking strength of concrete cylinders.

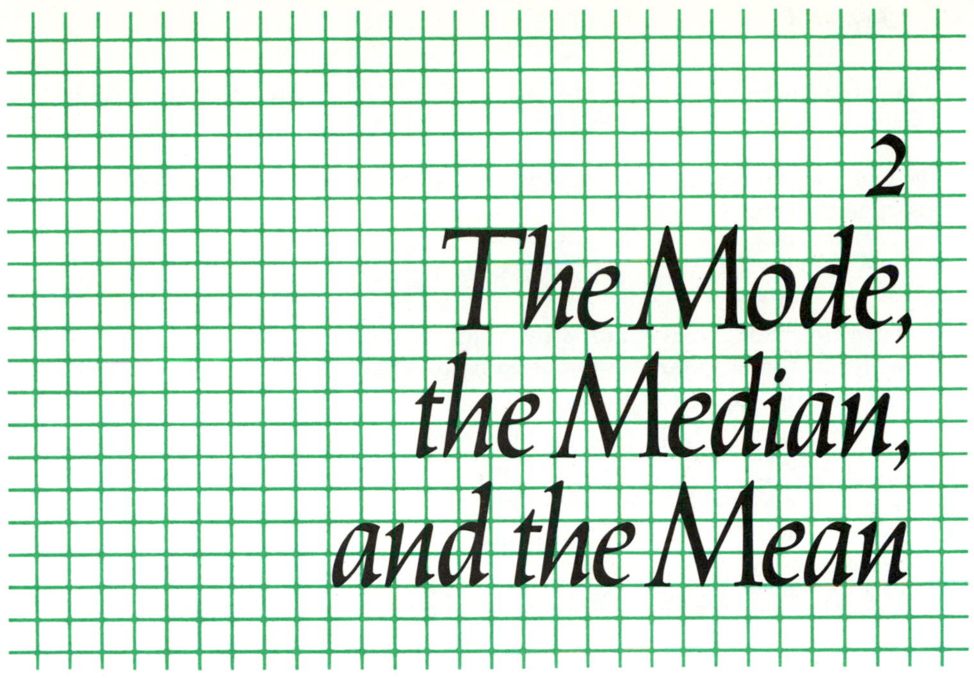

2 The Mode, the Median, and the Mean

2.1 Definitions

The most common statistical concepts to indicate the middle of a distribution are the **mean,** the **median,** and the **mode.** The best way to convey the meaning of these words is to illustrate them simply. Consider the following set of five terms or observations.

4
4
5
7
10

Definition 2.1 The mean is the value that is obtained by adding the terms and then dividing their sum by the number of terms.

In the example there are five terms and their sum is 30. The mean is 30 divided by 5, or 6.

$$\text{Mean} = \frac{30}{5} = 6$$

Definition 2.2 The mode is the value of the term that appears most frequently.

In the example, the mode is 4 since the term 4 appears twice and each of the other terms appears once.

Mode = 4

Definition 2.3 The median is the value of the term that is (or mean of the two terms that are) larger than or equal to half of the other terms and smaller than or equal to half of them.

In the example, the third term from the top, 5, is the median. The reason is that 5 is larger than two terms (4 and 4) and smaller than two terms (7 and 10).

Median = 5

For a distribution with an even number of terms like 4, 5, 5, 7, 10, and 11, there are two terms, 5 and 7, that are smaller than or equal to two terms (10 and 11) and larger than or equal to two terms (4 and 5). Their mean, ½(5 + 7) = 6, is the median of the distribution of six terms.

The mean is by far the most important of the three terms defined. To see why, consider the distribution of five terms again.

4	
4	Mean = 6
5	Median = 5
[7]	Mode = 4
10	

Suppose that the term inside the square were increased from 7 to 17. The terms would then be

4
4
5
17
10

The median is still 5 and the mode is still 4. But the mean of the new set of terms is now 8, not 6. The change has affected the mean but not the median or the mode. Certain changes would have affected the other two measures as well as the mean. For instance, adding 1 to the middle term would have increased the median to 6. However, the mean responds to *every* change whereas the median and the mode respond only to *some* changes. For this reason, **the mean is often described as "sensitive" and as "reflecting the entire distribution."** Its sensitivity seen in this way gives a hint of its importance. We shall see that the sensitivity of the mean may be a disadvantage to its representativeness when the distribution contains a few outlying terms at one extreme.

It is noteworthy that the mean of the original distribution was not actually a term in the distribution. That is, none of the five terms was a 6. The mean should be thought of as a statement about the distribution itself. The mean may or may not also be the same number as a term in the distribution.

For convenience, the terms in the distribution were presented in ascending order. That is, each term was equal to or larger than every term above it. Remember that any order might have been used, and, in practice, data are not ordered when collected. Here are some other orders.

10	7	10	[5]
4	[5]	[5]	4
[5]	4	7	10
7	10	4	4
4	4	4	7

Each of these columns represents the same distribution and in each case the median (in the square) is 5 by the definition given earlier. It is easy to make the mistake of thinking that the median in each case is the third number from the top. Remember that the definition does not say anything about placement on the page but specifies the median simply by its relative size.

There is little to say about identifying the mode. However, note that in some distributions there is no mode, because no one term occurs more than once.

3	2	9
5	5	3
2	7	2
0	3	1
8		0

A collection of terms may have more than one mode. Here is an example of a collection with two modes. It is said to have a *bimodal distribution*.

9
3
3
8
7
9
2

The two modes are 3 and 9, since each appears twice and no other number appears that often.

The mean, median, and mode are all types of averages, and there are others.

2.2 Two Properties of the Mean

We will now give our full attention to the mean. An important property of the mean is clear in a physical analogy. Imagine a long wooden plank with numbers on it at equal intervals from 4 to 10. Suppose five one-pound weights are placed on this plank in positions corresponding to the distribution of numbers already discussed: 4, 4, 5, 7, 10. This distribution is illustrated by placing two weights on 4, another on 5, another on 7, and another on 10 (see Fig. 2.1).

Fig. 2.1

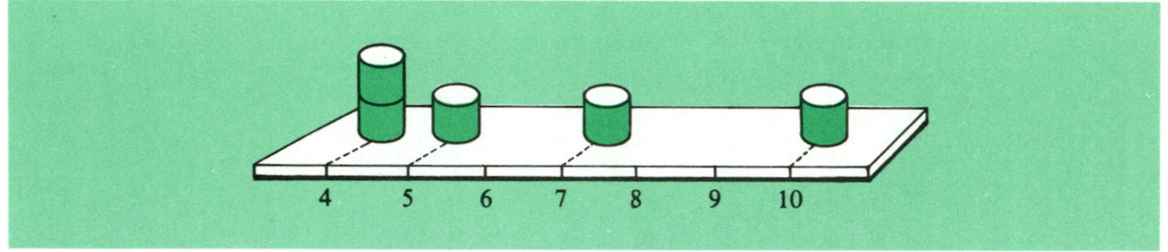

Now suppose we are given a fulcrum and told to balance the plank with its weights on this fulcrum, assuming the plank itself has no weight. After several attempts that fail (Fig. 2.2), we manage to find the balance point (Fig. 2.3).

Fig. 2.2

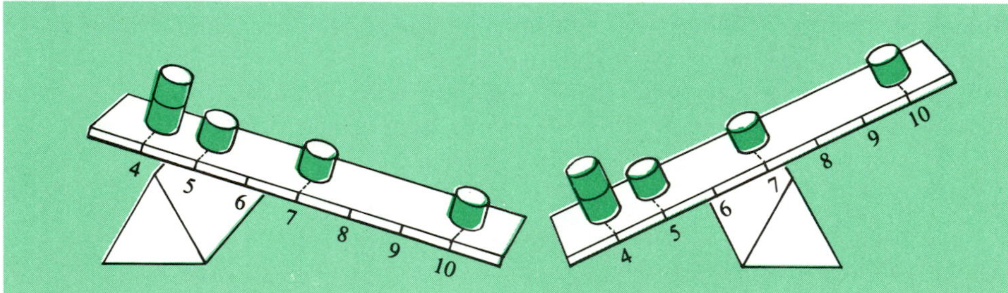

It turns out that the fulcrum has been placed under the mean of the terms. In Fig. 2.3 it is under the point 6. **The mean of any collection of terms is their balance point and, furthermore, the mean is the only balance point.**

Fig. 2.3

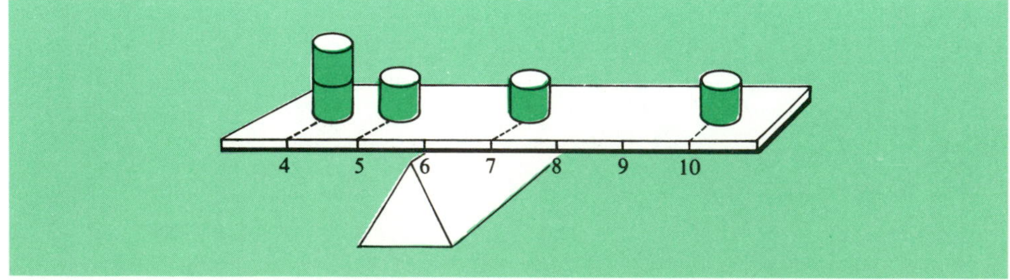

Consider the weights on the left side of the mean. Let us see how far from the fulcrum (mean) each of them is. The two weights on the extreme left are each 2 units away from the mean and the third is 1 unit away. The sum of these three distances is, therefore, 5. The weights on the right side also total 5 units away from the mean. It is this equivalence that balances the plank.

The sum of the distances of the weights from the mean is always the same on the left side as it is on the right side. The fact that the distances on either side of the mean "balance out" should be kept in mind. The way to obtain these distances is to subtract the mean from each of the terms. In Table 2.1 the five terms in the distribution are listed in the first column, and the mean, 6, has been written into the second column. The third column, called "Differences," has been found by subtracting the mean from each of the original terms. Instead of "differences," the word "distances" might have been used. Notice that some of the differences are negative numbers.

Table 2.1 Differences from Mean

Terms in Distribution	Mean	Differences
4	6	−2
4	6	−2
5	6	−1
7	6	+1
10	6	+4
		0 total of differences

When the differences are added, with attention paid to the plus and minus signs, the sum turns out to be zero. This makes sense if we think of each of the differences as a distance from the balance point, the mean. Differences with minus signs are results of weights on the left side of the mean and those with plus signs are results of weights on the right side.

The important facts to remember are that the sum of the differences from the mean equals zero and that the property of the differences totaling zero is *unique* to the mean. That the differences total zero means that no other number would have served as a balance point. For instance, in Table 2.2 the number 5 is tested as the balance point.

Table 2.2 Differences from Test Balance Point

Terms in Distribution	Test Balance Point	Differences
4	5	−1
4	5	−1
5	5	0
7	5	2
10	5	5
		5 total of differences

The sum of the differences is not zero. Using 5, the imaginary plank has fallen to one side.

The fact that the mean is the only number with the property described can help to find it. Just as one finds a balance point, in theory one may find the mean by trying different numbers until discovering the number that makes all the differences total zero.

For purposes of later reference, what has been said may now be summed up as a theorem.

THEOREM 2.1 In any distribution of terms, the sum of the differences from the mean equals zero. Conversely, if the sum of the differences from some number in a distribution equals zero, that number is the mean.

It should be noted that we have not proved Theorem 2.1 but have only illustrated it and given intuitive justification for it. We will often proceed in this way.

Now let us turn our attention to the set of differences that appeared in the third column of Table 2.1. These differences were

$$\begin{array}{r} -2 \\ -2 \\ -1 \\ 1 \\ 4 \end{array}$$

For our next theorem we square each of these differences, as shown in Table 2.3.

Table 2.3 Squares of Differences from Mean

Differences	Squares of Differences
−2	4
−2	4
−1	1
1	1
4	16
	26

The sum of the squares of the differences is 26. Note that these squares that we have just added were determined by the choice of the mean as a reference point. If we had chosen any reference point other than the mean, the sum of the squares of the differences from that point would have been larger than 26. In other words, to keep the sum of the squares of the differences from some point as small as possible we should choose the mean as the reference point.

What would have happened if we had chosen 5 as the reference point is shown in Table 2.4.

Table 2.4 Squares of Differences from Reference Point

Original Terms	5 as a Reference Point	Differences	Squares of Differences
4	5	−1	1
4	5	−1	1
5	5	0	0
7	5	2	4
10	5	5	25
			31

The sum of the squares of the differences has turned out to be larger than 26. We are now ready for a second theorem about the mean.

THEOREM 2.2 In any distribution of terms, the sum of the squared differences from the mean is less than the sum of the squared differences from any other point. Conversely, the point that minimizes the sum of the squared differences is the mean.

Obviously one makes no use of Theorem 2.2 in finding the mean. However, the "least squares" property of the mean has importance, as will be seen later.

2.3 How Changing the Terms Affects the Mean

Suppose now that the five terms in our distribution are the ages of the five Smith children, the mean age of a Smith child being 6 years. One might ask, "What will be the mean age of the Smith children 9 years from now?" The answer seems almost too obvious: In 9 years, the mean age of the Smith children will be 15 years. Obvious or not, this answer is correct.

Now, one more question. How did we arrive at this answer? The readers who did arrive at it will probably agree that their method was easier to use than to explain. To get the answer, we made use of a valuable property of the mean. It is essential to focus our attention on this property, for it will be helpful in less obvious problems.

The question was, what will be the mean age of the Smith children 9 years from now. One might have added 9 years to each of the terms in the distribution of ages of Smith children. In this way a new age would have been found for each child. The mean of these new ages would still have turned out to be 15. However, it is not necessary to use this method. We may correctly assume that when each child is 9 years older, the mean age of the children also increases by 9.

Our physical analogy suggests why this happens. Once more consider the five numbers 4, 4, 5, 7, 10 as weights on an imaginary plank. The fulcrum is the mean, 6. The plank, as usual, is considered to have no weight. Our problem tells us to consider each child as being 9 years older. To depict the ages, we move each weight 9 units to the right. The plank has been extended in Fig. 2.4 to show how the weights look in their new positions, with X's to denote their original positions.

Fig. 2.4

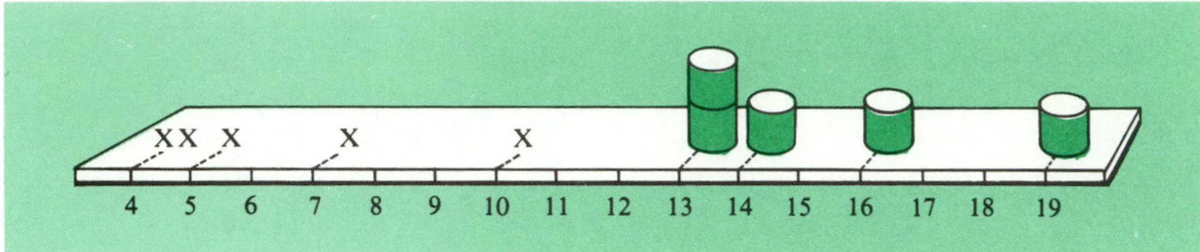

We may direct our attention to the portion of the plank that extends under the weights in their new positions. The new balance point is the new mean. This balance point is under the point 15 in Fig. 2.5.

Fig. 2.5

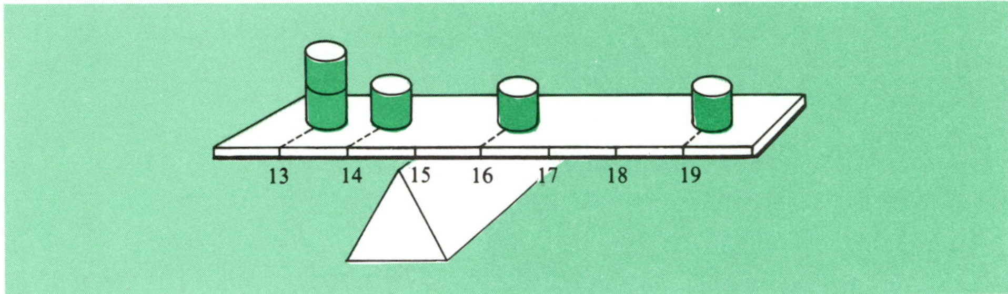

The mean has *shifted* along with the scores. It is exactly 9 units to the right of where it was. Thus the new mean is 9 units larger than the old mean. Notice that the scores have the same positions with respect to each other, even though each has been increased, and that the mean has the same relative position with respect to the scores as it formerly had.

Of course, there is nothing special about the number 9. Any one number that had been added to each of the scores would have preserved their relative positions while shifting them all. Whatever number had been added would have told exactly how much the mean had increased.

If a constant number had been subtracted from each of the scores, the weights would have been shifted to the left and the mean would have been decreased by that number. In general, when we decrease each term in a distribution by a constant amount, we are in effect decreasing the mean by that amount. Accordingly, we may say that the mean age of a Smith child a year ago was 5.

We will describe the addition of a given number to each term in a distribution as "adding a constant" and we will also have occasion to speak of "subtracting a constant." What has been said may now be summarized as a theorem.

THEOREM 2.3 If each term in a distribution is increased by a constant, the mean is increased by the same constant. If each term is decreased by a constant, the mean is decreased by the same constant.

Theorem 2.3 is not only intrinsically interesting but, more than that, we can "cash in" on it. Suppose we are after the mean of a distribution made up of awkwardly large numbers. It is often convenient deliberately to change these numbers by subtracting a given number or constant from each of them; thus arriving at a new distribution that yields an obvious mean. Our rule tells that this new mean is too small by exactly the amount of the constant we have subtracted from the original distribution. So we add back the constant to get the mean of the original distribution.

Suppose we are to find the mean of 92, 94, and 99. Subtract 90 from each term and find the mean of 2, 4, and 9. This mean is 5. Therefore, the mean of the original terms equals $5 + 90 = 95$.

The choice of the constant to be subtracted has no effect on the final result, since that constant is added back. Thus we might have subtracted 91 from each of the original terms 92, 94, and 99. The result would be a new distribution, 1, 3, and 8, which yields a mean of 4. Adding 91 to 4 again results in 95, the mean of the original terms.

Suppose that a statistician subtracted a constant exactly *equal to the mean*. That is, suppose that 95 were subtracted from each of the three terms. The result would be Table 2.5.

Table 2.5 Subtraction of Mean from Each Term

Original Terms	Constant Subtracted	Differences
92	95	−3
94	95	−1
99	95	4

This time the remaining differences have a sum of zero (Theorem 2.1) and the mean of the original terms is $0 + 95 = 95$. Notice how simple the arithmetic was in this last case.

When using the method of subtracting a constant to find the mean, it is usually preferable to subtract a constant that promises to be as close to the mean as possible. The nearer the constant is to the mean itself, the simpler the computations, because the differences are then small and easy to compute. It is a good idea to estimate what the mean will be before beginning to work, and then to use this estimate as the constant to be subtracted. The better the estimate, the more work saved, although the final answer is the same whatever the estimate.

The shortcut just recommended may seem roundabout, for admittedly the device of subtracting a constant to find the mean is seldom worthwhile when a distribution has only three terms. However, the method proves invaluable when the distribution is large.

Another technique sometimes used is to make the lowest term in the distribution the estimate, knowing in advance that it will not be the mean. The advantage of this technique is that it insures that none of the differences will be negative, and thus does away with one possible source of error, the inadvertent mishandling of minus signs. For instance, if we used the value 92 as our constant, we would obtain differences of

0, 2, and 7. Their mean being 3, we add 92 and 3, and once again get 95 as our computed mean of the distribution.

Next we ask what happens to the mean when each term in a distribution is multiplied by a constant. To begin with, observe what happens when each term in our distribution is multiplied by 2 (Table 2.6).

The effect has been to double the sum of the terms. That is, 60 is the new sum instead of 30. The multiplication by 2 in the new case led to the determination of a mean *twice* as large as formerly. By analogous reasoning, if we had multiplied each term by 3, the mean would have trebled its size. In general, when each term in a distribution is multiplied by a constant, the effect is to multiply the mean by the same constant.

Table 2.6 Multiplication of Terms by Two

	Terms	New Distribution When Terms are Doubled
	4	8
	4	8
	5	10
	7	14
	10	20
Sum	30	60
Mean	30/5 = 6	60/5 = 12

Finally, we can show that when each term in a distribution is divided by the same number, the mean becomes divided by that number. The process of dividing by a number is equivalent to that of multiplying by 1 over that number. For example, dividing each term by 3 is the same as multiplying each term by one-third. By our multiplication rule we know that when each term is multiplied by one-third, the mean becomes multiplied by one-third also. Thus dividing each term by 3 results in shrinking the mean to one-third of its original size. In general, when each of the terms in a distribution is divided by a constant, the mean is also divided by that constant.

To summarize, we can state the following theorem.

THEOREM 2.4 If each of the terms in a collection is multiplied by a constant, the mean becomes multiplied by the same constant. If the terms are divided by a constant, the mean becomes divided by that constant.

2.4 Some Applications of the Mean, the Median, and the Mode

The sensitivity of the mean frequently makes it a good measure of the size of all the terms in a group. Loosely speaking, the mean is often used as a measure of how high the terms in a group "are running." A frequent way to compare the sizes of the terms in two groups is to compare their means. Thus when we say that the mean number of home runs hit by players on the home team is larger than the mean number hit by players on the visiting team, we are saying that the home-team players are better home-run hitters as a group than the visiting-team players. Consider the statement that the average number of children per family in one country is 2.3, whereas in a second country it is 1.5. Very likely the mean has been used as the basis of comparison. The implication is that families tend to be larger in the first country mentioned.

The fact that the median (or middle term) is not affected by the size of the extreme terms makes it useful for certain purposes. The median provides important information, especially about a distribution in which a relatively small number of terms are extreme in one direction or the other. For instance, consider the distribution of the incomes of 6,000 citizens in a mining community. Virtually all of these citizens earn less than $10,000 per year. However, a few of them, the mine owners, earn as much as nearly all the others combined.

The median of this distribution of incomes gives a better picture of the economic level of the community than the mean does. The fact that the median is not influenced by the vastness of the incomes of a few makes it a representative measure. For instance, the median yearly income might be $9,400 per year. (That is, half of the citizens make less than $9,400 per year and half of them make more than this amount.) The mean, because it is swollen by the incomes of the few, may be $18,000. Thus it might turn out that only a tiny percentage of the people in the community earn as much as the mean income. The median is often a more revealing measure than the mean, especially when both are obtained from a distribution that contains a relatively small number of terms with values at one extreme.

The value of the modal term provides information to producers, designers, and storekeepers who aim their commodities at specific markets. For instance, a producer might wish to know the most frequent price being paid for men's watches in order to concentrate on making watches in the particular price range. A clothing salesman might wish to know the most popular sizes of men's suits so that he can purchase his stock accordingly. Naturally, many other factors enter into both of these illustrations. The point is merely that the mode indicates maximum frequency, which is often synonymous with maximum popularity.

2.5 Notation

The number of terms comprising a distribution is symbolized by the letter N. Thus for the following three distributions:

Distribution A		Distribution B		Distribution C	
5		5		5	
3	$N=3$	3	$N=4$	3	$N=5$
2		2		2	
		4		4	
				5	

The terms in a particular set may be given a letter value to distinguish them (usually the letter X or Y). For instance, we may use the letter X to stand for any term in Distribution A. When we specify a particular order for the terms, we may attach a subscript to indicate a particular term. For instance, suppose we are using the letter X to refer to any term in Distribution A and we think of them in the order indicated above. Then X_1 stands for the first term, 5. $X_2 = 3$ and $X_3 = 2$. Similarly, we may use the letter Y to refer to any term in Distribution B. Then $Y_1 = 5$, $Y_2 = 3$, $Y_3 = 2$, and $Y_4 = 4$.

The symbol Σ (Greek letter capital *sigma*) is used to indicate "the sum of." It tells us to add certain expressions. Remember that we used the letter X to denote a term in Distribution A. Thus $\Sigma X = 10$ and $\Sigma Y = 14$.

The symbol for the mean of a population is μ (Greek letter *mu*). Thus for Distribution A, $\mu = \dfrac{\Sigma X}{N} = \dfrac{10}{3}$. For Distribution B, $\mu = \dfrac{\Sigma Y}{N} = \dfrac{14}{4}$. In general,

$$\mu = \frac{\Sigma X}{N} \quad \text{for any distribution having terms denoted by } X. \qquad (2.1)$$

When we are considering several distributions simultaneously we use subscripts to avoid ambiguity. For instance, we can refer to the mean of Distribution A as μ_X and to the number of terms in Distribution A as N_x. Similarly μ_Y refers to the mean of Distribution B and N_Y refers to the number of terms in Distribution B. Thus:

$$N_X = 3; \quad \mu_X = \frac{\Sigma X}{N_X} = \frac{10}{3}$$

$$N_Y = 4; \quad \mu_Y = \frac{\Sigma Y}{N_Y} = \frac{14}{4}$$

As stated in the Preface, for this and subsequent chapters there are two sets of problems. Each set gives full coverage of the topics in the chapter.

Problem Set A

2.1 The five children of Mr. and Mrs. N. O. Fluoride had the following numbers of cavities discovered during their last family checkup: 3, 11, 5, 2, 11. Determine the mean, median, and modal numbers of cavities discovered for the children.

2.2 The first five morning customers through the checkout counter at Slim's Discount Foods presented the following amounts of cents-off coupons (values given in cents): 87, 15, 152, 0, 64. The first five afternoon customers presented the following amounts: 0, 53, 0, 0, 49. Find the mean, median, and mode for both the first five morning and first five afternoon customers.

2.3 Consumer studies show the costs of a head of lettuce on the first days of the previous 12 months were (in dollars) .35, .43, .45, .39, .50, .85, 1.05, .73, .50, .35, .33, .50.
   a. What was the mean price charged for a head of lettuce on the first days of each month during the last year?
   b. What price was charged most often?
   c. What was the median price?

2.4 Gas mileage improved by the following percentages on six of each of the three types of cars indicated below when bias-ply tires were replaced by steel-belted radials.

GM	*Ford*	*Chrysler*
4.7	4.0	6.0
4.9	4.0	6.3
3.2	6.2	7.2
5.0	5.8	4.1
5.7	4.3	4.1
6.0	4.1	4.7

Find the mean, median, and modal improvements in percent of miles added due to the tire change for (a) the six GM cars, (b) the six Fords, (c) the six Chryslers, and (d) the 18 cars considered as one population.

2.5 For the seven area high schools, the numbers of seniors failing the state proficiency test required for graduation were 37, 84, 17, 29, 48, 60, and 25. Find the average number of students per school who will fail to graduate because of the exam score. Which type of average did you use? Why?

2.6 Consider the following three distributions.

X	*Y*	*Z*
3	5	2
4	8	87
4	7	85
4	8	80
7	4	80
9	10	85

   a. Determine the mean, median, and mode for Distributions X, Y, and Z.
   b. If the numbers in Distribution X refer to stock numbers of fashions sold at an apparel store, which measure is the most revealing?
   c. If the values in the Y distribution are centigrade temperatures in a meat locker, which type of average would probably be most meaningful?
   d. Why would the median be the most suitable choice for measuring central tendency in Distribution Z?

2.7 Because of a programming error, the thousands of charitable contributions fed to a computer were each $10 too small. If the computer reports the mean amount contributed to be $38.47, what is the actual mean?

2.8 The ages at which 20 emphysema victims who were pack-a-day smokers were diagnosed are 68, 57, 35, 49, 57, 53, 70, 48, 59, 67, 63, 65, 48, 55, 63, 65, 41, 85, 60, and 57. Compute the mean and median ages at which the disease was diagnosed for this population.

2.9 The actual savings for 15 families from this year's mandatory state tax cut are (in dollars) 8.75, 9.35, 15.40, 1.25, 6.43, 8.11, 6.00, 5.95, 21.00, 2.74, 10.00, 6.50, 6.75, 5.25, 8.20. Find the mean savings per family for this group.

2.10 The numbers of acres destroyed in the last 16 major forest fires in a certain geographical section of the nation are

28,000	39,000	41,000	26,000
32,000	40,000	43,000	29,000
34,000	36,000	44,000	50,000
26,000	41,000	16,000	38,000

Find the mean number of acres destroyed per fire using Theorem 2.4 and dividing by a constant of 1,000.

2.11 Consider the following distribution: 100, 83, 88, 81, 83, 96, 105, 108, 78, 102, 97, 113, 126, 94, and 86.

a. Find the mean.
b. Verify that the sum of the differences from the mean is zero.
c. Find the sum of the squared differences from the mean.
d. Find the sum of the squared differences from 90. Be sure it is larger than your answer to part c.

2.12 Today's top local news story reports that one of the city's leading physicians has charged that, during the last six months, the following numbers of unnecessary surgical procedures were performed at an area hospital.

July	39
August	57
September	25
October	105
November	47
December	22

a. What is the mean for these six months?
b. If the doctor's accusations are true, and a mean of 40 unnecessary surgeries were performed monthly for the past year, what was the monthly mean for the first six months of the year?

2.13 The placement office for the school of business registers job offers in four categories. The numbers of offers on file at present are

Accounting	87
Management	45
Finance	23
Taxes	69

The placement office for health field majors lists jobs in six categories. At present there is a mean of 50 jobs per category.

a. Are more jobs registered in the business or health field office?
b. On the average, do individual business or health field majors have more jobs from which to choose? (Compare means.)

2.14 a. How many distributions have a mean of 50?
b. Find a distribution in which $X_1 = 49$, $\mu = 50$, the median is 50, and the mode is 50.

2.15 Suppose that you are buying a thermometer and there are many on display. You note that not all of these indicate the same temperature. How will you select one?

Problem Set B

2.16 Find the mean, median, and mode for each of the following three distributions.

Distribution A	*Distribution B*	*Distribution C*
99	−4	7.1
107	−2	8.2
34	−11	9.9
86	−13	5.1
127	24	6.0
69	−2	3.8

2.17 The weights of the horns of eight bighorn rams found in the Canadian Rockies are (in pounds) 28.5, 32.4, 30, 37, 25, 28.4, 30.5, and 31. Find the mean and median weights.

2.18 Sound House carries five brands each of low-, medium-, and high-priced stereo speakers. The power (in watts) required per speaker for each brand is given below.

Low	*Medium*	*High*
4	8	35
7	18	10
15	11	5
13	7	4
13	16	8

a. Find the mean and median power required for the brands in each category.

b. Find the mean and median numbers of watts required for the 15 brands the store carries.

2.19 If the first eight terms of a nine-term distribution with a mean of 14.3 are 83, 27, 3, 14.1, 2.6, 9, 8, and 51, what is the ninth term?

2.20 Ten mopeds tested for gas mileage yielded the following results in miles per gallon (mi/gal): 123, 85, 97, 92, 103, 114, 109, 91, 98, and 83.

a. Find the mean mi/gal for the ten tested mopeds.

b. If an antipollution device is installed on any moped, its gas mileage is decreased by 7.5 mi/gal. Find the mean mi/gal for the same 10 mopeds with the device installed.

2.21 The approximate sizes of the viruses that cause the following diseases are given below (in centimicrons).

Poliomyelitis	2.5
Yellow fever	52.5
Influenza	105.0
Chicken pox	137.5
Pharyngitis	75.0

a. Find the mean length of the viruses that cause the above five diseases.

b. The mean computed using the above five figures and the length of the virus that causes mumps is 80 centimicrons. Find the size of the mumps-causing virus.

2.22 The numbers of customers left powerless (for two or more hours) in the most recent blackouts in a three-state area are

400,000	81,000	385,000
38,000	137,000	92,000
95,000	185,000	89,000
215,000	200,000	43,000

a. Find the mean and median numbers of customers left powerless in the last 12 blackouts.

b. Suppose the last term, 43,000, is in error and should really be 83,000. Now what are the mean and median?

c. What property of the mean, lacking for the median, is illustrated by parts a and b?

2.23 Hospital occupancy figures show the following totals for number of unoccupied beds in the metropolitan area.

Jan.	385	Apr.	250	July	629	Oct.	297
Feb.	400	May	227	Aug.	600	Nov.	318
Mar.	270	June	573	Sept.	340	Dec.	635

a. Find the mean of the monthly totals.

b. If the construction of a new hospital would increase the number of unfilled beds by 25 percent as claimed by the hospital council, what would the new mean be?

2.24 The temperatures at 2:00 p.m. in the 15 buildings that comprise the "business heart" of Megapolis were recorded unofficially by the feature writer of the daily newspaper. The measures were (in degrees) 68, 75, 69, 72, 73, 75, 78, 82, 65, 73, 74, 75, 73, 78, and 75.

a. Find the mean, median, and modal temperatures for the 15 buildings.

b. What would the new mean be if each building lowered its temperature 5 degrees?

c. What would the new mean be if each building lowered its temperature 5 percent?

2.25 All state firms with 25 or more employees were recently surveyed. Among the questions asked was "Does your firm have a hire-locally-first policy?" The numbers of affirmative responses in the 10 largest cities were 12, 6, 9, 15, 29, 14, 3, 18, 20, 24.

A similar survey was conducted in a neighboring state. The mean number of affirmative responses for its 15 largest cities was 16.

a. What was the mean number of affirmative responses per city in the first state?

b. How many affirmative responses were recorded in the neighboring state's 15 largest cities?

c. Were more affirmative responses recorded in the first state's 10 largest cities or in the neighboring state's 15 largest cities?

2.26 One of the priorities of the new state administration is to encourage foreign tourists to visit the state. Records for the past eight years show that the following numbers of foreigners have vacationed in the state (per year).

200,500	155,500
185,000	145,000
190,000	187,000
210,000	165,000

a. What has been the mean number of foreign visitors per year during the last eight years?
b. By what percent would the mean yearly number have to be increased to reach the stated goal of 300,000 visitors from other lands per year?

2.27 Gravitational pull varies from place to place on the earth's surface. Consequently, Olympic record-keepers have established a table of correction factors for the javelin throw that includes the following (given in centimeters).

Helsinki	+14.15
Berlin	+ 9.00
Paris	+ 6.40
Los Angeles	− 4.13
Melbourne	− 1.60

If an athlete throws the javelin at the five locations listed, how much will the actual mean distance (in centimeters) differ from the mean the Olympic record-keepers would record? Which mean is larger?

2.28 The octane ratings of a dozen samples of Brand X gasoline are 87, 89, 88, 93, 93, 94, 88, 96, 90, 92, 91, and 91.
a. Find the mean rating for the 12 samples.
b. The octane ratings for 9 of 10 samples of Brand Y gasoline are 89, 87, 89, 94, 93, 93, 89, 92, and 95. What is the smallest that the 10th octane reading can be for the mean octane rating for Brand Y to exceed the mean rating for Brand X by a point or more?

2.29 Fifteen haulers presently provide trash service for the city. The numbers of residences served by the haulers are 6050, 4750, 8093, 3857, 6247, 5190, 3025, 5520, 4055, 6385, 6700, 4350, 4470, 3960, and 8030.
a. Find the mean number of residents served per hauler.
b. The haulers maintain that their equipment cannot continue to handle the large number of pickups required. The city agrees to contract with a new company to take over 7 percent of each route. How many residents will the new company serve?
c. What will the new mean per hauler be?

2.30 a. For the following population

44.7, 89, 33.2, −17, .07, 88, 109, 34.7, −24.2, 99

which is larger: $\Sigma(X - 45.65)^2$ or $\Sigma(X - 48.30)^2$? Why?
b. For the above population, find the sum of the differences from the mean and the sum of the squared differences from the mean.

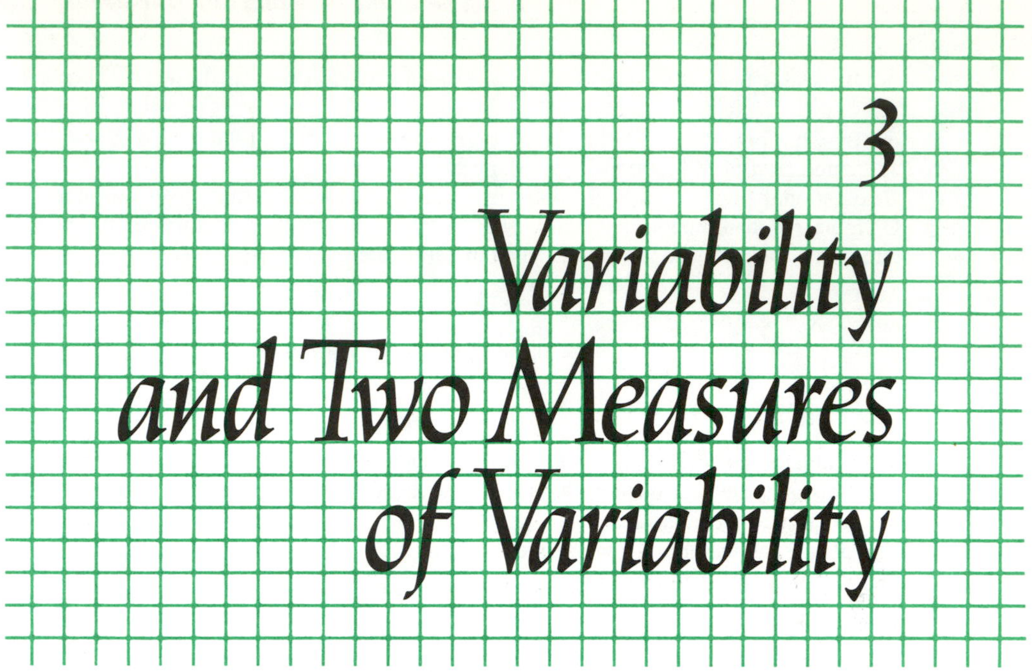

3 Variability and Two Measures of Variability

3.1 Variability

The terms in a distribution may vary widely or they may be very close together. To appreciate the difference, consider a distribution in which the terms are 8, 9, 10, 10, 13. We shall call it Distribution A. The representation of these terms as weights on a plank is shown in Fig. 3.1.

Fig. 3.1

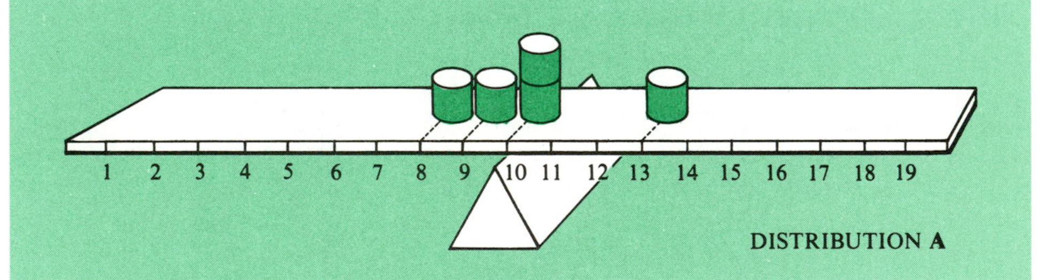

Now consider Distribution B made up of the terms 1, 5, 10, 16, 18. This distribution is illustrated in Fig. 3.2.

In Distribution A the weights appear closer together or *more clustered*, whereas in Distribution B they are relatively spread apart or *scattered*. Note that the means of the two distributions are the same. The statistician's way of describing the difference is to say that the terms in Distribution B are *more variable* than those in Distribution A.

Fig. 3.2

1 2 3 4 5 6 7 8 9 10 11 12 13 14 15 16 17 18 19

DISTRIBUTION **B**

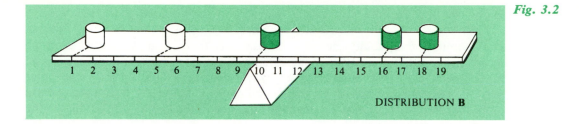

Variability is an extremely important idea to grasp. The property of variability refers to the relationship among all the terms in a distribution considered together. Thus a **change in any term changes the variability of the distribution.** Suppose, for example, we were to change the 18 to 20 in Distribution B. In the illustration this would mean moving the weight on number 18 two units to the right. The shift would scatter the weights even more than they were, with the result that their variability would be increased. On the other hand if the same weight had been shifted two units to the left—to 16, for instance—the variability would have been reduced.

Now suppose we were to change the number 10 to 11 in Distribution B. Would shifting the weight on 10 one unit to the right increase or decrease the variability of the distribution? The answer is not apparent, since we cannot see whether the shift increases or decreases the variability or "scatter," as we sometimes call it.

Frequently our eye cannot tell us which of two distributions is more variable. Compare Distribution C with Distribution D in Fig. 3.3.

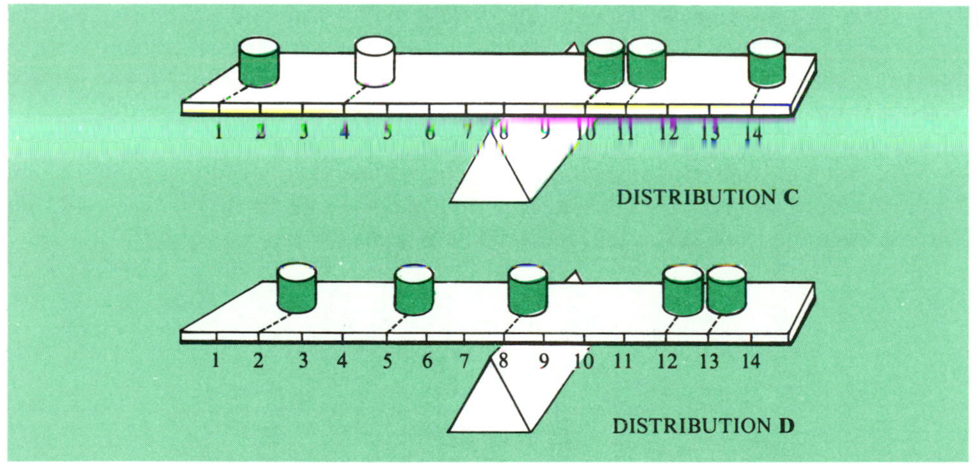

Fig. 3.3

How will we decide which distribution is more variable? Thus far we have not defined a measure of variability, so no decision is possible. Our next step is to define an appropriate measure.

3.2 The Variance

Our measure of variability must be some number that can be computed from the terms in any distribution. When computed, this number will tell us about the spread of the terms in the distribution. To satisfy our demands, the measure we are after must have certain properties. Our way of finding the measure will be to specify these properties in advance. Once we have specified them, they will guide us toward the measure we are seeking.

One property is that our measure must have a small value when it is computed from a distribution in which the terms are close together numerically. Its value must be large when the terms are widely scattered.

A second property is that the value of the measure should not be related to the number of terms in the distribution. Specifically, we would not want a measure that became large just because there were *many* terms in the distribution. Our measure should reflect only the similarity or the dissimilarity of the numbers themselves, and this has nothing to do with how many there are. We might have a thousand terms but, if they were all very similar, our measure should have a small value.

The third property is that our measure should be independent of the mean. We are interested only in how *scattered* the terms are. Distributions A and B in the previous section have the same mean even though Distribution B is more variable. Even if we were told that the mean was one million, we still could not say whether or not the weights representing the terms were far apart. The mean tells us nothing about the variability and its size should not influence our measure.

We now decide that when we say Distribution B is more variable than Distribution A, we mean that the weights in Distribution B *tend to be greater distances from the mean. Scatter* suggests wide distances from the fulcrum and *cluster* suggests that the weights are huddled around the fulcrum. Thus the measure of variability we are seeking should be some measure of how far the weights in our distribution are from the fulcrum or mean. Notice that we are not considering the numerical value of the mean itself, but merely the set of distances from the mean.

Now let us examine the distribution of five terms introduced in Chapter 2. This time the distances from each weight to the mean are indicated in Fig. 3.4. The distances to the left of the mean receive minus signs.

Fig. 3.4

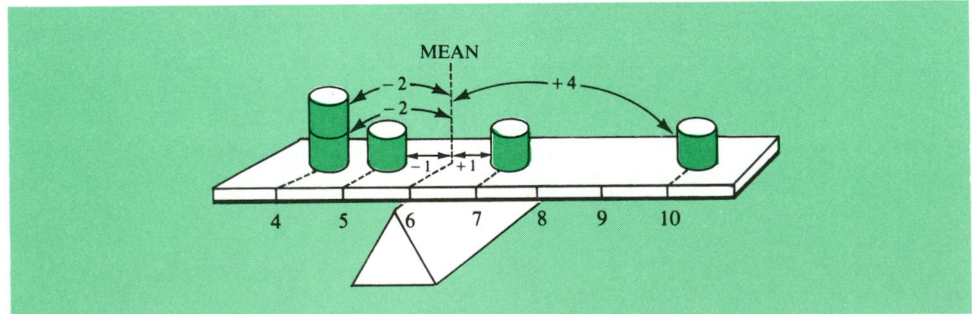

Our impulse might be to add up all the distances and to use their sum as an indication of variability. But the minus signs balance out the plus signs. We recall from Theorem 2.1 that no matter how long the distances are, their sum is always equal to zero.

To avoid the result of the differences balancing out to zero, we could treat the differences as though they were all positive. By summing the distances in this way and then taking their mean, an old-fashioned measure of variability, called the **average deviation,** was developed. However, the average deviation is seldom used.

Our way of escaping the minus signs is to square each of the differences—that is, to multiply each difference by its own value. We will profit by the fact that the square of any nonzero number is positive. Then, having squared each difference, we add up the squares. The computations performed on the distribution just illustrated are shown in Table 3.1.

Table 3.1 Computation of Squares of Differences from Mean

(1) Original Terms	(2) Mean	(3) (Original Term Minus Mean) Differences	(4) Squares of Differences
4	6	−2	4
4	6	−2	4
5	6	−1	1
7	6	1	1
10	6	4	16
			26

The column containing the squares of the differences includes no negative terms. May we now consider the sum of these squares, 26, as our measure of variability? Not quite. Suppose the number of terms in the distribution were increased from 5 to 11. That is, suppose 6 more weights were placed on the plank. The addition of these weights, unless we placed them all at the mean, would introduce new distances or differences. When these distances were squared and included, the sum of the squares would increase. Thus the *sum* of squares of distances reflects the *number* of weights or terms in the distribution as well as their variability. Since we want our measure not to be sensitive to the number of terms in the distribution, this sum is unsatisfactory as a measure of variability.

It is easy to "purge" our measure of this unwanted influence. We simply divide the sum of squares of the differences by the number of terms in the distribution. The quotient becomes our measure of variability. This measure is called the **variance.** Hence the variance for the distribution in Table 3.1 is $26/5 = 5.2$.

We have traversed a long road to find the variance, but our steps have not been difficult. In general, given a set of terms, our first step is to subtract the mean from each of them to obtain a set of differences. Then each difference is squared and the total of the squares is found. This total is then divided by the number of terms. The quotient, called the variance, becomes the measure of variability, for it satisfies all the demands we originally made.

Definition 3.1 The variance is the mean of the squared differences from the mean of the distribution.

We may now decide whether Distribution C or Distribution D is more variable. We find the variance of each in Tables 3.2 and 3.3.

Table 3.2 Finding the Variance of Distribution C

Terms	Mean	Differences	Squares of Differences
1	8	−7	49
4	8	−4	16
10	8	2	4
11	8	3	9
14	8	6	36
			114

Variance = 114/5 = 22.8

Table 3.3 Finding the Variance of Distribution D

Terms	Mean	Differences	Squares of Differences
2	8	−6	36
5	8	−3	9
8	8	0	0
12	8	4	16
13	8	5	25
			86

Variance = 86/5 = 17.2

Distribution C has a larger variance, so we say it is more variable.

The variance may be found by shortcut methods, which are indicated at the end of this chapter. However, the method we have given here is the only one that literally follows the definition. The method of computation that we have discussed must be grasped before we can understand the meaning of the measure.

3.3 The Standard Deviation

The variance enables us to tell which of a set of distributions is the most variable. But there is still another measure of variability called the **standard deviation.**

Definition 3.2 The standard deviation is the positive square root of the variance.

Admittedly we have as yet seen no purpose for this second measure of variability, but the standard deviation has various uses that will be discussed later. The standard deviation has all of the properties thus far ascribed to the variance. And the requirements that led to the development of the variance are met equally well by the standard deviation.

We may, for example, compare standard deviations to determine which of two distributions is more variable. Our conclusion will always be the same as if we had compared variances.[1] The reason is that if one distribution has a larger variance than another, it will also have a larger standard deviation. For instance, recall the distribution of Table 3.1 with its variance of 5.2. Suppose that Distribution B has a variance of 16. The former distribution has a standard deviation of 2.3, which is smaller than Distribution B's standard deviation of 4. The reader should verify that the standard deviation of Distribution C is 4.8 and the standard deviation of Distribution D is 4.1. Appendix I contains a table of square roots.

By now it should be clear that computing the standard deviation entails going one step beyond computing the variance. Yet the standard deviation is a much more familiar measure to most students of statistics, for it has more generally known applications. The standard deviation may be computed by the following four steps:

1. Subtract the mean from each term in the distribution to obtain a set of differences.
2. Square each difference and add the squares.
3. Divide the sum of the squared differences by the number of terms in the distribution.
4. Find the square root of the quotient just obtained; the positive square root is the standard deviation.

At this point we should think of the standard deviation as a line segment, or perhaps a ruler, of a given length. When the terms in the distribution are clustered, the ruler is relatively short (Fig. 3.5).

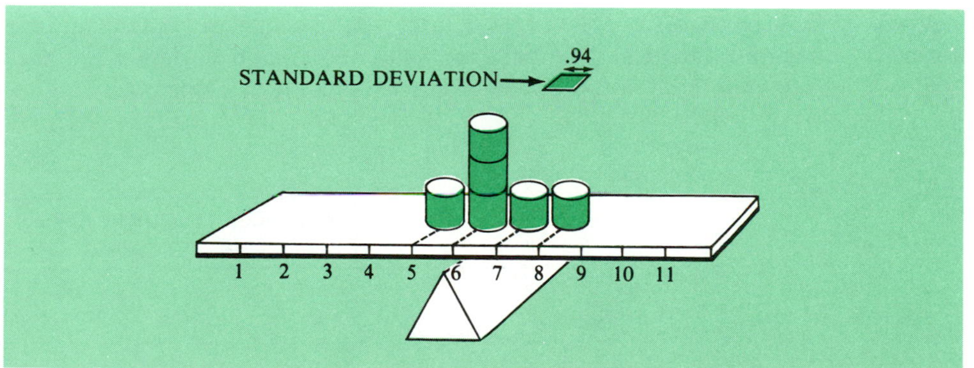

Fig. 3.5

[1]Note that neither the standard deviation nor the variance allows us to say that one distribution is a *certain number of times* more or less variable than another.

When the terms in the distribution are scattered, the line segment or ruler that applies to the distribution is relatively long (Fig. 3.6).

Fig. 3.6

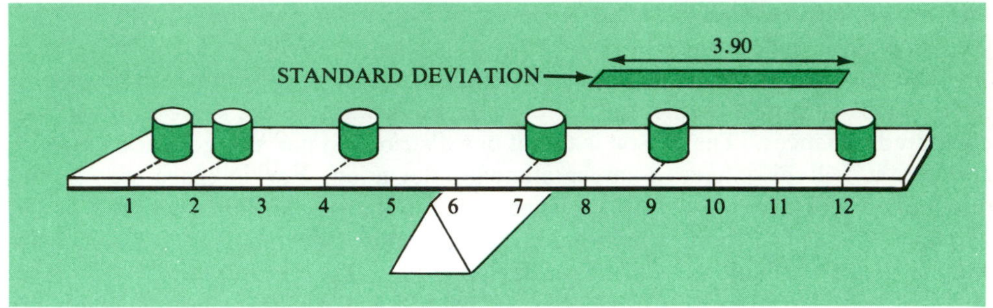

3.4 How Changing the Terms in a Distribution Affects the Standard Deviation and the Variance

We have already discussed the effect of adding a constant to each term in a distribution. In our picture analogy, the corresponding operation is to shift each weight some specified distance to the right. The weights in their new positions remain the same distance from each other and each weight preserves its original distance from the mean. Remember that the mean shifts along with the weights. Thus, adding a constant to each term leaves the variability of a distribution unchanged. Since the standard deviation and the variance measure this variability, it can be seen that adding a constant to every term does not change either the standard deviation or the variance.

For example, we may try adding 9 to each of the terms in the distribution 4, 4, 5, 7, 10. The new distribution has the terms 13, 13, 14, 16, 19. These two distributions were presented in Figs. 2.3 and 2.4. To verify that each has the same standard deviation and variance, it suffices to show that the sets of distances from the mean in the distributions are identical. The variance and the standard deviation are each derived entirely from these distances. The computations are in Table 3.4.

Table 3.4 Comparison of Distributions

Original Distribution			Distribution Obtained by Adding 9 to Each Term		
(1) Terms	(2) Mean	(3) Distances	(1) Terms	(2) Mean	(3) Distances
4	6	−2	13	15	−2
4	6	−2	13	15	−2
5	6	−1	14	15	−1
7	6	1	16	15	1
10	6	4	19	15	4

The sets of distances (Column 3) are the same in each distribution. We may now write as a theorem:

THEOREM 3.1 Increase or decrease of each term in a distribution by a constant amount does not alter the variability of the distribution, and therefore does not affect the variance or the standard deviation of the distribution.

On the other hand, multiplication of each term by a constant does affect the variability of a distribution. For instance, let us compare the illustration of our original distribution with that of the distribution obtained by doubling each of the terms (Fig. 3.7).

The new distribution is clearly more variable. Note that the distances between terms have doubled. The standard deviation of the new distribution turns out to be *twice* as large as that of the old distribution, a fact that the reader should verify. In general, when each term is multiplied by a constant, the standard deviation of the distribution becomes multiplied by the absolute value of that constant; for example, if each term is multiplied by 3, the standard deviation becomes three times as large as it was. If each term is multiplied by -2, the standard deviation becomes twice as large as it was.

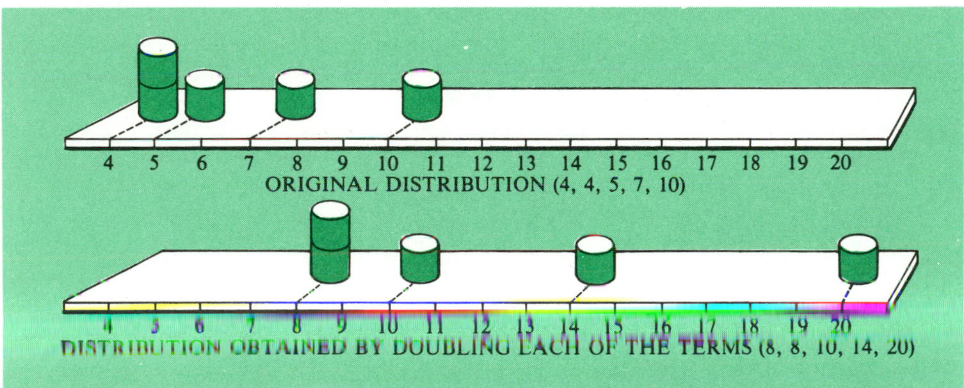

Fig. 3.7

Note that if the constant is a fraction between one and negative one, the effect is to diminish the standard deviation. Multiplication by a fraction pulls the weights closer together, whereas multiplication by a number greater than one or less than negative one spreads them out. Division by a constant effects a division of the standard deviation by the absolute value of the same constant.

Since the variance is the square of the standard deviation, any change in the standard deviation becomes squared in the variance. Thus multiplication of each term by a constant results in multiplication of the variance by the square of that constant; for instance, multiplication of each term by 3 results in multiplication of the variance by 9. Division by a constant results in division of the variance by the square of that constant. Division of each term by 3 effects a cutting down of the variance to one-ninth of its original size.

In sum, we may now write:

THEOREM 3.2 Multiplication of each term in a distribution by a constant results in multiplication of the standard deviation by the absolute value of that constant, and in multiplication of the variance by the square of that constant. The same holds when multiplication is replaced by division throughout.

3.5 Applications of the Variance

The variance is important as a measure of *heterogeneity*. For instance, the scatter of intelligence test scores in an elementary school class indicates the heterogeneity of the children with respect to intelligence. The variance of a single individual's different scores on a performance test given each day for 10 days indicates the unevenness of his or her performances. Think of two cities in each of which the mean yearly income is \$18,000. The city in which the distribution of incomes has the larger variance is the one in which the individuals differ more from each other in earning power. Similarly, the variance of the heights of trees is a measure of how different from each other these heights are. Note that the standard deviation might also have been used in any of the above contexts.

3.6 Notation

Suppose we are using the letter X to refer to the terms in a particular group. For instance, say that the group consists of three terms and that $X_1 = 1$, $X_2 = 3$, and $X_3 = 5$. Now suppose we square the value of each term, however many there are. For our particular group, $X_1^2 = 1$, $X_2^2 = 9$, and $X_3^2 = 25$. Now we add up the squares of the terms (for example, $1 + 9 + 25 = 35$). The sum of the squares of the terms in the group is symbolized ΣX^2. (In our particular group, $\Sigma X^2 = 35$.)

We have seen that the symbol denoting the sum of the terms themselves is ΣX. In our particular group:

$$\Sigma X = X_1 + X_2 + X_3 = 1 + 3 + 5 = 9$$

The square of the sum of the terms is symbolized by $(\Sigma X)^2$. In our particular group $(\Sigma X)^2 = (9)^2 = 81$.

Note the distinction between the **sum of the squares** of the terms denoted by ΣX^2, and the **square of the sum** of the terms denoted by $(\Sigma X)^2$. In the former case, the terms are squared individually and the individual squares are added up. In the latter case, the terms are added first and their sum is squared.

The order of operations is crucial. A graphic demonstration of the difference in a particular case will help us to remember the distinction. Watch what happens when all the terms are positive. We shall perform both operations on the group of three terms, 1, 3, and 5, this time using illustrations. We shall depict each term as a line segment designating its size. For instance, the first term would be a line segment 1 unit long, the second term a line segment 3 units long, and so on. We add the terms to find the sum of X by laying off the line segments end to end (Fig. 3.8).

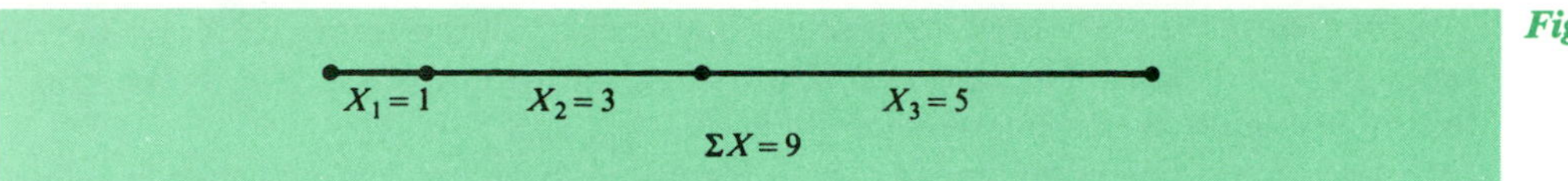

Fig. 3.8

Our next step is to square the value of each term. The square of a number is the (area of the) square built on the line segment representing that number. The squares of the three terms are indicated in Fig. 3.9.

To find ΣX^2 we now add up the areas of the three squares. $\Sigma X^2 = 1 + 9 + 25 = 35$.

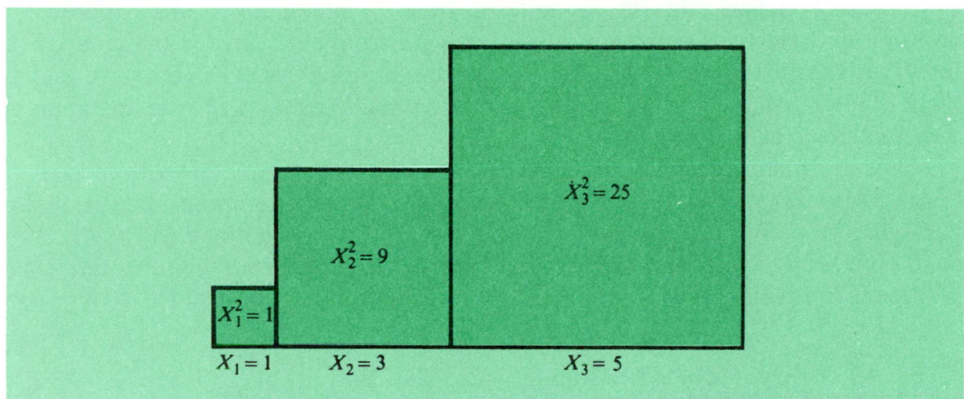

Fig. 3.9

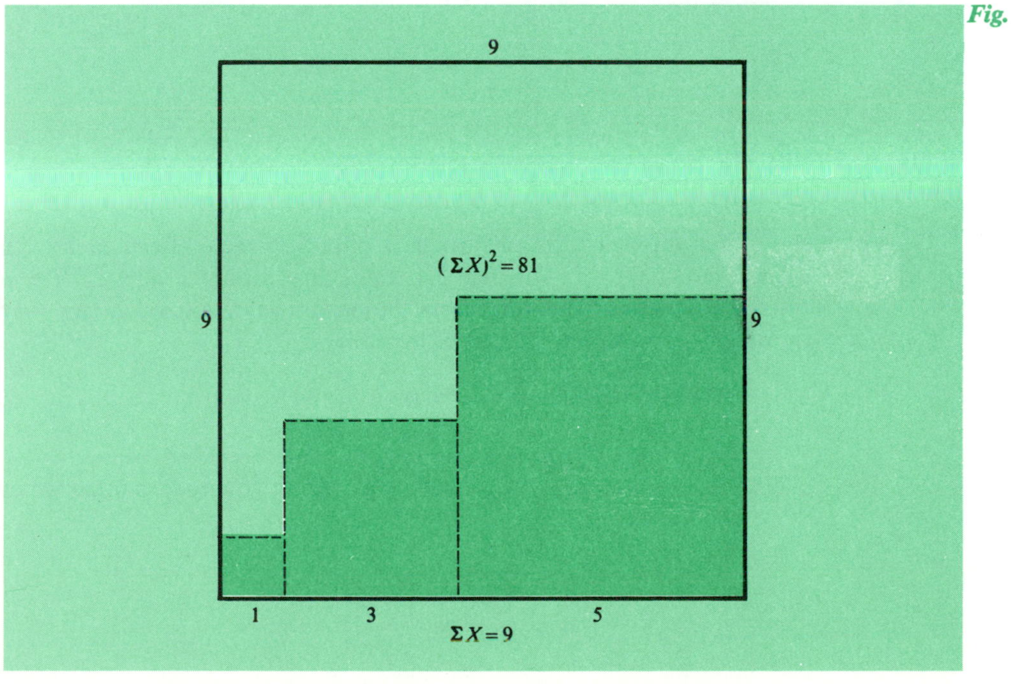

Fig. 3.10

Remember that we have represented the sum of all the terms, ΣX, by the line segment 9 units long. The *square of the sum* of all the terms, $(\Sigma X)^2$, is the area of the square built on this line segment (Fig. 3.10). $(\Sigma X)^2 = 81$.

The larger area is obtained by building up the baseline first and then building a square on it. Thus $(\Sigma X)^2$ is greater than ΣX^2 (when the terms are each positive). The purpose of this demonstration was merely to emphasize that the order of operations is important when there is squaring and adding to be done.

It is customary to use letters toward the end of the alphabet to refer to values of a variable and to use letters near the beginning of the alphabet to stand for constants. For instance, suppose the Greek letter α (read "alpha") represents a particular constant. Consider the operation of subtracting α from each term in the group. For instance, if the terms are X_1, X_2, and X_3 and we subtract α from each term, we get the three differences: $(X_1 - \alpha)$, $(X_2 - \alpha)$, and $(X_3 - \alpha)$. We may write the operation in three columns, as in Table 3.5.

We designate the sum of the differences obtained by subtracting α from each of the terms by $\Sigma (X - \alpha)$. Now suppose we square all the differences—that is, in the particular group of three terms we obtain $(X_1 - \alpha)^2$, $(X_2 - \alpha)^2$ and $(X_3 - \alpha)^2$. We have obtained the squared differences of each of the terms from the fixed reference point with the value α. The sum of these squared differences is designated $\Sigma (X - \alpha)^2$.

Table 3.5 Subtraction of a Constant

Term X	Constant α	Term Minus Constant $X - \alpha$
X_1	α	$(X_1 - \alpha)$
X_2	α	$(X_2 - \alpha)$
X_3	α	$(X_3 - \alpha)$
		$\Sigma(X - \alpha)$

The mean of a group is a constant, which is often subtracted from each of the terms. The mean is denoted by μ. Theorem 2.1 states that, for any group, $\Sigma (X - \mu) = 0$. The sum of the squared differences from the mean is symbolized $\Sigma (X - \mu)^2$. Thus the formula for the variance of a set of terms is

$$\frac{\Sigma (X - \mu)^2}{N}$$

The symbol for the variance is σ^2 (Greek small *sigma* squared), so that we can write

$$\sigma^2 = \frac{\Sigma(X - \mu)^2}{N}$$

The symbol for the standard deviation is σ (*sigma*), so that

$$\sigma = \sqrt{\frac{\Sigma (X - \mu)^2}{N}}$$

3.7 Computational Formula for the Variance

We have stated that:

$$\sigma^2 = \frac{\Sigma (X - \mu)^2}{N} \tag{3.1}$$

Two equivalent statements are

$$\sigma^2 = \frac{\Sigma X^2}{N} - \left(\frac{\Sigma X}{N}\right)^2$$

$$\sigma^2 = \frac{\Sigma X^2}{N} - \frac{(\Sigma X)^2}{N^2}$$

This is often written as

$$\sigma^2 = \frac{\Sigma X^2}{N} - \mu^2 \tag{3.2}$$

Formula (3.1) is usually referred to as the literal or *defining* formula for the variance, whereas Formula (3.2) is commonly called the *computational* formula.

The mean is likely to be a fraction even when the terms are each whole numbers, and then the differences obtained by using Formula (3.1) may be quite unwieldy. A

Table 3.6 Using the Computational Formula for Variance

X	X^2			
7	49	ΣX	=	61
8	64	N	=	5
15	225	ΣX^2	=	903
9	81	μ	=	12.2
22	484			
61	903			

tremendous advantage of computing the variance by Formula (3.2) is that there is no need to subtract the mean from each term in the group. When the original terms are whole numbers, Formula (3.2) enables the worker to compute the variance without using fractions.

When using Formula (3.2), it is necessary to set up only two columns in a table. Consider finding the variance of the distribution whose terms are 7, 8, 15, 9, and 22. The computations follow from Table 3.6.
According to the formula

$$\sigma^2 = \frac{903}{5} - (12.2)^2 = 31.76$$

The standard deviation is the square root of the variance or 5.64.

Problem Set A

3.1 Consider the Distribution X, which comprises the terms 0, 3, 9, 4, 7, and 1. Find the following.

a. ΣX
b. ΣX^2
c. $(\Sigma X)^2$
d. $\Sigma(X - 4)$
e. $\Sigma X - 4$
f. $\Sigma(X - 4)^2$
g. $\Sigma(X - \mu)$
h. $\Sigma(X - \mu)^2$
i. $\Sigma X^2 + 3$
j. $\dfrac{\Sigma(X - \mu)^2}{N}$

3.2 Compute the variances and standard deviations of Distributions W, Y, and Z using Formula (3.1).

W	Y	Z
3	4	24
7	15	27
8	17	28
	84	30
		31
		40

3.3 Compute the variances and standard deviations of the three distributions in Problem 3.2 using Formula (3.2).

3.4 Find the variance and standard deviation of each of the following three distributions by any method.

X	Y	W
7	−3	1.75
8	−2	2.81

X	Y	Z
9	1	3.90
11	0	4.00
15	7	2.36
8	8	
25		

3.5 An inventory control manager identifies 10 boxes of a certain brand of tissue. As each box is sold, the number of days it remained on the shelf is noted. The results are 1, 1, 3, 4, 5, 5, 5, 6, 7, and 9.
  a. Find the variance and standard deviation.
  b. If standard deviation is measured in days, what would variance be measured in?

3.6 Joggers Anonymous, a private athletic club, has eight members who are physicians. The doctors' physical fitness programs call for them to jog the following numbers of miles per week: 3.5, 4.75, 5, 8, 10.5, 15, 5.5, and 20.
  a. Compute the mean, variance, and standard deviation.
  b. Suppose each jogger increases his or her distance by one mile per week. Find the new mean, variance, and standard deviation.

3.7 Fifteen sufferers from migraine headaches required the following numbers of milligrams of the active ingredient in a new remedy for relief.

325, 350, 270, 280, 295, 380, 400, 250, 420, 415, 315, 365, 385, 415, 505

  a. What is the variance and standard deviation of these 15 terms? (Hint: Subtracting a constant from each term will give smaller scores to work with.)
  b. Had each person required 50 milligrams less, what would the variance and standard deviation have been?

3.8 A chemist monitoring the presence of a toxic substance in the city's drinking water records measures (in parts per billion) of 49, 67, 63, 69, 75, 52, 48, 43, 75, 74, 73, 62, 69, 70, 72, 49, 50, and 55. Compute the mean, variance, and standard deviation of these readings.

3.9 A dozen large banks report that sales of gold (in ounces) for the week were 685, 857, 973, 495, 453, 892, 1173, 733, 1244, 797, 852, and 971.
  a. Find the mean, variance, and standard deviation for gold sales at the 12 banks for the week in question by subtracting 400 from each term and using Theorems 2.3 and 3.1.
  b. If prices had gone up as expected, sales would have been down approximately 10 percent. What would the mean, variance, and standard deviation have been had each term been decreased by 10 percent? (Hint: This is the same as multiplying each term by .9.)

3.10 For the past 20 months the numbers of fires listed as arson on police reports have been 83, 48, 79, 66, 53, 94, 88, 87, 95, 94, 97, 83, 108, 67, 90, 85, 77, 65, 135, and 94.
  a. Find the variance and standard deviation for this group of terms.
  b. Proponents of a long-range plan assert that appropriate measures can cut the number of incidents of arson by 25 percent. Assuming the assertion to be true, if the plan had been in effect for the past 20 months, thus cutting each term by approximately 25 percent, about what would we expect the variance and standard deviation of the terms to be?

3.11 Consider the job-proficiency scores of the 17 applicants for the position of head counselor at a training center: 23, 25, 18, 29, 43, 35, 32, 45, 29, 27, 24, 26, 39, 35, 49, 38, and 37.

a. Find μ, σ^2, and σ.
b. Add 10 to each score and determine the new mean, variance, and standard deviation.
c. Divide each original term by 5. Find the new mean, variance, and standard deviation.
d. Increase each proficiency score by 30 percent. Compute the new mean, variance, and standard deviation.

3.12 a. Can the variance of a population be less than the standard deviation? If so, when?
b. Can the variance of a population be negative? If so, when?
c. Can the variance of a population be zero? If so, when?
d. Can the variance of a population equal the standard deviation? If so, when?

3.13 The numbers of repairs necessary over a five-year period for 25 automobiles of the same make are given here.

37	43	21	28	29	17	42	33
25	26	29	24	37	35	32	32
18	15	24	39	24	29	18	41
12							

How many of the above terms are within (a) one, (b) two, and (c) three standard deviations of the mean?

3.14 Consider a Distribution X whose mean is μ_X and whose standard deviation is σ_X. Suppose Distribution Y is obtained from Distribution X by adding 3 to each of its terms. Suppose Distribution Z is obtained from Distribution Y by multiplying each term in Y by 5. Is each of the following statements true or false?

a. $\mu_X = \mu_Y + 3$
b. $\mu_X = \mu_Z$
c. $\sigma_X = \sigma_Y$
d. $5\sigma_X^2 = \sigma_Z^2$
e. $\sigma_Y = \sigma_Z$
f. $\mu_Y = \mu_X + 3$
g. $\mu_Z = 5(\mu_X + 3)$
h. $\sigma_X = \sigma_Y + 5$
i. $\sigma_X = \sigma_Y - 5$
j. $\sigma_Y = 5\,\sigma_Z$
k. $\sigma_Y = \dfrac{\sigma_Z}{5}$
l. $\sigma_Z^2 = 25\,\sigma_Y^2$

3.15 Consider Distribution W, which is obtained by subtracting 2 from each term in Distribution V. Distribution U is then obtained by dividing each term in W by 3. Determine whether each of the following statements is true or false.

a. $\mu_V = \mu_W + 2$
b. $\mu_U = 3\mu_W$
c. $\sigma_V^2 = \sigma_U^2$
d. $\mu_V = \dfrac{\mu_U}{3}$
e. $\sigma_W = \sigma_V$
f. $\sigma_U = \dfrac{\sigma_V - 2}{3}$
g. $\sigma_U^2 = 9\sigma_W^2$
h. $\mu_U = \dfrac{\mu_V - 2}{3}$
i. $9\,\sigma_U^2 = \sigma_W^2$
j. $\sigma_V^2 = 9\sigma_U^2$
k. $9\sigma_V^2 = \sigma_U^2$
l. $\sigma_W = \sigma_U$

Problem Set B

3.16 Compute the variances and standard deviations of the three distributions given below using Definitions 3.1 and 3.2.

Distribution A	*Distribution B*	*Distribution C*
5	14	0
10	23	7
8	17	4
3	18	19
11	8	
5		

3.17 Find the variance and standard deviation of each of the three distributions in Problem 3.16 using Formula (3.2).

3.18 Find the variances and standard deviations of the four distributions given here using any method.

X	*Y*	*Z*	*W*
14	88	−7	−2.1
18	89	−8	3.7
19	63	−11	−1.1
27	55	−25	3.0
36	47	−14	−4.7
11	23	0	
12	108		
15			

3.19 Suppose Distribution T is made up of five terms: 17, 23, 11, 11, and 7. Determine the following.

a. ΣT

b. ΣT^2

c. $(\Sigma T)^2$

d. $\Sigma T + 5$

e. $\Sigma (T + 5)$

f. $\Sigma (T + 5)^2$

g. $\Sigma T^2 + 5$

h. $\frac{1}{2} \Sigma T + 5$

i. $\frac{1}{2}(\Sigma T + 5)$

j. $\Sigma (T - \mu)^2$

3.20 The shortage of fossil fuels has been a boon to the wood-burning stove industry. Black Stack, Ltd., sells wood-burning stoves through eight of its New England outlets. The numbers of stoves sold during the last fall quarter at the stores were 68, 87, 108, 99, 157, 75, 146, and 85. Find the variance and standard deviation of these terms.

3.21 Ten standard homeowner insurance policies from different companies were compared with the companies' standard policies for the previous year. In each case the newer policy was easier to understand and contained fewer words. The word reductions (in numbers of words) for the 10 policies were 897, 513, 400, 1057, 615, 299, 753, 1184, 387, and 350.

a. Determine the variance and standard deviation for the 10 terms.

b. If each policy were shortened by an additional 150 words, what would the variance and standard deviation be?

3.22 A dozen interior decorators prepared dormitory rooms for viewing by freshmen during summer preenrollment. The total cost to outfit each room was clearly displayed for the students' benefit. The 12 total costs were (in dollars):

355	275	530	250
247	625	208	175
205	227	187	211

a. Find the mean, variance, and standard deviation of the distribution.
b. If inflation increases costs uniformly by 1 percent by the time school starts, what would the new mean, variance, and standard deviation be expected to be?

3.23 A state welfare agency supports 15 credit counseling bureaus across the state. Each bureau counsels approximately the same number of people. The mean age (in years) of those seeking help with credit problems is provided here for each bureau.

33.7	35.4	36.9	30.7	33.9
32.6	34.4	30.9	32.6	32.5
37.3	36.2	31.0	32.7	32.8

a. What is the variance and standard deviation of these ages? (Hint: Subtract 30 from each term.)
b. Find the average (mean) deviation of the above terms from their mean.

3.24 Plow-Em-Under Associates recently purchased 16 farms in the nation's breadbasket. The sizes of the farms (in acres) are as follows:

350	275	210	450	425	600
520	340	480	480	510	525
315	375	220	680		

a. How much do the sizes of the 16 farms vary as measured by variance and standard deviation? (Hint: Subtract a constant from each term.)
b. Assuming workers are deeded 50 acres of each farm in exchange for cultivating the land, find the variance of the sizes of the plots held by the Associates.
c. Assuming workers are deeded 15 percent of each farm to cultivate, find the variance of the sizes of the plots held by the Associates.

3.25 The numbers of shoplifters under the age of 19 who have appeared before Judge H. Jurie for the past 18 months are 27, 18, 17, 13, 29, 37, 36, 39, 47, 52, 28, 31, 20, 31, 17, 23, 21, and 16.

a. Indicate the variability in the number of young shoplifters appearing from month to month over the past year and a half by finding the variance and standard deviation of the above terms.
b. Judge Jurie feels that 65 percent of those who appear before her on shoplifting charges should be charged with more serious offenses. If each term above were decreased by 65 percent, what would the variance and standard deviation be?

3.26 The large number of recent complaints concerning the comfort of employees assigned to the Liberal Arts Classroom Building has prompted the maintenance department to monitor temperature levels in the building's 13 classrooms. The results indicate the following temperatures (in degrees Fahrenheit).

69	67.5	73	73.5	69.5	75	78
78	79	79	83	81	77	

a. Find the variance and standard deviation of the 13 classroom temperatures.
b. Which of the following procedures would create the least variable climate in the building: increasing each temperature by 5 degrees, reducing each temperature by 10 percent, or increasing each temperature by 35 percent?
c. What two considerations (concerning the mean and standard deviation) are involved in establishing satisfactory room temperatures?

3.27 Each member of the music faculty contributes the same amount to the coffee fund each month. The total number of cups consumed, however, varies from teacher to teacher as indicated by the totals for one week given here.

14	25	46	15	24	18	23
29	10	5	15	8	17	17
15	30	11	12	10	25	

a. Find the variance and standard deviation for the number of cups consumed for the week in question.
b. Assuming everyone cuts his or her coffee drinking exactly in half next week, find next week's variance and standard deviation.
c. If everyone increases his or her coffee drinking next week by five cups, what will the variance and standard deviation be?

3.28 Consider the following terms, which indicate miles walked or ridden to school by the 17 first-graders in an experimental school designed to integrate the community culturally and economically.

3.2	.1	6.8	9.0	2.1	4.4	7.8
2.1	5.5	2.3	14.0	3.7	12.3	8.7
6.2	9.7	11.2				

How many children travel distances within (a) one, (b) one and a half, and (c) three standard deviations of the mean?

3.29 Can the standard deviation of a distribution ever equal the mean? Why or why not?

3.30 Consider Distributions X and Y.

X: 20, 17, 16, 15, 19, 19, 18, 12, 13, 13, 17, 16, 14, 14, 16, 15, 18
Y: 3, 11, 9, 12, 9, 8, 16, 13, 12, 11, 6

If the data from X and Y are combined into one distribution, verify that its variance is

$$\sigma^2 = \frac{N_X\sigma_X^2 + N_Y\sigma_Y^2 + N_X(\mu_X - \mu)^2 + N_Y(\mu_Y - \mu)^2}{N_X + N_Y}$$

where μ is the mean of the combined distribution.

3.31 Show that Formulas (3.1) and (3.2) are algebraically equivalent.

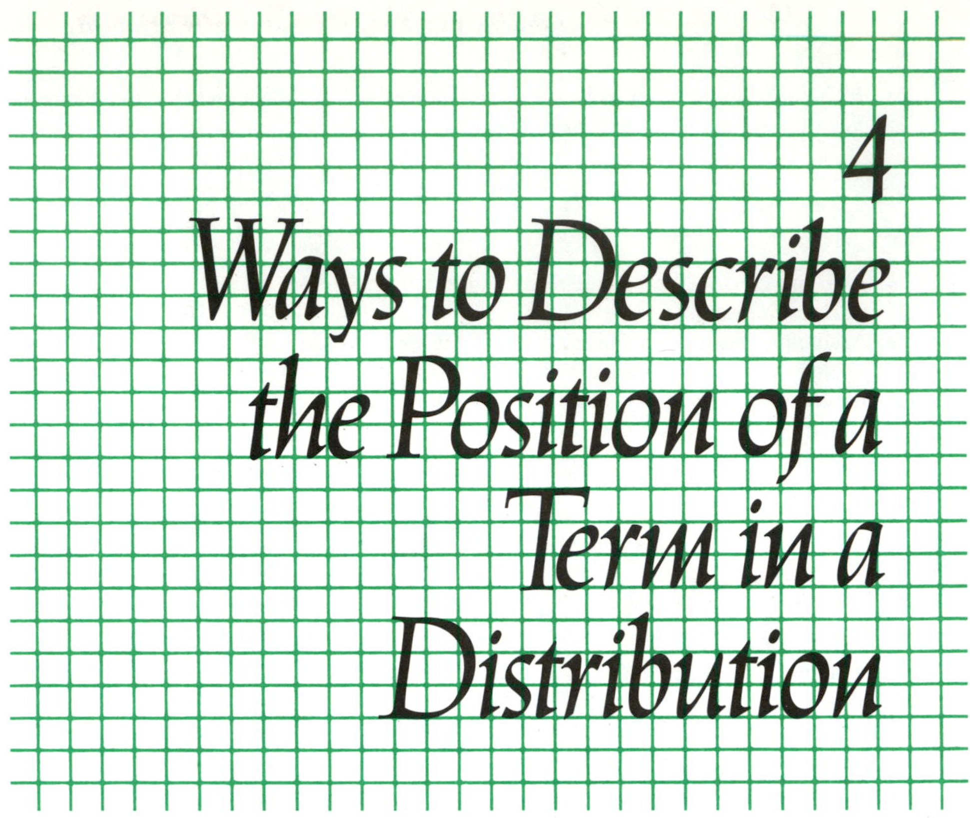

4 Ways to Describe the Position of a Term in a Distribution

4.1 Percentile Ranks and Percentiles

To describe the position of a term in a distribution is often useful. For instance, the terms may be the numerical grades of students in a class and we may wish to communicate how a particular student is doing. We may say something like "John got the third highest mark in the class," or "Arthur got next to the lowest mark." These two statements each locate a term by telling how many others there are on one side of it.

However, such information does not enable us to determine how successful Arthur was or how successful John was. There might have been three students in the class or a hundred. We must know how many students there were to determine how well these particular students fared.

To indicate the status of a term it does not suffice to announce the number of terms above or below it. The total number of terms in the distribution must also be taken into account.

We are going to take a different approach, which enables us to disregard the actual number of terms in the distribution.

We shall now consider how to locate the position of a term in a distribution by assigning to it a value called a **percentile rank.** To begin with a specific example, suppose that Bill is one of 10 students in a class and got a mark higher than exactly five of his fellow students. So far we have accounted for six students, Bill and the five others whom he surpassed. The remaining four students got marks higher than Bill's.

Here is the percentage breakdown. Fifty percent of the marks (5 out of 10) were lower than Bill's. Bill's mark itself (1 out of 10) comprises 10 percent of the marks in the distribution. Forty percent of the marks (4 out of 10) are higher than Bill's mark. These percentages are illustrated in Fig. 4.1.

Fig. 4.1

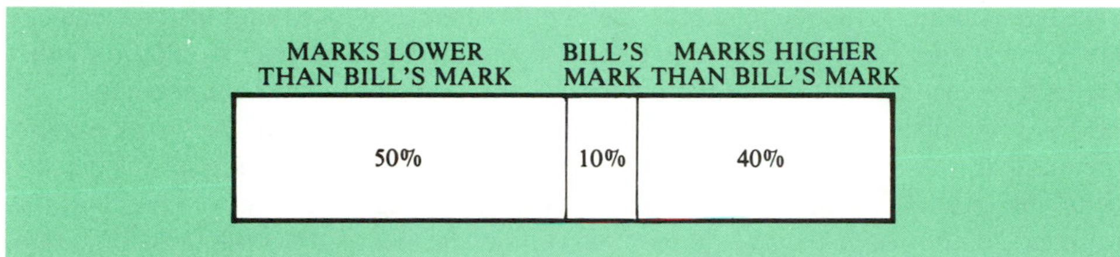

Now we are going to represent Bill's place in the percentage picture by a single line. We choose the midpoint of the segment representing his mark and draw a vertical line through it, as shown in Fig. 4.2.

Fig. 4.2

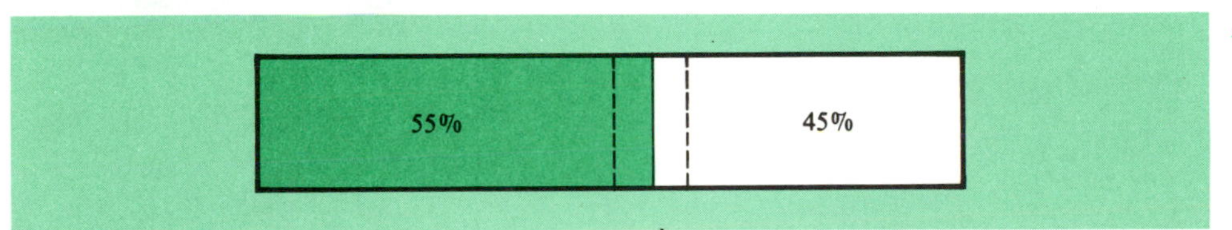

Fifty-five percent of the terms in the distribution are below this point and 45% of them are above it.

We say that the percentile rank of Bill's mark is the percent of terms in the distribution below the line we have just found to represent it. The percentile rank of Bill's mark is 55.

Definition 4.1 The percentile rank of a term in a distribution is the sum of the percent of terms below it and one-half of the percent that the particular term comprises.

As another illustration, suppose that Jim, one of 50 students in his class, received a higher mark than 17 of his classmates. We wish to find the percentile rank of Jim's mark. We begin by noting that 34% of the marks in the distribution are lower than Jim's. Jim's mark comprises 2% of the distribution of marks.

Thus, the percentile rank of Jim's mark is 34 plus one-half of 2, or 35.

Note that Bill's mark has a percentile rank of 55 whereas Jim's mark has a percentile rank of 35. We may now say that Bill did better in his class than Jim did in his. Percentile ranks have enabled us to compare the two students even though factors such as the severity of the two teachers and their methods of marking may have been different in the two classes. In general, **percentile ranks enable some comparisons to be made between terms in different distributions.** Judging by percentile ranks, Bill's accomplishment in his class is superior even though Jim happened to surpass more students in his particular class. Bill's performance was superior since he surpassed a larger fraction of his competitors than did Jim. However, we are assuming that the levels of competition are equivalent in the two classes. In practice this assumption is often not justified. Note that we have paid no heed to the *numerical* values of the terms themselves. That is, we have not considered the actual values of the marks that Bill and Jim received.

It is common practice to express percentile ranks as whole numbers between and including 0 and 99. A percentile rank of 100 assigned to a number would imply that the terms exceeded by the number and half of the terms equal to the number make up 100% of the distribution. Since this leaves 0% for the other half of the terms equal to the number, the rank 100 is never assigned. If the sum of the percent of terms below a term and one-half of the percent that the term comprises is not a whole percent, it is rounded to the nearest whole percent to obtain the percentile rank.[1] To illustrate, suppose that the world's 20 largest cities are scored by an economic index that reflects standard of living as in Table 4.1.

Table 4.1 Standard of Living Index

City	Index	City	Index
A	2.8	K	9.8
B	3.9	L	5.3
C	4.6	M	9.8
D	5.3	N	2.7
E	10.2	O	3.9
F	9.8	P	7.7
G	7.7	Q	7.6
H	13.0	R	10.1
I	2.1	S	8.4
J	.3	T	8.3

To find the percentile rank of any city we must first arrange the terms in ascending order as follows:

.3 2.1 2.7 2.8 3.9 3.9 4.6 5.3 5.3 7.6
7.7 7.7 8.3 8.4 9.8 9.8 9.8 10.1 10.2 13.0

Since there are 20 terms, each term must comprise 5% $\left(\frac{100\%}{20 \text{ terms}}\right)$ of the distribution. To find the percentile rank of city A's index of 2.8, we add the percent of terms

[1]In the case where the sum is between 99.5 and 100, 99.7 for instance, the percentile rank assigned to the term is 99, not 100.

below 2.8, which is $3(5) = 15$ to one-half the percent 2.8 comprises, which is $\frac{1}{2}(5) = 2.5$. The result, 17.5, is rounded to the nearest whole percent, 18. So we may conclude that among the world's 20 largest cities, city A has a percentile rank of 18 when the standard of living index is considered. The calculations for city F would be

$$14(5) + \tfrac{1}{2}(3 \cdot 5) = 70 + 7.5 = 77.5$$

City F's percentile rank is 78.

The percentile rank of a term is often more meaningful than its actual value. For instance, suppose you are told that Betsy Jones received a score of 113 on the Kent-Smith Aptitude Test for carpenters and then are asked whether you think Betsy Jones will be a good carpenter. You cannot possibly answer without more information. But if you are told instead that Betsy Jones' score is at the 94th percentile in a distribution of scores of carpentry students, you may say, judging from the test, that she is a good prospect for the trade of carpentry.

The word **percentile** refers directly to a term in the distribution or to a value intermediate to two terms. The term with a percentile rank of 35 is said to be at the 35th percentile. The term with a percentile rank of 74 is said to be at the 74th percentile, and so on. In Chapter 6 we shall see how to determine the value of any specified percentile in a distribution.

4.2 *z* Scores

Percentile ranks leave some questions unanswered. For instance, suppose each of the two students to be compared is the most outstanding in his respective class. One may dwarf his competitors, whereas the other's superiority in his class may be slight. The percentile ranks of the two students might be the same, giving us no clue to this difference. In this section we shall see a new way to describe the position of a term in its distribution. This new method, unlike the one just described, is sensitive to differences. We will pay no attention to percentile ranks. Instead, our way of locating a term will be to indicate how far it is from the balance point (the mean) and on which side of the mean it is.

Consider the distribution of six terms, 4, 6, 7, 10, 15, 18, with its mean of 10. By subtracting the mean from any term, we can determine the distance of that term from the mean. The sign of the distance indicates whether the term is above or below the mean. The term 15, for instance, is five units above the mean, so we may give it a "distance score" of +5. The term 7 is three units below the mean, so its "distance score" becomes −3. In the same way, we can convey the location of each of the terms in any distribution.

Suppose that the terms in the distribution given are the test marks obtained by the six students in a class. We now write the students' names, their marks, and their "distance scores" (Table 4.2).

Each student's distance score indicates the position of the student's mark in the distribution. However, we cannot use distance scores to compare students in two different classes. One reason is that, in another class, the marks may run differently

Table 4.2 Distance Scores from Mean

	Class A		
Student	**Mark**	**Mean**	**Distance Score**
Robert	4	10	−6
Arthur	6	10	−4
Bert	7	10	−3
Edward	10	10	0
Richard	15	10	5
Tom	18	10	8

(Mean = 10. Standard Deviation = 5.)

and a few points above or below the mean may have a very different meaning. For instance, consider the marks in Class B (Table 4.3).

Here the marks are much more scattered than in Class A. The distance scores of the various students tend to be larger. For instance, Marty has obtained a distance score of +25, which is considerably larger than any of the distance scores obtained by students in Class A. Yet Marty was not even the best student in his class. Furthermore, the distance score of +8, which is very outstanding in Class A, would be run-of-the-mill in Class B.

Table 4.3 Distance Scores from Mean

	Class B		
Student	**Mark**	**Mean**	**Distance Score**
Frank	20	50	−30
James	30	50	−20
Tony	35	50	−15
Bill	50	50	0
Marty	75	50	25
David	90	50	40

(Mean = 50. Standard Deviation = 25.)

In order to compare a student in one class with a student in the other we cannot simply compare their distance scores. We must do something to make up for the fact that distance score points may be "cheap" in one distribution and "expensive" in another. Suppose we wish to compare Marty's performance (in Class B) with Tom's (in Class A). Marty's distance score is +25 and Tom's is only +8, but we cannot conclude that Marty did better since the two students are in different classes. If we ask who did better, we are in effect asking: Was Marty's distance score more outstanding in Class B than Tom's distance score was in Class A? Therefore, to proceed we must find a way to determine how outstanding a distance score is in its own distribution.

Note that when the distance scores in a distribution are large as a group, the standard deviation of the distribution is large. When the distance scores are small, the standard deviation is small. The standard deviation, after all, is a measure of how far from the mean the terms are (or how large the distance scores are as a group). It follows that a student whose mark has a distance score of, say, +10 may not be outstanding if the standard deviation of marks in his class is large, for this means that some other distance scores in the distribution are also large. But a student whose mark has a distance score of +10 might be quite outstanding. If the standard deviation of marks in his class is small, this indicates that some other distance scores in the class are small. Since the standard deviation tells us about the size of the other distance scores, to appraise a distance score we should consider the size of the standard deviation of the distribution from which it came. If we were appraising the height of a 6-foot aborigine we would think of him one way if his compatriots were all giants and another way if they were all dwarfs.

The extent to which a distance score is large compared with the standard deviation of the distribution indicates the extent to which that distance score is outstanding. For instance, if a distance score is twice as large as the standard deviation, then the distance score is outstanding; if the distance score is less than the standard deviation, then it is commonplace. Remember that *the sign (+ or −) of the distance score simply tells whether the original term was above or below the mean.*

We may now wonder how outstanding Tom was in his class. What we actually ask is: How does Tom's distance score compare with the standard deviation of marks in his class? Tom's distance score is +8 and the standard deviation is 5. So we answer that Tom's distance score is 8/5 as large as the standard deviation of his class.

Definition 4.2 The ratio—distance score divided by standard deviation—is called a *z* score.

We say that Tom's z score is $8/5 = 1.6$. To find Marty's z score, we divide his distance score of +25 by the standard deviation of marks in his class (Class B). This standard deviation is 25, so Marty's z score is 1.0. Since Tom's z score is higher, we may now say that Tom was more outstanding in his class than Marty was in his.

To repeat, **the *z* score of a term tells how many standard deviations that term is above or below the mean of its distribution.** For instance, if a term has a z score of 1.2, the term is 1.2 standard deviations above its mean. A term with a z score of −.3 is 3/10 of a standard deviation below the mean of its distribution. One important

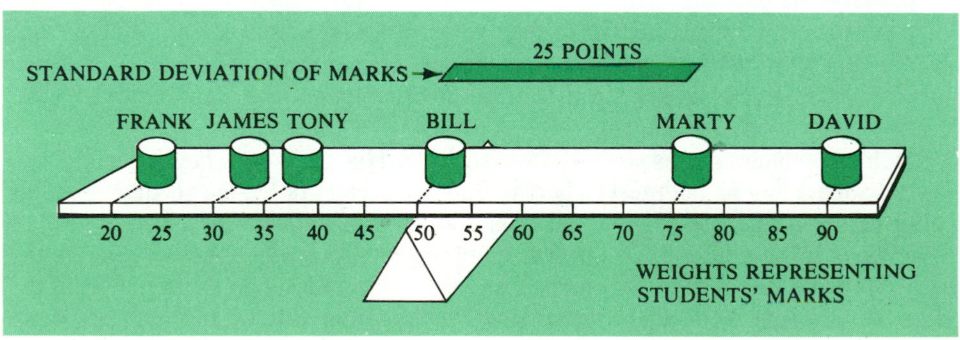

Fig. 4.3

use of z scores is to compare terms in different distributions. Unlike distance scores, z scores are not affected by the "cheapness" or "expensiveness" of points. The reason is that a z score locates a term while taking into consideration the variability of the entire distribution.

We pointed out that the standard deviation is like a ruler that reflects the scatter of the terms in the distribution (Fig. 4.3).

The z score of a term tells how many ruler lengths the particular term is from the mean. For instance, the z score of James' mark is −.80 (Fig. 4.4). The z score of Marty's mark is 1.0.

Fig. 4.4

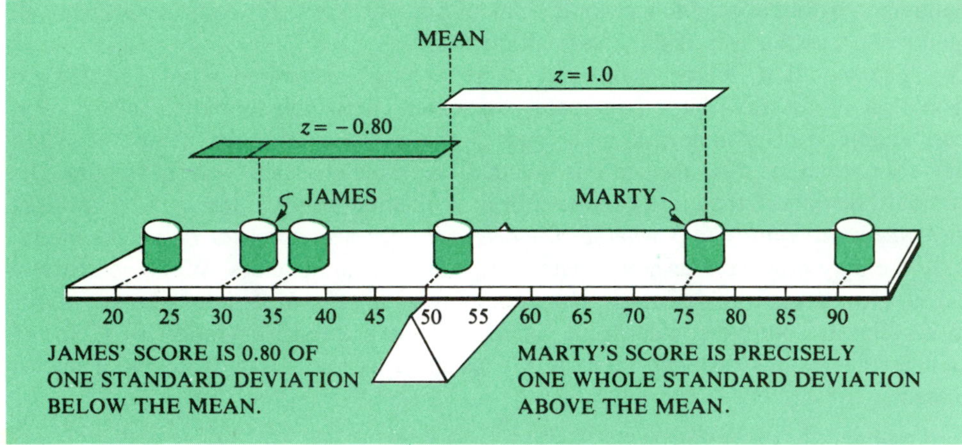
MEAN
z = 1.0
z = −0.80
JAMES
MARTY
20 25 30 35 40 45 50 55 60 65 70 75 80 85 90
JAMES' SCORE IS 0.80 OF ONE STANDARD DEVIATION BELOW THE MEAN.
MARTY'S SCORE IS PRECISELY ONE WHOLE STANDARD DEVIATION ABOVE THE MEAN.

4.3 z Score Averages

z scores are sometimes useful for finding averages when differing units of measurement are in use. Suppose a teacher gives two tests, A and B, to a class. The highest possible mark on Test A is 100 and the highest possible mark on Test B is 10. One student, Sue, gets a 76 on Test A and an 8 on Test B. Another student, Bob, gets an 83 on Test A but completely fails Test B with a mark of 1. The teacher gives Sue the higher grade, reasoning that she has done adequately on both tests whereas Bob has passed only one of them.

Our problem begins the next day when Bob returns with his grade card, complaining bitterly: "My two test marks, 83 and 1, give me a mean of 42. Sue's two marks were 76 and 8, so her mean is also 42. But Sue got the higher mark. It isn't fair."

Bob's complaint seems somewhat reasonable. His mean is as high as Sue's. Yet Sue did adequately on both tests whereas Bob completely failed the second one. Our feeling that Sue deserved a better mark is not substantiated by a simple comparison of their means. As a matter of fact, it was inappropriate to average the two tests in the first place because score points on one test were worth more than score points on the other. To compute the mean in the usual way is possible only when the score points being averaged have approximately the same *worth*. Bob surpassed Sue by seven points on Test A, but Sue came out seven points higher on Test B, where each test point counted considerably more. In a sense, Bob picked up his seven-point advantage when points were *cheap*, whereas Sue's seven-point superiority indicated a markedly superior performance. By comparing means, Bob gave himself an unfair advantage, since he was, in effect, counting each point he had earned on Test A as heavily as each point he had lost on Test B. Bob was like a man with \$2.08 claiming to be as rich as a man with \$8.02.

Let us suppose that the 100-point test had a mean of 75 and a standard deviation of 10, whereas the 10-point test had a mean of 7 with a standard deviation of 1.2. Now we can determine each student's test scores in terms of z scores. Bob's z scores are $\frac{83 - 75}{10} = +.8$ and $\frac{1 - 7}{1.2} = -5$, which have a mean of $\frac{.8 - 5}{2} = -2.1$. Bob's mean score, then, is 2.1 standard deviations *below* the mean. Sue's z scores, on the other hand, are $\frac{76 - 75}{10} = +.1$ and $\frac{8 - 7}{1.2} = .83$, which have a mean of $\frac{.1 + .83}{2} = +.47$. Sue's work is .47 standard deviation above the mean. When viewing the scores in this light, there is no doubt that the teacher was justified in awarding Sue the higher grade.

By converting the marks of each student into z scores, it is possible to obtain equal units of measurement despite the fact that the original units of measurement may have been different. The reason is that a given z score has the same meaning in any distribution. Thus, instead of comparing direct means, Bob's mean z score was compared with Sue's mean z score. In general, when different terms are based upon different units of measurement, the ordinary average loses its meaning and it is advisable to convert the terms into z scores before averaging them.

4.4 No Inherent Relationship between z Scores and Percentile Ranks

z scores and percentile ranks do not have an inherent equivalence; thus it is impossible to determine either one from the other unless other information (to be discussed later) is given. In theory, a term with a particular percentile rank may have nearly any z score. Furthermore, two terms in different distributions with identical z scores may have markedly different percentile ranks. Let us consider the distribution of the five terms 2, 3, 6, 8, 11, with its mean of 6 (Fig. 4.5).

Two terms or weights (2 and 3) are to the left of the mean, so they have negative z scores. Two others (8 and 11) are to the right of the mean, so they have positive z scores. One term (6) is equal to the mean, so its z score is zero.

The five terms are listed in Table 4.4 with their z scores and percentile ranks.

The term 8 has a percentile rank of $70 = 3(20) + ½(20)$. The z score of this term is .61. The fact that this z score is positive tells us that the term is above the mean.

Fig. 4.5

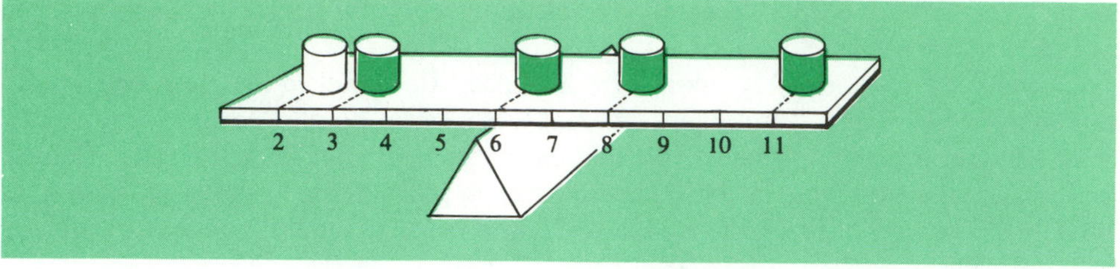

However, it is possible for a term to have a percentile rank of 70 and still be below the mean (that is, to have a negative z score). For example, consider the distribution of the five terms 0, 1, 2, 3, 14, with its mean of 4 (Fig. 4.6).

Table 4.5 shows the terms, their z scores, and their percentile ranks.

Table 4.4 Comparison of z Scores and Percentile Ranks

Terms	z Scores	Percentile Ranks
2	−1.21	10
3	−0.91	30
6	0.00	50
8	.61	70
11	1.52	90

Table 4.5 Comparison of z Scores and Percentile Ranks

Terms	z Scores	Percentile Ranks
0	−.78	10
1	−.59	30
2	−.39	50
3	−.20	70
14	1.96	90

The term 3 has a percentile rank of $70 = 3(20) + ½(20)$. But it has a negative z score because it is below the mean. In even more extremely "unbalanced" distributions, a term may have a percentile rank as high as 90 and still be below the mean.

Fig. 4.6

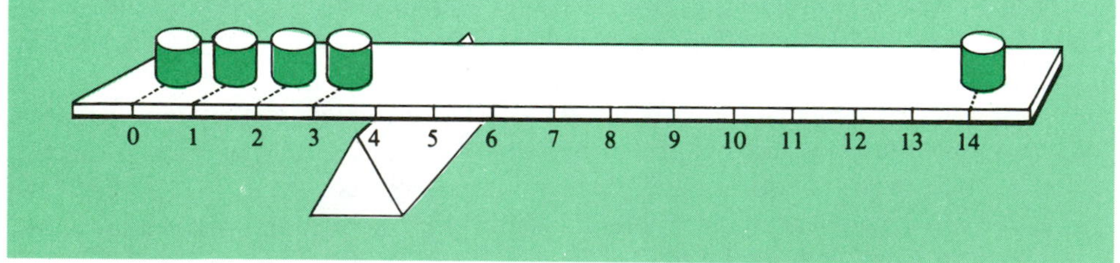

To summarize: There are no hard and fast rules for a relationship between z scores and percentile ranks. Each locates a term in a different way. The percentile rank of a term tells what percent of terms in the distribution are below it, but says nothing about its relationship to the mean. The z score of a term indicates how it stands in relationship to the mean of the distribution but does not indicate what percent of terms are above or below it.

Admittedly, in nearly every distribution that the reader is likely to meet, a term with a percentile rank as high as 70 or 80 does have a positive z score. Also, when the z score of a term is as high as 1.0, that term is generally larger than at least half of the remaining terms. But the exact relationship between percentile ranks and z scores depends upon the distribution.

4.5 The Mean and Standard Deviation of a Set of z Scores

Recall that the z score of a term is found in two steps. The first is to subtract the mean from the term to find its distance score. The second is to divide the distance score by the standard deviation of the distribution.

To demonstrate several properties of z scores, let us find the z scores of all the terms in a distribution. Suppose we are given the distribution 4, 7, 8, 10, 11, 20, which has a mean of 10 and a standard deviation of 5. In Table 4.6 the original terms have been listed in Column 1. The mean, 10, has been subtracted from each of them, leaving the distance scores listed in Column 2. These distance scores have each been divided by the standard deviation, 5, and the quotients or z scores are listed in Column 3.

Table 4.6 Finding z Scores

(1) Original Terms	(2) Distance Scores	(3) z Scores
4	−6	−1.2
7	−3	−.6
8	−2	−.4
10	0	0.0
11	1	.2
20	10	2.0

The distribution of distance scores (in Column 2) has been obtained by subtracting the mean from each of the terms in Column 1. In other words, the constant, 10, was taken from each term in Column 1 to yield the set of numbers in Column 2. We have seen that subtracting a constant from each term in a distribution reduces the mean by that constant (Theorem 2.3). Therefore, the mean of Column 2 equals the mean of Column 1 minus 10. Thus the mean of Column 2 is $10 - 10 = 0$.

Since subtracting a constant does not change the standard deviation of a distribution (Theorem 3.1), the standard deviation of Column 2 is the same as that of Column 1. Thus the standard deviation of Column 2 equals 5.

Now that we know the mean and the standard deviation of Column 2, we may temporarily forget Column 1.

Now we are after the mean and the standard deviation of Column 3, the column of z scores. Remember that we divide each term in Column 2 by 5 to get the terms in Column 3. We have seen that the process of dividing the terms in a distribution by a constant has the effect of dividing both their mean and their standard deviation by the same constant (Theorems 2.4 and 3.2). Thus the mean of the terms in Column 3 equals the mean of the terms in Column 2 divided by 5, and the standard deviation of the terms in Column 3 equals the standard deviation of the terms in Column 2 divided by 5. The mean of the terms in Column 3 is zero and the standard deviation of the terms in Column 3 is one. By the same reasoning, it can be shown that in general:

THEOREM 4.1 The mean of a distribution of *z* scores is zero and the standard deviation is one.

Note that the steps we took did not depend upon any coincidental factors in the distribution we began with. We might have begun with any set of terms listed in Column 1. Since the set of distance scores was obtained by subtracting the mean from each of them, these distance scores inevitably had a mean of zero and a standard deviation equal to that of the original terms. The process of dividing these distance scores by the standard deviation produced a mean that still was equal to zero, since zero divided by any nonzero number is zero. Thus the mean of the z scores had to be zero. Since we divided the terms in Column 2 by their own standard deviation, the effect was to divide the standard deviation of the terms in Column 2 by itself. This inevitably yielded a new standard deviation of one. Thus we might have started with any set of terms, and still we would have been able to show that their z scores have a mean of zero and a standard deviation of one.

4.6 Standard Scores or T Scores

Since z scores contain decimal points and since they may be either positive or negative, it is easy to make errors when doing computations with them. For some purposes it is convenient to change a distribution of z scores into a new distribution consisting entirely of *positive whole numbers*. The first step toward doing this is to multiply each z score by 10 and to round off each resulting term to the nearest whole number. By this process, a z score of 1.23 becomes converted to 12, and a z score of −.67 becomes converted to −7. The terms thus obtained are free from decimal points, although some of them are negative.

Now by adding 50 to each of the terms, still another distribution may be created in which all the terms are positive whole numbers. These terms are called **standard scores** or ***T* scores.** (In theory it is possible for a standard score or T score to be negative, but a negative standard score would correspond to a z score of less than -5.00. Later on we shall see that such z scores are virtually nonexistent.

A standard score or T score is conventionally obtained from a z score by multiplying the z score by 10, rounding off, and then adding 50. For instance, a z score of 1.23 becomes a standard score of 62, and a z score of $-.67$ becomes a standard score of 43. Standard scores or T scores may be used in many of the same ways as z scores. For example, the performances of students in different classes may be compared by comparing their standard scores. **The mean of a distribution of standard scores or T scores is 50 and its standard deviation is 10.**

4.7 Applications

Percentiles and percentile ranks are useful measures in many fields in which scores are obtained. For instance, college entrance committees usually consider the percentile ranks of applicants' marks rather than the numerical mark values themselves. The reason is that a good average in one preparatory school is apt to be a poor one in the next. The percentile rank of a student's average indicates how the student did with respect to his or her classmates, whereas the numerical value of the student's average reflects the stringency of the marking system in his or her school as well as the level of the student's own performance.

Economists sometimes speak of the percentile rank of an individual's income in the community or in the country. They might speak also of the percentile rank of production by a city in its particular state.

z scores play an important part in a multitude of procedures, as we shall see. However, they seldom serve as descriptive measures or ends in themselves. When data are to be presented, z scores are often converted into standard scores or T scores. Note that a single standard score conveys considerable information. Thus, a student who is told that his or her standard score on a particular test is 60 may infer that his or her mark is one standard deviation above the class mean. As we shall see, when certain other information is present, one can make rather precise inferences from knowledge of a standard score.

4.8 Notation

The letter P with a subscript is used to denote a particular percentile. Recall that a percentile is a term or value that has a specified percentile rank. Thus P_{10} stands for the tenth percentile (the term whose percentile rank is 10). P_{90} is the value of the ninetieth percentile (the term whose percentile rank is 90), and so on.

The percentile rank of a term X is designated by PR_X. If the term 45 has a percentile rank 80, we write $PR_{45} = 80$.

The z score of any particular term in the distribution is obtained by subtracting the mean from the term and dividing the obtained difference by the standard deviation of the distribution. For instance, the z score of the term X, designated z_X, is found by the following formula:

$$z_X = \frac{X - \mu}{\sigma} \tag{4.1}$$

This formula can be rearranged by solving for X yielding an equally useful formula, which is

$$X = z_X\sigma + \mu \tag{4.2}$$

The standard score of the term X, which we shall symbolize T_X, is equal to 50 plus 10 times the z score of the term. Thus,

$$T_X = 50 + 10z_X \tag{4.3}$$

When we substitute the value of z_X given in (4.1) in Formula (4.3)[2], we get a longer expression for the standard score of a term in a distribution.

$$T_X = 50 + 10\left(\frac{X - \mu}{\sigma}\right) \tag{4.4}$$

Problem Set A

4.1 The five major industries in Megapolis have had their share of labor disputes during the past decade. Below are listed the number of disputes for each industry.

Clothing manufacturing	22
Energy production	14
Banking and finance	9
Entertainment	8
Health	5

Find the percentile rank of each industry.

4.2 Find the percentile ranks of 15, 30, and 40 in each of the following distributions.

X	Y			Z			
15	2	17	30	0	25	15	18
20	7	23	39	7	40	29	103
30	9	29	40	7	40	40	22
35	15	30		3	32	52	30
40	15	30		9	15	30	25
45				18	17	35	15

[2]Note that the formula does not reflect the fact that T scores must be rounded to whole numbers.

4.3 One way of rating a television show is to indicate the percentage of the viewing audience tuned to the program. For all shows on a certain network the mean rating is 18 and the standard deviation is 5. Find the z scores of programs on that network that capture the following percents of the audience: (a) 7; (b) 25; (c) 34; (d) 10; (e) 20.

4.4 Find the T scores for each of the five programs in Problem 4.3.

4.5 The percents of the U.S. RDA (United States recommended daily allowance) of vitamin B_6 provided by 20 brands of unsweetened breakfast cereal are 25, 23, 21, 23.5, 27, 19.25, 20.33, 18.5, 28.7, 14, 33.9, 35, 27, 21.5, 23, 25, 17.5, 19.25, 23, 24.6. Find the percentile ranks of the brands that supply (a) 23, (b) 25, (c) 14, (d) 33.9, and (e) 20.33% of the U.S. RDA.

4.6 The results on the last exam given in Dr. Data's statistics class had a mean of 63 and a standard deviation of 20. Find the z scores and T scores of the following: (a) 45; (b) 52; (c) 65; (d) 73; (e) 90.

4.7 Swindler Sam sells 15 makes of domestic and imported automobiles. This year the heights of the 15 car types are (in inches)

42.5	47	50	49.5	47
49	43.5	51	45	45.5
48.5	47	50	48	46

a. Find the mean and standard deviation of the above terms.
b. Find the z scores of 45, 42.5, 48.5, and 51.
c. Find the T scores of 45, 42.5, 48.5, and 51.

4.8 The editor of *Consumer Crusader*, A.D. Vocate, has rated the 29 Freedom Party state senators on their efforts to keep state taxes from rising. The scores given the senators are

29	59	83	92	53	65	72	83	71	79
73	62	27	35	47	60	65	79	87	76
68	85	91	39	47	58	73	73	75	

Find the percentile rank of the senator who (a) is third from the top, (b) received the lowest score, (c) received a score of 73, and (d) received a score of 62.

4.9 If the heights of a new strain of tomato plant are distributed with a mean of 2 feet and a standard deviation of 4.5 inches, find the heights of plants that have the following z scores: (a) 0; (b) 1; (c) 1.85; (d) -2.34; (e) -5.

4.10 Salesperson Sarah sells 50 units on day one when the mean number of units sold is 65 and the standard deviation is 12, and 18 units on day two when the mean is 15 with a standard deviation of 3. Salesperson Saul sells 45 units the first day and 21 units on the second day. Which salesperson is the most successful on the basis of these two days' sales?

4.11 There are 41 cars in the Hodgeson Factory parking lot. The numbers of people who rode to work in the 41 cars are 1, 1, 1, 4, 1, 1, 1, 2, 1, 1, 1, 3, 2, 1, 1, 2, 1, 1, 4, 1, 1, 1, 3, 1, 1, 2, 1, 1, 2, 2, 3, 1, 1, 1, 1, 1, 1, 2, 1, 5, 1. Find the percentile rank of each car carrying only one person to work.

4.12 There are 16 brands of car ramps currently on the market. The weight ratings are as follows (in pounds):

3,500	7,300	6,500	8,500
3,700	6,000	6,600	4,300
8,000	5,000	4,000	3,900
5,500	4,500	3,500	8,100

a. Find the mean and standard deviation of the weight ratings. (Divide by a constant first!)
b. What are the z scores of the four ratings in the first column?

Suppose that a new metal alloy made it possible for each rating to be increased 500 pounds.

c. What would the new mean be?
d. What would the new standard deviation be?
e. How would z scores be affected? Explain.

4.13 Thirty-five acne medications were tested by hundreds of teenagers across the country. They were identified only as compound one, compound two, and so on. On the basis of the responses, the compounds are listed (by number) in order of effectiveness from least to most effective. Circled compounds tied for the positions they occupy.

1.	1	8.	23	15.	2	22.	18	29.	9
2.	22	9.	32	16.	17	23.	25	30.	29
3.	7	10.	14	17.	11	24.	34	31.	12
4.	35	11.	31	18.	27	25.	8	32.	24
5.	15	12.	26	19.	19	26.	10	33.	5
6.	6	13.	16	20.	30	27.	20	34.	13
7.	28	14.	21	21.	4	28.	3	35.	33

Find the percentile ranks of compounds

a. 6 b. 26 c. 11 d. 4 e. 20 f. 12 g. 33

4.14 In a distribution of 50 terms, a certain term has a percentile rank of 45. There are two other terms equal to it. How many terms does P_{45} exceed?

4.15 Show that a distribution of T scores must always have a mean of 50 and a standard deviation of 10.

Problem Set B

4.16 Consider Distributions W, U, and V.

W		*U*				*V*	
0	−3	6	11	7	9	4	9
2	−5	0	12	7	16	3	14
7	−14	−3	15	8	12	102	13
8	8	−7	12	3	11	11	8
4	7	2	−1	4	1	7	7

a. In Distribution W find the percentile ranks of 0, 7, and 4.
b. In Distribution U find the percentile ranks of −3, 0, and 6.
c. In Distribution V find the percentile ranks of 7, 11, 102, and 13.

4.17 The pollution indexes in Roverville for the first 14 days of July were 49, 52, 57, 88, 85, 73, 65, 77, 62, 80, 53, 50, 49, and 45.

a. Find how the third day of the month ranks among the first 14 days by finding its percentile rank.
b. Which day has a percentile rank of 96?

4.18 The level of Lake Ihde has a mean of 4 feet above the spillway with a standard deviation of 3.5 feet. Negative values indicate the level is below the spillway. Find the z scores of the following levels.
a. 5 feet
b. −7 feet
c. 24 inches
d. −1.5 yards

4.19 a. Find the T scores of the levels given in Problem 4.18.
b. Find the levels that have z scores of 1.55 and −2.13.

4.20 The city transit authority operates eight buses. The average number of minutes that each bus has been late over the last six weeks has been calculated. Here are the results.

Bus	*Average Number of Minutes Late*
A	3
B	15
C	12
D	2
E	20
F	0
G	4
H	3

Determine the percentile rank of each bus. Give buses with *lower* averages *higher* rankings.

4.21 Last year the county administered a state-supported program to aid qualifying families in paying utility bills. A month-by-month breakdown of the number of families aided is given below.

Month	*Families Aided*
January	59
February	73
March	48
April	30
May	8
June	10
July	45
August	60
September	55
October	50
November	50
December	60

a. Find the mean and standard deviation of the number helped per month.
b. Find the z scores for February, August, and May.
c. Find the T scores for February, August, and May.

4.22 Declining enrollment and escalating costs have forced staff cuts at each of the 16 public schools in the Middleberry District. The cuts, by school, are

13	11	14	7	8	5	3	2
14	3	7	11	7	5	9	11

a. What percent of the schools released more than 10 people?
b. Find the percentile ranks of the schools in the district that cut 11, 3, and 14 positions.

4.23 Thirteen brands of a specific type of adhesive can be purchased in this country. The labels claim that the holding strengths (in lb/in^2) are 385, 373, 481, 916, 418, 516, 630, 720, 415, 500, 525, 400, and 825.
a. Find the mean and standard deviation of these terms. (Subtract a constant from each term to yield smaller numbers.)
b. Find the z scores of the brands claiming holding strengths of 373, 415, 500, and 400.

4.24 Three hundred anglers entered the LaVoy County fishing tournament. The mean weight of strings caught was 15 pounds with a standard deviation of 7 pounds. Find the weights of strings with z scores of
a. -1.4
b. .9
c. 2.11
d. How likely is it that a string would have a weight whose z score is -3?

4.25 If Richter Scale readings of the previous 17 earthquakes are as given below, find the percentile ranks of (a) the median value, (b) the value closest to the mean, and (c) the modal value.

3.7	3.9	4.2	7.1	6.3	3.7	3.8	4.5	7.2
5.3	5.6	3.7	2.9	5.1	5.0	4.6	5.1	

4.26 Thirty-five female students took part in a biorhythms study. Each rated herself on various traits daily on a scale from 1 to 10. The ratings on intellectuality for the 15th day are provided here.

1	7	8	4	3	9	6	7	5	4
3	1	4	8	7	9	10	3	8	9
5	4	8	7	6	4	9	8	9	8
1	9	10	9	4					

a. Calculate the mean and variance.
b. Calculate the z scores for the terms 2, 4, and 9.
c. Calculate the T scores for 2, 4, and 9.

4.27 "Slick" Stevens studies oil slicks. The width measurements of the most recent 25 he has examined have been (in miles):

5.0	7.5	8.4	3.9	6.0	15.0	12.5
5.5	14.0	21.5	3.8	7.0	9.0	11.2
4.9	5.5	8.0	7.0	3.5	14.4	11.0
20.0	7.0	12.6	12.0			

a. Find the percentile ranks of the measurements 3.8, 14.4, and 5.5.
b. If each term is increased by 5, what are the percentile ranks of 3.8, 14.4, and 5.5?
c. If each term is increased by 10 percent (multiplied by 1.1), what are the percentile ranks of 3.8, 14.4, and 5.5?

d. How does adding a constant to each term and multiplying each term by a constant affect percentile ranks?

4.28 The weight changes of the 40 clients participating in a reducing program at Fannie Famine's Figure Factory are as follows:

0	+3	−2	−5	+4	−7	−10	+3	+6	0
−1	−2	−1	0	−6	+3	+8	−1	0	+4
+1	+2	0	−1	−15	+1	+3	0	0	+5
+10	−5	0	+1	−1	−1	−2	−3	0	+7

a. Find the mean and standard deviation of the weight changes.
b. Find the z scores of −6, 0, and +3.
c. Find the percentile ranks of −6, 0, and +3.

4.29 Give an example of a distribution in which the largest term smaller than the mean has a percentile rank of 95.

4.30 Verify that the mean and standard deviation of the z scores of the terms in the following distribution are 0 and 1 respectively.

3 8 −2 1 4

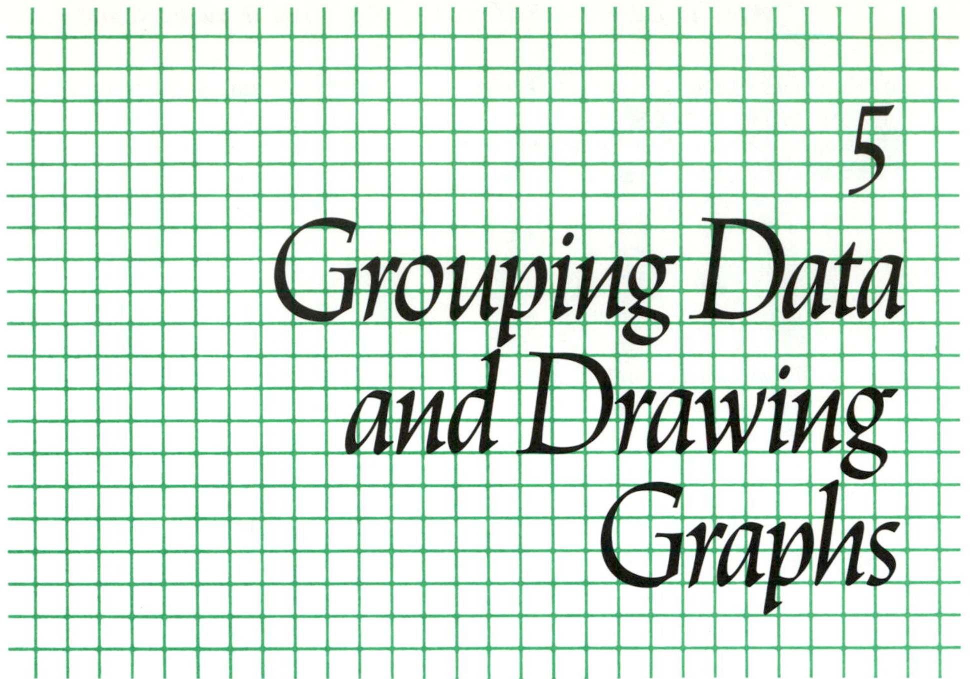

5.1 Discrete and Continuous Variables

Recall that the terms in a distribution may be regarded as the values assumed by some variable for the members of a certain population. Every distribution thus depends on two things: a population and a *variable*.

For instance, consider the distribution of body temperatures of the 15 patients in Ward B of County General Hospital. The population consists of the 15 patients. The variable is body temperature. As we take the temperature of one person after another, the variable assumes different values, specifically 15 different values, the temperatures of the 15 people.

As another example, think about the Gross National Products for the years 1950–1980. There are 31 years (including 1950), so this time the population has 31 members. The variable is the GNP. As we move from year to year, the GNP takes on new values. The distribution in question has 31 terms, one standing for the GNP of each of the 31 years in the population.

To treat a distribution properly we must take into account the kind of variable with which we are dealing. Variables are classified as either **discrete** or **continuous**.

Discrete variables are, in general, those variables that may assume values that can be listed or can be placed in order of first, second, third, and so on. A good example is the variable, scores on a 100-point true–false test. The values that the variable can assume can be listed quite easily: 0, 1, 2, 3, . . . 100. This makes 101 different values. We could list them all. If it is possible to list every single value a variable might assume, the variable is discrete.

Sometimes there are so many values that it is impossible to write them all down. Such is the case with the number of people viewing the World Series on television.

Each year a new value is assumed by the variable. Next year, theoretically, any number of people within range might watch. This means any positive whole number is a possibility: 1, 2, 3, 4, 5,... 1,000,000,... Even though we cannot list all of the values, we can clearly see that they can, theoretically, be placed in order of first, second, third, and so on. Whenever it is possible to put in order all the values a variable may assume, that is, in the position of first, second, third, and so on, the variable is also discrete.

Several further examples of discrete variables should prove instructive. The number of "yes" votes by the members of a 10-person committee assumes one of 11 possible values at each observation. We can list them all: 0, 1, 2, 3, 4, 5, 6, 7, 8, 9, 10. The variable, number of people married each hour at City Hall, is also discrete. It may assume any of the values 2, 4, 6, 8, 10,... and so on. Even though it is impossible to list them all, the position of any value on the list can be determined. The value 100, for example, is the 50th term.

The values of a discrete variable are often categories. For instance, marital status is discrete since all of its values may be listed: single, married, divorced, and widowed. Army status is also a discrete variable as is toothpaste preference, since there are a limited number of categories or responses that may serve as values.

Variables not discrete are classified as continuous. The values of a continuous variable cannot be listed, nor can the position of any possible value be determined to be first, second, third,... For instance, the variable, age, is continuous. Even if we consider only those ages between 5 and 10 years, we cannot possibly list them all. For one child the variable may assume the value 4.54 and for another 4.55. Between these two ages are an infinitude of other values that may also be assumed (such as 4.541, 4.542, and 4.5499). Variables such as height, weight, humidity, and temperature are also continuous. Between any two values that may be assumed, an infinitude of other possible values may be found. Between 1 pound and 2 pounds, or 1 foot and 2 feet, or 1 degree and 2 degrees are numerous other values. **In general, if between any two values the variable assumes, there are other values that the variable could also assume, the variable is continuous.**

A useful distinction is to think of continuous variables as *measurable* and discrete variables as *countable*. Height and weight are measured whereas attendance and product preference are obtained by counting.

Often it is necessary to treat a truly continuous variable as discrete for experimental purposes. Intelligence and special abilities are generally thought to be continuous, but we score them using such tests as the Wechsler Adult Intelligence Scale, which allows for only 158 possible scores. Likewise, height may be expressed in whole inches and weight in half-pounds, thus creating discrete variables from continuous ones.

5.2 Tabulating and Graphing the Values of a Discrete Variable

The distribution of values of a discrete variable may be conveniently summarized in what is called a **frequency table.** To begin with, we shall define the word *frequency*.

Definition 5.1 The frequency of any value that a variable may assume in a distribution is the number of times that value appears in the distribution.

As an example, suppose that among employees in a factory 10 are single, 20 are married, 15 are divorced, and 5 are widowed. The frequency of single workers is said to be 10, the frequency of married workers is 20, and so on.

Table 5.1 is a frequency table of the distribution of marital status in the factory. The possible values of the variable, marital status, are listed in the first column and the frequencies of these values are listed in the second column.

Table 5.1 Distribution of Marital Status

Marital Status	Frequency
Single	10
Married	20
Divorced	15
Widowed	5

The terms in the distribution are illustrated as weights on a plank in Fig. 5.1.

Note that the four sections on the plank corresponding to the four possible values are separated so that the weights in different piles do not touch each other. The frequency of each value is indicated by the number of weights piled up at that value—that is, the 10 weights piled up over the value "single" indicate that there are 10 single people in the distribution, and so on.

Fig. 5.1

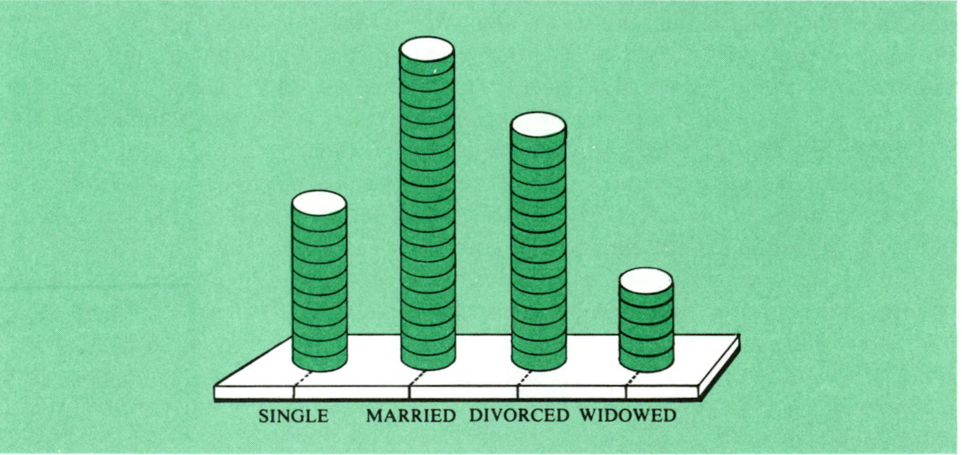

We may simplify the representation by drawing a front-view diagram of the weights. The plank is represented by a horizontal line and the weights are represented by rectangles. A vertical axis serves to indicate the number of weights in each pile (Fig. 5.2).

Now we are ready to simplify the representation into its most usual form. Fig. 5.3 is called a **bar graph**.

Fig. 5.2

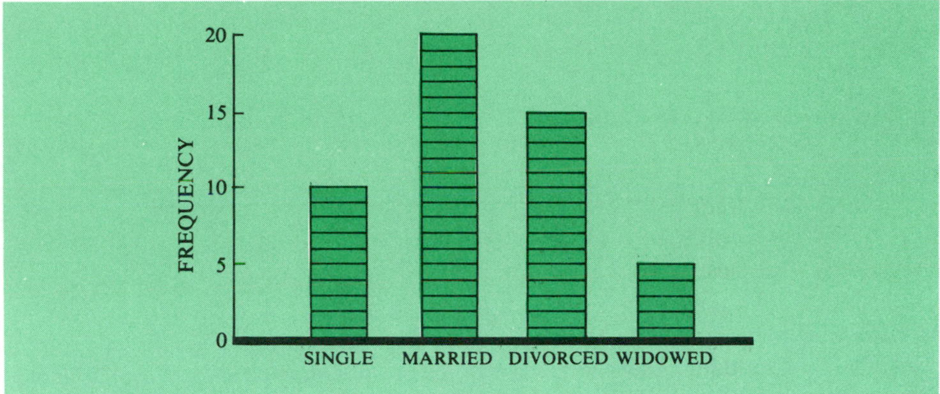

Fig. 5.3

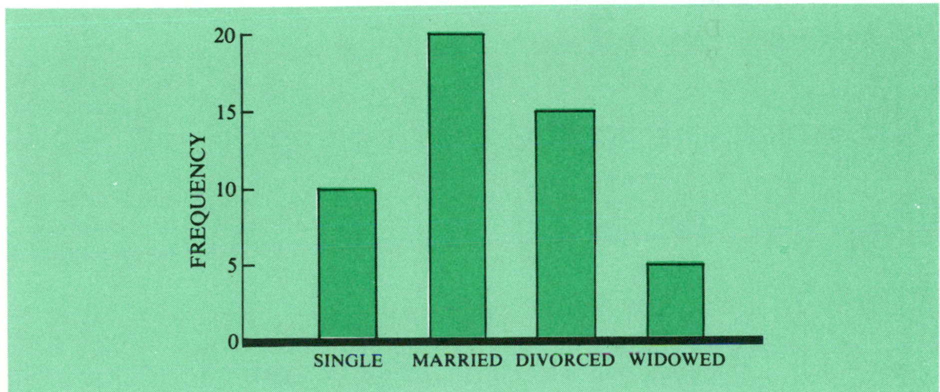

The bar graph is the most convenient and popular kind of graph used to depict the distribution of values of a *discrete* variable. The first step toward creating a bar graph is always to draw the frequency table of the distribution. Then the bar graph is drawn from this table. Note that there are spaces between the different bars.

5.3 Rounding Off

In practice, the values of a continuous variable must *always* be rounded off. Often rounding the values of a discrete variable is advantageous. Measurements are never made with perfect accuracy and, even if they were, the task of working with exact values would be prohibitive. The process of rounding off facilitates many computational procedures and it is often serviceable to round off values of a discrete variable as well as a continuous one.

When we round off values of a variable, we are in effect "forcing" the variable to assume one of a limited (or finite) number of values. For instance, suppose we round off heights to the nearest inch in the interval 4 feet 11½ inches up to but not including 6 feet ½ inch. We are forcing the variable, height, to assume one of 13 rounded-off values. These values are 5 feet, 5 feet 1 inch, 5 feet 2 inches, and so on. We picture these rounded-off values as spots along a plank (Fig. 5.4).

Fig. 5.4

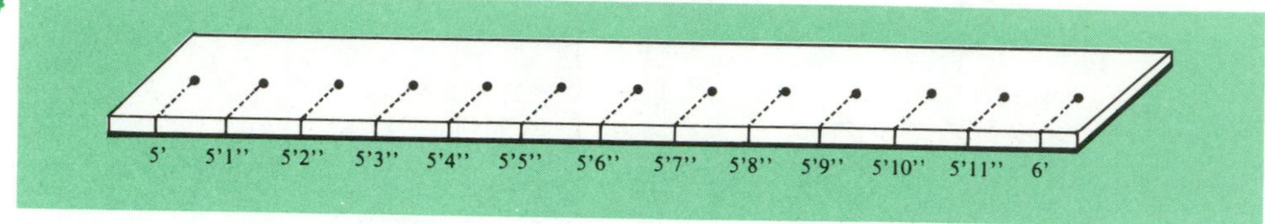

Rounding off a term means giving it the value of the spot or "rounded value" on the plank to which it is closest. Each rounded-off value on the plank is the midpoint of an interval that includes observations extending ½ inch on each side of it. Altogether, there are 13 intervals, each an inch wide, stretched across the plank (Fig. 5.5).

Fig. 5.5

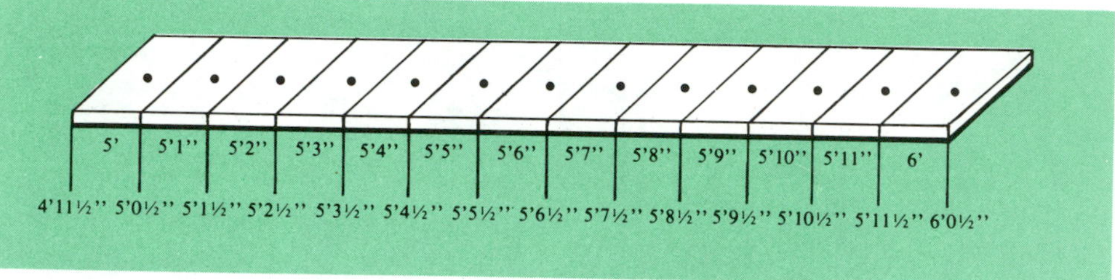

When we round off a term, we are in essence locating it in one of the 13 intervals and then assigning it the value of the midpoint of that interval. Thus the term 5 feet 5¾ inches falls in the indicated interval (Fig. 5.6). We assign it the value of the midpoint of that interval, which is 5 feet 6 inches. In general, the process of rounding off the terms in a distribution involves grouping the terms into intervals and then assigning the terms in each particular interval the value of the midpoint of that interval. Techniques of working with rounded-off values in order to simplify computations are called **grouping methods**.

Fig. 5.6

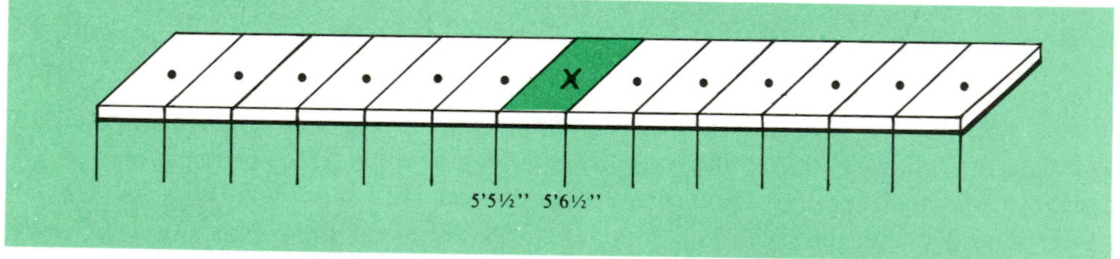

Note that the width of the intervals used indicates the degree to which the terms are being rounded off. In our example, the intervals are each one inch wide, reflecting the fact that heights are being rounded off to the nearest inch.

5.4 Grouping Data—The Discrete Case

We shall now consider the formal procedure of putting a mass of data into intervals for purposes of drawing graphs and simplifying computing procedures. Values from either discrete or continuous variables may be used to form what is called a grouped frequency table. In effect, we are considering a formal procedure for putting the terms in a distribution into intervals with midpoints that are convenient numbers with which to work.

A *five-step procedure* is generally used to group a mass of observations and to tabulate them. First let us consider the method used for discrete variables.

1. The first step is to compute the range. This is the difference between the largest and smallest terms in the distribution. As an illustration, suppose the data are the marks on a 100-word spelling test given to 1,000 fourth-grade children in a community. The highest mark turns out to be 100 and the lowest mark to be 3. The range of marks is $100 - 3 = 97$.

2. The next step is to decide upon the number of intervals to be used. There are no hard and fast rules as to the number of intervals to be used. We shall, however, suggest some guidelines that should make the task easier for the beginner. Do note, however, that different authors and researchers may use other methods that are perfectly acceptable. In other words, *the end product is not unique.*

Suppose that N represents the total number of terms to be grouped and n represents the number of intervals that will contain these terms. One good rule of thumb is to use the largest n that will satisfy

$$2^n \leq N \tag{5.1}$$

In our case we need to find the largest n such that $2^n \leq 1{,}000$. Since $2^9 = 512$ and $2^{10} = 1{,}024$, we should use about 9 intervals. The word "about" needs to be stressed. If 8 or 10 intervals prove to be more convenient as we go along, we reserve the right to use the more convenient number. Table 5.2 provides the approximate number of intervals to be used for up to 4,095 terms.

Table 5.2 Selecting Number of Intervals

N	n
16–31	4
32–63	5
64–127	6
128–255	7
256–511	8
512–1,023	9
1,024–2,047	10
2,048–4,095	11

3. Once the number of intervals has been chosen, in this example 9, the size of each interval must be determined. The approximate size is found by dividing the range by the number of intervals. In the example, $97/9 = 10.78$. It is of great convenience to make the width of the intervals an odd number so that the midpoint of each interval will be a whole number. Thus, in this case, we would probably decide to make the intervals 11 units wide since 11 is the closest odd number to 10.78. The expanse of 9 intervals, laid end to end, is 99 units. It is essential that the expanse of intervals be at least one more than the range. Note that for the spelling test scores, there are 98 whole numbers between and including 3 and 100. That is one more than the range, 97. So far we have specified the width for our intervals and decided how many to use.

4. The next step is to specify the endpoints and the midpoints of the intervals. In our illustration, we will make the highest interval 90–100. Note that when we state the endpoints of an interval we intend to include these endpoints in the interval. Thus, the interval 90–100 is 11 units wide since it contains 11 whole numbers. The next highest interval becomes 79–89. We keep on constructing intervals until we find the lowest interval, which is 2–12.[1]

The nine intervals and their midpoints are shown in Table 5.3. Remember that each interval is 11 points wide since both endpoints are included. We have constructed nine consecutive equal-sized intervals and have indicated their midpoints. The midpoints may be found by finding the means of the interval endpoints.

Table 5.3 Grouped Distribution Pattern

Intervals	Midpoints
79–89	84
68–78	73
57–67	62
46–56	51
35–45	40
24–34	29
13–23	18
2–12	7

5. To complete the grouped frequency table, we tabulate the frequencies of the different intervals. In other words, we tabulate the number of terms falling into each interval and record these numbers. For instance, the grouped frequency table of the distribution of children's scores on the spelling test might look like Table 5.4.

Table 5.4 tells us that 78 children got spelling marks between 90 and 100 (including the marks of 90 and 100). It tells us that 140 children got marks between 79 and 89 (including the marks of 79 and 89), and so on.

[1]We could have made the highest interval 91–101 and worked down to 3–13. There are no definite rules as to where to begin since the grouped frequency table for a set of data is not unique.

We can now proceed as if the distribution were made up of 78 terms with a value of 95, 140 terms with a value of 84, and so on. That is, we can assign to all the terms

Table 5.4 Grouped Frequency Table—Discrete Case

Intervals	Midpoints	Frequencies
90–100	95	78
79–89	84	140
68–78	73	192
57–67	62	120
46–56	51	140
35–45	40	80
24–34	29	119
13–23	18	81
2–12	7	50
		1,000

in each interval the value of the midpoint of that interval. Doing this is, in essence, rounding off the terms to the nearest 11-point units, just as we rounded off heights to the nearest inch in the previous section. We will see how to compute various measures by group methods in Chapter 6.

Having constructed a grouped frequency table using data from a discrete variable, we may exhibit the results using a bar graph (Fig. 5.7). Note that the interval endpoints and midpoints have been indicated along the horizontal axis.

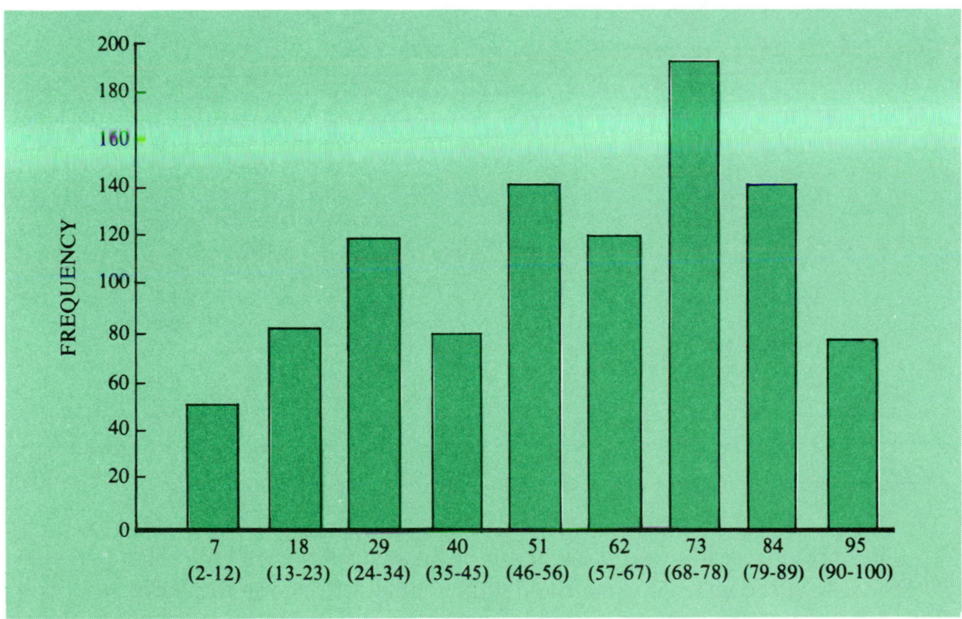

Fig. 5.7

5.5 Grouping Data—The Continuous Case

Next we shall consider the grouped frequency table for values assumed by a continuous variable and the methods for drawing graphs using the table. To be specific, consider again the population of 1,000 children. This time, however, instead of looking at their spelling scores, consider the percent of the United States recommended daily allowance (U.S. RDA) of a specified vitamin that each received on a certain day. Suppose that the highest percent is 100 and the lowest percent is 3. Be aware that values other than whole numbers may appear in the distribution, and the variable is continuous.

The first three steps of the five-step procedure for obtaining a grouped frequency table are the same as for the discrete case. So the range would be $100 - 3 = 97$. Again we decide upon nine intervals of width 11.

In the case of a continuous variable we must be able to place every conceivable number in the range (between 3 and 100 in our example) into some interval. This means the lower boundary of an interval *must coincide* with the upper boundary of the interval immediately below it. In the continuous case, the width of an interval is the difference between its upper and lower boundaries. It is usually most convenient to use boundaries halfway between two consecutive whole numbers.

The boundaries of our top interval would be 89.5 and 100.5. Its width is $100.5 - 89.5 = 11$. Values such as 90.356 and 98.326 would be placed in this interval. The next lower interval would be 78.5–89.5. We can find the boundaries of all the intervals. They are indicated in Table 5.5 along with the frequencies for each interval.

Table 5.5 Grouped Frequency Table—Continuous Case

Boundaries	Midpoints*	Frequencies
89.5–100.5	95	119
78.5–89.5	84	81
67.5–78.5	73	50
56.5–67.5	62	140
45.5–56.5	51	132
34.5–45.5	40	145
23.5–34.5	29	160
12.5–23.5	18	75
1.5–12.5	7	98
		1,000

*The midpoint may be found by finding the mean of the boundaries of each interval.

We are now ready to draw a graph of the distribution of children's percents of U.S. RDA of the vitamin using the data in Table 5.5. Along the horizontal axis of Fig. 5.8 we note the boundaries of the successive intervals and also their midpoints.

We use the vertical axis to indicate frequencies as we did when we drew the bar graph. Once again we draw bars over the intervals to indicate the frequencies or number of scores in each interval. The bar over each interval extends from its lower boundary to its upper boundary so that adjacent bars just touch each other. The heights of the various bars in Fig. 5.8 indicate the frequencies of the various intervals.

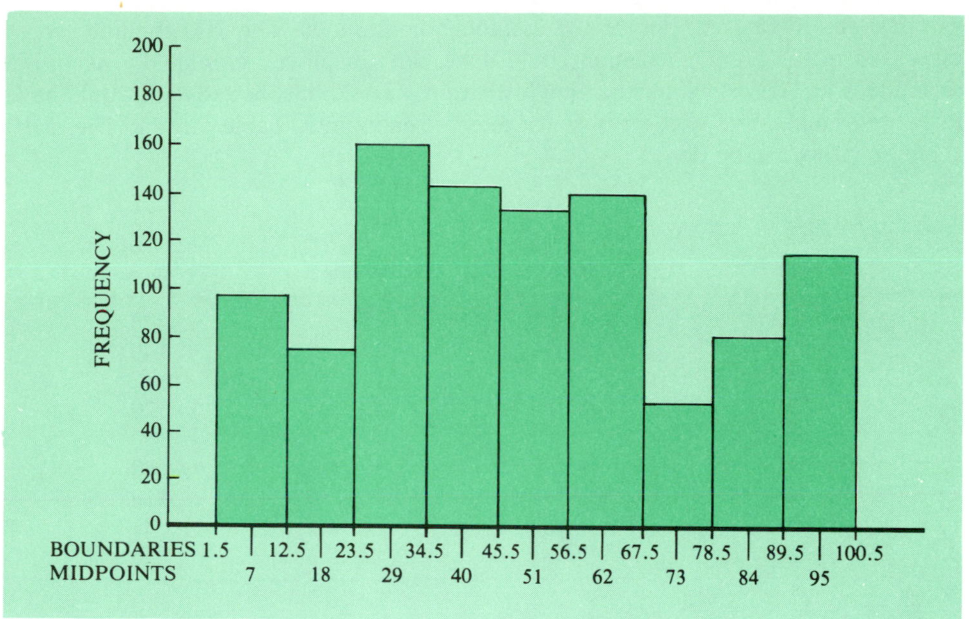

Fig. 5.8

We now remove the vertical lines separating successive bars in order to indicate that we are graphing a continuous variable. The graph in Fig. 5.9 is called a **histogram.**

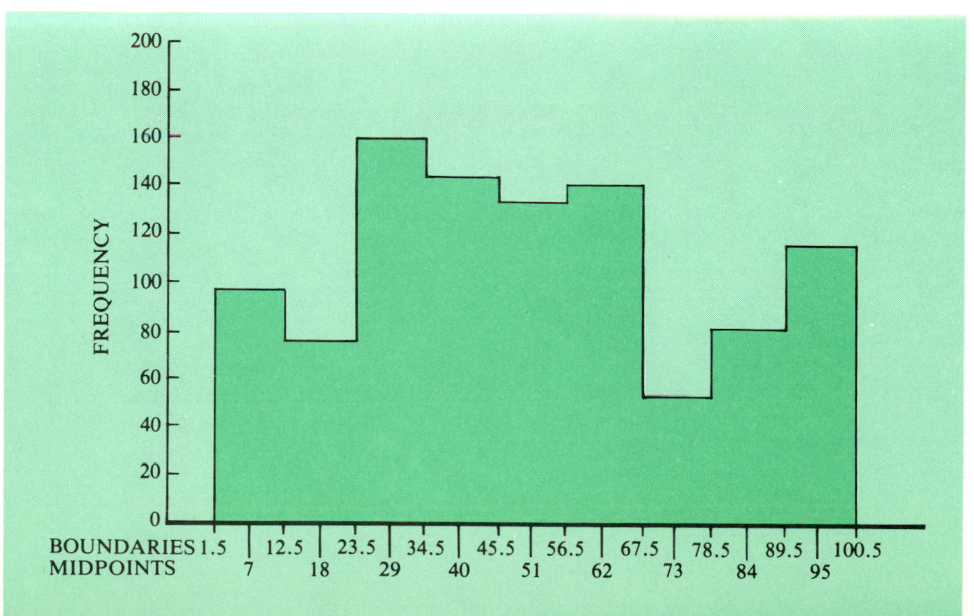

Fig. 5.9

The histogram, which is a graph of a continuous variable, resembles the bar graph in various ways. Values are recorded along the horizontal axis in each case and frequencies are indicated by the heights of the various columns. However, the bars are quite separate in a bar graph, indicating that the underlying variable is discrete. Successive bars are "run together" in a histogram to indicate that the underlying variable is continuous.

Discrete data are often treated as though continuous. The 100 spelling scores discussed in the previous section could have been displayed graphically using the techniques just described for drawing histograms. To do this, boundaries would have to be determined and used in place of interval endpoints. Table 5.6 and Fig. 5.10 show how this can be done.

Table 5.6 Spelling Scores

Intervals	Boundaries	Midpoints	Frequencies
90–100	89.5–100.5	95	78
79–89	78.5–89.5	84	140
68–78	67.5–78.5	73	192
57–67	56.5–67.5	62	120
46–56	45.5–56.5	51	140
35–45	34.5–45.5	40	80
24–34	23.5–34.5	29	119
13–23	12.5–23.5	18	81
2–12	1.5–12.5	7	50

Fig. 5.10

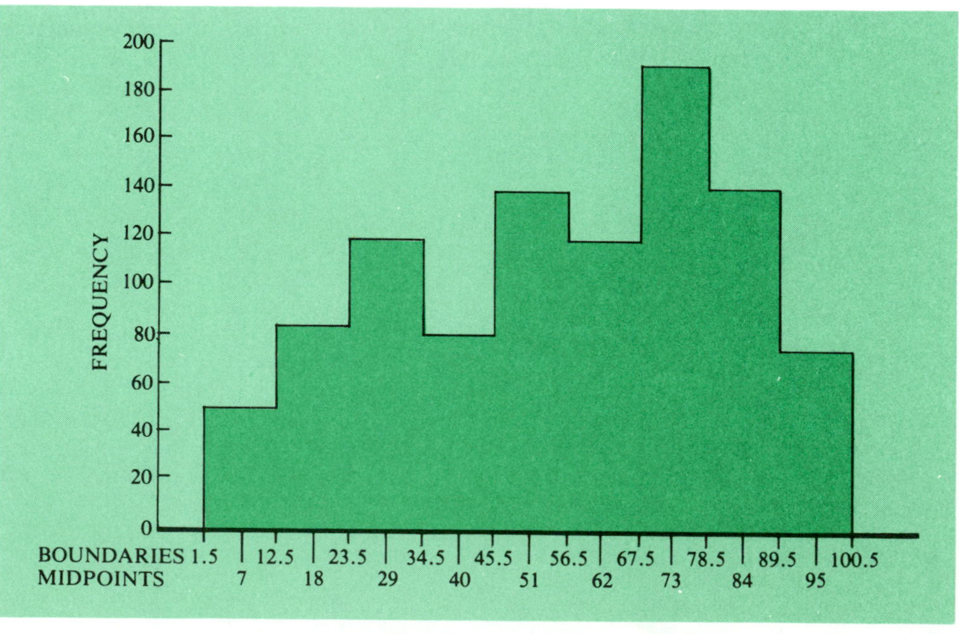

A second way to graph values of a continuous variable is by means of a **frequency polygon.** The frequencies of the various intervals are indicated by points drawn over the midpoints of the intervals and these points are connected by straight lines as shown in Fig. 5.11.

Fig. 5.11

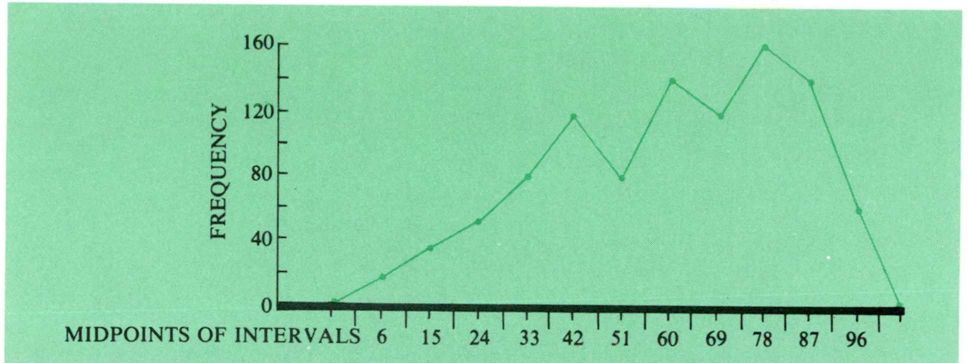

5.6 The Curve as an Approximate Graph

We shall now compare several histograms that depict an identical distribution. This time we shall consider for our illustration the distribution of heights of all American adult males between 5′ 3½″ and 5′ 11½″. The histogram of these heights rounded off to the nearest inch is shown in Fig. 5.12.

Fig. 5.12

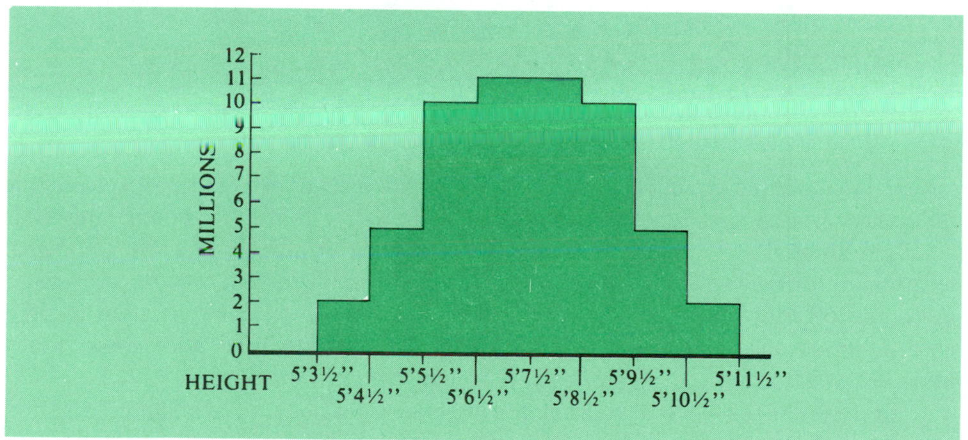

This histogram tells us that 2 million men are between 5′ 3½″ and 5′ 4½″. It tells us that 5 million men are between 5′ 4½″ 5′ 5½″, etc. Twelve and a half percent of the area is to the left of 5′ 5½″ and 50% is to the left of 5′ 7½″. We may conclude from the graph that one-eighth of the men are shorter than 5′ 5½″ and that one-half of them are shorter than 5′ 7½″.

Now let us look at another graph of the same distribution. The intervals in the histogram in Fig. 5.13 are half inches.

Fig. 5.13

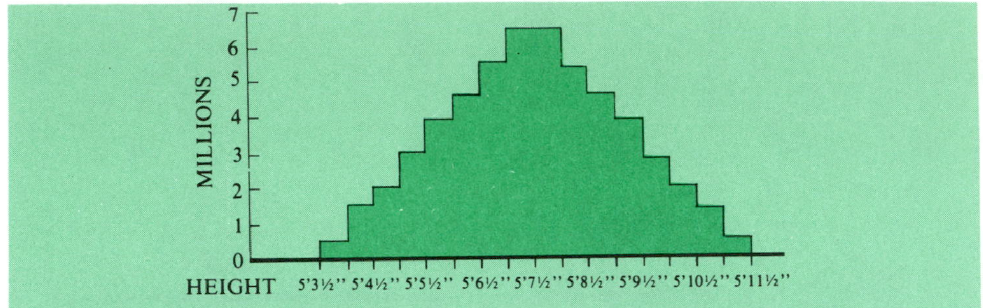

The effect of narrowing the intervals has been to increase the number of bars and to cause successive bars to change more gradually. As we break down the intervals further, the changes in height from each bar to the next become more and more gradual. The histogram begins to look like a curve. In Fig. 5.14 we see how it looks when the intervals are 1/10 of an inch long.

Fig. 5.14

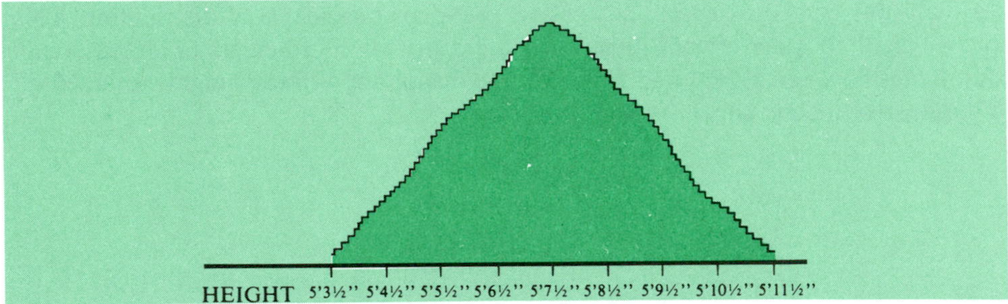

If we are free to choose intervals as small as we like, we can reduce the difference in area between the histogram and the curve that it is approaching to as small an amount as we like. However, there must be a sufficiently large number of terms in the distribution for us to do this. The reason is that we are creating a multitude of intervals as we cut down their size, and it must be possible to pour enough terms into each interval for the approximation to hold. Otherwise, many intervals will be empty or nearly empty, leaving breaks in the graph.

There are times when decreasing the intervals does not lead to an approximation of a curve as described. But such times are exceptional. In any event, we shall later discuss conditions under which the approximation is guaranteed to occur.

The purpose of this section was simply to give the reader an intuitive explanation of how a curve can be used to represent a distribution of values of a continuous variable. Thus a highly accurate graph of the distribution of heights that we have been discussing would look like Fig. 5.15.

Fig. 5.15

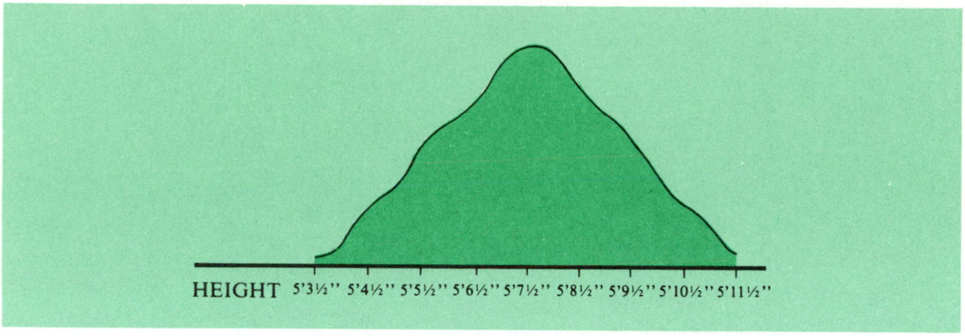

5.7 The Meaning of Area

The interpretation of the area under a graph is extremely important. Later on we will see why. The best way to illustrate is to imagine that we have just come across a bar graph that was published without a vertical axis to indicate the number of terms in the various piles (Fig. 5.16).

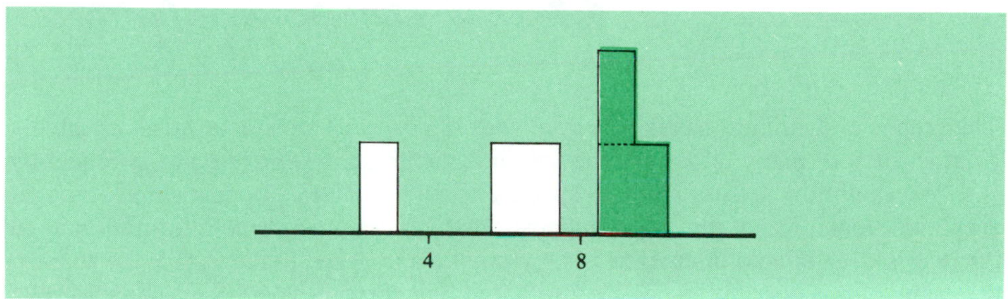

Fig. 5.16

A careful examination of this graph reveals that the total area of the columns to the left of the number 8 is equal to the total area of the columns to the right of 8. (The area of the light columns equals the area of the shaded columns.) We now ask what this tells us about the distribution being described.

Without a vertical axis, we have no idea how many weights are being represented or how large the weights are supposed to be. However, we do know that half of them were placed on the plank below 8 and half of them above 8. In other words, half of the terms are below 8 and half of them are above 8.

It is entirely possible that the original distribution looks like one or the other of the two parts of Fig. 5.17.

In either event our inference is correct. It happens that one-sixth of the area is below 4, so we may conclude that one-sixth of the terms in the distribution are less than 4. In general, **we may interpret the fraction of area over an interval as the fraction of the terms in the distribution that fall in that interval.** This interpretation does not depend upon our knowing how many terms there are in the whole distribution.

Fig. 5.17

5.8 Data That Account for the Shape of a Graph

The graphs of distributions take many shapes. Some of these shapes arise repeatedly. That is, they fit many different types of data. Other shapes arise quite infrequently.

We shall now discuss certain aspects of data that suffice to determine how their graph will look. To set the stage, suppose that we are given some information about the distribution of ages of citizens of Daytonville and asked to determine the shape of the graph of this distribution. We are told that one person's age has a z score of -2.0 and that other ages have z scores of -1.0, 0, 1.0, and 2.0. If we attempt to draw the graph of the distribution using this information, our first move might be to construct a horizontal axis and then to note where the five z scores appear. But once this is done, we can proceed no further. We do not yet have enough information to determine the shape of the distribution.

Now suppose we are given the following five additional facts:

1. The age with a z score of -2.0 is the smallest term in the distribution.
2. The age with a z score of -1.0 exceeds 40% of the terms.
3. The age with a z score of 0 exceeds 70% of the terms.
4. The age with a z score of 1.0 exceeds 80% of the terms.
5. The age with a z score of 2.0 is the largest term in the distribution.

This information contributes a great deal. We can now build a histogram from it. The base of the first rectangle is the segment going from $z = -2.0$ to $z = -1.0$ (Fig. 5.18).

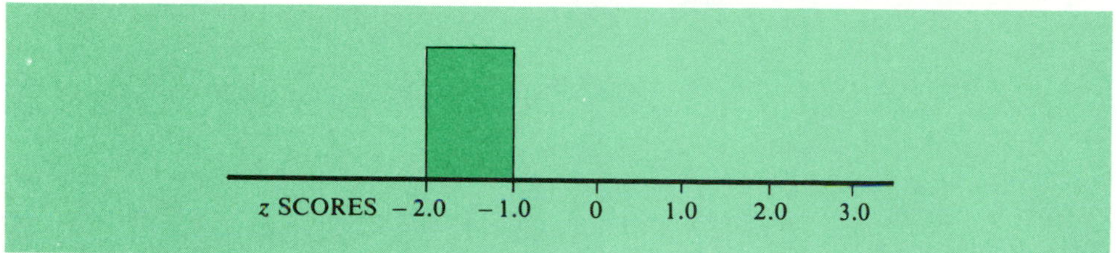

Fig. 5.18

There are no scores to the left of $z = -2.0$ because it is the smallest term.

Since 40% of the ages in the distribution have z scores between −2.0 and −1.0, we know that the area of the rectangle we have drawn comprises 40% of the area of the entire graph.

We know also that 70% of all the ages must have z scores less than 0. Thus 30% of the area of the entire graph must be in the rectangle based on the interval $z = -1.0$ to $z = 0$. This rectangle must be three-fourths as high as the first one.

By this same reasoning, 10% of the area must be in the next rectangle and 20% in the last rectangle. We obtain the graph in Fig. 5.19.

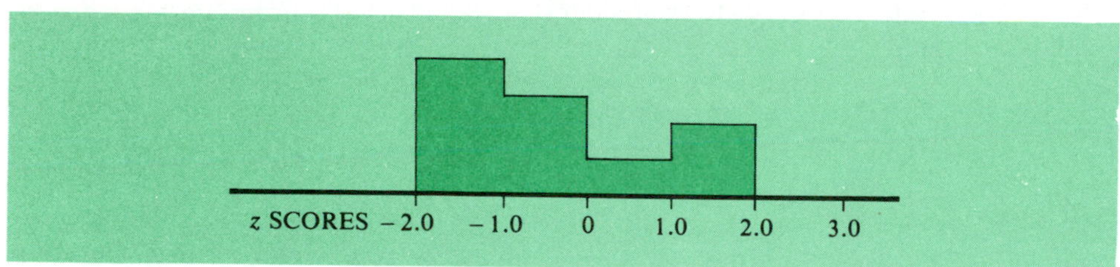

Fig. 5.19

We have not been given sufficient information to reconstruct the graph with a high degree of accuracy. However, given an elaborate table, one can construct a detailed graph.

The point is that **a graph is determined by a table giving z scores and the percents of terms exceeded. That is, from a detailed set of such pairings we can draw the shape of the graph.** The pairings tell us how the area is divided up over different portions of the base line.

We might say that a table of pairings like the one given is a description of a picture. The picture, of course, is the graph.

If two graphs differ in shape, the tables that correspond to them must differ. If two graphs have identical shapes, the tables that correspond to them must be identical. For instance, the graph of the distribution of men's weights in City A is shown in Fig. 5.20.

The graph of incomes of teenagers in City B happens to have an identical shape (Fig. 5.21).

Since the two curves are identical, a single table of z scores and the corresponding percents of terms exceeded describes both distributions. Thus the weight in City A, which has a z score of 0, exceeds 60% of the terms. Similarly, in City B the teen whose income has a z score of 0 earns more than do 60 percent of his or her

Fig. 5.20

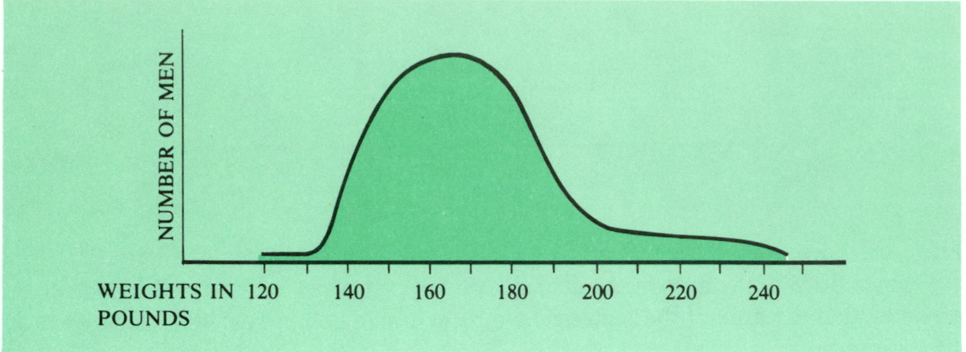

competitors. The percents that go with each particular z score are the same in each distribution, reflecting the fact that the shapes of the two graphs are the same.

We have seen how to construct a graph from this information. The purpose of this section was simply to show that there is an equivalence between *a graph* on the one hand and *a table of* z *scores and the percents of terms exceeded* on the other. Ordinarily it might not occur to the reader that such a set of pairings corresponds to a particular graph.

Fig. 5.21

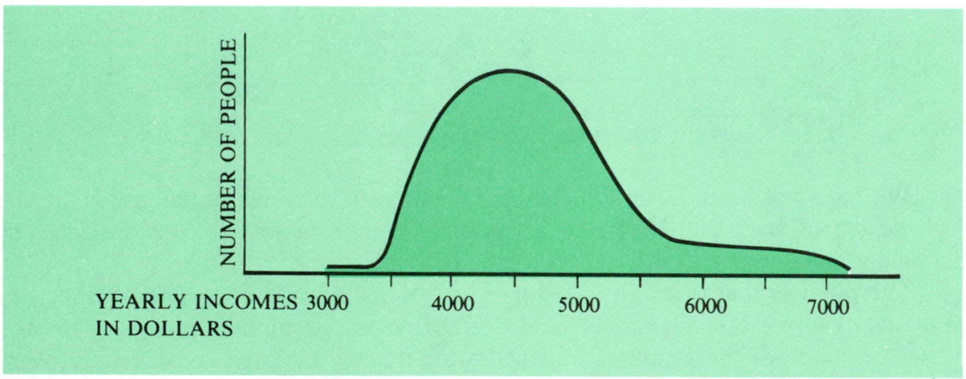

Incidentally, the person drawing the graph decides in advance how much space to give to it. For instance, successive z score units might have been set much closer to each other than they were in Fig. 5.19. The graph would have looked more compact but essentially it would not have been different. The reader should not be fooled into thinking that two graphs are different when their dissimilarities are due wholly to differences in spacing of units along one or both axes.

5.9 Applications

Bar graphs and histograms are extensively used to represent virtually all kinds of data. Histograms are used more frequently, but when two distributions are represented on the same axes it is often advantageous to use frequency polygons, since they are less likely to coincide over sections of the graph.

As we shall see, it is a relatively easy matter to compute the various measures already discussed once a mass of data has been "poured" into the intervals of a grouped frequency table. The experienced researcher can conjure up the shape of a distribution by looking at the frequency table, and often plunges into computing the various measures without stopping to draw a graph.

Problem Set A[2]

5.1 Classify each of the following variables as discrete or continuous.

a. time
b. velocity
c. number of children
d. width
e. attendance
f. distance
g. age at next birthday
h. hourly wage to nearest cent
i. weight
j. dress size

5.2 Phil's Film Factory sells eight types of camera film. Sales figures for the week before Christmas indicate the following numbers of rolls purchased by type:

Film Type	*Number Sold*
A	253
B	185
C	77
D	59
E	105
F	37
G	15
H	20

a. Identify the variable. Is it discrete or continuous?
b. Identify the population.
c. Represent the data graphically.

5.3 Round off each of these numbers as directed.

a. 499.13 pounds to the nearest pound
b. 114.79 millimeters to the nearest millimeter
c. 8 feet 11 inches to the nearest yard
d. 1,378 errors to the nearest thousand errors
e. age 111 months to the age at last birthday

5.4 This distribution of 75 terms shows the number of patients treated for drug overdose at each of the regional hospitals during the year. Construct a grouped frequency table.

[2] There are no unique answers to the questions in this set calling for histograms, frequency polygons, and grouped frequency tables.

5	9	12	23	52	83	115
5	9	12	23	55	83	123
5	9	13	25	55	84	130
5	10	13	29	55	85	
6	10	13	33	56	86	
6	11	13	35	57	86	
7	12	13	38	59	87	
7	12	13	38	63	99	
7	12	13	38	72	102	
8	12	15	43	72	107	
8	12	17	47	72	109	
9	12	18	52	72	109	

5.5 Drivers with more than three noninjury-causing traffic violations are required to attend a defensive driving class to prevent revocation of driving privileges. The sizes of the most recent 36 classes have been 15, 23, 49, 73, 12, 81, 11, 14, 72, 73, 65, 52, 41, 20, 19, 16, 43, 15, 17, 29, 37, 35, 29, 37, 11, 17, 53, 49, 83, 14, 22, 71, 70, 15, 63, and 52.
   a. Is the variable involved continuous or discrete?
   b. Construct a grouped frequency table and represent the distribution using both a bar graph and a histogram.

5.6 The weights of catches (in tons) for the 800 boats in the shrimp fleet are as follows:

Catch (in tons)	*Frequency*
.5–1.5	55
1.5–2.5	34
2.5–3.5	22
3.5–4.5	150
4.5–5.5	125
5.5–6.5	115
6.5–7.5	40
7.5–8.5	102
8.5–9.5	97
9.5–10.5	60

Represent the data on a histogram.

5.7 Oceanview has 39 service stations. The total hours of operation each day are provided for each station.

8.00	7.25	5.00	24.00	20.00
18.75	8.00	6.00	7.50	9.00
8.00	16.00	12.50	20.00	22.00
8.00	9.50	7.00	18.00	9.00
16.50	16.00	5.50	5.00	12.50
6.00	10.00	6.00	7.00	5.00
8.00	12.00	15.75	9.75	7.25
8.50	9.50	12.00	12.25	

   a. Construct a grouped frequency table for this distribution. Indicate boundaries and midpoints.
   b. Using the table obtained in part a, represent the data graphically using a histogram.

5.8 A gerontologist has devoted the last 25 years to studying the effects of antiradiation drugs on the life span of dogs. The life spans of 25 purebred dogs administered the drug proved to be (in years) 7.8, 9.5, 15, 17.5, 16.8, 17.3, 6.2, 12.9, 11, 10, 12.3, 24, 21, 17.1, 8.9, 7.6, 9.4, 15.2, 14, 13.9, 8.3, 9.7, 15.2, 12, and 10.5.

a. Construct a grouped frequency table for the data.
b. Using the grouped frequency table, draw a frequency polygon.

5.9 Sketch a histogram of a distribution using the following information about six of its terms. Be as accurate as the information will permit.

z Score	*Percent of Terms Exceeded*
−3.0	0
−2.0	15
−1.0	35
0.0	55
1.0	75
2.0	95

No term has a z score as large as 3.

5.10 The sports programming coordinator of a major network has requested her administrative assistant to compile the lengths of the last 50 televised professional football games. The data collected show the following lengths (in minutes): 103, 107, 95, 110, 115, 123, 96, 107, 115, 111, 112, 97, 90, 125, 123, 127, 95, 102, 107, 115, 93, 108, 110, 111, 115, 139, 116, 105, 97, 144, 98, 104, 114, 119, 129, 133, 122, 121, 111, 127, 118, 115, 123, 118, 94, 132, 93, 106, 114, 115.

Group the data into a frequency table and represent it on a histogram.

5.11 Below are the z score and percentile rank pairings of seven terms from a distribution of values assumed by a continuous variable. No term in the distribution has a z score as big as 4.

z Score	*Percentile Rank*
−3.0	0
−2.0	10
−1.0	15
0.0	20
1.0	25
2.0	35
3.0	95

a. Sketch the histogram with as much accuracy as the information permits.
b. Name two things that would increase the accuracy of the drawing.

5.12 According to the assessor's office, the sizes (in hundreds of square feet) of the 77 homes in the Fairfax addition are

16.83	17.30	23.90	15.21	18.75	19.31	15.95
25.72	18.40	15.00	19.30	22.00	21.70	16.45
23.43	19.30	17.41	17.42	18.75	18.30	28.00
16.02	15.10	16.32	13.05	15.47	16.81	16.11
18.00	28.70	18.20	25.20	15.90	29.30	30.04
18.10	19.40	17.93	15.50	19.00	19.60	19.76
16.50	16.70	16.03	19.51	16.30	28.40	15.75
27.20	14.30	16.00	23.93	30.00	15.00	23.00
14.70	14.25	17.11	19.50	15.21	17.15	17.80
15.50	18.00	16.50	15.00	16.00	14.03	18.03
21.08	18.50	18.70	24.10	15.50	17.08	21.40

Represent the data using a grouped frequency table and a histogram.

5.13 **The area under the curve between 98 and 99 represents 23% of the total area. How many terms are in the distribution?**

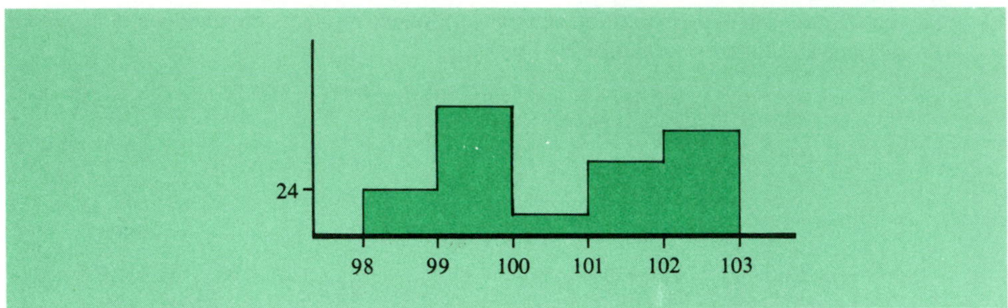

5.14 Some of the pairings of z scores and percents of terms exceeded are given here for a very special distribution.

z Score	*Percent Exceeded*
−3.0	.13
−2.5	.62
−2.0	2.28
−1.5	6.68
−1.0	15.87
−0.5	30.85
0.0	50.00
0.5	69.15
1.0	84.13
1.5	93.32
2.0	97.72
2.5	99.38
3.0	99.87

a. Sketch a histogram using the above information.

b. Make an educated guess as to what the curve would look like if an infinitude of additional pairings were provided.

5.15 A private religious-affiliated adoption agency placed 635 children under the age of 10 in homes last year. Only 45 had not celebrated a first birthday. One hundred had yet to celebrate a second birthday, 213 a third, 385 a fourth, 473 a fifth, 519 a sixth, 570 a seventh, 615 an eighth, and 624 a ninth. As already mentioned, none of the 635 children had reached his or her 10th birthday. Construct a grouped frequency table and histogram.

Problem Set B[3]

5.16 a. Roll an ordinary die 50 times and record the results.
b. Use a bar graph to represent these results.
c. Identify the variable involved in this situation and classify it as either continuous or discrete.

5.17 Classify the following variables as continuous or discrete.

a. voltage	f. favorite soft drink
b. number of voters	g. calories
c. type of variable	h. enrollment
d. age	i. perimeter
e. diameter	j. strength

5.18 Find the midpoints of the following intervals: (a) 82–97; (b) 87.5–93.5; (c) −4–15; (d) 139.5–474.5.

5.19 Organize the following values assumed by the discrete variable, number of dogs impounded during the week, into a grouped frequency table.

18	15	25	6	19	8	18	32
7	23	14	33	18	11	19	14
17	27	17	22	27	12	29	28

5.20 Know-It-All Encyclopedias, Inc., employs 45 regional representatives on either a full- or part-time basis. The third quarter totals (in number of sets sold) for the 45 salespeople are as follows:

9	27	80	32	16
87	35	79	37	37
14	22	16	45	34
20	6	52	11	43
81	19	94	83	91
50	27	12	72	18
72	32	17	40	21
84	11	24	15	27
123	34	21	63	30

a. Construct a grouped frequency table for the data.
b. Draw a bar graph and a histogram from the table.

5.21 The following shows the number of turkeys in each weight class delivered to one of its outlets in the metropolitan area by a large turkey distributor. Draw a histogram and a frequency polygon that illustrate the data.

Weight (in pounds)	*Frequency*
23.5–25.5	15
21.5–23.5	25
19.5–21.5	60

(continued)

[3]There are no unique answers to the questions in this set calling for histograms, frequency polygons, and grouped frequency tables.

17.5–19.5	70
15.5–17.5	60
13.5–15.5	10
11.5–13.5	7
9.5–11.5	5

5.22 A county agent sent questionnaires to all landowners in the region. Each farmer was asked to indicate the percent of land owned that was on a planned crop rotation program. The 30 forms returned by farmers reported the following percents.

65.5	73.8	70.5	72.9	70.0
88.4	89.4	75.7	54.5	65.0
59.0	45.7	76.8	45.9	72.0
73.0	53.8	65.4	52.0	83.0
75.0	62.9	63.2	56.0	51.5
52.5	68.4	65.0	83.0	87.0

a. Is the variable involved continuous or discrete?
b. Construct a grouped frequency table.

5.23 To evaluate its direct mail advertising campaign, the chairman of the board has the following figures collected. Each shows the number of sales for one campaign. The numbers are in order for convenience.

49	61	70	75	79	86	111	121
49	61	70	75	83	86	112	131
50	61	70	75	85	94	112	133
56	61	71	75	85	103	113	135
56	63	71	77	86	103	118	135
56	64	73	77	86	104	118	136
56	68	73	77	86	105	120	138
57	69	73	78	86	111	121	140
61	69	74					

a. Construct a grouped frequency table.
b. Draw the histogram that follows from the table.

5.24 Use the following pairings to sketch the histogram of the distribution with as much accuracy as the information permits. No term has a z score as large as 3.

z Score	*Percent Exceeded*
−3.0	0
−2.0	10
−1.0	50
0.0	60
1.0	85
2.0	90

5.25 Consider the following histogram prepared by the school nurse to indicate the weights of kindergartners at Adams Elementary School.
How many kindergartners weigh
a. between 35.5 and 37.5 pounds?

b. less than 39.5 pounds?
c. more than 41.5 pounds?
d. between 35.5 and 45.5 pounds?

What percent of the kindergartners weigh
e. between 37.5 and 39.5 pounds?
f. less than 43.5 pounds?

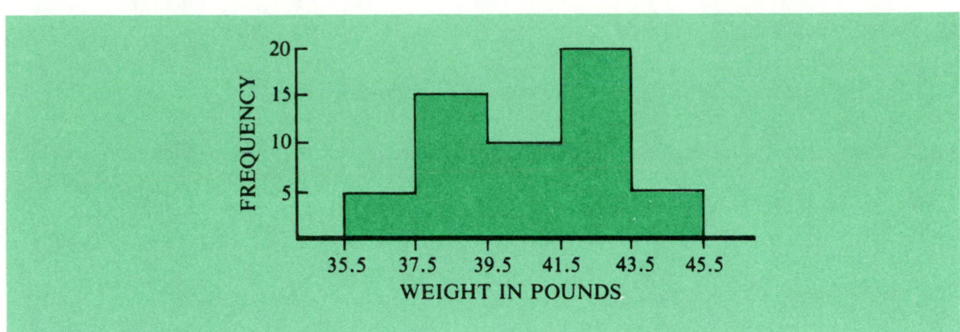

5.26 A meteorologist has been studying the weather patterns in an urban area. The mean weekly temperatures for a one-year period are as follows (in degrees Fahrenheit):

−3	40	69	89	43
0	47	75	86	41
8	47	75	75	32
5	45	70	73	20
15	52	80	75	14
11	59	86	65	15
13	60	89	65	10
20	70	82	60	4
15	72	81	52	
30	73	93	50	
35	71	90	49	

Construct a grouped frequency table and frequency polygon to show the distribution of mean weekly temperatures.

5.27 From the pairings below sketch the histogram. No term has a z score of 3.5 or higher.

z Score	*Percent Exceeded*
−2.5	0
−2.0	5
−1.0	40
0.0	60
1.5	75
2.0	85
2.5	90

5.28 Qwik Kab Taxi Company has a fleet of 44 cabs. The gallons of gas used for the week are given here for each taxi.

125.4	120.6	127.8	129.6	121.3
120.7	127.9	132.4	120.5	127.5
122.8	142.1	128.5	122.8	127.2
124.5	126.3	129.6	145.3	125.0

(continued)

137.1	137.8	132.1	123.7	125.0
139.4	139.4	130.4	133.7	132.0
143.2	133.2	130.0	128.3	141.0
140.0	132.4	123.7	139.4	138.6
133.3	144.6	129.3	129.4	

Construct a grouped frequency table and histogram for these data.

5.29 From the information below, sketch the histogram of the distribution. Three exceeds all z scores.

z Score	*Percent of Terms Exceeded*
−4.0	0
−3.5	9
−3.0	27
−2.5	48
−1.5	54
0.0	72
1.5	75
2.0	84
2.5	93

5.30 What are the advantages and disadvantages of using many or few intervals to group data?

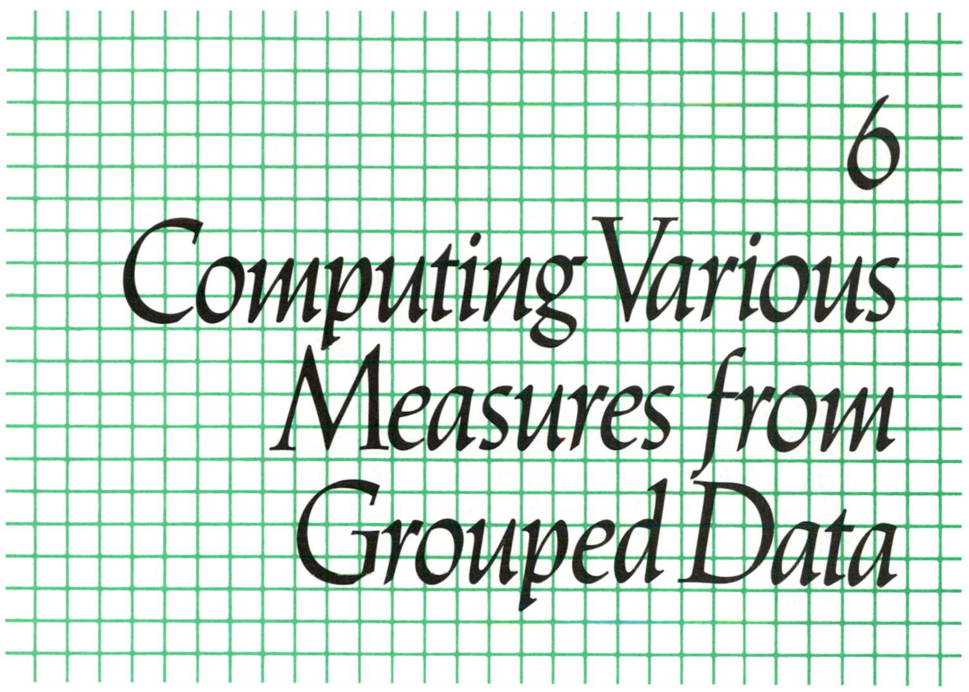

6 Computing Various Measures from Grouped Data

6.1 Introduction

In Chapters 2, 3, and 4, we discovered the meaning of certain measures of average value, variability and location, and we learned how to compute these measures for distributions of given data. In Chapter 5 we saw the desirability of grouping a large mass of data into intervals to form a grouped frequency distribution and we learned how to represent such a distribution graphically by the use of histograms and frequency polygons. In the present chapter we shall see how to combine these previous ideas and compute measures for grouped data. Our previous graphical analysis will prove useful, and we shall also consider an additional graphical procedure.

6.2 The Median and Percentiles from a Histogram

We recall that the median is a number that is larger than or equal to half of the terms in a distribution and smaller than or equal to half of them. It is also the 50th percentile. We saw that the median need not be the same number as a term in the distribution. It is extremely important to keep these facts in mind when discussing the median of a distribution of grouped data where, as we have seen, the individual terms of the original distribution have lost their identity.

For reference we reproduce here the grouped frequency table (Table 6.1) and histogram (Fig. 6.1) for the children's spelling scores. So far we know only that the *median* is larger than half of these 1,000 scores and smaller than half of them. To find it, we must find a number on the horizontal axis of Fig. 6.1 that is larger than 500 of the scores. Looking at the frequency table and adding the frequencies, we have 50 in the lowest interval, 50 + 81 = 131 in the lowest two intervals combined, 131 + 119 = 250 in the lowest three intervals combined. Continuing in this way we find 330 in the lowest four intervals, 470 in the lowest five, and 590 in the lowest six intervals. In other words, there are 470 scores less than the boundary 56.5 and there are 590 scores less than the boundary 67.5. The median must be a number larger than 56.5 and smaller than 67.5, but closer to 56.5.

Table 6.1 Grouped Frequency Table

Intervals	Boundaries	Midpoints	Frequencies
90–100	89.5–100.5	95	78
79–89	78.5–89.5	84	140
68–78	67.5–78.5	73	192
57–67	56.5–67.5	62	120
46–56	45.5–56.5	51	140
35–45	34.5–45.5	40	80
24–34	23.5–34.5	29	119
13–23	12.5–23.5	18	81
2–12	1.5–12.5	7	50
			1000

The area above the interval from 56.5 to 67.5 represents the 120 scores that were grouped into that interval by rounding them off to the value 62. In order to find the median we must find a point on the horizontal axis at which to split the bar by a vertical line so that 30 scores out of the 120 are to the left of that line. This will put 500 scores to the left of the line and our point on the axis will then be at the median. In drawing in such a line we must put 30/120 or 1/4 of the area of the bar to the left of the line and the remaining 3/4 to the right. It should be noted that the fact that the top of the bar is horizontal or parallel to the base has an important implication for us at this point. It means that in considering the histogram and the median we are assuming that the 120 scores (which lost their individual identities when they were all rounded to the value 62) are distributed evenly throughout the interval.[1] With this consideration in mind a vertical line has been drawn in with dashes on the histogram. It

[1]It is standard statistical procedure to make this assumption in the computation of the median.

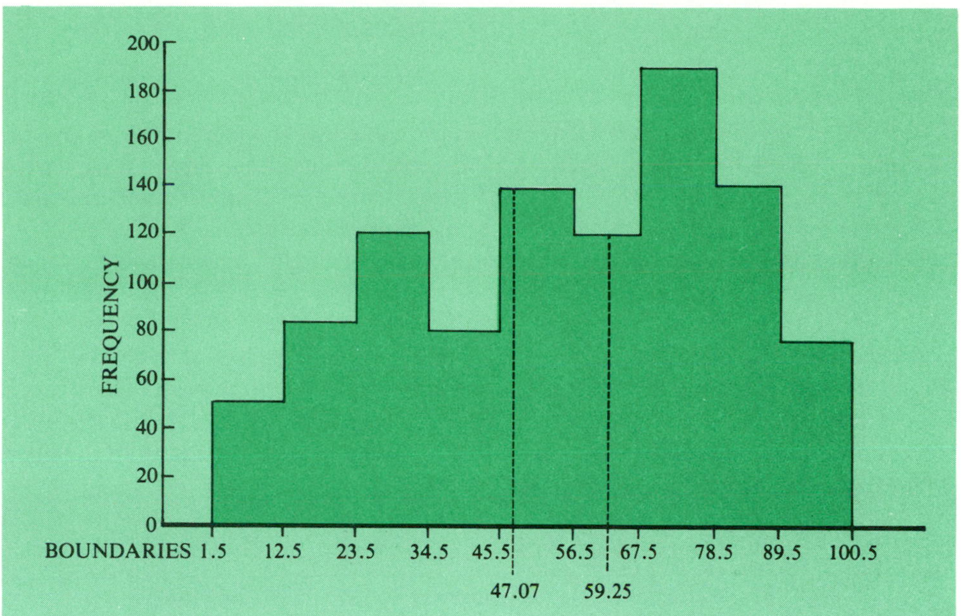

Fig. 6.1

crosses the horizontal axis at the score value of 59.25. This means that the median of the frequency distribution is about 59.25. Note that half of the total area of bars on the histogram is to the left of the dashed line and half is to the right.

The histogram can be used in the same way to obtain percentiles for a grouped frequency distribution. We recall that a percentile is also a term in the distribution or is a score value intermediate to two terms. For example, the 35th percentile is a score that exceeds 35% of the scores (including half of those at that score). If we wish to find the 35th percentile for the spelling distribution we must find a number on the horizontal axis such that 35% of the scores are smaller than that number. Or stated differently, now we wish to draw in a vertical line such that 35% of the area is to the left of that line and 65% is to the right of it.

Note first that 35% of 1,000 scores represents 350 scores. Looking back to where we added successive interval frequencies, we found 330 scores in the lowest four intervals combined and 470 in the lowest five intervals combined. This means that the 35th percentile is between 45.5 and 56.5. We must put 20/140 or 1/7 of the area of the bar for that interval to the left of the line we draw and 6/7 to its right. Such a line has been drawn in with dots in Fig. 6.1. It crosses the horizontal axis at about the value 47.07. This means that the 35th percentile of the frequency distribution is about 47.07. Note that 35% of the total area of bars on the histogram is to the left of the dotted line and 65% is to the right. The value of any percentile can be found from the histogram by the same procedure as we have used here. The reader will recognize that the median is just another name for the 50th percentile and that its estimation from the histogram followed the same pattern.

6.3 Computation of the Median and Other Percentiles

The method for computation of any percentile, including the median, for a grouped frequency distribution is simply a numerical version of the procedure that we used in the preceding section. We shall go over the two examples again in some detail and refine the method so that reference to the histogram is unnecessary.

First, it will be convenient to introduce what is called a "cumulative less than" column in our frequency table. We shall refer to this simply as the "less than" column. In this column we record the sums we obtained when we added the frequencies of successive intervals in our previous work. This has been done in Table 6.2 in the last column. The column is constructed from the bottom up. There are 50 scores less than 12.5, 50 + 180 = 131 scores less than 23.5, 131 + 119 = 250 scores less than 34.5, and so on. In other words, the "less than" refers to the *upper* boundary of each interval.

Table 6.2 "Cumulative Less Than" Frequencies for Spelling Scores

Boundaries	Frequencies	Less Than
89.5–100.5	78	1,000
78.5–89.5	140	922
67.5–78.5	192	782
56.5–67.5	120	590
45.5–56.5	140	470
34.5–45.5	80	330
23.5–34.5	119	250
12.5–23.5	81	131
1.5–12.5	50	50

In seeking the median we want that point where the "less than" reading is 500. The reason is that 500 represents half of our total of 1,000 cases. Our table shows that this point will be between 56.5 and 67.5. But how far must we go starting from 56.5 and proceeding towards 67.5 before reaching this point? At 56.5 we have 470 scores, so we need 500 minus 470 or 30 scores out of the 120 scores in the interval from 56.5 to 67.5. In the previous section we discussed splitting the bar of the histogram so that 30/120 or 1/4 of its area was to the left of the dashed line that we drew in. But we noticed there the importance of the fact that the tops of the bars in the histogram are parallel to the horizontal axis. We are actually just finding the point on the horizontal scale (Fig. 6.2) that is 1/4 of the distance from 56.5 to 67.5. Since the total distance is 11, we want to go to the point that corresponds to the number that is 56.5 plus 1/4 of 11. That is, the median is $56.5 + (1/4)(11) = 56.5 + 11/4 = 56.5 + 2.75 = 59.25$. This means that 500 spelling scores are less than 59.25 and 500 are greater.

Now for the 35th percentile we must find that point where the "less than" reading

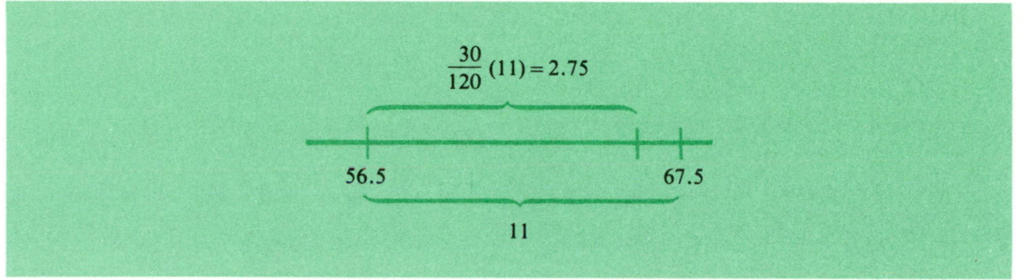

Fig. 6.2

is 35% of 1,000, or 350. Our "less than" column in Table 6.2 shows that it will be in the interval from 45.5 to 56.5; we must then determine the stopping point as we proceed from 45.5 to 56.5 along the horizontal scale (Fig. 6.3). We have 330 scores up to 45.5, and we need 350 minus 330 or 20 scores out of the 140 scores that have been grouped into the interval that we are considering. This means that we need to cover 20/140 or 1/7 of the 11 units of distance from 45.5 to 56.5. If we do, we will arrive at the point that represents the 35th percentile, often designated as P_{35}. Hence,

$$P_{35} = 45.5 + (1/7)\,(11) = 45.5 + 1.57 = 47.07$$

This means that 35% of the scores are less than 47.07 and 65% are greater.

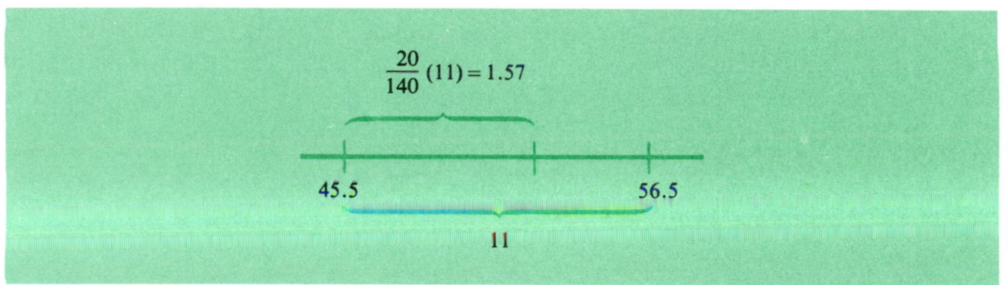

Fig. 6.3

For those who prefer symbolic statements, we could summarize our method for finding percentiles by the formula:

$$P_n = L + (s/F)(i) \tag{6.1}$$

Here we are finding the nth percentile, P_n. We refer to the "less than" column and find that P_n falls in the interval having L for its lower boundary. We still need s scores in order to arrive at the point on the horizontal scale (Fig. 6.4) that corresponds to the percentile. The distance from the lower boundary L to the upper boundary $L + i$ is i and there are F scores in the interval from L to $L + i$. For instance, reference to the preceding example gives us $P_{35} = 45.5 + (1/7)(11)$, which we had before.

Fig. 6.4

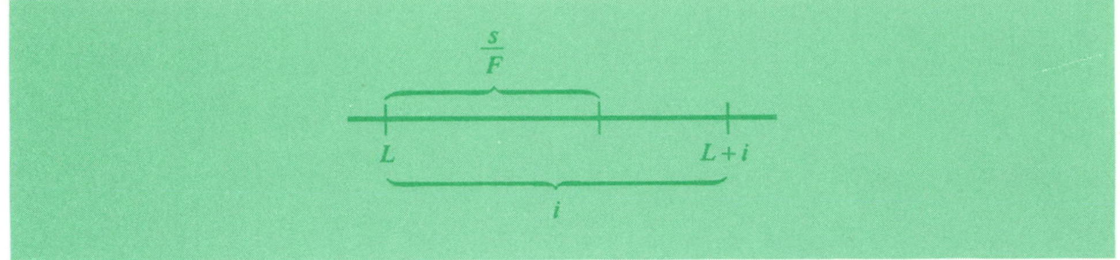

6.4 Quartiles and Deciles

Two terms frequently used in professional literature are *quartile* and *decile*. The first quartile, or Q_1, is the value corresponding to that point on the scale of scores such that 25% (or one-quarter) of the scores are smaller than it and 75% are greater. In other words it is another name for P_{25}. The second quartile, or Q_2, is another name for the median or P_{50}. Similarly, the third quartile, or Q_3, is another name for P_{75}. A student whose spelling score is among the lowest one-fourth is properly spoken of as being "in the lowest quarter" of the distribution but a student is not *at* the first or lower quartile unless his score falls at the point on the scale of scores such that exactly 25% of the scores are less than his. Similarly, a student is *at* the upper quartile, Q_3, if his score exceeds exactly 75% of the scores; but a pupil is "in the upper quarter" if his score has a percentile rank anywhere in the interval from 75 to 100.

In the same way, deciles refer to tenths of a population. The first decile, or D_1, is the value corresponding to that point on the scale of scores such that 10 percent of the scores are smaller than it. It should be clear than $D_1 = P_{10}$, $D_2 = P_{20}$, and so on.

6.5 Percentile Ranks from Grouped Data

We have just seen how to find a term in a distribution that exceeds a certain percent of the scores. Now we turn our attention to finding the percent of terms exceeded, given a particular score. In other words, we shall be determining percentile ranks of scores. In the example involving spelling scores, a parent might wish to know the rank of a child who scored 70.

Because the tops of the bars in the histogram are parallel to the horizontal axis, we noted that the values in each interval may be assumed to be evenly distributed throughout the interval. Therefore, we shall forego a discussion of the histogram and consider the horizontal axis. Specifically, we shall be concerned with the interval in which the score 70 falls, 67.5–78.5 (Fig. 6.5).

The approximate location of the term 70 has been indicated, as have the endpoints and the interval width, 11. Reference to the "less than" column of Table 6.2 reveals that 590 of the 1,000 scores fall below 67.5. Now we need to determine the number of scores between 67.5 and 70, which we may add to 590 to obtain the total number below 70.

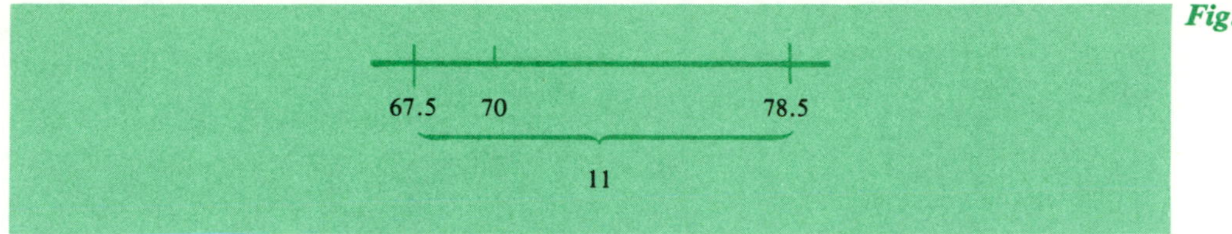

Fig. 6.5

The distance between 67.5 and 70 is $70 - 67.5 = 2.5$. Therefore, 2.5 of the 11 units or about 23% ($2.5/11 = .227$) of the 192 terms in the interval fall between the lower boundary 67.5 and 70 (Fig. 6.6).

Fig. 6.6

That is, about 44 terms (23% of 192) fall in the interval as indicated in Fig. 6.6. Now it is clear that $590 + 44 = 634$ of the 1,000 scores or 63.4% of the scores fall below 70. The percentile rank of 70 is therefore 63.

Again for those who prefer symbolic statements, we could summarize the method for finding percentile ranks as follows:

$$PR_X = \left(\frac{C + \dfrac{X - L}{i} F}{N} \right) 100 \tag{6.2}$$

Here we are finding the percentile rank of the term X, which falls in the interval whose lower boundary is L, whose width is i and whose lower boundary exceeds C scores. The interval contains F scores of the total N scores in the distribution.

Reference to the example just completed yields

$$PR_{70} = \left[\frac{590 + \dfrac{70 - 67.5}{11} 192}{1000} \right] 100 = 63.36$$

Having found this value, we round to the nearest percent, giving us an answer of 63.

6.6 The Cumulative Curve

We shall now consider the "less than" column that we added in Table 6.2 for the computation of percentiles and percentile ranks. The cumulative "less than" graph

Fig. 6.7

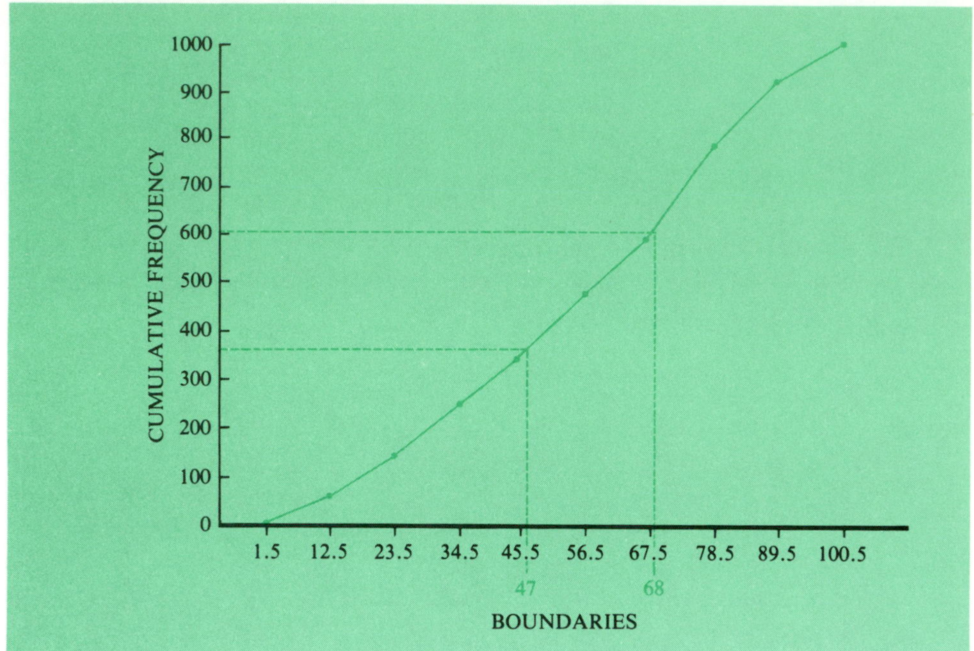

for the spelling scores is shown in Fig. 6.7. Remembering that the "less than" refers to the upper boundary of each interval, we have plotted a point for each value in the "less than" column of Table 6.2 and we have placed these points directly above the upper boundaries of the corresponding intervals. In addition we added a point at the extreme left of the curve to bring the curve down to the baseline and indicate that there are zero scores less than 1.5, the lower boundary of the lowest interval.

The cumulative "less than" curve for a grouped frequency distribution may readily be used to obtain both percentiles and percentile ranks. Suppose we want the 60th percentile, P_{60}. Since there are 1,000 scores, we go up the vertical scale to the point corresponding to 60% of 1,000, or 600 scores. The dashed line is drawn parallel to the base until it hits the curve. Then it is dropped perpendicular to the base and seems to strike the scale of scores at about 68. Hence, $P_{60} = 68$. Had we started with a score of 68 and desired its percentile rank, we would have drawn the parts of the line in the reverse order and would have estimated the frequency on the vertical scale to be about 600. This would represent 600/1,000 of the scores, and when divided this gives 60%, or a percentile rank of 60.

The dotted line in Fig. 6.7 shows the estimate of P_{35} to be 47 or the percentile rank of 47 to be 35. This is the example that we have done twice previously. As the examples show, the cumulative "less than" curve (often called an *ogive*) can be very useful if several percentiles or percentile ranks must be estimated for the same frequency distribution.

6.7 The Mode for Grouped Data

There is little to be said about the determination of the mode for data that have been placed in a grouped frequency distribution. The reader will recall from Chapter 2 that a distribution of actual scores may have two or more modes—that is, the mode is not always uniquely determined. The same sort of situation prevails for grouped data. First, we have to decide what number to identify as the mode. It is customary to identify the interval having the greatest frequency as the **modal interval** and to call its midpoint the **crude mode.** Thus, in the spelling scores example (Table 6.1) the modal interval is the interval 68–78, since its frequency, 192, is greater than that of any other interval in the table. The crude mode is 73, the midpoint of the modal interval. Had there been another interval with frequency 192, we would have had two modal intervals and two crude modes. The crude mode is not used in further computations for grouped frequency distributions.

6.8 The Mean for Grouped Data—Direct Method

In Chapter 2 we learned how to calculate the mean of a distribution by direct application of its definition. We also discussed some important properties of the mean and saw how we could "cash in" on one of these properties in order to shorten the calculation of the mean when the data involved large numbers. This was the property that states that if a constant is subtracted from each term in a distribution, the mean is decreased by the same constant. We also saw that dividing each term by a constant has the effect of dividing the mean by that same constant. In this (and the next) section we shall see how to calculate the mean from a grouped frequency table. Again we shall first do this by direct application of the definition and then we shall see how to shorten the work by making use of the two properties of the mean that were just mentioned.

We have seen that a simple method of rounding off terms is to put them in a grouped frequency table. When this is done, the terms in each particular interval get rounded off to the value of the midpoint of that interval. For instance, according to Table 6.1, the spelling scores of 50 children are in the interval from 2 through 12. The midpoint of this particular interval is 7. When we compute the mean for this grouped frequency distribution we must have one value to use for all of the 50 scores in this interval. The logical choice is the value at the midpoint, so we proceed as if 50 children had spelling scores of exactly 7. Similarly, 81 terms in the distribution are in the interval from 13 through 23, so we proceed as if there are 81 terms that have the value 18, and so on. The reader may well object at this stage that we are not being consistent because earlier in this chapter, when we computed percentiles for grouped data, we made the assumption that the terms in an interval are evenly distributed throughout that interval. This assumption was based on the observation that the tops of the bars in the histogram are horizontal. A little reflection, however, should remove the reader's objection, for if the terms in an interval are, indeed, distributed evenly throughout the interval, then their mean will be at the midpoint of that interval. It is this very midpoint that we are now choosing when faced with the necessity of using one value to represent all of the terms in the interval.

When we round off terms, we are in effect creating a new distribution. We admit only certain possible values, those that are the midpoints of the successive intervals. The frequencies of the intervals become the frequencies of these values. The process of rounding off entails increasing slightly the values of some of the terms and decreasing slightly the values of others. The reason is that the terms that happen to fall in a particular interval may include some that are above its midpoint and others that are below it; but we proceed as if all the terms have the value of the midpoint. Note that the width of the intervals tells exactly how much rounding off we are doing. The wider the interval, the more we may be changing the terms from their original values.

First we shall compute the mean of a distribution by the direct method. Consider the grouped frequency distribution of the spelling marks of the 1,000 fourth grade children (Table 6.1). Our first step is to obtain the sum of these 1,000 spelling marks or terms in the distribution. Remember that we are now considering the distribution of rounded-off terms. The highest 78 terms in this grouped distribution each have the same value, 95. Thus the sum of these highest 78 terms is $(78)(95) = 7,410$. The next highest 140 terms have each been rounded off to the value of 84. Thus the sum of these next 140 terms is $(140)(84) = 11,760$. In the same way we can find the sum of the terms in each of the nine intervals. To get the sum of the rounded-off terms in any particular interval, we multiply the midpoint of that interval by the number of terms included, that is, the frequency of the interval.

We shall now add to Table 6.1 a column indicating the sum of the terms in each interval. This is done in Table 6.3. In this table we use the letter X to refer to the midpoints of the intervals and the letter F to refer to the frequencies of the intervals. Thus the sum of the terms in each interval is denoted by FX.

Table 6.3 Computation of Mean

(1) Intervals	(2) Midpoints X	(3) Frequencies F	(4) Sum of Terms FX
90–100	95	78	7,410
79–89	84	140	11,760
68–78	73	192	14,016
57–67	62	120	7,440
46–56	51	140	7,140
35–45	40	80	3,200
24–34	29	119	3,451
13–23	18	81	1,458
2–12	7	50	350
		$\Sigma F = 1,000$	$\Sigma FX = 56,225$

Column (4) tells us the sum of the terms in each interval. To get the sum of all the terms in the distribution, we add up all of the values in Column (4). The sum of these values is denoted by ΣFX and is found to be 56,225. In other words, the sum of the rounded-off spelling marks of all the 1,000 children is 56,225.

Remember that when the terms are listed individually and we consider each term as an individual value of X, then the sum of the terms is denoted ΣX. Now that we are using the grouped frequency method, the sum of the terms is ΣFX, where the X values are the midpoints of intervals rather than individual terms. Thus the two symbols ΣX (used in Section 2.5) and ΣFX (used in this section) are equivalent except for errors caused by the grouping. The difference in notation merely means that we have summed up all of the terms in the distribution by different methods.

The frequencies of the different intervals are indicated in Column (3) of Table 6.3. In practice, we add up these frequencies to determine how many terms there are in the distribution (test scores). Here the sum of the frequencies, denoted ΣF, is 1,000. In other words, we are working with 1,000 test scores. It should be clear that $\Sigma F = N$.

It follows that the formula for the mean of a distribution, computed by the group method, is

$$\mu = \frac{\Sigma FX}{\Sigma F} \tag{6.3}$$

Thus for the data given in Table 6.3,

$$\mu = \frac{\Sigma FX}{\Sigma F} = \frac{56{,}225}{1{,}000} = 56.225$$

6.9 The Mean for Grouped Data—Coded Method

Although the computation just completed by Formula (6.3) was perfectly straightforward, the reader may have been chagrined by the fairly large numbers involved. Recall that on page 19 we were faced with a similar situation and we made the work simpler by applying Theorem 2.3 and subtracting a constant from each score. That constant was then added back to the mean of the new scores. We saw that the choice of a constant to be subtracted had no effect on the final result. It was advantageous to choose a constant that appeared to be as near the mean as possible, since this led to calculations with smaller numbers. Now let us see how we can apply this same method to our grouped data on spelling scores. Since we are working only with the midpoints, we will need to subtract the same constant from each midpoint.

The midpoints and frequencies have been entered in Columns (1) and (2) of Table 6.4. We shall choose 51 as the constant to subtract from each midpoint. This value is the midpoint of one of the middle intervals. Our choice is dictated by the large frequency in the interval with 51 as its midpoint. Now the new score for the value 51 itself becomes zero. The fact that a large frequency, namely 140, will now be multiplied by zero will reduce our computations considerably. The value 51 has been subtracted from each of the midpoints and the set of new midpoints thus obtained appears in Column (3) of Table 6.4. These new midpoints have been labeled X'. Note that $X' = X - 51$.

Table 6.4 Computation of Mean by Shorter Method

(1) Midpoints X	(2) Frequencies F	(3) New Midpoints X'	(4) Sum of New Terms FX'
95	78	44	3,432
84	140	33	4,620
73	192	22	4,224
62	120	11	1,320/13,596
51	140	0	0
40	80	−11	−880
29	119	−22	−2,618
18	81	−33	−2,673
7	50	−44	−2,200/−8,371
			$\sum FX' = +5{,}225$

Now we must find the mean, using the X' midpoint values, and then add back the 51 to that new mean. We proceed as in the previous computation, except that we now use the X' values. Column (4) of Table 6.4 shows the product FX' for each interval. We need to remember that the product of a positive number and a negative number is a negative number. The sums of the positive and negative values in Column (4) are shown separately. The sum of all the terms after subtraction of the constant is denoted by $\sum FX'$ and for the data in Table 6.4 it is found to be 5,225. Then the mean of the new (X') scores is $\sum FX'/\sum F$. We shall denote this by μ'. Thus we have $\mu' = \sum FX'/\sum F = 5{,}225/1{,}000 = 5.225$. Now we must add back the 51 to this mean of the new scores, that is, $\mu = \mu' + 51$. This gives us $\mu = 56.225$, exactly the same as when we computed the mean without subtracting a constant from each midpoint score. It is always true that this shortcut method will give a result identical to that obtained by the longer method when both are applied to the same grouped distribution. It will be instructive for the student to do the computation of Table 6.4 over again, this time using 40 as the constant to be subtracted from each midpoint value. Although the intervening numbers will be different, the final result should again be the same.

The reader may have noticed that all of the X' values in Column (3) of Table 6.4 are multiples of 11. This occurs because the distance between successive midpoints is 11. The fact that these numbers are all multiples of 11 enables us to devise a further shortcut in the computation by applying Theorem 2.4. We shall divide each of these numbers by 11 and then find the mean of the new numbers. The theorem tells us that dividing each number by 11 will have the effect of dividing the mean by 11. Hence, we shall then multiply the new mean by 11. This will give us only the mean of the X' values, so we will still need to add 51 to get the mean of the original spelling scores.

In Table 6.5, the X, X', and F columns from Table 6.4 have been recorded as Columns (1), (2), and (3), respectively. Column (4) contains the results of dividing each X' value by 11. These values have been denoted by U. We then proceed as in the previous computations except that we now use the U values. Column (5) shows the product FU and gives the sum of the U scores for each interval. The sum of all of the 1,000 U scores for the distribution is the sum of Column (5) and is denoted by ΣFU. This is found to be +475, as shown in the table. Then the mean of these newest or U scores is $\Sigma FU/\Sigma F$. This may be denoted by μ''. For the data in the table we then have

$$\mu'' = \Sigma FU/\Sigma F = 475/1{,}000 = .475$$

Now we must multiply back by 11 so that $\mu' = 11(.475) = 5.225$, the mean of the X' scores, as before. Finally, $\mu = 5.225 + 51 = 56.225$, the mean of the original grouped spelling scores and the same result as obtained before.

Table 6.5 Computation of Mean Using Coded Scores

(1) X	(2) X'	(3) F	(4) U	(5) FU
95	44	78	4	312
84	33	140	3	420
73	22	192	2	384
62	11	120	1	120/1,236
51	0	140	0	0
40	−11	80	−1	−80
29	−22	119	−2	−238
18	−33	81	−3	−243
7	−44	50	−4	−200/−761
		$\Sigma F = 1{,}000$		$\Sigma FU = +475$

Now, admittedly, if we had to go through all of the above computations our method of using U scores would not be much of a shortcut from our original computation of the mean for the grouped spelling scores. However, a little reflection will show the reader that the U values in Column (4) of Table 6.5 could be written in at once without the intervening step of the X' scores in Column (2). We simply choose 51 as the original midpoint score to be replaced by a new or U score of zero. This particular midpoint (51 in our example) becomes our reference point. We then enter 0 in the U column opposite 51 and successively add 1 to get the U scores for the intervals having midpoints larger than 51. This gives 1, 2, 3, . . . and so on for these U scores. Likewise, we successively subtract 1 to get the U scores for the intervals having midpoints smaller than 51. This gives −1, −2, −3, . . . and so on for these U scores. The U scores or **coded scores,** as they are often called, can thus be entered directly in Column (4) of Table 6.5. It must be remembered that we are subtracting

51 from each midpoint value and then dividing the result by 11 when we write the U scores. Then once we have found the mean of the U scores (denoted by μ''), the mean of the original scores must be found by first multiplying back by the 11 and then adding back the 51. Symbolically $\mu = 11\,\mu'' + 51$ or $\mu = 11\,(\Sigma\,\text{FU}/\Sigma\,\text{F}) + 51$. To generalize this for distributions other than spelling scores, let us denote the reference point (51 in our example) by R and the distance between successive midpoints (11 in our example) by i (since it is the same as the width of the interval from boundary to boundary). The general formula is then

$$\mu = i\,\frac{\Sigma\,FU}{\Sigma\,F} + R \tag{6.4}$$

As an example of direct coding of scores, let us calculate the mean for the IQ data shown in Table 6.6. Here we have entered the midpoints, X, in Column (3); the coded scores are shown in Column (4), which indicates that we have decided to use 105.5 as the reference point. Note that in this case $i = 10$, so that the entry of coded scores in Column (4) really amounts to subtracting 105.5 from each midpoint or X value and then dividing the result by 10. The reader should verify this for the individual midpoints. We then compute Column (5) by multiplying the F and U values for each row. We find that $\Sigma FU = -109$. Then using Formula (6.4) we have

$$\begin{aligned} \mu = i(\Sigma\,FU/\Sigma\,F) + R &= 10\,(-109/250) + 105.5 \\ &= 10\,(-.436) + 105.5 \\ &= -4.36 + 105.5 = 101.14 \end{aligned}$$

The reader should note that the $-.436$ represents the mean of the U or coded scores and that we then multiplied back by the 10 and added back the 105.5. It will be instructive for the reader to do the work for Table 6.6 over again, this time using 95.5 as the reference point. Note that the final result is the same (101.14) although the intervening numbers are different.

Table 6.6 Calculation of Mean of IQ Scores Using Coded Scores

(1) IQ Interval	(2) F	(3) X	(4) U	(5) FU
121–130	20	125.5	2	40
111–120	40	115.5	1	40/ +80
101–110	77	105.5	0	0
91–100	55	95.5	−1	−55
81–90	40	85.5	−2	−80
71–80	18	75.5	−3	−54/−189
	$\Sigma\,F = 250$			$\Sigma FU = -109$

6.10 The Variance and the Standard Deviation for Grouped Data—Direct Method

The reader should recall the definition of the variance (as given in Chapter 3), and that the standard deviation was defined as the positive square root of the variance. It was noted that one useful formula for finding the variance is the following.

$$\sigma^2 = \frac{\Sigma X^2}{N} - \left(\frac{\Sigma X}{N}\right)^2 \qquad (6.5)$$

In Formula (6.5), σ^2 stands for the variance, ΣX stands for the sum of the terms, ΣX^2 stands for the sum of the squares of the terms, and N stands for the total number of terms involved.

Several important properties of the variance and the standard deviation were also discussed in Chapter 3. One of these was that the subtraction of a constant from each term in a distribution has no effect on either the variance or the standard deviation. Another was that division of each term in a distribution by a constant has the effect of dividing the standard deviation by the absolute value of that constant and of dividing the variance by the square of that constant. In this section and the next we shall follow the same plan that we used in the preceding two sections. We shall first see how to carry out the computation for grouped distributions using a modification of Formula (6.5). Then we shall see how to shorten the work by making use of the two properties that we have just mentioned.

Once again we shall use a single value to represent all of the scores in each interval of the distribution. The logical choice is again the midpoint, for the same reasons discussed in the preceding section. We are, therefore, assuming that all of the scores in each interval are located at the midpoint of that interval. For example, the 50 spelling scores in the interval from 2 through 12 in the distribution of Table 6.1 are all considered to have the value 7, the midpoint of that interval.[2]

Now to illustrate the computation by means of a modification of Formula (6.5), let us again use the distribution of children's spelling scores. We will need the data in Table 6.3, and these have been reproduced in Table 6.7. We will also need the values of X^2 in order to find σ^2. As before, we will use X to represent the midpoints of the intervals; the sum of the terms in each interval is given by the corresponding entry in the FX column. The sum of all of these values in Column (4) is $\Sigma FX = 56{,}225$. Column (5) shows the square of each X value in Column (2) and Column (6) shows the value of the sum of the squares of the terms in each interval. For example, each term in the 2–12 interval (last row) is 7 and there are 50 such terms. We have $X = 7$, so $X^2 = 49$ and the sum of all these 50 terms is $FX^2 = (50)(49) = 2{,}450$. In the first

[2]A correction factor is sometimes introduced to make up for the error occasioned by the rounding off done when computing the variance as described. However, the correction itself is usually trivial when the distribution contains a large number of cases, and we need not consider it here.

row $X = 95$, so $X^2 = 9{,}025$ and the sum of the 78 squares is $FX^2 = 78\,(9{,}025) = 703{,}950$. The Column (6) numbers for the other rows may be obtained in like fashion—by multiplying the corresponding numbers in Columns (3) and (5). Some readers may recognize that $FX^2 = (FX)(X)$ and hence that for the last row $FX^2 = (350)(7) = 2{,}450$. This shows that Column (5) is unnecessary and that the values in Column (6) may be obtained by multiplying the corresponding values in Columns (2) and (4). As a further illustration of this, note that for the first row $(FX)(X) = (X)(FX) = (95)(7{,}410) = 703{,}950 = FX^2$.

Table 6.7 Calculation of the Variance

(1) Intervals	(2) Midpoints X	(3) Frequencies F	(4) Sum of Terms FX	(5) Squares X^2	(6) Sum of Squares FX^2
90–100	95	78	7,410	9,025	703,950
79–89	84	140	11,760	7,056	987,840
68–78	73	192	14,016	5,329	1,023,168
57–67	62	120	7,440	3,844	461,280
46–56	51	140	7,140	2,601	364,140
35–45	40	80	3,200	1,600	128,000
24–34	29	119	3,451	841	100,079
13–23	18	81	1,458	324	26,244
2–12	7	50	350	49	2,450
		$\Sigma F = 1{,}000$	$\Sigma FX = 56{,}225$		$\Sigma FX^2 = 3{,}797{,}151$

The sum of the squares of all 1,000 terms in the distribution is then the sum of Column (6) and is denoted by $\Sigma\, FX^2$. Hence, $\Sigma\, FX^2 = 3{,}797{,}151$, in our example. The formula for the variance of a distribution as computed by the direct method is then

$$\sigma^2 = \frac{\Sigma\, FX^2}{N} - \left(\frac{\Sigma\, FX}{N}\right)^2 \tag{6.6}$$

Thus for the data given in the table we have

$$\begin{aligned}\sigma^2 &= (3{,}797{,}151/1{,}000) - (56{,}225/1{,}000)^2 = 3{,}797.151 - (56.225)^2 \\ &= 3{,}797.151 - 3{,}161.251 = 635.9\end{aligned}$$

Therefore the variance is 635.9 and the standard deviation $\sigma = \sqrt{635.9} = 25.22$.

6.11 The Variance and the Standard Deviation for Grouped Data—Coded Method

By now the reader may well be dismayed at the prospect of having to carry out computations that involve squaring and adding such large numbers. However, just as before, the shortcut of using coded scores may be employed. To begin with, we may subtract some particular constant from each score in the distribution—that is, from each midpoint value listed. Theorem 3.1 tells us that to do so will have no effect whatsoever on the variability of the distribution, so that we shall obtain the same values of σ and σ^2 as if we had not bothered to subtract the constant. It follows that having subtracted a constant from each midpoint will not even necessitate our making a readjustment by adding it back to compute the standard deviation or the variance.

The reader will recall that the final step in setting up the shortcut for the mean was to divide the new midpoint values by the width of the interval in order to obtain the final coded or U scores. We shall do precisely the same thing when computing the variance. Theorem 3.2 tells us that this will have the effect of dividing the standard deviation by that same amount and hence of dividing the variance by the square of that amount. We will then have to "multiply back" in order to get the final values for these measures.

To see the application of this coded score method to the spelling score data, we have reproduced four columns of Table 6.5 as the first four columns of Table 6.8. These columns show the original midpoint values, X; the frequencies, F; the coded scores, U; and the sums of the coded scores for each row, FU. Remember that the U scores were obtained by subtracting 51 from the corresponding X scores and then dividing by 11.

Two new columns have been added to the table: Column (5) shows the square of

Table 6.8 Computation of Variance and Standard Deviation Using Coded Scores

(1) X	(2) F	(3) U	(4) FU	(5) U^2	(6) FU^2
95	78	4	312	16	1,248
84	140	3	420	9	1,260
73	192	2	384	4	768
62	120	1	120	1	120
51	140	0	0	0	0
40	80	−1	−80	1	80
29	119	−2	−238	4	476
18	81	−3	−243	9	729
7	50	−4	−200	16	800
	$\Sigma F = 1{,}000$		$\Sigma FU = 475$		$\Sigma FU^2 = 5{,}481$

the coded score for each row, and Column (6) gives the sum of these squared scores for each row. In obtaining Column (5) keep in mind that the product of two negative numbers is a positive number and, in particular, the square of a negative number is a positive number. The values in Column (6) may be obtained by multiplying the corresponding entries in Columns (2) and (5). Thus in the first row $FU^2 = (F)(U^2) = (78)(16) = 1{,}248$, in the second row $FU^2 = (140)(9) = 1{,}260$, and so on. The reader may also recognize that $FU^2 = (FU)(U) = (U)(FU)$ and hence the Column (5) values are unnecessary and the Column (6) values may be obtained by multiplying the corresponding values in Columns (3) and (4). Thus in the first row $FU^2 = U(FU) = (4)(312) = 1{,}248$, in the second row $FU^2 = (3)(420) = 1{,}260$, and so on. Again we must remember that the product of two negative numbers is a positive number; for instance, in the last row $FU^2 = U(FU) = (-4)(-200) = 800$.

The sum of the squares of all 1,000 terms in the coded or U score distribution is the sum of Column (6) and is denoted by ΣFU^2. Hence, $\Sigma FU^2 = 5{,}481$ in our example. We now proceed as in the previous computation except that we use the U values. Let σ_U^2 denote the variance of the U values. Then we have

$$\sigma_U^2 = \frac{\Sigma FU^2}{N} - \left(\frac{\Sigma FU}{N}\right)^2 \tag{6.7}$$

Thus for the data in Table 6.8 we have

$$\sigma_U^2 = \frac{5{,}481}{1{,}000} - \left(\frac{475}{1{,}000}\right)^2 = 5.481 - (.475)^2 = 5.481 - .226 = 5.255$$

Then the standard deviation of the U values $\sigma_U = \sqrt{5.255} = 2.292$.

As stated earlier, Theorems 3.1 and 3.2 tell us that the only adjustment needed to get the values of σ^2 and σ for the original distribution is to multiply back to undo the effect of having divided by 11. By Theorem 3.2 we must multiply σ_U by 11 to get σ. We have, then, $\sigma = 11\sigma_U = 11(2.292) = 25.21$, the same result that we got with uncoded scores.[3] By squaring σ, we get the value of σ^2. $\sigma^2 = (25.21)^2 = 635.5$.[3] Note that this value of the variance is what we also obtained when we worked with uncoded scores. It will be instructive for the student to do the computation over again, this time using 40 as the reference point for the coded scores. This should again illustrate that the choice of reference point has no effect on the final result.

For those who prefer a completely symbolic statement, we can combine Formula (6.7) with the multiplying back and obtain

$$\sigma = i\sqrt{\frac{\Sigma FU^2}{N} - \left(\frac{\Sigma FU}{N}\right)^2} \tag{6.8}$$

where we have again used i for the width of the interval in the distribution.

[3]The small difference is due to the rounding error.

It should be noted that Table 6.8 [even with Column (5) omitted] contains all data needed for computation of the mean as well as the variance and standard deviation of the spelling marks using coded scores. The reader will have noted that $(\Sigma FU)/N$ appears in both computations. This quantity is the mean of the coded scores. For grouped frequency distributions the coded scores approach is generally the fastest and easiest to apply when these measures are needed.

Problem Set A

A prominent physician recently devised a scale to measure a person's dependency on drugs. The measure, which ranges from 1 to 30, takes into account not only daily dosage and type of drug required, but also a subject's willingness to obtain the drug despite the cost and consequences. The dependencies of 100 young persons beginning a rehabilitation program (some willingly, some not) are measured. The results that follow are to be used to answer Questions 6.1–6.4.

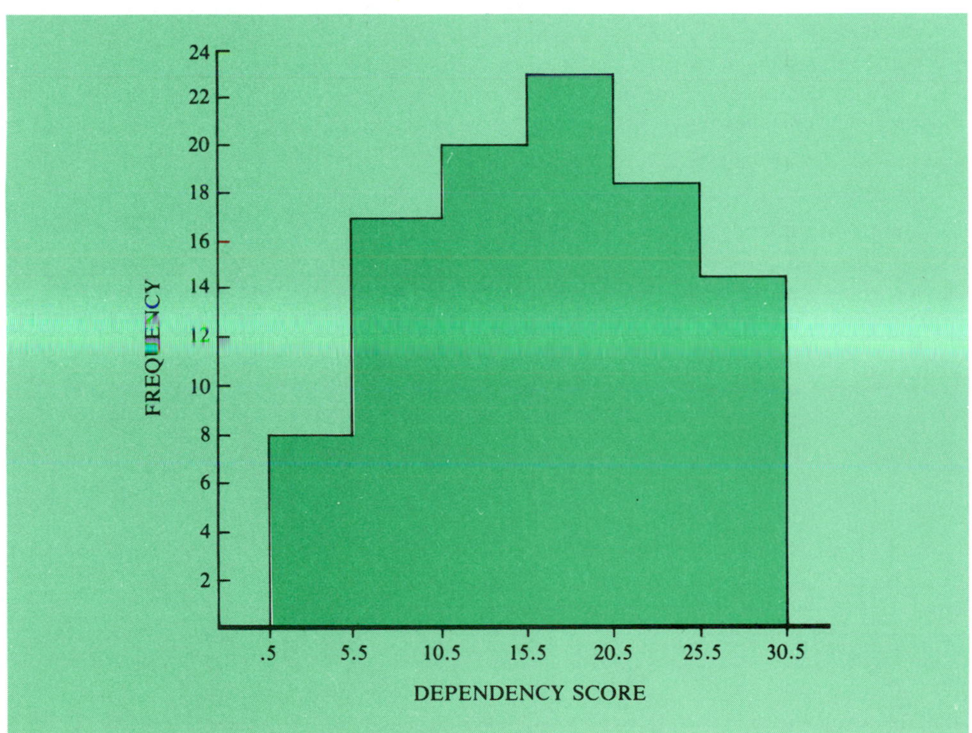

6.1 Construct the grouped frequency table for the distribution.
6.2 Compute the median dependency score, P_{35}, D_{10}, and Q_3.
6.3 Find the percentile ranks of the following dependency scores.

a. 10.5 b. 4 c. 22

6.4 a. Compute the mean and standard deviation using the direct method.
b. Compute the mean and standard deviation using coded scores by subtracting a constant of 18.

For Problems 6.5–6.7 use the following grouped frequency table, which shows the numbers of deaths due to heart failure per 1,000 men who took varying amounts of aspirin per day. The subjects were all 45 years old when the 10-year study began.

Daily Dosage (in milligrams)	*F*
1,750–2,100	3
1,400–1,750	7
1,050–1,400	15
700–1,050	14
350–700	21

6.5 Compute the following from the distribution of daily aspirin dosages for the men who died.

a. P_{15} b. $PR_{1,500}$ c. the median d. Q_3

6.6 a. Draw the histogram and cumulative curve for the distribution of dosages.
b. Locate the median on both the histogram and cumulative curve.

6.7 a. Find the crude mode, mean, and standard deviation directly.
b. Use coding to find the mean and standard deviation.

6.8 Which of the following curves could not serve as a cumulative frequency curve? Why not?

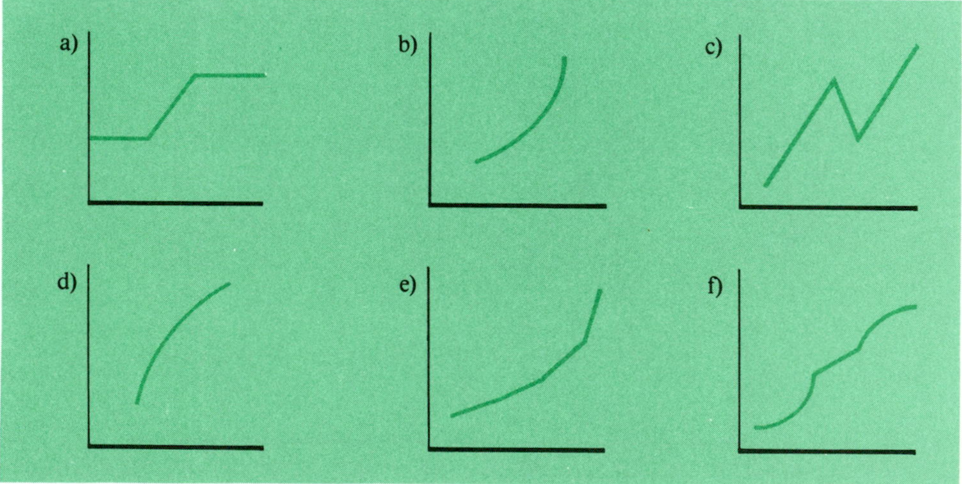

The following situation is to be used to complete Problems 6.9–6.12.

Five hundred test homes throughout the country each had a 42-inch by 36-inch window facing south. The heat flow was measured at each site. Then a roof overhang was added to each house and the heat flow measured again. The percents of *heat flow reduction* are supplied for the homes in the following grouped frequency table.

Percent Reduction	*F*
51.5–58.5	9
44.5–51.5	41
37.5–44.5	59
30.5–37.5	105
23.5–30.5	123
16.5–23.5	62
9.5–16.5	71
2.5–9.5	30

6.9 Compute Q_2, D_1, and P_{45} for the distribution of percents of reduction of heat flow.

6.10 Find the percentile ranks of (a) 9, (b) 36, (c) 49, and (d) 53.7.

6.11 a. Draw the cumulative curve for the distribution.
b. Use the curve to estimate Q_2, D_1, P_{45}, PR_9, PR_{36}, PR_{49}, and $PR_{53.7}$. Compare your results with your answers to Problems 6.9–6.10.

6.12 a. Compute the mean and variance of the distribution by the direct method.
b. Compute the mean and variance using coded scores and subtracting a constant of 27.

Problems 6.13–6.15 refer to the following results from a study conducted by a local religious congregation to determine the age distribution of the group.

Age	*F*
90–100	3
80–90	30
70–80	53
60–70	117
50–60	190
40–50	210
30–40	325
20–30	75
10–20	115
0–10	103

6.13 a. Compute the mean age of the congregation and its standard deviation.
b. What is the most common age of the parishioners?

6.14 a. Find the ages that represent P_{15}, P_{30}, P_{75}, and P_{90}.
b. Find the percentile ranks of persons who are 10.5, 43.7, and 62 years old.

6.15 a. Draw the cumulative curve and find the percentiles and percentile ranks called for in Problem 6.14.
b. Draw the histogram that represents the data.

Problem Set B

Deel Delivery Service has been in business for 25 years. During that time they have been forced to replace 30 delivery trucks. The lengths of service for the replaced vehicles are given here. They are to be used to answer Problems 6.16–6.18.

Years of Service	*F*
16.5–20.5	1
12.5–16.5	6
8.5–12.5	6
4.5–8.5	12
.5–4.5	5

6.16 Compute the mean, median, and mode for the distribution of years of service.

6.17 a. Find the variance and standard deviation of the terms.
b. What are the z scores of values 6.5, 8.75, and 16?

6.18 Draw a cumulative frequency curve and use it to estimate the percent of values that are at least 5. What percent are at least 10?

For Problems 6.19–6.21 use the cumulative frequency curve provided here for the prison terms of the 150 inmates in the state's maximum security facility who are less than 21 years of age.

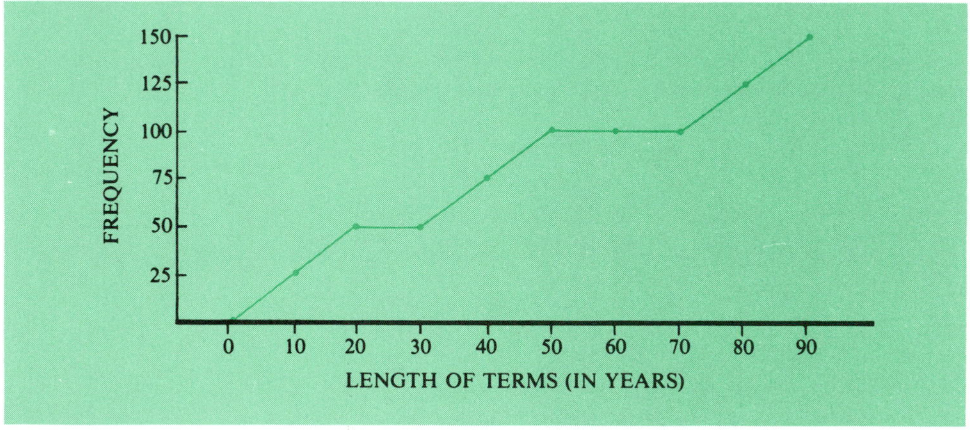

6.19 None of the inmates under 21 years of age have sentences of what lengths?

6.20 Find the median sentence and the sentences that represent P_{75}, P_{40}, and D_6.

6.21 Find how the prisoners with the following terms rank among their peers by finding their percentile ranks.

a. 15 years b. 516 months c. 76 years d. 2 years

A committee of educators and parents has been attempting to uncover the cause of the decline in achievement scores at Roosevelt Elementary School. The committee sent a questionnaire and survey to each pupil's home. The 330 forms returned indicated the following numbers of hours of television watched per school day by Roosevelt students. Use the results to work Problems 6.22–6.25.

Hours Watched	*F*
5–6	2
4–5	15
3–4	38
2–3	125
1–2	97
0–1	53

6.22 Find the following values and ranks.

a. P_{16} b. $PR_{2.75}$ c. P_{88} d. PR_4 e. $PR_{4.3}$ f. D_6

6.23 Draw the cumulative curve and histogram for the data.

6.24 a. Compute the mean and standard deviation directly.

b. Compute the mean and standard deviation by the coded method. Subtract a constant of 2.5.

6.25 Suppose each child cuts his or her TV viewing time in half. What should the new boundaries be to maintain six intervals? What would the new mean and standard deviation be?

The Tireless Tire Company employs 2,782 workers. Its president, Ty Tired, is concerned over increasing absenteeism. He orders a report of the numbers of workdays missed by his employees during the past six months. The results below will be needed to answer Problems 6.26–6.29.

Number of Days Missed	*F*
27–29	5
24–26	59
21–23	63
18–20	97
15–17	262
12–14	197
9–11	152
6–8	368
3–5	692
0–2	887

6.26 Compute the mean, median, and mode of the distribution.

6.27 Compute the distribution variance and standard deviation directly and by coding.

6.28 a. Find the percentile rank of the president's son, Tyler Tired, II, who missed 13 days of work.

b. Find the ranks of the president's nephews who missed 19, 23, and 25 days.

6.29 Verify your answers to Problem 6.28 by drawing a cumulative curve and locating the four percentile ranks.

6.30 Construct the grouped frequency table and histogram of the distribution that has the following cumulative curve.

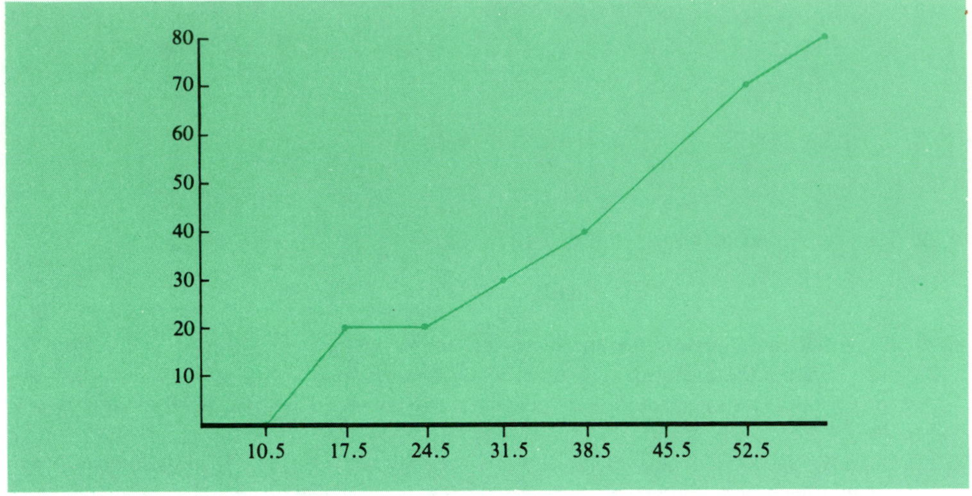
80
70
60
50
40
30
20
10
10.5
17.5
24.5
31.5
38.5
45.5
52.5

Review I

The material we've been considering so far falls under the general category of descriptive statistics. That is, we have been examining ways to describe a distribution of values. Our emphasis has been both on the terms comprising the distribution (on their placement and how to represent them) and on a whole distribution considered as an entity in itself.

Before going on to new topics, the reader should know the procedures for finding percentile ranks, z scores, T scores, and percentiles, and understand their meaning. Finally, there should be an understanding of how to draw and interpret the pictorial representations discussed: bar graphs, histograms, frequency polygons, and cumulative frequency curves. For further work full grasp of these concepts will be necessary.

The mean, median, and mode are indicators of a distribution's center. By far the most important of these is the mean, which may be thought of as a balance point. Although the mean gives a good deal of information about the distribution, it tells nothing about how scattered the terms are. The standard deviation and the variance tell us this. It may be helpful to think of the standard deviation as a yardstick that measures distance from the mean.

The positions of individual terms in a distribution may be indicated by their percentile ranks or their z scores. Each type of description has its own advantages. The percentile rank locates a term by showing only its position relative to other terms. The z score, on the other hand, shows the number of standard deviations a term is above or below the mean. Thus it takes into consideration the magnitude of a term's difference from the mean and compares it with the variability of the distribution.

Pictorial representations of distributions (such as bar graphs, histograms, frequency polygons, and cumulative frequency curves) offer convenient and useful ways to present data. If for every term in a distribution two facts are known, its z score and the percent of terms it exceeds, a precise picture or graph of the distribution can be drawn.

So far we have dealt with distributions having a finite number of terms. In the next chapters we shall extend many of the concepts just discussed to infinite or theoretical distributions.

The exercise sets that are included in each of the four Reviews (following

Chapters 6, 10, 14, and 19) are intended to provide additional practice in problem-type recognition as well as in problem solving.

Problem Set A

I.1 Calculate the mean, median, mode, variance, and standard deviation for the three distributions, X, Y, and Z, provided here.

X	*Y*	*Z*	
7	18	19	14
4	22	17	19
3	4	8	27
8	84	40	21
1	23	37	18
9	22	29	12
	22	3	11
		49	

I.2 The terms 8, 17, and 25 each appear in each of Distributions U, V, and W below. Find the percentile ranks of 8, 17, and 25 in each distribution.

U	*V*				*W*				
8	7	8	17	21	0	8	9	24	17
17	7	8	17	22	1	5	17	25	17
25	8	9	21	25	2	4	13	18	8
					7	2	25	17	8
					8	25			

I.3 Consider this distribution of the noise levels of five factory machines (measured in octave bands):

70 77 73 60 59

a. Find the mean, median, and mode if they exist.
b. What is the variance and standard deviation of the terms?
c. Find the z score and percentile rank of the term 60.

I.4 Reported auto theft monthly totals for the past year are 77, 107, 99, 152, 80, 93, 55, 75, 68, 93, 84, and 73.

a. Find the mean, median, and modal monthly totals for the year.
b. Compute the standard deviation using both the literal and computational formulas.
c. Determine the z scores of 77, 68, and 93.
d. Find PR_{77}, PR_{68}, and PR_{93}.

I.5 The weights (in pounds) of eight calves being purchased to raise for slaughter are 507, 602, 490, 540, 535, 525, 675, and 476.

a. Find the mean, variance, and standard deviation of the weights.

b. Suppose each weight is increased by 50 pounds. What are the new μ, σ^2, and σ?
c. Suppose each weight is increased by 50% (multiply each term by 1.5). Find the new μ, σ^2, and σ.
d. Suppose, due to illness, each calf loses 10% of its weight. Find the new μ, σ^2, and σ.

I.6 The ages of 15 children suffering from enuresis who are being screened at Clark Clinic for multihandicapping conditions are provided here.

3.50	6.00	7.00	4.25	8.50
5.50	5.00	4.75	3.75	6.50
6.00	7.00	7.25	6.75	7.25

a. Determine the mean age.
b. Find the standard deviation of the distribution.
c. If the children are chosen to participate in a five-year study, what will be the mean, variance, and standard deviation of their ages at the end of the study?

I.7 Verify that the sum of the term distances from the mean is zero using the distribution of ages in Problem I.6.

I.8 For Distribution W whose 10 terms are 11, 8, 17, 3, 14, 3, 8, 15, 19, and 27, find

a. ΣW
b. ΣW^2
c. $(\Sigma W)^2$
d. μ_W
e. $(\Sigma W - \mu_W)^2$
f. $\Sigma W^2 - \mu_W^2$
g. $\Sigma (W - \mu_W)$
h. $\Sigma (W - \mu)^2$
i. $[\Sigma (W - \mu_W)]^2$
j. $\Sigma W + \mu_W$

I.9 The mean age of 1,000 single homeowners granted home loans by a certain mortgage company is 37.5. The standard deviation of the group ages is 6.3 years.
a. Find the z scores of ages 27, 35, 42, and 53.5.
b. Find the T scores of ages 27, 35, 42, and 53.5.
c. Find the ages whose corresponding z scores are -1.5, -2.75, and 1.2.
d. Find the percentile rank of Single Sara who is older than 845 of the group members and is the only person her age.

I.10 The efficiency ratings of two types of engines are measured on two obstacle courses. The first course has a history of allowing scores with a mean of 20 and standard deviation of 4, whereas the second course has allowed a mean and standard deviation of 25 and 7, respectively. Engine A scores 21 and 30 on the two courses, and engine B scores 25 and 26. Which engine should be considered most efficient on the basis of these tests?

I.11 Classify the following variables as continuous or discrete.

a. weight gained
b. number who died
c. favorite color
d. light intensity
e. noise level
f. number of births

g. reading skill h. shoe size i. dress length
j. batting average

I.12 The Kingman Corporation makes large belts for conveyor systems. The quality control manager, Mr. Raymond, periodically selects 100 belts to test for elasticity. The deviations from the ideal are recorded. The last check yielded the following results given in grouped form.

Deviation	*Frequency*
1,104.5–1,155.5	9
1,053.5–1,104.5	14
1,002.5–1,053.5	30
951.5–1,002.5	25
900.5–951.5	15
849.5–900.5	7

a. Compute μ and σ directly.
b. Compute μ and σ using coded scores.
c. Find the median.

I.13 There is often a considerable time span between the discovery of a drug and its introduction to the market. Testing, government regulations, and production-marketing activities account for much of this time. A study of 500 drugs available today, some by prescription and some over-the-counter, shows the time between development and introduction for each drug. A summary of the results follows.

Time (in years)	*Frequency*
21.5–24.5	11
18.5–21.5	70
15.5–18.5	52
12.5–15.5	84
9.5–12.5	17
6.5–9.5	44
3.5–6.5	85
.5–3.5	137

a. Find the interval midpoints and cumulative frequencies for the grouped frequency table.
b. Draw the histogram and cumulative frequency curve.
c. Calculate P_{50}, P_{35}, $PR_{7.8}$, and PR_{16}. Also estimate these on the cumulative frequency curve.

I.14 a. Find the mean for the distribution given in Problem I.13 both by the direct method and by using coded scores.
b. Find the standard deviation for the distribution in Problem I.13 directly and by coding.

I.15 Use the following pairings to draw a rough sketch of the distribution's curve. The pairings are for five different terms from a distribution whose terms all have z scores below 3.

z Score	*Percent of Terms Exceeded*
−4.5	0
−3.0	30
−1.0	50
0.0	70
1.5	85

What does area under the curve represent?

Problem Set B

I.16 The production coefficients (number of successful contacts divided by total number of contacts) of the eight company representatives in the Tulsa office have been determined to be .47, .83, .65, .72, .89, .77, .75, and .71.

a. Determine the mean and median coefficients for the Tulsa representatives.
b. Determine the variance for the eight coefficients.
c. Assuming the coefficients of the company representatives nationwide have a mean of .68 and a standard deviation of .15, find the z score (in the national distribution) of the smallest Tulsa coefficient, the largest Tulsa coefficient, and the mean Tulsa coefficient.

I.17 If Distribution Y has six terms, 4, 8, 16, 29, 7, and 8, find the following.

a. the mode
b. P_{50}
c. μ_Y
d. ΣY
e. ΣY^2
f. $(\Sigma Y)^2$
g. $\Sigma (Y - \mu)$
h. $\Sigma (Y - \mu)^2$
i. $\dfrac{\Sigma (Y - \mu)^2}{N_Y}$
j. $\Sigma (Y + 5)^2 + 7$

I.18 According to the *Daily News,* the following totals indicate the numbers of marriages in June and December for the past 10 years.

June:	107	92	84	73	121	88	79	62	85	91
December:	43	45	29	35	39	37	51	47	42	38

In which of the two months was the number of marriages more variable?

I.19 The mean rainfall in Calico County is 19 inches per year. The standard deviation is 2.75 inches.

a. Find the z scores of the years with 21.7, 15.3, and 12 inches of rain.
b. Find the T scores of the years mentioned in part a.

c. What would the rainfall need to be for a year to have a z score of 3, -1.8, $-.37$, and 1.75?

I.20 Professor U. I. Fale gives his statistics classes the same exams year after year. The past has shown that the class averages on Exam I have a mean of 75 and a standard deviation of 3. On Exam II the class averages have a mean of 65 and the standard deviation is 8. This semester Professor Fale teaches three sections of statistics, A, B, and C. On Exam I class A averages 71, B averages 77, and C averages 76. On Exam II the averages of sections A, B, and C are 71, 70, and 65, respectively. Which class performed best?

I.21 The tensile strengths (in pounds per square inch) of 15 metal disks are

65,000	53,800	62,100
72,000	46,500	67,000
53,300	81,700	71,400
42,100	39,300	73,400
65,900	38,800	35,000

a. Find the mean tensile strength.
b. Determine the standard deviation by first dividing by 100.
c. How does dividing each term by 100 affect the mean and standard deviation of a distribution?

I.22 The numbers of voters who favor candidate Lettie U. Down are provided here for the 20 voting districts of a state society.

81	71	49	62
77	70	59	67
29	60	47	25
68	54	81	49
54	53	73	50

a. Find the mean and median of the distribution.
b. If Ms. Down's support increases by approximately 12% in each district, what are the new mean and median numbers of support per district?
c. What additional piece of information is necessary to make the mean number of supporters per district meaningful?

I.23 Use the histogram at the top of page 119, which shows the waiting times during a one-week period for patients in a hospital emergency room, to answer the following questions.
a. What percent of the patients waited 15 minutes or longer?
b. What percent waited less than 5 minutes?
c. To determine the total number of patients treated in the emergency room during the study period, what information would need to be added to the graph?

I.24 Twenty-five people delivered scrap aluminum to the reclamation collection site last Saturday. The pounds sold by each person are as follows:

27	13	25	19	22	20	47
15	18	15	29	23	20	21

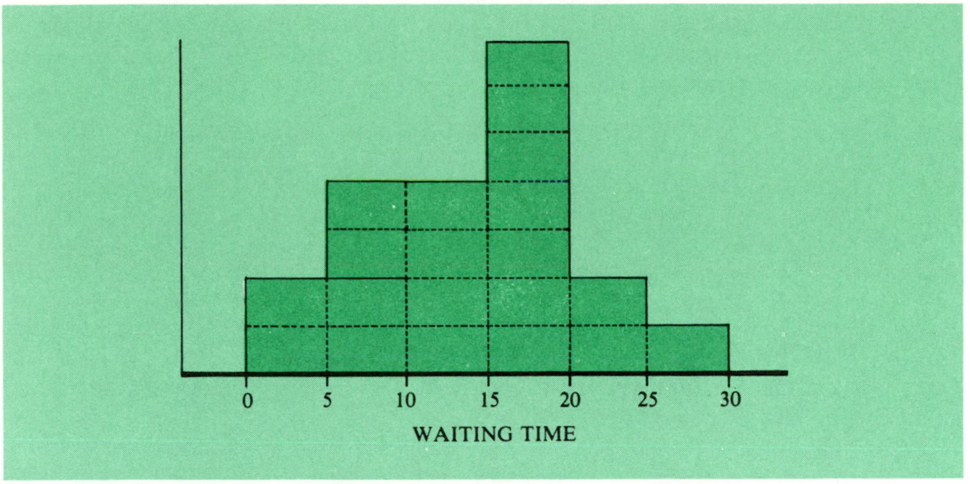

45	32	31	23	10	12	15
13	25	18	31			

a. Find the mean weight per person of the aluminum collected by the 25 people.
b. Find the standard deviation of the 25 weights.
c. Had each person collected 10 more pounds, what would the mean and standard deviation have been?
d. Had each person collected 20% more aluminum by weight, what would the mean and standard deviation have been?

I.25 A faculty physical exercise test is given in three parts (I, II, III). In each, a certain task is to be completed as quickly as possible. The means and standard deviations for three age groups are given below for each part of the three part test. (All values are in minutes.)

Age	I	II	III
40–50	$\mu = 15 \quad \sigma = 5.0$	$\mu = 28 \quad \sigma = 7.0$	$\mu = 21 \quad \sigma = 5.9$
30–40	$\mu = 12 \quad \sigma = 3.5$	$\mu = 25 \quad \sigma = 5.5$	$\mu = 19 \quad \sigma = 5.0$
10–30	$\mu = 10 \quad \sigma = 2.0$	$\mu = 21 \quad \sigma = 4.0$	$\mu = 15 \quad \sigma = 3.0$

A 25-year-old instructor scored 9, 22, and 18.
A 35-year-old associate professor scored 13, 29, and 17.
A 45-year-old full professor scored 20, 25, and 17.
Whose performance was the most outstanding? (Hint: Find the mean of three z scores for each teacher.)

I.26 Pat Block, graduate student, has been researching the reaction times of crustaceans subjected to a certain stimulus. The times of 60 of the creatures are grouped in the following table.

Reaction Time (in seconds)	*Frequency*
24.5–29.5	9
19.5–24.5	10
14.5–19.5	19
9.5–14.5	15
4.5–9.5	7

a. Find the mean reaction time directly and by coding.
b. Find the standard deviation directly and by coding.
c. Find P_{15}, D_3, and Q_3.
d. Determine PR_{10} and PR_{22}.

I.27 a. Complete the grouped frequency table in Problem I.26 by adding midpoints and cumulative frequency entries.
b. Draw the histogram and cumulative frequency curve.
c. Use the cumulative frequency curve to substantiate your answers to parts c and d of Problem I.26.

I.28 School District Number Four has 1,053 households in the district that it may tax. The budget committee of the school board completed the following table for use in preparing next year's budget.

Household Income	*Frequency*
70,500–77,500	19
63,500–70,500	11
56,500–63,500	47
49,500–56,500	43
42,500–49,500	0
35,500–42,500	55
28,500–35,500	127
21,500–28,500	165
14,500–21,500	289
7,500–14,500	193
500–7,500	51

a. Draw a frequency polygon that represents the data.
b. Find the mean income. Use coded scores and subtract a constant of 32,000.
c. Find the standard deviation using coded scores.

I.29 Use the data presented in Problem I.28 to solve the following.
a. Find the interval midpoints and cumulative frequency entries for the grouped frequency table.
b. What is the median income?
c. Find P_9, Q_3, and D_4.
d. Find $PR_{30,000}$ and $PR_{8,700}$.

I.30 One hundred seniors were surveyed at each of 800 high schools. Among questions asked was "Do you feel that divorce laws should be liberalized?" The grouped frequency table below summarizes the affirmative responses.

Number of Affirmative Responses	*Number of Schools*
200–224	200
175–199	8
150–174	105
125–149	187
100–124	123
75–99	105
50–74	55
25–49	17
0–24	0

a. Calculate the mean number of affirmative responses directly. Check by using the coded method.
b. Calculate the standard deviation directly. Check by using the coded method.
c. Find P_{50}, Q_3, and D_2.
d. What is the crude mode?

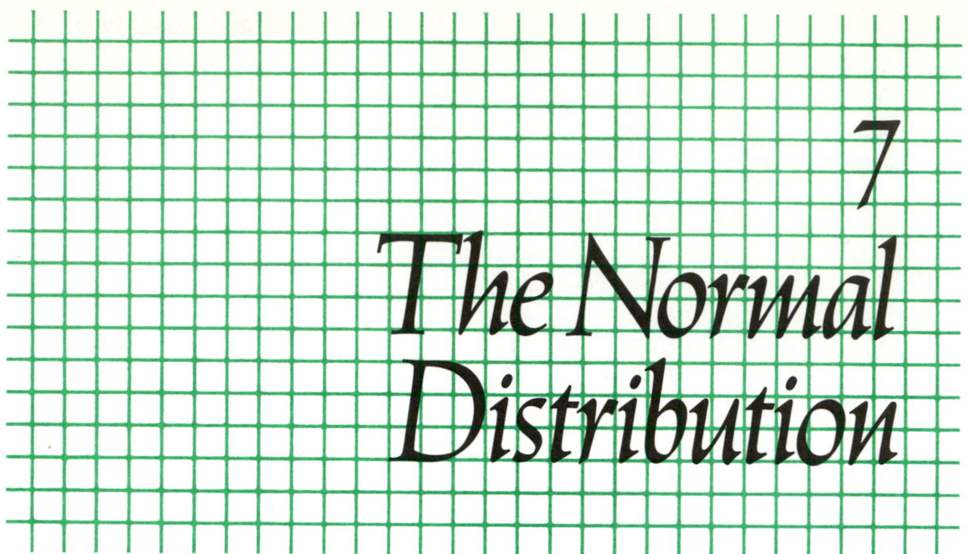

7 The Normal Distribution

7.1 Theoretical Distributions

We have seen that as we add more and more terms to a distribution we may often depict its shape with increasing accuracy by means of a curve. Under some conditions, as we continue collecting terms, the effect of new terms on the shape of the distribution becomes increasingly negligible. After a while, the shape of the distribution may be said to "settle down" and it becomes possible to consider properties of the distribution without specifying the number of terms included in it. We need specify only that the number of terms is "large enough."

We shall use the phrase **an infinitude of terms** to mean so many terms that the effect of adding more by the same process need not concern us. A **theoretical distribution** is one composed of an infinitude of terms. In other words, a theoretical distribution is one that has "settled down" and we may talk about its properties and its shape as if they were fixed. In a theoretical distribution, each term comprises a negligible percent of the distribution and we say that the percentile rank of a term is merely the percent of terms below it (usually given to the nearest whole number of percent).

7.2 The Normal Distribution and the Normal Curve

We shall now give our attention to a particular theoretical distribution called the **normal distribution** and the graph that depicts it, the **normal curve**. The normal distribution is extraordinarily important from both a theoretical and a practical standpoint. Much of the development of statistics—indeed, perhaps most of it—has sprung from consideration of the normal distribution. The normal distribution is of prominent importance in virtually all fields where data are gathered.

The graph in Fig. 7.1 is the normal curve. z score values are plotted along the baseline. Consecutive z score units are equal distances apart.

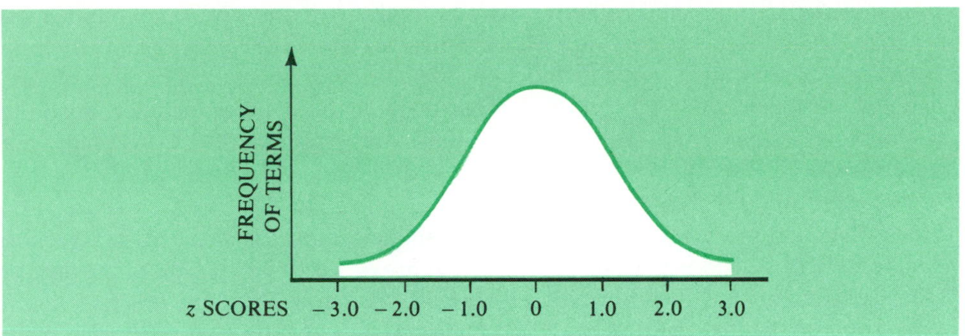

Fig. 7.1

The exact shape of the normal curve depends somewhat upon how far apart we decide to set the z score units along the baseline. For instance, the normal curve looks somewhat different when we set the z score units closer together (Fig. 7.2). The spacing of units along the vertical axis is the same in Fig. 7.2 as in Fig. 7.1.

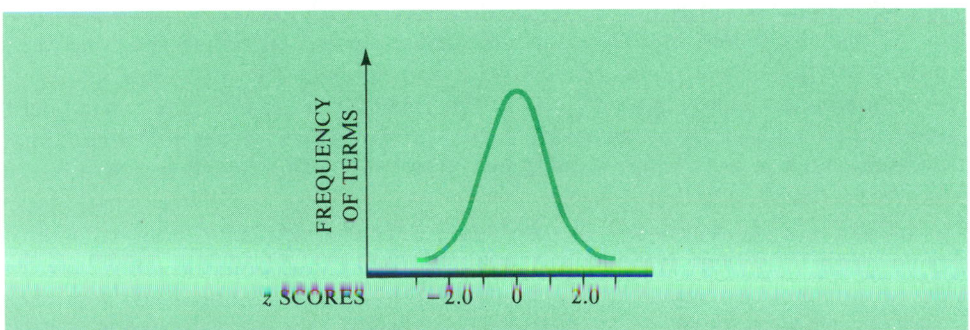

Fig. 7.2

Several properties of the normal distribution are reflected in the normal curve, no matter how it is drawn. For instance, the normal curve is symmetric and its highest point is over the value $z = 0$, facts that the reader may verify by looking at either Fig. 7.1 or 7.2. We shall point out more of these properties in Section 7.3.

The normal distribution and the normal curve are precisely and completely defined by a mathematical formula. Unfortunately, we cannot define the normal distribution in the usual way without presupposing a knowledge of calculus. Instead we shall present a somewhat imprecise but roughly equivalent definition of the normal distribution, to which we shall repeatedly refer.

We saw in a previous chapter that a table listing z score values along with the percents of the distribution exceeded by the terms with those z score values constitutes a numerical description of a distribution. We were able to draw the graph of a distribution from such a table with no further information.

There exists a particular set of relationships between the z scores of terms and the percents of terms exceeded such that, when these relationships hold, the distribution is said to be a normal distribution.

The particular set of relationships that defines the normal distribution is given in Appendix II. This table is known as a normal distribution table. A number of z scores ranging from -4 to $+4$ are listed. When we know the z score of a term in the normal distribution and would like to know what fraction of the total number of terms is below the given term, this normal distribution table tells us. For instance, according to the table, in the normal distribution the term with a z score of -1.0 is bigger than approximately 159/1,000 of the terms in the distribution. In other words, the term with a z score of -1.0 is at about the 16th percentile. The term with a z score of 0 is bigger than half of the terms in a normal distribution; thus it is at the 50th percentile. The term with a z score of 1.5 is bigger than approximately 933/1,000 of the terms in the distribution. Thus the term with a z score of 1.5 is roughly at the 93rd percentile.

The normal distribution may be roughly defined as the distribution in which the relationships listed in Appendix II hold. The statement that a distribution is normal means that the relationships between the z scores of the terms and the percents of terms exceeded in the distribution are as indicated in the table. In short, if the relationships are as indicated in the table, the distribution is normal at least to the extent that the table, being incomplete, is able to identify the distribution. Admittedly we have given an approximate characterization of the normal distribution rather than a precise definition of it.

Appendix II is necessarily incomplete and imprecise. Obviously, no table can be extensive enough to include all possible z scores; even the most elaborate tables provide only an approximate definition. The fractions in the table are carried out to the nearest thousandth, but they still are not precise. In any event, our approximate definition will prove to be serviceable for characterization of the normal distribution, as we shall see.

Given the information in the table, the reader can proceed to draw an approximate graph of the normal distribution. Actually the result would be a histogram rather than a smooth curve, but if that reader were carefully to "smooth out" the tops of the bars of the histogram after drawing it, the resulting figure would be virtually identical with the normal curve. Whether this picture would resemble the normal curve in Fig. 7.1 or the one in Fig. 7.2 would depend upon how the z score units along the horizontal axis were spaced.

We have defined the normal distribution by referring to a table of relationships rather than by referring to the normal curve. The curve of the distribution characterized by the table is the normal curve whether it looks relatively flat, as in Fig. 7.1, or peaked, as in Fig. 7.2. Appendix II is our best substitute for the formula that perfectly defines the normal distribution and the normal curve. A more extensive table would naturally define the normal distribution more precisely.

7.3 Four Properties of the Normal Distribution

We shall now consider four properties of the normal distribution and note their influences on the normal curve (as drawn in Fig. 7.3). We shall refer to the vertical center line in Fig. 7.3 as the **vertical axis**. The vertical axis is at the value $z = 0$ where the curve reaches its maximum point.

Fig. 7.3

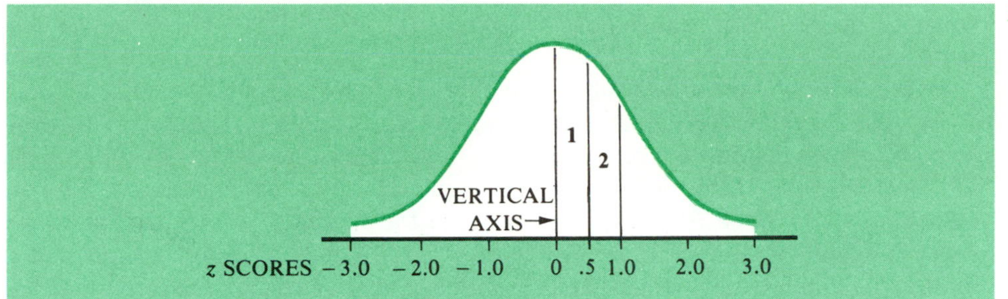

1. One property of the normal distribution is that the terms tend to cluster around the point $z = 0$. That is, as the curve moves in either direction away from the vertical axis, it drops, revealing the fact that there are fewer and fewer terms. For instance, there are more terms with z scores between 0 and .5 than there are terms with z scores between .5 and 1.0. In Fig. 7.3 the area in sector 1 (which signifies the fraction of terms with z scores between 0 and .5) is larger than the area in sector 2 (which signifies the fraction of terms with z scores between .5 and 1.0). The further we go from the vertical axis in either direction, the fewer terms we find.

2. A second property is that the normal curve is symmetrical around the vertical axis. The height of the curve over any z score value is exactly the same as its height over the negative of that value. For example, the curve is the same height over the point $z = 1.0$ as it is over the point $z = -1.0$. The curve is the same height over the point $z = -2.57$ as it is over the point $z = 2.57$.

It follows that half of the terms are to the left of the vertical axis and have negative z scores and half of them are to the right of the vertical axis and have positive z scores. Also there is the same fraction of terms in any sector as in its corresponding sector on the opposite side of the vertical axis. For instance, it turns out that about 15% of the terms have z scores between .5 and 1.0. The same percent of the terms have z scores between $-.5$ and -1.0 (Fig. 7.4).

Fig. 7.4

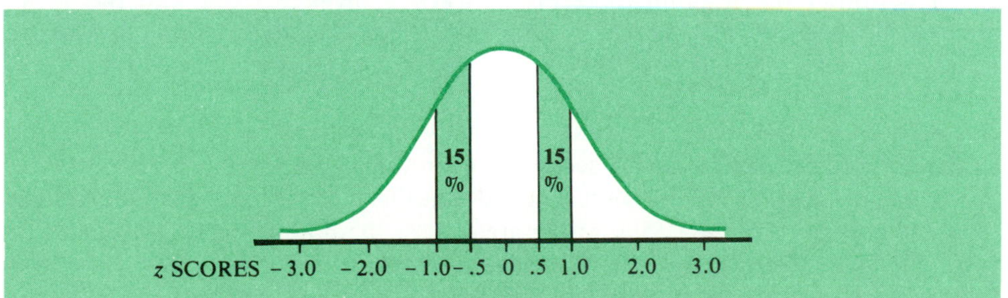

In Section 7.4 we shall make use of the fact that each sector contains the same fraction of terms as its symmetrically opposite sector. We shall use this fact to draw a simplified and convenient table of the normal distribution.

3. The sizes of the terms in the normal distribution are not bounded in either direction. In theory there exist terms with z scores of all magnitudes. There are terms with z scores as small as $-1{,}000$ and z scores as large as 1,000.[1] Such terms are incredibly rare and in practice we do not consider them. However, we take cognizance of them in the normal curve by not dropping the curve completely down to the baseline, no matter how far out we extend it in either direction.

4. The mean, median, and mode of the normal distribution all have the same value. The term that has a z score equal to 0 has all three properties (see Fig. 7.3). The term with the z score of 0 is at the mean since it is the balance point of the distribution. The same term is the mode since the normal curve is highest where $z = 0$, indicating that more terms have the z score of 0 than any other z score value. Finally, the property of symmetry tells us that exactly half of the area is to the left of the vertical axis, where $z = 0$, and half of the area is to the right of it. Thus half of the terms have z scores less than 0 and half of them have z scores larger than 0. Accordingly, the term with the z score of 0 is also the median term.

7.4 Making Inferences about Terms Using the Normal Distribution Table

Because the normal distribution is symmetric, we can put the information in Appendix II into a more convenient form. We need only describe the relationships for half of the distribution and the reader can use the table whether working with positive or negative z score values. The table in Appendix III gives the fraction of terms that have z scores between $z = 0$ and various z score values. Appendix III carries areas or fractions of terms out to four decimal places, one more place than Appendix II. Generally, we shall use Appendix III for all calculations. Different values of z scores are listed, with the first two digits in the left-hand column and the third digit at the top of one of the other columns. To illustrate the use of Appendix III we shall begin by considering four types of problems dealing with the normal z score distribution.

1. Finding the fraction of terms between a given z score and the mean ($z = 0$). Suppose we wish to find in Appendix III the fraction of terms that have z scores between $z = 0$ and $z = .70$. We look for the entry that is in the row for $z = .7$ and that is also in the column for .00. We do this because $.70 = .7 + .00$. We find this entry to be .2580. Loosely speaking, slightly more than one-quarter of the terms in the normal distribution have z scores between 0 and .70.

Examine Fig. 7.5. We have just stated that the sector bounded by the vertical axis and $z = .70$ contains .2580 of the area under the normal curve. Appendix III also tells us the fraction of terms that have z scores between $z = 0$ and $z = -.70$. Due to the symmetry of the curve, the same fraction of terms are in this sector as in

[1]Thus the formula for the normal distribution divulges to the mathematician the exact fraction of terms in the normal distribution that have z scores between 1,000 and 1,001.

the sector we have considered. That is, the fraction of terms with z scores between $z = -.70$ and $z = 0$ is .2580. The sector bounded by $z = -.70$ and the vertical axis is also delineated in Fig. 7.5.

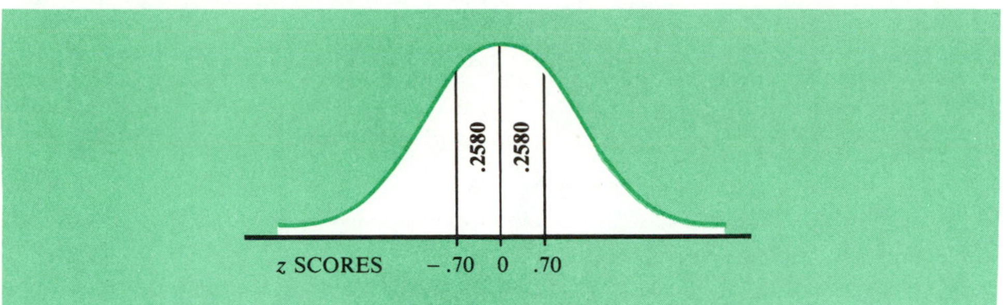

Fig. 7.5

2. Finding the fraction of terms between a given positive z score and a given negative z score. Suppose we wish to determine the fraction of terms in the normal distribution that have z scores between $z = -1.96$ and $z = 1.96$. We first read in Appendix III that the fraction of terms with z scores between $z = -1.96$ and $z = 0$ is .4750. This is found in the row for $z = 1.9$ and the column with .06 at the top, because $1.96 = 1.9 + .06$. That is, the fraction of area in the sector bounded by $z = -1.96$ and the vertical axis is .4750 (Fig. 7.6). Similarly the fraction of terms with z scores between 0 and 1.96 is .4750.

The fraction of terms in the two sectors together is the fraction of terms with z scores between -1.96 and 1.96. Therefore, the fraction of terms with z scores between -1.96 and 1.96 is $.4750 + .4750 = .9500$ (Fig. 7.7).

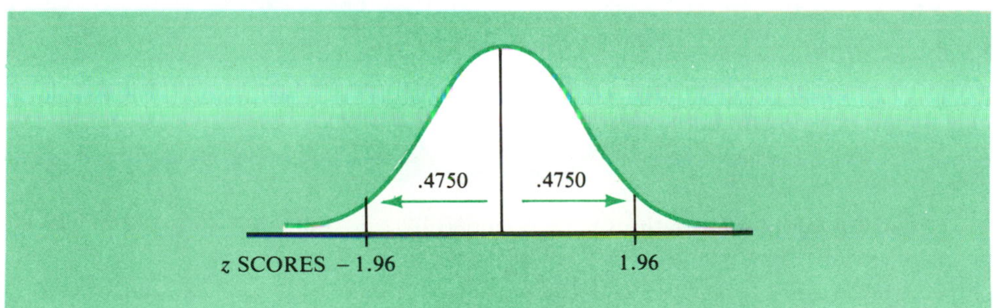

Fig. 7.6

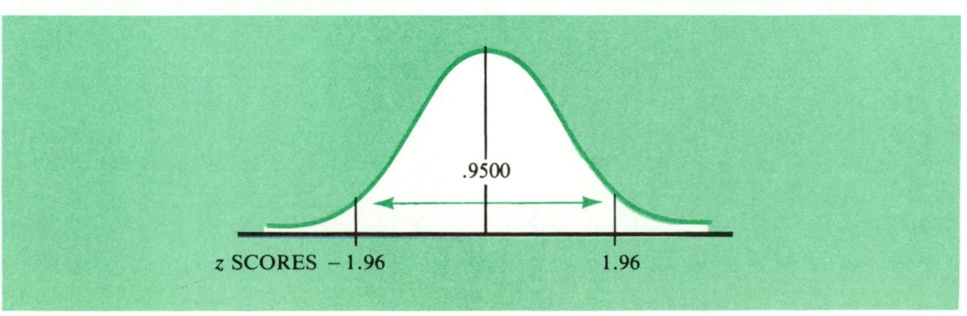

Fig. 7.7

In general, when we want to find the fraction of scores between a positive and a negative z score, we add the areas in the two sectors bounded by $z = 0$ and the two given z scores (Fig. 7.8).

Fig. 7.8

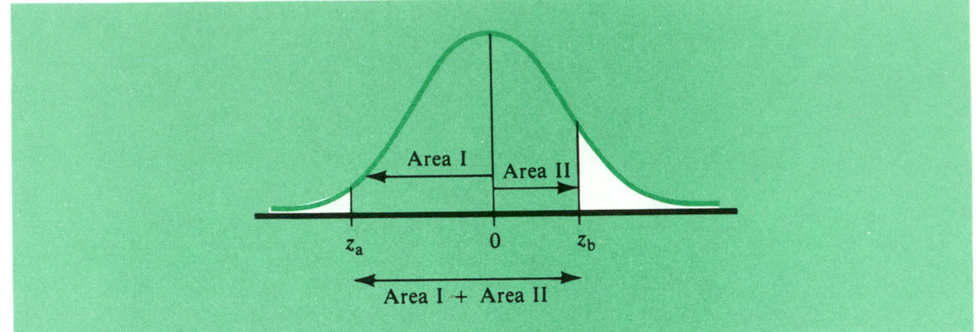

3. Finding the fraction of terms between two positive or two negative z scores. For instance, suppose the problem is to determine the fraction of terms in the normal distribution that have z scores between $z = .50$ and $z = 1.0$. We compute the area of the specified sector by subtracting the area of one sector from that of another. According to Appendix III, the fraction of area in the sector bounded by $z = 0$ and $z = 1.0$ is .3413 (Fig. 7.9).

Fig. 7.9

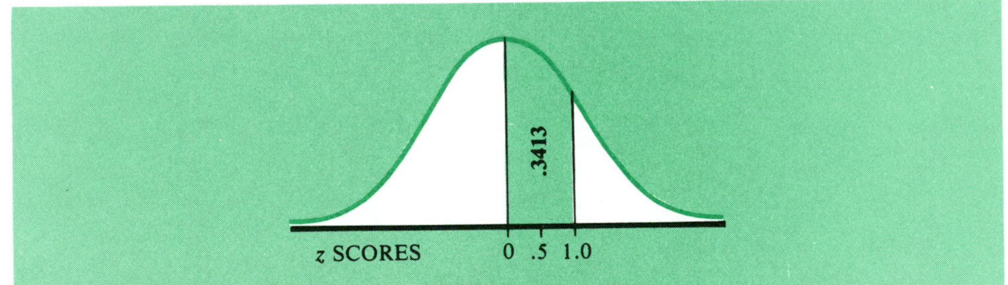

The fraction of area in the sector bounded by $z = 0$ and $z = .50$ is .1915 (Fig. 7.10).

Fig. 7.10

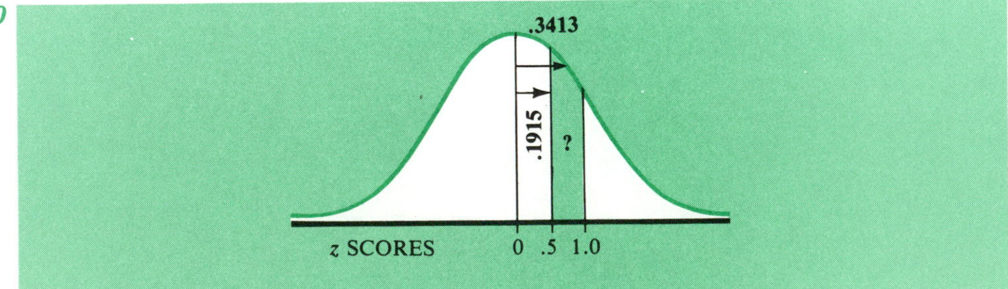

The fraction of area in the sector $z = .5$ to $z = 1.0$ may be found by subtracting the area of the second sector from that of the first. Thus the fraction of area in the sector we are after is $.3413 - .1915 = .1498$. In other words, .1498 of the terms in a normal distribution have z scores between $z = .50$ and $z = 1.0$.

In general, when we want to find the fraction of z scores between two scores with the same sign, we subtract the areas in the two sectors bounded by $z = 0$ and the two given z scores (Fig. 7.11).

Fig. 7.11

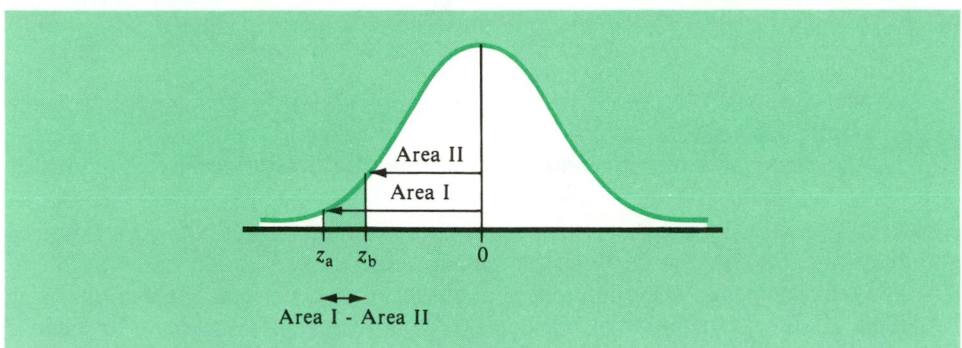

4. Finding the fraction of terms above or below a given z score. Suppose our task is to find the fraction of terms in the normal distribution that have z scores below 1.55. According to Appendix III, the fraction of area in the sector bounded by $z = 0$ and $z = 1.55$ is .4394 (Fig. 7.12). To the area .4394, which we can see from Fig. 7.12 is to the left of $z = 1.55$, we must add .5000, which is also clearly to the left of $z = 1.55$. Therefore, below $z = 1.55$ we find $.5000 + .4394 = .9394$ of the terms.

Fig. 7.12

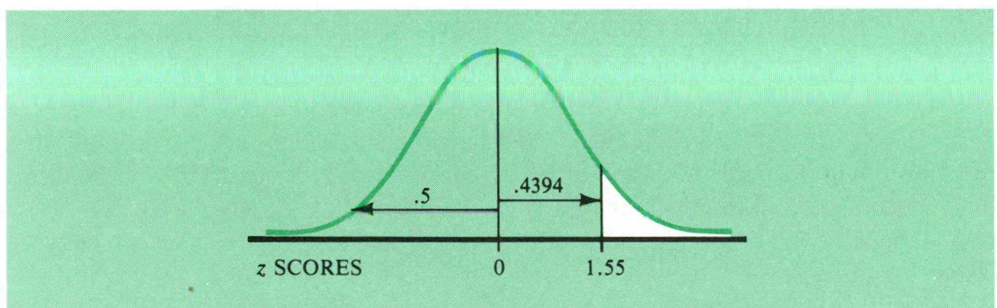

As another illustration, let us consider the problem of finding the fraction of terms above the term whose z score is .47. From Appendix III we learn that .1808 of the terms lie in the sector bounded by the mean and $z = .47$ (Fig. 7.13). We now use the information that half (.5000) of the terms are above the mean ($z = 0$). We must clearly subtract the area found in Appendix III from .5000, yielding .3192.

The reader should verify that finding the fraction of terms with z scores greater than $-.18$ requires adding the area of a sector to .5000 and finding the fraction of terms with z scores below -1.89 requires subtracting the area of a sector from .5000.

Fig. 7.13

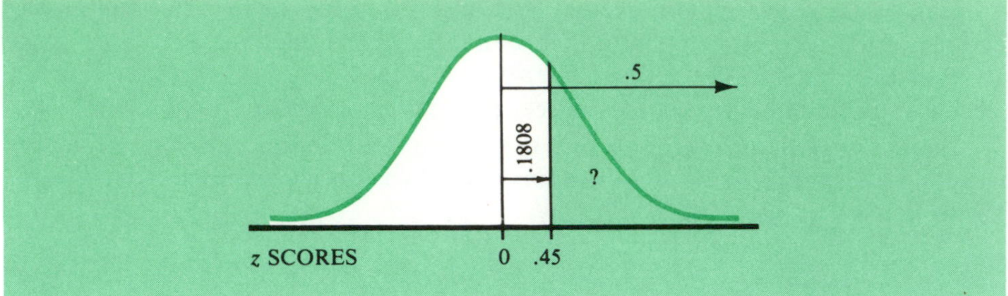

The answers to these two problems are .5714 and .0294, respectively, which the reader should also verify.

The fractions of terms in various sectors may be determined by the methods described. Appendix III gives enough information so that by adding sectors or subtracting them, as the situation requires, the reader can "isolate" the sector desired and determine the fraction of terms contained in it.

We will now make some inferences about terms in a normal distribution, using Appendix III. The distribution of IQ scores of American adults is normal. Its mean is 100 and its standard deviation is 15 points (when scores are obtained from the Wechsler Adult Intelligence Scale).[2]

Remember how we get *z* scores (Section 4.2). The IQ score of 115 is one standard deviation above the mean and thus has a corresponding *z* score of 1.0; the IQ score of 130 has a corresponding *z* score of 2.0; the IQ score of 85 corresponds to a *z* score of −1.0, and so on.

To determine the position of an IQ score in the distribution, we first convert the IQ score into a *z* score. Then we use either Appendix II or Appendix III to determine the fraction of terms in the normal distribution which have smaller *z* scores. We use Appendix III in this section, since we shall make exclusive use of it in the chapters that follow.

To begin with, suppose we are to determine the percentile rank of the IQ score of 115. The IQ score of 115 corresponds to a *z* score of 1.0. Our problem becomes that of determining the percentile rank of the *z* score of 1.0. Note that 50% of the terms have *z* scores less than zero (Fig. 7.14).

According to Appendix III, roughly 34% of the terms have *z* scores between 0 and 1.0 (Fig. 7.15).

When added, the two sectors of the distribution comprise roughly 84% of it (Fig. 7.16). Thus the term with a *z* score of 1.0 is higher than approximately 84% of the terms in the distribution. In effect, we have said that the IQ of 115 is higher than approximately 84% of the IQs in the entire distribution.

[2]In practice, we consider a distribution normal even when it does not follow the table at the extreme tails. Virtually all actual distributions are bounded. For instance, the distribution of IQ scores has no terms below zero, and no score can be above 157, which is the upper limit of the Wechsler Adult Intelligence Scale. So long as a distribution "behaves normally" between $z = -3.0$ and $z = 3.0$, we shall consider it normal.

Fig. 7.14

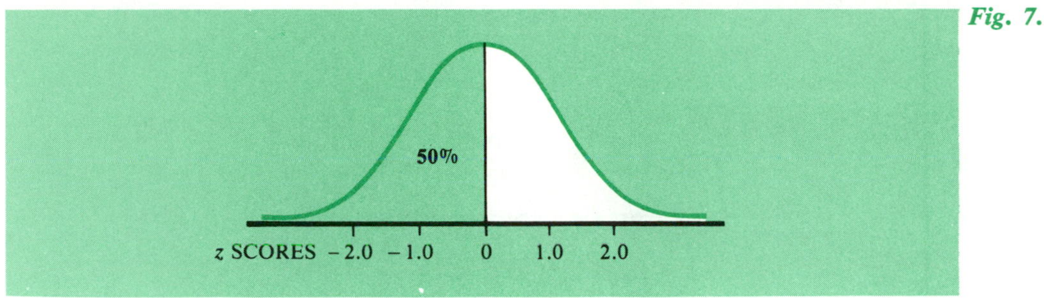

Fig. 7.15

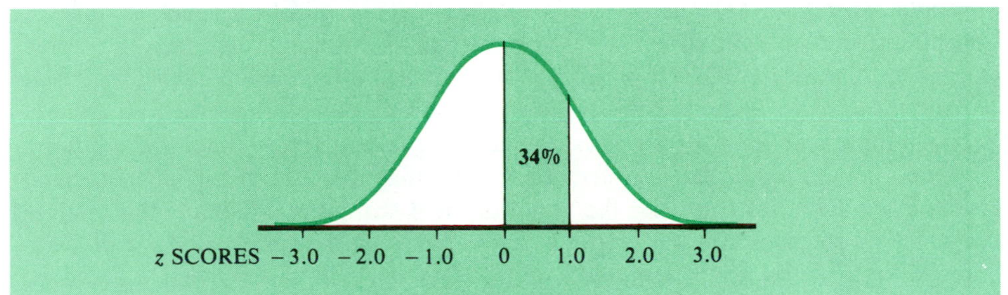

Fig. 7.16

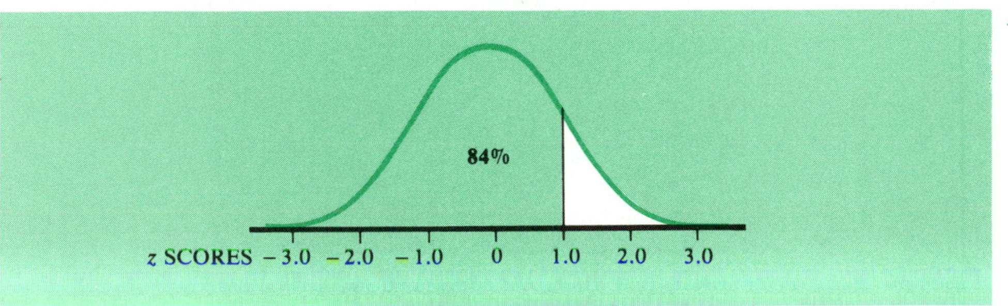

Perhaps the reader can guess how to use Appendix III to locate the position of an IQ score that is below the mean.

For instance, our task might be to determine the percentile rank of the IQ of 70. As usual, the first step is to convert the IQ score into a z score. The IQ score of 70 corresponds to a z score of -2.0. Appendix III tells us that roughly 48% of the terms in a normal distribution have z scores between 0 and -2.0 (the same as between 0 and 2.0). Thus the term with the z score of -2.0 is surpassed by roughly 48% + 50% of the terms in the distribution (see Fig. 7.17). To put it another way, the term with a z score of -2.0 is at the second percentile. Thus the person with an IQ of 70 surpasses only about 2% of the people in the entire distribution.

Somewhat more complex problems relating terms in a normal distribution arise but, if what has been said is clear, the reader should have no trouble handling them. *The first step is always to convert the original scores (or terms of another kind) into z scores*. Once the conversion has been made, one of the procedures already discussed may be used. For instance, the problem might be to determine the fraction

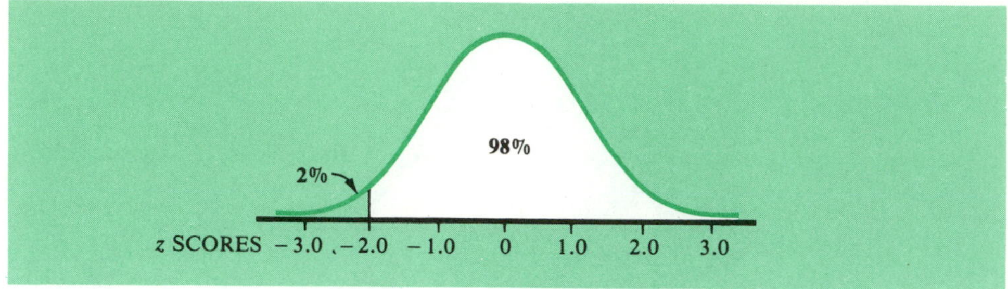

Fig. 7.17

of people who have IQs between 90 and 110. This problem becomes that of determining the fraction of terms in the normal distribution that have z scores between $-.67$ and $+.67$. **It is advisable to draw the normal curve for each problem and to delineate the relevant sectors on it before proceeding.** It usually becomes apparent which sectors must be added or subtracted to reach the solution.

Once we are told that a distribution is normal and are given its mean and standard deviation, we can draw the graph of the distribution indicating the original scores (or more generally, the original terms) along the baseline, along with the z scores. For instance, the distribution of IQs has a mean of 100 and a standard deviation of 15 points, so that the normal curve of IQs looks like that in Fig. 7.18.

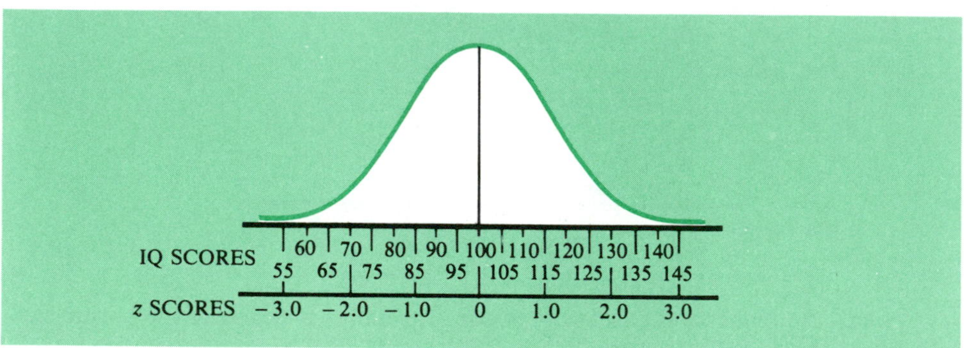

Fig. 7.18

Note that the IQ of 100 is the mean, the median, and the modal IQ of the distribution. The IQs tend to cluster around 100; in other words, we find fewer and fewer IQs as we move further and further from 100 in either direction. The symmetry of the distribution tells us that as many people have IQs of 70 (that is, 69.5 to 70.5) as of 130 (that is, 129.5 to 130.5), the same number of people have IQs of 95 as those who have IQs of 105, and so on.

7.5 The Distribution of Shots

We shall now consider a completely different kind of situation in which the normal distribution is found. In Chapter 8 are specified conditions, found in nature, that "force data" into the normal distribution shape. The purpose of this section is simply to give the reader more experience in working with the normal distribution and a greater appreciation of its generality.

Suppose a rifle-shooting tournament is held in which male contestants are presented with the task of hitting a telephone pole a given distance away. The rules are that each marksman is allowed a single shot. Each marksman has to wait in a nearby building until it is his turn to shoot so that he cannot witness his predecessors at the firing line. An electronic device is set up to record how far each bullet is from the pole when it passes it. For instance, the device may record that one bullet passes 7 feet to the right of the pole and another 3 feet to the left of it. The score given to each marksman indicates where his bullet passes the pole; for example, the marksman whose shot goes 7 feet to the right of the pole gets a score of + 7. The one who fires 3 feet to the left of the pole gets − 3. Anyone who hits the pole gets a score of zero, which guarantees him at least a tie for the prize. We shall assume that the conditions are such that a "miss" is as apt to occur on one side of the pole as on the other.

The contest is starting and it is our job to record the outcome. The first marksman steps to the line and fires. Our device tells us that his bullet passes 3 feet to the right of the pole. His score is + 3 and we record it by placing a tiny rectangle on a horizontal axis that we have drawn specially for the occasion (Fig. 7.19).

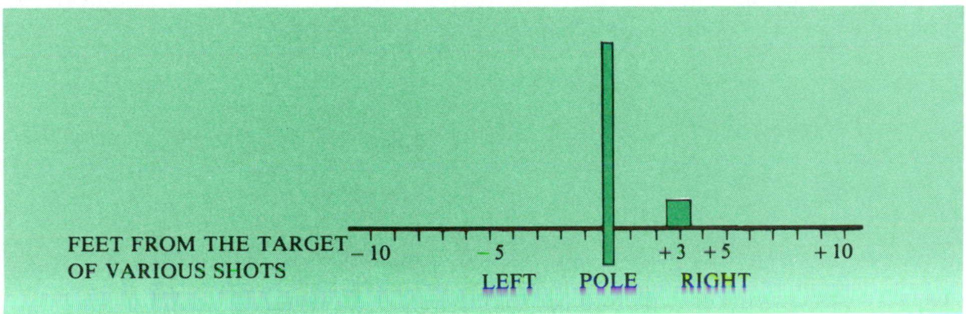

Fig. 7.19

The sound of the shot is a signal for the next person to emerge from the building and step to the firing line. Bang! His shot goes 5 feet to the left of the pole and again we record it (Fig. 7.20). After several shots our picture looks like Fig 7. 21. We may suppose that the contest goes on indefinitely but it is time for us to leave. Our picture

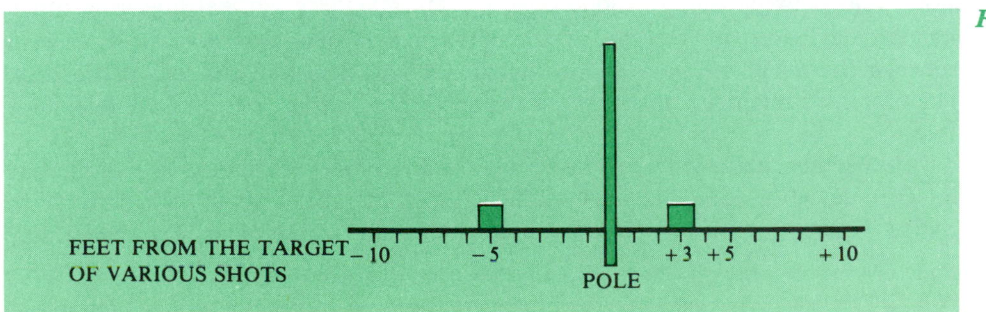

Fig. 7.20

Fig. 7.21

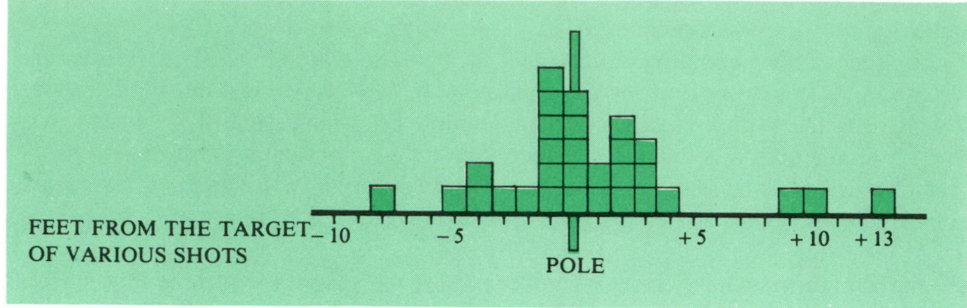

has come to look like Fig. 7.22.[3] Note that virtually as many shots went to the left of the telephone pole as to the right of it and that hits are more numerous near the pole than farther away on either side. The approximate representation by a curve becomes more and more accurate as more shots are fired. As you may have guessed, this curve is the normal curve and the distribution of shots is the normal distribution.[4] The standard deviation of the distribution reflects the abilities of the marksmen as a group. A small standard deviation would indicate that the shots tend to cluster around the telephone pole and a large standard deviation would indicate that they are relatively scattered. To be specific, we shall now suppose that the standard deviation of the shots is 4 feet. The distribution of shots looks like Fig. 7.23.

Fig. 7.22

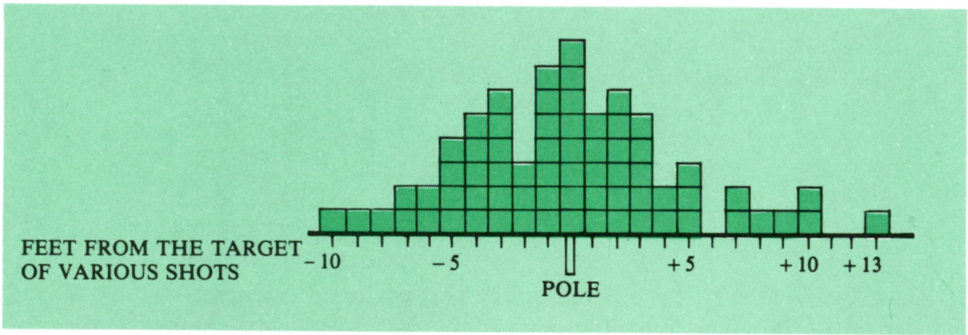

Appendix III tells us that in a normal distribution about 19% of the terms lie within a half of a standard deviation to the right of the mean. Thus about 19% of the shots went between the telephone pole and an imaginary vertical line a half of a standard deviation (or 2 feet) to the right of the pole. Similarly, 19% of the shots went between the telephone pole and an imaginary line 2 feet to the left of the pole. Therefore, we might say that 38% of the shots were within 2 feet of the telephone pole.

Altogether, about 68% of the terms in the normal distribution are within one standard deviation of the mean on one side or the other. Thus about 68% of the shots landed within 4 feet of the telephone pole on one side or the other. The reader may

[3]In this illustration the scores are rounded off to the nearest unit (foot).

[4]We shall consider the principle "explaining" the normality of the distribution of shots in Chapter 8.

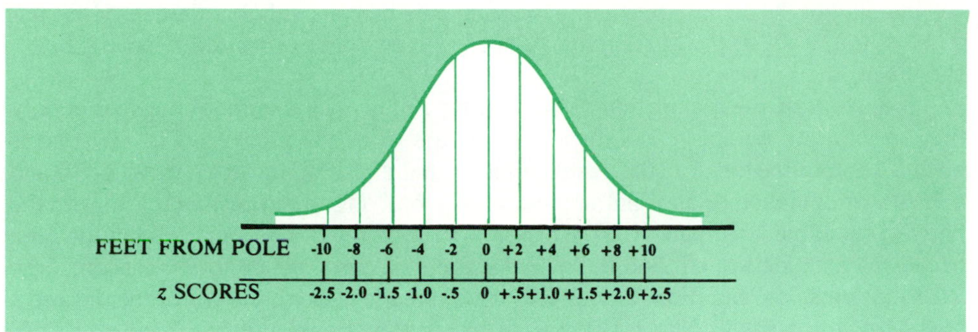

Fig. 7.23

verify that about 95% of the shots landed within 8 feet of the pole. Note that we need know only the values of the mean and standard deviation of a normal distribution in order to make inferences about any of the individual terms, using Appendix III.

7.6 Determining Scores from Percentile Ranks

Another type of problem that we can solve using the table of normal curve areas in Appendix III is essentially the opposite of those previously illustrated in this chapter. Up to this point the problems we have considered provided us with a score or term and required our finding an area. Now we shall be given the area and be required to find the score.

Suppose that we wish to know what z score has a percentile rank of 35; that is, we wish to find P_{35}. This means that 35% of the area is to be to the left of that score and 15% is to be in the sector between the desired z score and $z = 0$ (Fig. 7.24). (Remember that Appendix III is a "between" table. We must always translate the information we have to work with into some statement about area between the mean and some z score.)

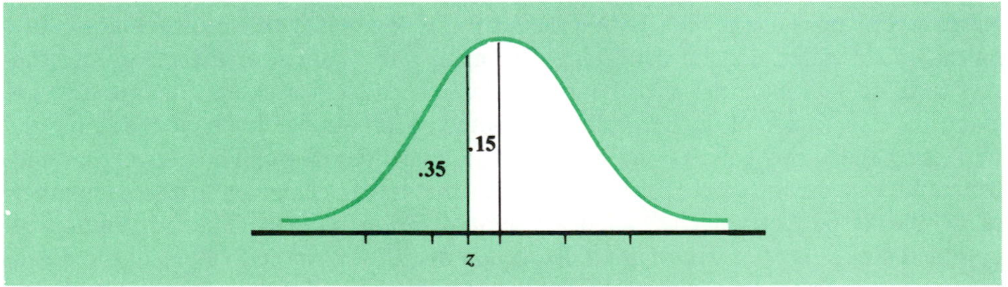

Fig. 7.24

We look in the body of the table in Appendix III this time and our answer will be a z score read from the margins. The fraction given in the body of the table that is nearest to .15 is .1517. This corresponds to a z score of .39. We obtain .39 by adding the heading of the column in which .1517 appears to the z value at the left of its row. That is, $0.3 + .09 = 0.39$. Therefore, since we know that the desired z score is below the mean, the answer is $z = -.39$.

To illustrate the application of this last type of problem, let us suppose that only the top 10% of the applicants for a position are to be interviewed. The test that is given to determine the top 10% results in a normal distribution with a mean of 80 and a standard deviation of 5. What is the cut-off score? Now we must again refer to the body of the table in Appendix III. This time we seek the z score for a percentile rank of 90, from which we will find the actual score that corresponds to this z score. Fig. 7.25 illustrates the situation. Since 90% of the area is to be to the left of the desired z score, we must have 40% of the area between that score and the mean at $z = 0$. Therefore, we look in the body of the table for the number closest to .40. It is .3997, and corresponds to a z score of 1.28 as read from the margins. This means 1.28 standard deviations. Since one standard deviation in our given distribution is 5 score units, we must multiply 1.28 by 5 to get to our desired cut-off point. The result tells us that the cut-off point is 6.40 score units above the mean. Hence it is $80 + 6.40$ or 86.4.

Fig. 7.25

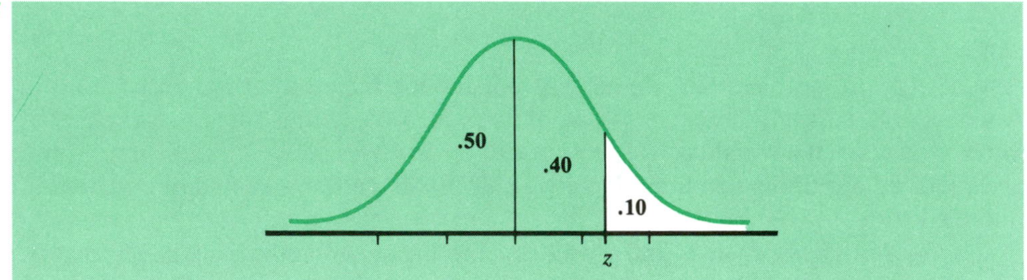

7.7 Applications

Researchers make use of the fact that the positions of terms with various values in the normal distribution can be determined from only the values of the mean and the standard deviation of the distribution. In effect, the researchers can provide an elaborate description of the normally distributed observations they have made simply by stating these two values—the mean and standard deviation. Thus educators and psychologists, among others, often obtain tests that have been given to a large number of people as part of what is called a standardization process. The distribution of obtained scores is often approximately normal and the computed values of the mean and standard deviation of this distribution are published. The educator or psychologist can use these two values to determine precisely the place of an individual score in the distribution. Their procedure is to convert the individual score into a z score and then to locate this z score in the normal distribution, using the table in Appendix III.

Problem Set A

7.1 Find the percent of the terms in a normally distributed population with z scores between the following pairs of z scores.

a. 0 and 1.4
b. 0 and −2.5
c. 2.2 and 1.5
d. −.5 and −.6
e. 1.4 and −.9
f. 2.03 and 1.87
g. −1.54 and −1.36
h. 2.14 and −1.33
i. 0 and 8
j. 2.58 and −1.81

7.2 Find what fraction of the area under the normal curve lies
a. to the right of $z = .5$.
b. to the right of $z = -1.3$.
c. to the right of $z = 1.27$.
d. to the left of $z = .9$.
e. to the left of $z = 2.05$.
f. to the left of $z = -1.32$.
g. within one standard deviation of the mean.
h. within 2.5σ of the mean.
i. within 2σ of the mean and is left of $z = 0$.
j. within $.5\sigma$ of the mean and is right of $z = .1$.

7.3 Find the percentile ranks of the terms in a normally distributed population that have the following z scores.

a. 1.0
b. 1.5
c. −2.0
d. −2.5
e. 1.2
f. 1.23
g. −3.04
h. .87
i. −.94
j. .07

7.4 Find the z scores of the terms in a normally distributed population that have the following percentile ranks.

a. 95
b. 80
c. 76
d. 63
e. 52
f. 48
g. 33
h. 29
i. 18
j. 3

7.5 Find the z scores in a normally distributed population that represent

a. P_{97}
b. P_{65}
c. Q_1
d. D_{30}
e. Q_3
f. P_{45}
g. P_{15}
h. P_{84}
i. D_1
j. P_{53}

7.6 Find the z value(s) in a normal distribution such that

a. one-fourth of the terms lie between it and the mean (two answers).
b. 95% of the terms are exceeded by it.
c. 10% of the terms are larger than it is.
d. the middle half of the terms lie between them.
e. the middle 18% of the terms lie between them.
f. the middle 95% of the terms lie between them.
g. the middle 99% of the terms lie between them.

7.7 Find the shaded areas in the following diagrams.

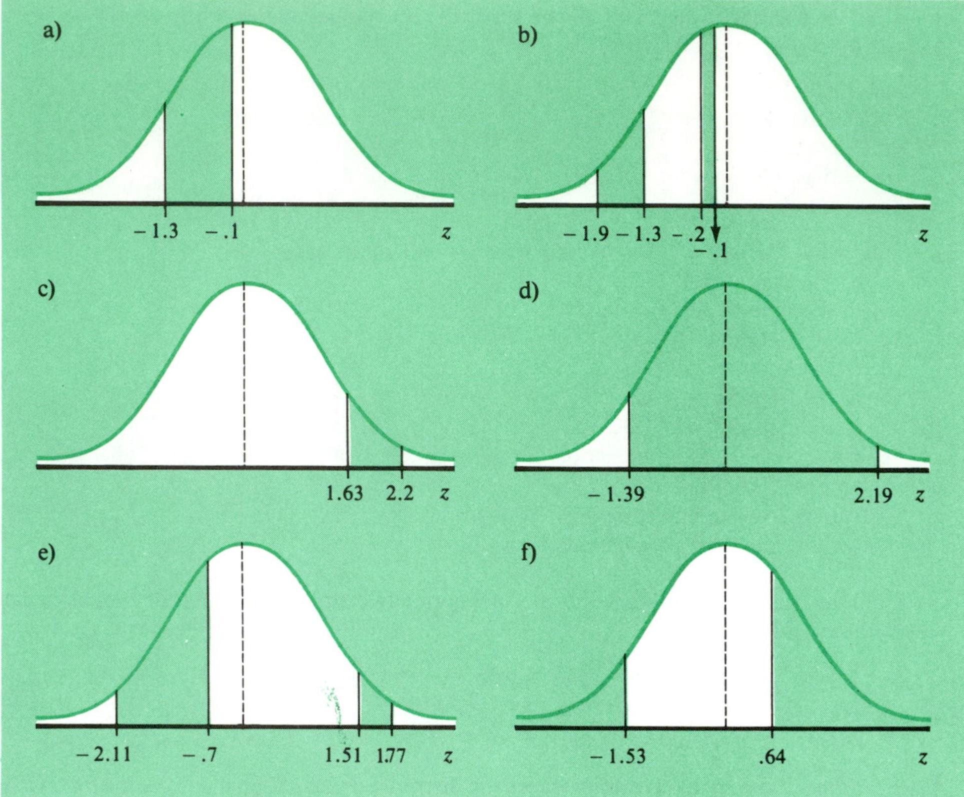

7.8 For a distribution of 1,000 test scores, the mean is found to be 500 and the standard deviation is 100. Assume that the distribution is approximately normal and answer the following questions: (a) What percent of the students had scores less than 750? (b) What is the percentile rank of a score of 475? (c) What is the score exceeded by 60% of the students? (d) Find an interval, centered at the mean, that includes 60% of the scores. (e) There are only 10 scores less than what score?

7.9 The time lengths of films produced by a French film company are found to be normally distributed with a mean length of 92 minutes and a standard deviation of 23 minutes. Draw the normal curve depicting the distribution of film lengths. Find what percent of films last (a) less than 65 minutes, (b) more than 77 minutes, (c) less than 112 minutes, (d) more than 120 minutes, (e) between 69 and 115 minutes, (f) between 92 and 100 minutes, (g) between 100 and 108 minutes.

7.10 The attention span of 4-year-olds is known to be normally distributed with a mean of 12 minutes and a standard deviation of 3 minutes. Find

a. the attention span of the child whose attention span exceeds those of 85% of all 4-year-olds.
b. the fraction of 4-year-olds who will sit still for a 10-minute story.
c. the fraction of 4-year-olds with attention spans between 9 and 13 minutes.

7.11 Find the value(s) labeled z in each of the following diagrams.

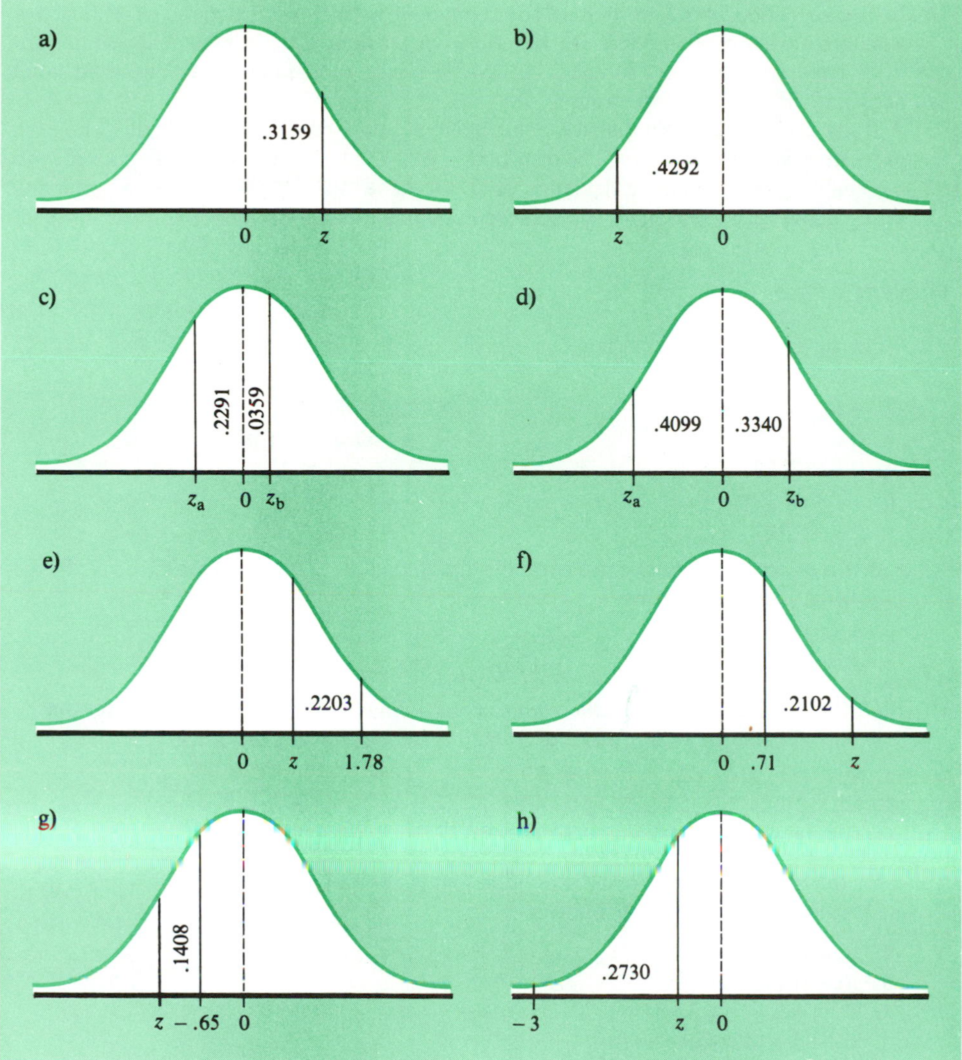

7.12 A vending machine dispenses soda in paper cups. The amounts vary normally with a mean $\mu = 8$ ounces and a standard deviation $\sigma = .5$ ounce. What percent of the cups dispensed contain (a) 7 or fewer ounces? (b) more than 9 ounces? (c) between 7.5 and 10.5 ounces? (d) Find the amounts contained in the 10% of the cups containing the most soda.

7.13 Municipal Court judges in a given state spend a mean of 4 hours and 15 minutes per day on the bench with a standard deviation of 45 minutes. (a) What fraction of the time would you expect these judges to spend less than 3 hours per day on the bench, assuming the distribution of times to be approximately normal? (b) Again assuming normality, find a time interval whose center is 4 hours 15 minutes that tells the lengths of 50% of all the days put in by the judges being considered.

7.14 In a distribution of scores that is approximately normal with a mean of 100 and a standard deviation of 50, there are 165 scores larger than 200. (a) How many scores are to be expected between 170 and 210? (b) How many scores are to be expected to be negative? (Hint: Use proportions.)

7.15 A study is made to obtain the well-being index of numerous college graduates who earn more than $20,000 per year. The distribution is found to be approximately normal with a mean of 5 and a standard deviation of .7. Of those studied, 580 people had an index rating above 6.0. How many subjects were included in the study?

Problem Set B

7.16 Find the areas under the normal curve in the sectors bounded by the following pairs of z scores.

a. 0 and −11	f. 3.01 and 1.62
b. 0 and −.7	g. −1.47 and −1.06
c. 0 and 2.35	h. 2.44 and −.45
d. 1.7 and 1.6	i. −.43 and 1.48
e. −2.3 and −.4	j. −.17 and −.28

7.17 Find the z scores of the terms in a normally distributed population that are exceeded by the following fractions of terms.

a. .8340	c. .2877	e. .7939	g. .0256
b. .3446	d. .6331	f. .1170	h. .5714

7.18 Find the z scores and percentile ranks of the following terms in a distribution that is normally distributed with a mean of 90 and a standard deviation of 20.

a. 90	d. 62	g. 99.8	j. 105.7
b. 85	e. 55	h. 40.3	
c. 103	f. 115	i. 77.5	

7.19 Find the fraction of terms in a distribution that is normally distributed with a mean of 50 and a standard deviation of 5 that is

a. greater than 48.
b. greater than 57.39.
c. less than 44.
d. less than or equal to 44.
e. within 1.65 standard deviations of 50.
f. within 2.33 standard deviations of 50.
g. between 42 and 53.
h. between 41.5 and 57.6.

7.20 Find the term in the distribution that is normally distributed with a mean of 100 and a standard deviation of 10 which represents

a. P_{13}	d. P_{75}	g. D_7	j. P_{59}
b. D_3	e. P_{80}	h. D_9	
c. Q_2	f. P_{17}	i. P_{35}	

7.21 Find the shaded areas in each of the following diagrams.

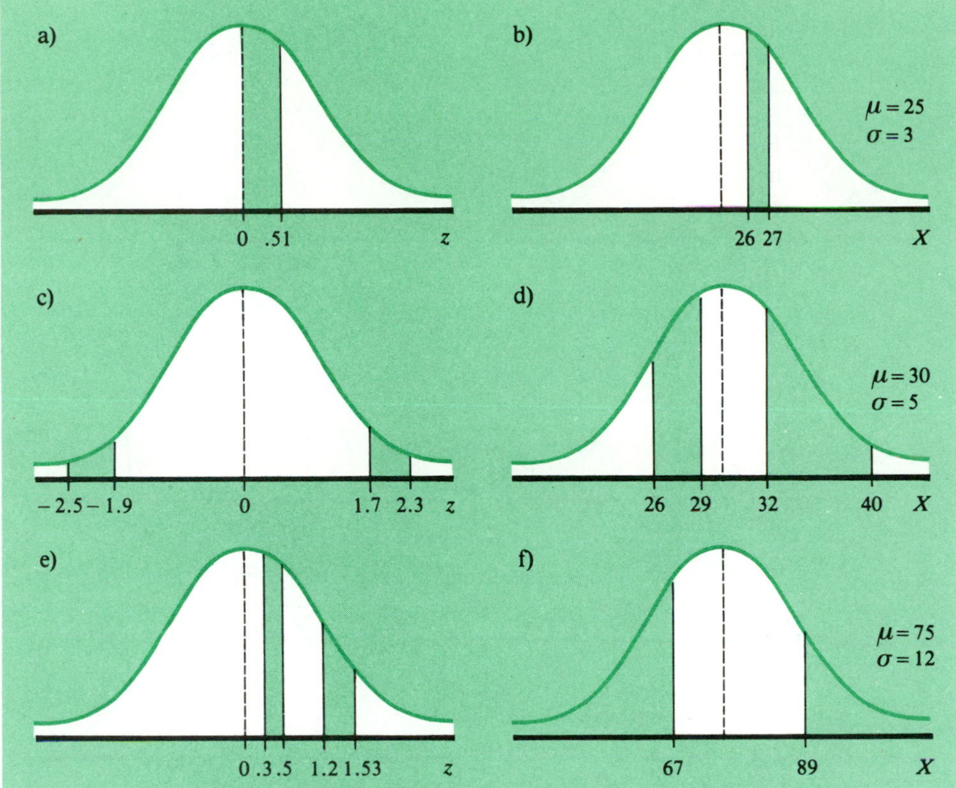

7.22 LeRoy's Auto Mall specializes in clean, late-model used cars. He has sold 3,000 cars since opening his business eight years ago. Their speedometer readings have been roughly normally distributed with a mean of 20,000 miles and a standard deviation of 5,000 miles. Find

a. the percentile rank of the reading 28,000 miles.
b. the number of readings that have been less than 17,500 miles.
c. the fraction of the readings that have been between 15,000 and 25,000 miles.
d. an interval centered at the mean that includes 40% of the readings.
e. the cut-off point for the largest 15% of the readings.
f. the cut-off point for the top 10 readings.

7.23 The recovery time for a particular contagious disease is normally distributed with a mean of 22 days and a standard deviation of 2 days.

a. What percent of the afflicted recover in less than three weeks?
b. The 48 pupils in Mr. Kelly's physical education class have all been exposed to the disease and are expected to contract it. About how many should be expected to be ill between 20 and 25 days?
c. What is the smallest number of days a person can be ill and yet be ill longer than 90% of those afflicted?

7.24 Find the values labeled X or z in the following diagrams.

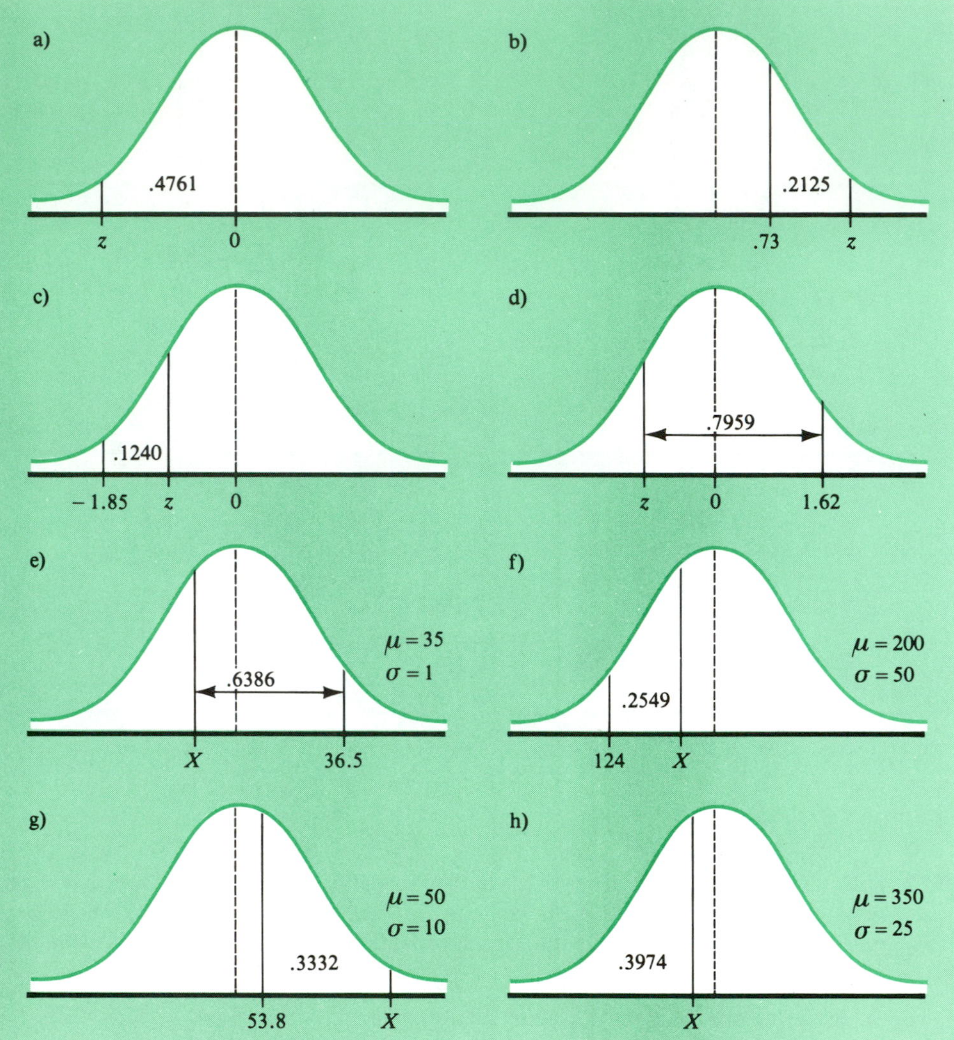

7.25 Court records in a small resort community show that the number of miles over the speed limit for convicted speeders is normally distributed with a mean of 15 mph and a standard deviation of 4 mph.

a. About what percent of convicted speeders were found to be going between 14 and 18 mph over the posted limit?
b. If 200 speeders are arrested, about how many of them would be expected to have exceeded the speed limit by more than 20 mph?
c. What speed is exceeded by only 5% of the speeders if the posted limit is 45 mph?

7.26 A certain machine manufactures steel balls. The diameters of the balls are normally distributed with a mean of 0.3456 inch and standard deviation 0.0078 inch. Find the percent of balls with diameters (a) between 0.3400 and 0.3500 inch, (b) greater than 0.3560 inch, (c) less than 0.3232 inch.

7.27 Professor Trish has just finished grading final exams and now must determine final grades. She knows from experience that the scores will be approximately normal in distribution. The mean grade proves to be 74 and the standard deviation 9.

a. If the middle 35% are to be graded C, find the cut-off points for C.

b. Laura's average is an 87. If the top 10% receive A's, can Laura expect an A?

7.28 Past experience has shown that the mean life of Whiz-Matic irons is 72 months with a standard deviation of 10 months. The product is sold with an unconditional four-year guarantee. The lives are normally distributed.

a. What fraction of the irons should be expected to be returned for failing to satisfy the guarantee?

b. This year 100 irons were returned for not meeting the guarantee. Assuming that all four-year-old irons not still functioning were returned, how many irons were sold four years ago?

7.29 Certain commodities are graded by weight and normally distributed; 20% are called standard, 50% large, 20% super, and 10% colossal. If the mean weight is 0.92 ounce with a standard deviation of 0.08 ounce, what are the limits for the weights of the supers?

7.30 a. The test scores of a group of students are normally distributed with a mean of 74.8. If 17% of the scores are at least 79.5, what is the standard deviation of these test scores?

b. The test scores of a group of students are normally distributed with a standard deviation of 15. If 20% of the scores are less than 65, what is the mean of these scores?

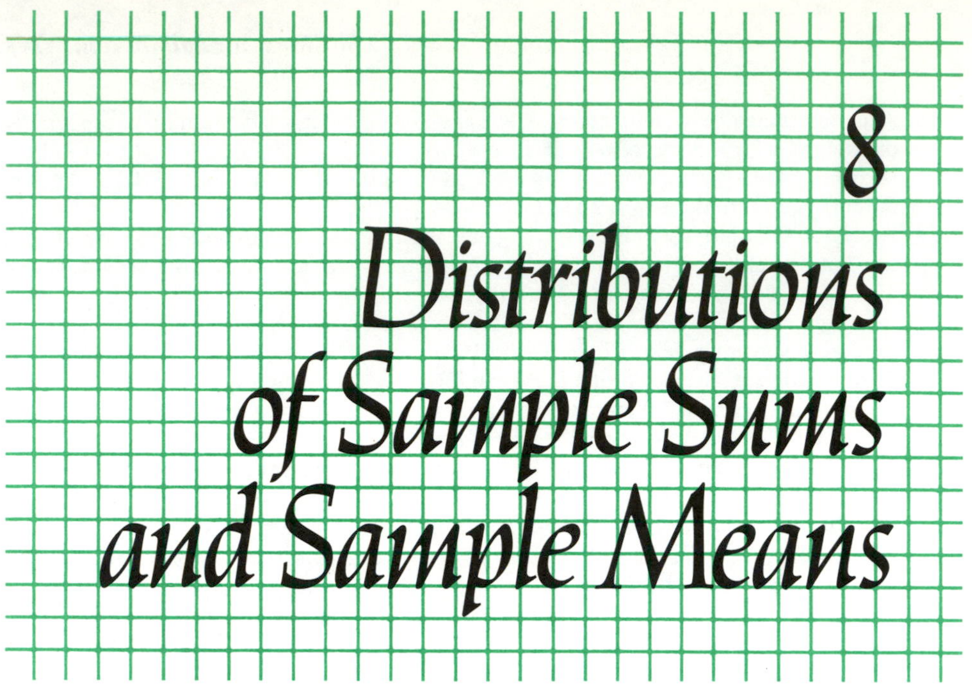

8 Distributions of Sample Sums and Sample Means

8.1 The Central Limit Theorem

The reader may wonder how it is that the curve that depicts the incidence of bullets around a target also describes the distribution of IQs in the United States. The heights of mature trees are also distributed normally and might have been used as illustrative observations. In fact, we are scarcely restricted to subject matter when looking for illustrations of the normal distribution.

This chapter introduces a highly theoretical discussion leading to an explanation of why the normal distribution arises in so many contexts. For the sake of discussion, we are going to pretend that it is our job instead of nature's to make up the values of a multitude of terms. In making up these values, we shall use the same procedure that nature often employs when she produces a distribution of normally distributed terms.

To clarify its details we shall discuss the procedure as if we are carrying it out ourselves. In Section 8.2 we will see how nature makes use of the procedure. As scientists our role is that of observing the normal distributions that she produces. We will postpone examining the relevance of the procedure for the time being and begin simply by discussing its technical aspects.

Assume that to begin with we have access to an infinitely large population of terms and that we can choose at random as many terms from this population as we wish. *To proceed we need know nothing about the shape of the distribution of terms in this infinite population*. We assume only that it has some unknown mean and standard deviation and that we have access to its terms. Theorem 8.1 describes the procedure for gathering terms from the infinite population and combining them in such a way as to create a normal distribution.

THEOREM 8.1 (The Central Limit Theorem) Suppose that a multitude of equal-sized random samples is gathered from the same infinite population. The sum of each sample is computed and the sums of the different samples are put together to form a new distribution. It follows that the new distribution is approximately normal. (An assumption is that the random samples yielding the sums are large enough.[1])

Note the assumption that we have access to the terms in an infinite population about which we know nothing. The procedure is to gather equal-sized random samples from this population and *to contribute the sum of each sample as a single term in a new distribution.* Theorem 8.1 states that under these conditions the new distribution will be approximately normal even when the original distribution is not normal. The approximation gets better as sample size increases, but is almost always satisfactory if either the samples contain at least 30 terms or the population being sampled from is normal.

The following illustration should make Theorem 8.1 clear. Picture a giant wooden crate filled with an inexhaustible supply of tiny slips of paper. On each slip a number is written. For simplicity, assume that only the numbers 0, 1, and 2 are included and that each of these numbers appears on roughly one-third of the slips. (Strictly speaking, we need not make either of these assumptions.) Remember that the supply of slips of each denomination is infinite. That is, there are so many that we can withdraw any number of slips of any denomination and still we need not be concerned with having changed the balance among those remaining in the crate.

The wooden crate has a small opening in its corner through which a slip can be released. It also contains an apparatus that shuffles the slips so thoroughly that every one of them stands an equal chance of being released when we decide to draw one. To draw a slip we press a button that automatically shuffles the slips before releasing one. Thus we can obtain a slip at will, but when we do, every slip in the crate is an equally likely candidate to be the one that is released. Accordingly, our observations are independent and our sample is random.

We are now ready to gather equal-sized samples. We decide to gather samples of 200 slips (that is, terms) each.

We begin by taking 200 slips, one after the other. The number on the first one is a 2; the number on the second one is a 0; on the third, a 2; and so on. The total of all the numbers on the 200 slips (comprising the first sample) turns out to be 210. In other words, the sum of the terms in the first sample is 210. This sum becomes the first term in the new distribution about to be created. We denote the value of 210 on our "scorecard."

The sum of the next 200 slips turns out to be 194. The process is repeated for a

[1]Nearly always the samples are large enough if they contain at least 30 terms each. Even this assumption is unnecessary if the infinite population sampled from is known to be normal. The proof of Theorem 8.1 is beyond the scope of this book.

multitude of samples of 200 slips each. Theorem 8.1 tells us that the distribution of sums of samples approaches the normal distribution as more and more sums are contributed. Fig. 8.1 shows one stage of the process.

Fig. 8.1

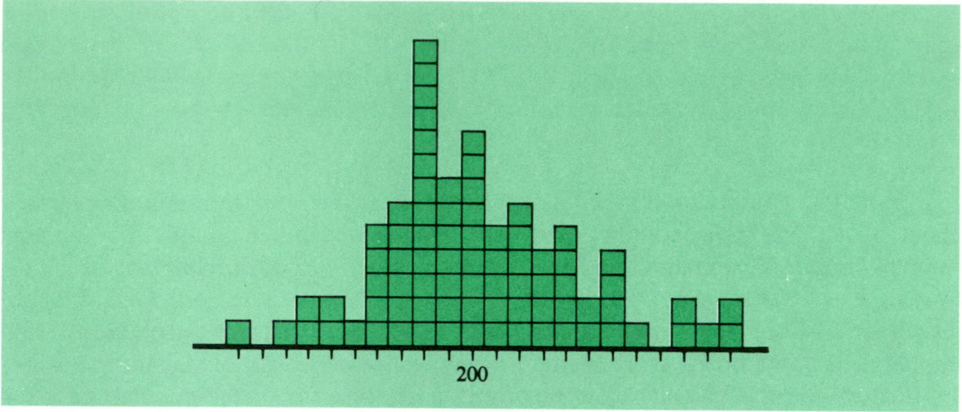

The reader should make sure it is clear how Theorem 8.1 predicts the normality of the distribution in Fig. 8.1. The surprising thing about Theorem 8.1 is that nothing is stated about the population from which the samples are drawn.[2] The distribution of the sums of the samples is normal regardless of the shape of the distribution from which the samples are drawn. However, when the "parent" population is extremely unbalanced, it is necessary to take giant-sized samples to ensure that the sums of these samples will be normally distributed.

8.2 How Nature Uses Theorem 8.1

Theorem 8.1 explains why the normal distribution arises in so many different contexts. The method we used in sampling with the slips of paper seems to be a favorite procedure of nature. In each situation presented, the numbers that comprise a normal distribution may be viewed as sums of equal-sized samples of independent observations.

Consider again the bullets that fell into a normal distribution surrounding the telephone pole. The place where each bullet struck was, in effect, the sum of a number of influences. For instance, the way a particular marksman stood may have directed his bullet 2 feet to the right of the target (+2). A wind may have veered his bullet a foot to the left (−1). A cloud passing overhead changing the light may have had no effect whatsoever (0).[3] The marksman's proneness might be to fire "first shots" a foot to the left (−1), and an involuntary tension may have jerked his aim 2 feet to the right (+2). A shadow may have shifted his aim 1½ feet to the right (+1½).

[2]An exception is that the original distribution must contain more than just one value repeated infinitely many times.

[3]This influence would have to be included if its effect on any other shots was different from zero.

These and other influences operate simultaneously on a particular marksman and they are different for different marksmen. A marksman's score, indicating where his bullet finally went, is the sum of a sample of influences. To be concrete, let us say that each marksman's score was the sum of 70 major influences. Therefore, the score given to each shot may be viewed as the sum of a sample of 70 terms and corresponds to the sum of the numbers on 70 slips. Thus the scores obtained by the different marksmen may be viewed as the sums of different equal-sized samples. The distribution of shots (or sample sums) was normal, as would have been predicted by Theorem 8.1.

The normality of people's intelligence levels can be explained by the theorem. Each person's intelligence level may be viewed as a sum of small contributions from various sources. An individual's actual intelligence is the sum of contributions from hereditary factors and other factors like nutrition, opportunity, emotional makeup, and interest. For example, the contribution to an individual's intelligence from his or her grandmother might correspond to one IQ point. The contribution to the person's intelligence from nutrition might be worth two points. According to this view, the intelligence level of each person is the sum of a vast number of contributions. Thus the distribution of intelligence levels of a multitude of people is normal by Theorem 8.1.

The distribution of the heights of trees is normal for essentially the same reason. The height of a tree may also be viewed as the sum of a vast number of small contributions. Its heredity and environment both contribute to its growth in numerous ways. **Similarly, whenever the terms in a population are each subject to the same vast number of small and relatively independent influences, then the distribution of the terms is normal.**[4]

In Chapter 9 we shall consider a very special kind of distribution of sample sums called the binomial. Distributions of sample sums differ from one another for various reasons. One distinguishing factor is the parent population that gives rise to the new distribution. When the parent population contains only two values (repeated infinitely many times), the resulting distribution of sample sums is called binomial. A more in-depth study of the binomial must be preceded by a discussion concerning probability.

8.3 The Central Limit Theorem for Sample Means

Let us now consider a variation of Theorem 8.1. To begin with an illustration, we shall use the crate analogy again. Suppose as before that we draw a multitude of random samples (of 200 numbered slips each) from the crate. In Section 8.1 we made

[4]When there is a large number of terms in each sample, the distribution of sample sums is normal even though not all the observations (or influences) are mutually independent. In other words, the demand for absolute independence becomes diminished when the samples yielding their sums are large. Undoubtedly many of the influences on the marksman's aim were somewhat related when he fired, and many of the influences on intelligence are interrelated. Yet the distribution of shots and that of intelligence levels are each normal.

up a distribution out of the sums of these various samples. Theorem 8.1 stated that the distribution of these sample sums is approximately normal.

This time suppose that we find the mean of each sample instead of its sum. Now the various sample means are put together to form a new distribution. Theorem 8.2 states that the distribution of sample means is approximately normal.

THEOREM 8.2 (The Central Limit Theorem for Means) Suppose that a multitude of equal-sized random samples is gathered from the same infinite population. The mean of each sample is computed and the means of the different samples are put together to form a new distribution. It follows that the new distribution is approximately normal. (An assumption is that the random samples yielding the means are large enough.[5])

Theorem 8.2 follows directly from Theorem 8.1. The mean of each sample is its sum divided by the number of terms included in the sample. Remember that each sample must have an identical number of terms. (This number was 200 in the crate analogy.)

Let us say that there are N terms per sample. Thus if the sums of five samples happen to be 200, 400, 400, 800, and 1,200, then the corresponding sample means are $200/N$, $400/N$, $400/N$, $800/N$, and $1{,}200/N$. Specifically, where $N = 200$, the means of the five samples are

$$\frac{200}{200} \quad \frac{400}{200} \quad \frac{400}{200} \quad \frac{800}{200} \quad \frac{1{,}200}{200}$$

or, more simply, 1, 2, 2, 4, 6. One might say the set of means has been obtained from the set of sums by "proportional shrinking." Thus the distribution of means is a small-scale version of the distribution of sums. The graph of the sums of the five samples is shown in Fig. 8.2, and the graph of their means is shown in Fig. 8.3.

Fig. 8.2

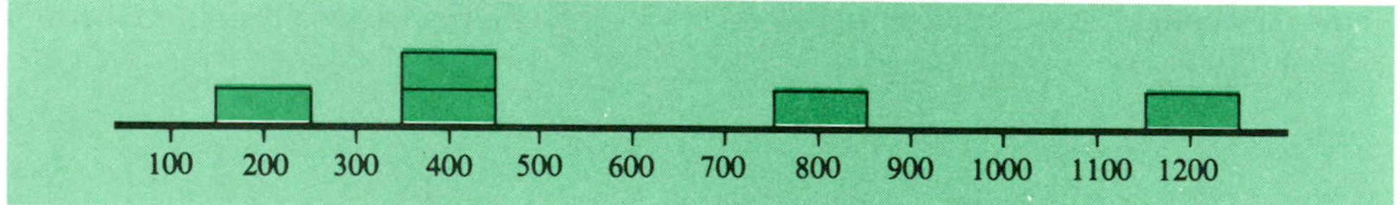

Fig. 8.3

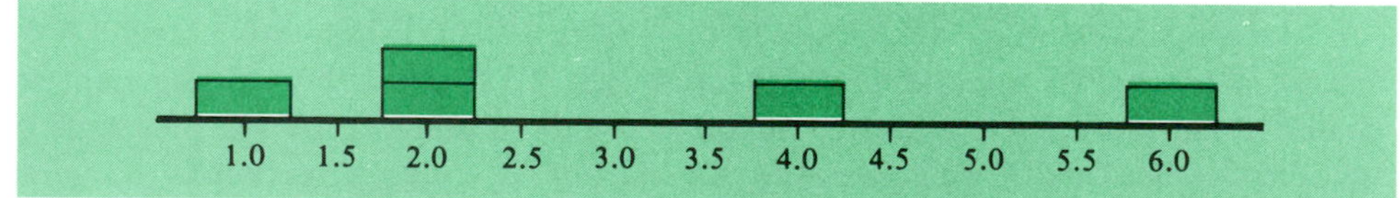

According to Theorem 8.1, as more and more samples are gathered, the distribution of sample sums approaches the normal distribution, as shown in Fig. 8.4.

[5]See the footnote on page 145.

Fig. 8.4

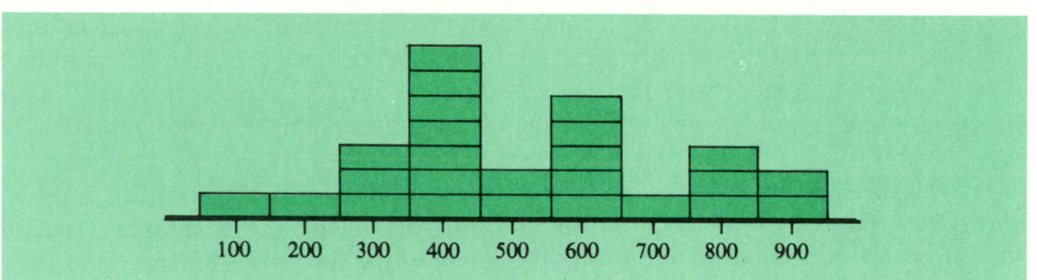

The distribution of sample means continues to mirror precisely the distribution of sample sums. Thus the distribution of sample means also becomes normal (Fig. 8.5).

Fig. 8.5

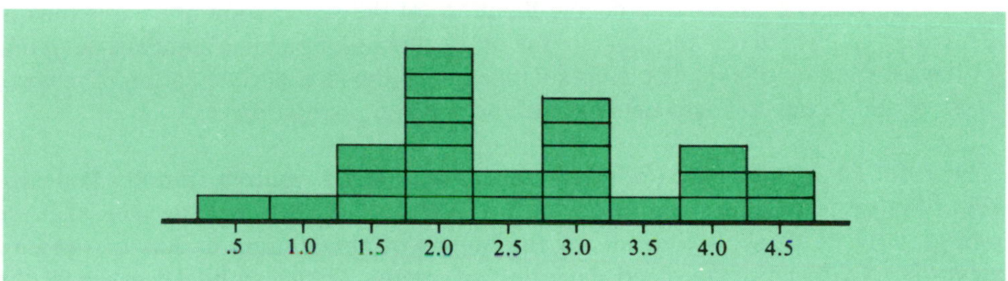

We have demonstrated Theorem 8.2 but we have not proven it rigorously. The reader who has mathematical facility might deduce Theorem 8.2 from Theorem 8.1. It would have to be shown that dividing all the terms in a distribution by the same constant does not change either the z scores or the percents of terms exceeded by any of the terms. Thus the operation of dividing all the terms by N leaves the pairings of z scores and percents of terms exceeded unchanged. The set of pairings fully defines the shape of a distribution. Since the distribution of sample sums is normal by Theorem 8.1, the distribution of sample means is also normal.

8.4 Properties of the Distribution of Sample Means

Our main focus will now be on the distribution of sample means. It is helpful to view this distribution as a secondary distribution. Specifically, starting with any population, suppose that equal-sized random samples are drawn from it and that the means of these samples become the numbers or elements that form a new distribution. It is the new distribution thus formed that is called the distribution of sample means.

From now on it will be assumed that this new distribution is infinite. That is, an infinite number of samples are drawn from the original population so that an infinite number of sample means comprise the new distribution.

There exist important relationships between the mean and standard deviation of the original population and the mean and standard deviation of the distribution of sample means created from this population. The mean (or balance point) of the distribution of sample means is the same as the mean of the original population.

THEOREM 8.3 Suppose an infinitude of equal-sized random samples is drawn from the same infinite population and the means of these samples are put together to form a new distribution. The mean of the new distribution (composed of sample means) has the same value as the mean of the original population.

We call such a new distribution *the sampling distribution of means.* A sampling distribution is a theoretical distribution of some particular value, like a mean, as it would be obtained from an infinite number of samples.

The symbol $\mu_{\bar{X}}$ is used to denote the mean of the distribution of sample means. Thus $\mu_{\bar{X}} = \mu$, where μ is the mean of the original population. The standard deviation of the distribution of sample means shall be called the standard deviation of means.[6] It is related to the standard deviation of the original population.

THEOREM 8.4 Suppose an infinitude of equal-sized random samples is drawn from the same infinite population. Denote the number of terms in each of these samples by N. The distribution of the means of these samples has a standard deviation (called the standard deviation of means or the standard error of the mean). The standard deviation of means equals the standard deviation of the original population divided by the square root of N.

The symbol for the standard deviation of the means is $\sigma_{\bar{X}}$. Thus, Theorem 8.4 states that

$$\sigma_{\bar{X}} = \frac{\sigma}{\sqrt{N}} \tag{8.1}$$

where σ is the standard deviation of the population from which the samples are drawn and N is the size of the samples.

Consider the population of IQ scores of adults in New York City. The mean of this population is 100 ($\mu = 100$). The standard deviation of this population is 15 points ($\sigma = 15$). Suppose we gather an infinitude of random samples of 10 scores each from this population and we put together the means of these samples to comprise a distribution. Since the parent population is made up of IQ scores, and therefore, is known to be normal, we may sample using 10 terms and still be assured by Theorem 8.2 that the distribution of sample means is at least approximately normal.

According to Theorem 8.3, the distribution of sample means has a mean (or balance point) at the value 100. Theorem 8.4 enables us to find the standard deviation of the distribution of sample means.

$$\sigma_{\bar{X}} = \frac{\sigma}{\sqrt{N}} = \frac{15}{\sqrt{10}} = 4.7$$

[6]This standard deviation of means is often called the *standard error of the mean.*

Fig. 8.6 shows the graph of the distribution of individual IQ scores. The graph of the means of random samples of 10 scores each is also shown.

Fig. 8.6

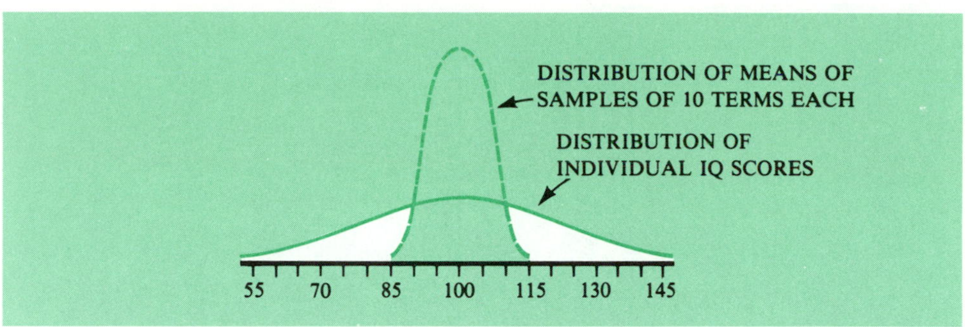

The graph of the distribution of sample means is the more compact, reflecting the fact that the standard deviation of means is 4.7 points whereas the standard deviation of the original population is 15 points. Viewed in another way, the sample means cluster around the balance point more than do the individual IQs. Thus the means of different samples resemble each other more closely than do the individual IQ scores.

The resemblance between the means of samples may be ascribed to a "balancing process" within each sample. Within any sample there are apt to be both high scores and low scores. Usually these scores tend to "average each other out" so that the sample mean turns out to be somewhere near 100, the value of the population mean. Thus it would be very rare to find a sample with a mean as low as 80 or as high as 120. The members of the sample would have to perform either consistently poorly or consistently well, and, since the members in each sample are randomly chosen, this is unlikely. Whereas individual scores are apt to differ markedly from the population mean, the sample means are less apt to spread out.

We may view the situation another way. Theorem 8.4 implies that **the standard deviation of a distribution of means of samples is always smaller than the standard deviation of the original distribution from which the samples are drawn.** Thus, the means of samples always resemble each other more closely (and cluster closer to the balance point) than do the individual terms in the population from which the samples are drawn.

The size of the samples yielding the means determines how closely the means will cluster around the balance point. **The larger the samples are, the more closely their means cluster together.** We can see in Formula (8.1) that, as the value of N increases, the value of $\sigma_{\bar{X}}$ decreases. In other words, when the number of terms per sample is large, the standard deviation of the sample means becomes small.

Three distributions are shown in Fig. 8.7. The first (Curve A) is the distribution of individual IQ scores with a mean of 100 and a standard deviation of 15 points. The second (Curve B) is the distribution of means of samples of 10 terms each drawn from the IQ population. This distribution of sample means has a mean of 100 and a standard deviation of 4.7 points. It is clearly more compact than the distribution of individual scores. Finally, the distribution of the means of samples of 100 scores each is also shown (Curve C). The mean of this distribution is 100 but its standard deviation is only 1.5 points. In other words, the means of samples of 100 scores each tend to be very much like each other.

Fig. 8.7

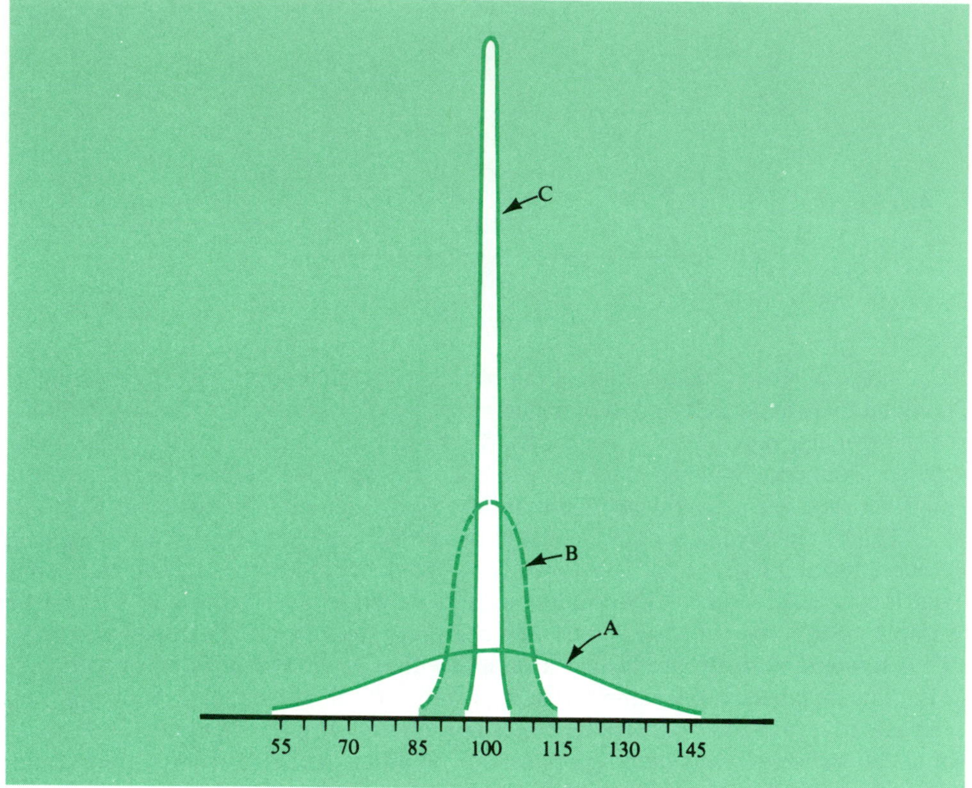

8.5 Locating a Sample Mean in the Distribution of Sample Means

Suppose that a vast number of equal-sized random samples are drawn from the same population and that the means of these samples form a normal distribution. We have seen how to compute the z score and the percentile rank of a single term in a normal distribution. We shall now consider how to compute the z score and the percentile rank of a single sample mean in the normal distribution of sample means. (Note that we are assuming that the samples are large enough to make the distribution of their means normal.)

To begin with, suppose we know the values of the mean and standard deviation of the population of individual terms (from which the samples are drawn). We have access to only one of the samples and we compute its mean. Without seeing any of the other samples, we are to determine the z score and the percentile rank of our particular sample mean in the distribution of sample means.

Before illustrating the method, we shall outline it briefly. Remember that we are given only the mean and the standard deviation of a population. We are told that many equal-sized random samples have been drawn from this population. We know the mean of one of these samples and wish to determine its z score and its percentile rank in the distribution of sample means.

Remember, Theorem 8.2 tells us that the distribution of sample means is normal. The first step is to compute the mean and standard deviation of the distribution of sample means. Our particular sample mean is a single term in this normal distribution of sample means. We compute the z score of our sample mean in its distribution by a variation of the standard method used to compute the z score of a term. Since we are locating the sample mean in the distribution of sample means, we use as a yardstick the standard deviation of means. The z score of our particular sample mean is the number of yardsticks it is above or below the balance point of the distribution of sample means.

At this point, we have computed the z score of our particular sample mean in the normal distribution of sample means. It should be clear that the sample mean is simply a term in a normal distribution. Now that we know the z score of this term we are able to find its percentile rank by referring to the normal distribution table.

We shall now illustrate in detail the method of finding the z score and the percentile rank of a sample mean in its distribution of sample means. Once again we begin with the population of individual IQs, which we know is approximately normal. We will say it has a mean of 100 ($\mu = 100$) and a standard deviation of 15 points ($\sigma = 15$). An infinite number of random samples of 25 cases are each drawn from this population ($N = 25$) and their means comprise a distribution. However, we have access to only one sample and have found that its mean is 106. We shall denote the mean of a sample by $\bar{X}$ so that $\bar{X} = 106$. The problem is to find the z score and the percentile rank of the sample mean of 106 in the distribution of sample means.

The distribution of sample means which has been created is normal (by Theorem 8.2) and has a mean of 100. By Formula (8.1) the standard deviation of the distribution of sample means turns out to be 3 units.

$$\sigma_{\bar{X}} = \frac{\sigma}{\sqrt{N}} = \frac{15}{\sqrt{25}} = 3$$

Therefore the distribution of sample means looks like Fig. 8.8.

The sample mean of 106 is located in Fig. 8.8. Our next step is to find the z score for the sample mean of 106. Remember how we found the z score of an individual's IQ—we used the standard deviation (of 15 points) as a yardstick; we then determined how far the individual's score was from the balance point and divided that distance by the length of the yardstick.

Our attention is now focused on the distribution of sample means. Therefore, we will use an analogous yardstick, the standard deviation of means. The z score of a particular sample mean is obtained by first finding the distance from the balance point to the particular sample mean. We then divide this distance by the length of the yardstick we are using. The quotient we get, which tells how many yardsticks the

Fig. 8.8

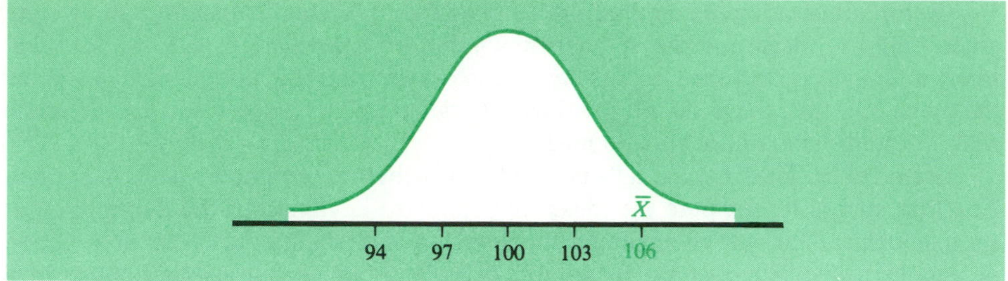

particular sample mean is from the balance point, is the z score of the particular sample mean.

The sample mean of 106 is 6 points above the balance point of its distribution. The standard deviation of means is 3 points. Thus the sample mean is above the balance point a distance that is twice the length of the standard deviation of means, as illustrated in Fig. 8.9. In other words, the z score of the sample mean is 2.

$$z = \frac{\bar{X} - \mu_{\bar{X}}}{\sigma_{\bar{X}}} = \frac{106 - 100}{3} = 2$$

Fig. 8.9

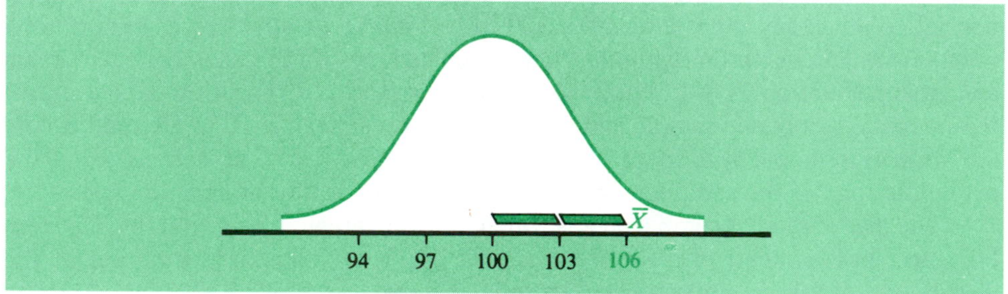

Finally, we can determine the percentile rank of the sample mean of 106 from its z score. Appendix III tells us that, in a normal distribution, the term with a z score of 2 has a percentile rank of 98. Thus the sample mean of 106 is higher than approximately 98% of all the sample means.

Before considering a second problem, it is worthwhile restating the main principles in what has been said. The supposition is that a vast number of equal-sized random samples are drawn from the same population. The means of these random samples form a normal distribution, called the distribution of sample means. Our knowledge of the original population enables us to determine the values of the mean and standard deviation of the distribution of sample means. Once we know the values of the mean and standard deviation of the distribution of sample means we can readily compute the z score of any particular sample mean. We are then able to determine the percentile rank of that sample mean from its z score.

Incidentally, from now on when we speak of the z score or the percentile rank of a sample mean, it will be assumed that the particular sample mean is being considered as a member of the theoretical distribution of equal-sized random samples drawn from the same population.

To illustrate the procedures again, suppose that the population of IQs among individuals of army age has a mean of 100 and a standard deviation of 15 points. After induction, the vast number of draftees are randomly assigned to squadrons containing 400 soldiers each.

The mean IQ of those in a particular squadron, Squadron B, is 99.25. Our problem is to determine the relative standing of Squadron B among all the squadrons. Specifically, we ask what percent of all the squadrons have (mean) IQs below 99.25.

We are given that the distribution of individual IQs has a mean of 100 and a standard deviation of 15 points. The squadrons are each random samples of 400 IQs drawn from this population. Therefore, the distribution of squadron means has a balance point at 100 (Theorem 8.3). The standard deviation of means is found by Formula (8.1).

$$\sigma_{\bar{X}} = \frac{\sigma}{\sqrt{N}}$$

$$\sigma_{\bar{X}} = \frac{15}{\sqrt{400}} = \frac{3}{4}$$

In sum, the distribution of squadron means has a mean of 100 and a standard deviation of 3/4. (The relevant yardstick is 3/4.) The distribution of squadron means is shown in Fig. 8.10.

Fig. 8.10

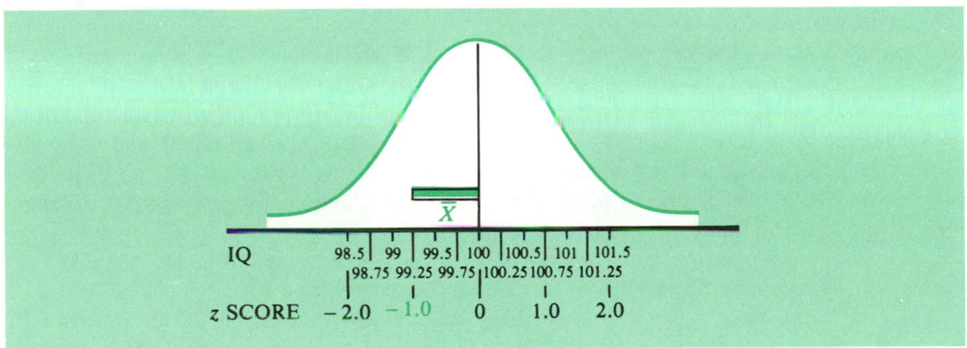

The mean of Squadron B, 99.25, is located in the illustration. The *z* score of the mean of Squadron B may be determined from two facts:

1. The mean of Squadron B is .75 point below the balance point.
2. The standard deviation of means, which is the relevant yardstick, is 3/4 of a point.

The *z* score of the mean of Squadron B is $\frac{-.75}{.75} = -1$. The distribution of squadron means is normal (since it is a distribution of large equal-sized random samples from the same population).

Appendix III tells us that in a normal distribution the term with a z score of -1 is larger than approximately 16% of the other terms. Thus the mean of Squadron B is at the 16th percentile. In effect we have compared a particular sample with the other samples by finding the z score of its mean and then converting this z score into a percentile rank. In the chapters to follow, we shall have many occasions to locate a particular sample mean in the distribution of sample means.

Problem Set A

8.1 Which of the following distributions are normal or approximately normal? Justify your answers.
   a. the diameters of thread spools produced by a certain machine
   b. the results obtained on successive tosses of a fair coin
   c. the mean ages of samples of 100 randomly selected Canadians
   d. the total weights of boxes of 100 cookies produced by the same method
   e. the incomes of a certain type of doctor, which are distributed with a mean of $50,000 and a median of $55,000
   f. the mean lifetimes of equal-sized large samples of a certain type of lightbulb
   g. the speeds of automobiles in Centerville

8.2 Let X be the distribution obtained by successive tosses of a fair die and Y be the distribution formed by summing successive groups of 36 tosses of the die. Suppose Z is the distribution formed by finding the means of successive groups of 50 values from X and W is the distribution formed by finding the means of successive groups of 100 values from X.
   a. Which of the Distributions X, Y, Z, and W are normal? Why or why not?
   b. What can be said about the means of Distributions X, Z, and W?
   c. Place the standard deviations of Distributions X, Z, and W in increasing order.

8.3 Consider a parent population that is distributed with a mean of 50 and a standard deviation of 5 but is not normal.
   a. Find the mean and standard deviation of a distribution of means of random samples of size 36 taken from the parent population.
   b. What proportion of the sample means would be expected to be less than 49?
   c. What proportion would be expected to be greater than 47?

8.4 The mean amount of nicotine per cigarette in Light-Up brand is 20 milligrams with a standard deviation of 1.80 milligrams. The amounts vary normally. The cigarettes are packaged 20 to a pack.
   a. Find the mean and standard deviation of the mean amounts of nicotine per cigarette for a vast number of packs.
   b. What percent of the packs would have mean amounts per cigarette below 20.5 milligrams?
   c. What percent of the cigarettes would contain less than 20.5 milligrams?
   d. What percent of the packs would have mean amounts per cigarette between 19 and 21 milligrams?
   e. How many milligrams would we expect 30% of the package means to exceed?

8.5 A marine biologist is studying a type of whale whose weights are normally distributed with a mean of 4.78 tons and a standard deviation of .20 ton.

a. What percent of the whales weigh less than 4.6 tons?

b. What percent of the samples of six whales would have a mean weight of less than 4.6 tons?

8.6 A musical aptitude test was given to all third-grade classes in New York City. The mean score on this test was 75 and the standard deviation of children's scores was 15 points. Suppose that the classes are composed of 36 children each, so that in effect there are a vast number of samples of 36 scores each. (a) In what percent of the classes would we expect the class mean to be as high as 80? (b) In what percent would the mean be as low as 72? (c) Answer parts a and b using samples of 49 scores each. (d) Answer parts a and b using samples of size 64. (e) How does the size of the samples used affect a distribution of sample means?

8.7 Hay fever remedy Ah-Choo relieves the well-known symptoms of the malady for a mean of 12 hours with a standard deviation of 2.5 hours. What fraction of the allergists prescribing the drug to 49 patients would note a mean relief time of (a) 13 hours or more? (b) between 11.5 and 12.25 hours?

8.8 The bus trip from Bathville to John City takes a mean of 85 minutes with a standard deviation of 15 minutes and is normally distributed. Bunny Hop Transit schedules eight buses daily between the two points.

a. What percent of the daily mean times should exceed an hour and a half?

b. One-fourth of the daily means are less than what time?

8.9 Lobsters off the coast of Florida in the waters harvested by Lefty Boatman have a mean weight of 2 pounds with a standard deviation of 6 ounces. The weights are normally distributed. Lefty captures 16 of the crustaceans in each trap. What percent of the traps contain lobsters with a mean weight in excess of 35 ounces?

8.10 Ruth's Poodle Palace schedules eight poodles for grooming each day. The times required to groom the dogs are normally distributed with a mean of 55 minutes and a standard deviation of 10 minutes. What percent of the time does Ruth work more than a 7-hour day (420 minutes)? (Hint: This means the average (μ) per dog must be more than 420/8 minutes.)

8.11 The average adult has completed a mean of 10.75 years of formal education with a standard deviation of 1.5 years.

a. Find the mean and standard deviation for a distribution of means of random samples of size 100.

b. Find the percent of random samples of 100 adults that would be expected to have a mean number of completed years above 11.

c. Find the number of years that we would expect to be exceeded by 20% of the sample means.

8.12 A school psychologist administers an anxiety test to all students when they begin the first grade. The test scores are known to be normally distributed with a mean of 40 and a standard deviation of 6. First graders are put into classes of 18 pupils each. What percent of the first-grade teachers receive classes with mean anxiety scores (a) as low as 39? (b) between 39.5 and 40.5?

8.13 The amount of insulation per home in a well-defined northern region is distributed with a mean of 500 pounds and a standard deviation of 75 pounds. For purposes of an energy study the homes are randomly assigned to groups of size 50.

a. Find the insulation level (in pounds) that 85% of the group means fail to meet.

b. Find the sample mean that represents P_{35}.

c. Find an interval centered at the mean that would be expected to include half of the sample means.

8.14 At the plant where Cherry Slurp soda is bottled, machines fill "12-ounce" bottles with a mean of 12.1 ounces and a standard deviation of .2 ounce.

a. The amounts in the individual bottles can be considered approximately normal. Why?
b. What percent of the six-packs actually contains bottles with a mean of 12 ounces or more?

8.15 Consider a population consisting of only the scores 3, 6, 8, 11, and 15. (a) Find the mean and standard deviation of these scores. (b) Find all the possible samples of size two (that is, 3 and 6, 6 and 3, 3 and 3, etc.) that can be drawn from this population, *with replacement*, and list these (there are 25 such samples). (c) Find the means of all the samples of size two. (d) Find the mean of the sample means. (e) Find the standard deviation of the sample means. Compare the mean of all the samples of size two and the standard deviation with the results in (a).

Problem Set B

8.16 Which of the following distributions are normal or approximately normal? Why or why not?

a. the ages of all school children in North America
b. the heights of all graduating seniors in South Carolina
c. numerous means of equal-sized large samples selected randomly from a nonnormal distribution
d. numerous means of equal-sized small samples selected randomly from a nonnormal distribution
e. numerous means of equal-sized small samples selected randomly from a normal distribution
f. total passenger weights of full 80-passenger jet planes
g. weights of U.S. postage stamps

8.17 Consider a normal distribution with a mean of 500 and a standard deviation of 25.

a. Find the mean and standard deviation of a distribution of means of randomly selected samples of size 16.
b. What percent of the means are smaller than 495?
c. What percent of the terms in the original population are smaller than 495?

8.18 A "2-pound box" of a leading brand of chocolate chip cookie contains 100 cookies. All such cookies produced weigh a mean of .34 ounce with a standard deviation of .04 ounce, and the distribution is normal.

a. Find the mean and standard deviation of the mean weights per cookie for numerous boxes.
b. What percent of the boxes contain cookies with a mean weight of more than .335 ounce?
c. What proportion of the cookies weigh more than .335 ounce?
d. What percent of the boxes have mean weights per cookie that equal or are exceeded by .33 ounce?
e. Find an interval centered at .34 ounce into which we would expect 75% of the box means to fall.

8.19 A scientist is studying a rare organism known to exist in a remote mountain stream. Each cubic centimeter of water contains a mean of 400 organisms with a standard deviation of 100. Each day the scientist or an assistant collects 40 water specimens of a cubic centimeter each.

a. If the experiment is conducted for many, many days, what percent of the days can the scientist expect to have a mean of 425 or more organisms per specimen to study?

b. What mean number of organisms per specimen would a third of the daily means fail to exceed?

8.20 The mean age at which women give birth to a first child has risen to 24.7 years with a standard deviation of 2.4 years.

a. In what percent of the samples of size 100 would we expect the sample mean to be as low as 24 years?

b. In what percent of the samples of size 75 would we expect the sample mean to be as low as 24 years?

c. In what percent of the samples of size 30 would we expect the sample mean to be as low as 24 years?

d. Explain why the answers to parts a, b, and c vary as they do.

8.21 Sleeping time per 24-hour period for full-time college students is normally distributed with a mean of 7.35 hours and a standard deviation of 1.80 hours. For a sleep study a vast number of students are recruited to record their sleeping times each day for a month. Each student contributes 30 times to form a distribution whose terms are randomly grouped into samples of size 25 by a computer that then supplies the mean for each sample group of 25. What percent of the sample means (a) exceed 7 hours? (b) are less than or equal to 8.25 hours? (c) are between 7.5 and 8 hours?

8.22 An improved diet plan causes chickens to gain a mean of 1.5 pounds more than the usual diet does; and the standard deviation is only .5 pound. The added weights vary normally. The grower sells the chickens in lots of 100 each to various packaging and marketing companies.

a. What percent of the lots have mean weights per bird that are in excess of 1.6 pounds more than they would have been had traditional chicken feed been supplied?

b. What percent of the birds weigh in excess of 1.6 pounds more than if traditional feed had been supplied?

c. What percent of the lots have mean weights per bird that are between 1.45 and 1.55 pounds more than they would have been had the new diet not been used?

8.23 A certain type of fuse will function until a mean of 18.5 amperes passes through it (with a standard deviation of 3 amperes); then it will burn out. The burn-out times are normally distributed. The fuses are sold in boxes of six each. If the claim on the box states the fuse can handle up to a mean of 19 amperes, what percent of the boxes fail to live up to the claim?

8.24 Suppose that the heights of American soldiers are normally distributed with a mean of 5 feet 7½ inches and a standard deviation of 3½ inches. (a) What percent of samples of size 350 have a mean as great as 5 feet 8 inches? (b) What percent of samples of size 50 have a mean as small as 5 feet 6 inches? (c) Find the height that we would expect to be exceeded by the means of only 10% of the samples of size 100.

8.25 Thirty preschoolers in each of 500 child-care centers across the country are given a maturity test through a federally funded study. The mean and standard deviation of the individual scores prove to be 55 and 7 respectively. About how many centers have children whose scores have a mean of (a) 53 or more? (b) more than 56? (c) 56 or more? (d) between 54 and 56?

8.26 The final exam scores in Dr. I. M. Tuff's beginning course in statistics distribute themselves normally with a mean of 76 and a standard deviation of 11. Dr. Tuff's contracts stipulate that he teach classes of 50 students each.

a. If C grades range from 70 to 80, what proportion of Dr. Tuff's students receive a C?
b. What percent of his classes have a mean grade of C?
c. If 2% of the classes have means above a C, what is the upper cut-off point for a C grade?

8.27 The deadline for classified ads in the *Bakke Bugle* is 1:00 p.m. A mean of 65 ads is received each day during the hour before the deadline (12:00–1:00). The standard deviation is 19. The typesetter has requested monthly reports to help in meeting crucial deadlines. If nothing changes, what fraction of the monthly reports giving daily means of the number of ads received during the noon hour will vary between 60 and 70? (Use 30 days per month.)

8.28 The number of misdemeanors reported on the subway in Megapolis varies normally with a mean of 208 per month and a standard deviation of 46. Sergeant Weeks files a monthly summary.

a. What percent of the summaries should cite more than 250 misdemeanors?
b. What fraction of the quarterly reports would have a monthly mean between 200 and 250?

8.29 The product control manager at Kitchen Magic, Inc., periodically selects 25 plastic mixing bowls at random from the production line to test for quality. The bowl diameters are normally distributed with a mean of 8 inches and a standard deviation of .35 inch.

a. What percent of the manager's samples have mean diameters between 7.9 and 8.1 inches?
b. Find an interval centered at 8.1 that includes 10% of the sample means.

8.30 Let X be the distribution of incomes in a large country. Tax returns are sent to numerous regional receiving stations throughout the country where they are checked for errors. Each regional station receives 10,000 returns. Let Y be the distribution of means for the regions. The returns are then forwarded to numerous sectional receiving stations. Each sectional station receives returns from 100 regional stations. Let Z be the distribution of means for the sectional stations.

a. Identify Distributions X, Y, and Z on this diagram.

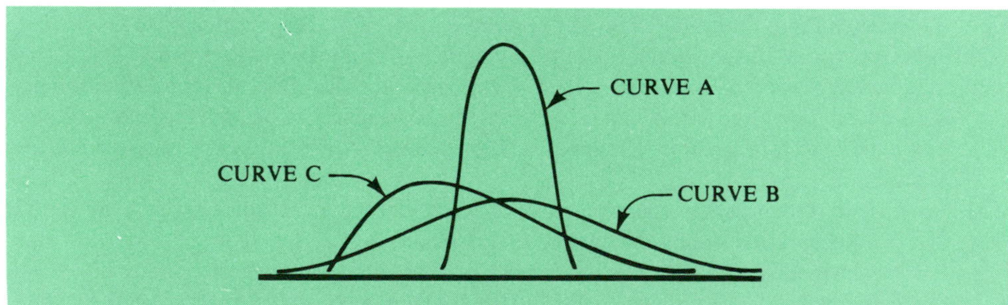

Answer true or false for the following.

b. $\mu_X = \mu_Y = \mu_Z$

c. Distributions Y and Z are normal.

d. $\sigma_X < \sigma_Y < \sigma_Z$

e. $\sigma_Y = 100\,\sigma_X$ [Hint: Use Formula (8.1).]

f. $\sigma_Y = 10\,\sigma_Z$

g. $\sigma_X = 1{,}000\,\sigma_Z$

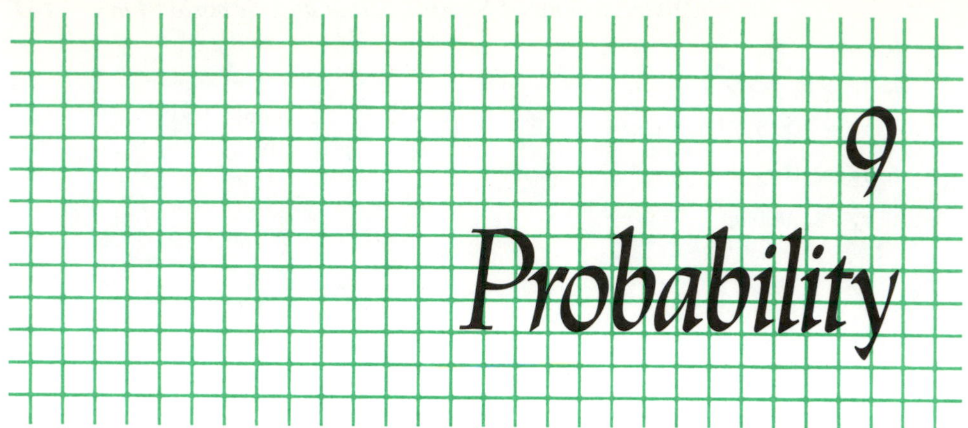

9.1 Probability

Probability is a topic that has long fascinated philosophers and mathematicians. The concept of probability is subtle; as a matter of fact, a century after two mathematicians named Pascal and Fermat founded the science of probability, no one knew quite how to define it. Probability seemed to imply that things "partly happen and partly don't," whereas in the world things either happen completely or they don't happen at all.

Fortunately, the science of probability was fruitful enough to survive until a satisfactory rationale was developed for it. The science of probability has received considerable attention, and today there are several ways of defining probability that are adequate for most purposes. Probability theory has flowered in the 20th century and it now enriches many applied fields.

On some level we all sense the meaning of probability, even though we do not have a handy definition of it. Children seem to have some understanding of probability, because they show surprise at improbable happenings. Even white rats manifest a sense of probability; they act very differently when their rewards or punishments are made more or less probable by psychologists.

Probability must be viewed in the context of an experiment or procedure that permits *definable outcomes.* For instance, the flip of a coin is an experiment that permits two outcomes, a head and a tail. A race is an experiment in which we may specify as many outcomes as there are possible winners. If we are interested in the exact order in which the participants finish, we must consider each possible order of finishing as an outcome that may occur.

The weather on a Sunday afternoon is in this sense also an experiment. If we are interested in whether it will rain we may specify two outcomes, rain and no rain. Remember that the outcomes of an experiment must be defined and recognizable and that the experiment must inevitably lead to some outcome. The outcomes will be defined in such a way that no two of them can possibly occur together.

The drawing of a slip from the wooden crate, as described in the preceding chapter, is an experiment. If the numbers on the slips are 0, 1, and 2, then these numbers are the three possible outcomes. The drawing of a *sample* of slips to determine its mean may also be viewed as a single experiment. The mean of the sample may be considered the outcome of this experiment. In the same way, selecting a person at random and determining his or her height is an experiment that theoretically permits an infinitude of outcomes corresponding to the possible heights that may be obtained. Choosing a random sample of people and finding the mean height in this sample is also an experiment with an infinitude of possible outcomes.

The word *probability* is always used with reference to the outcome of an experiment. The exact nature of the experiment and the specified outcome must be made clear before the probability of the outcome can be found. The probability of an outcome occurring is a *fraction* or *proportion.*

Definition 9.1 The probability of an outcome is the fraction of times that the outcome would occur if the experiment were repeated indefinitely.

Suppose the experiment is the flip of a coin, which has two possible outcomes, a head and a tail. The probability that a head will turn up is defined as the proportion of times that heads would appear if the coin was flipped over and over again an infinitude of times.

Picture someone flipping a coin into the air over and over again. It is our job to record the number of heads and the number of flips of the coin. We construct a fraction that has as its numerator the number of heads already obtained and as its denominator the total number of flips of the coin. To be concrete, suppose that a head comes up four times in the first 10 flips. At this point the fraction is 4/10 or .40. The flipping continues and after 100 flips suppose that the number of heads is 53. The fraction is now 53/100 or .53. The repetitions of the experiment go on and on; that is, the flipping of the coin continues indefinitely.

Table 9.1 provides a record of the number of heads obtained after various numbers of flips of the coin. The different proportions of heads are also indicated.

Table 9.1 Heads Obtained in Coin Experiment

Heads	4	53	139	237	480
Flips	10	100	300	500	1,000
Proportion of heads	.40	.53	.46	.47	.48

Early in the game, the proportion of heads obtained is greatly influenced by the outcome of each flip. For instance, the appearance of one more head during the first 10 flips would have raised the proportion of heads after 10 flips from .40 to .50. However, as the number of flips increases, the importance of a few flips becomes less and less. The proportion of heads gains a kind of stability, and after a while even long runs of heads or tails scarcely influence it.

Mathematicians like to say that the proportion of heads *settles down* at some value, and this value is defined as the probability that a head will turn up. If the coin is a fair one, the value is .50. If the coin is biased in favor of heads, the fraction is more than .50. If the coin is biased against heads, the fraction is less than .50.

Let us suppose the coin being investigated has a tail on each side. Now as the repetitions of the experiment continue, the numerator—indicating the number of heads obtained—remains equal to zero. The denominator—indicating the number of repetitions of the experiment—keeps increasing. No matter when we stop, the proportion of heads stays equal to zero. Similarly, **whenever any outcome is impossible, its probability is zero.**

On the other hand, suppose the coin has a head on both sides. The number of times a head turns up will always equal the number of repetitions of the experiment. Thus the numerator of the fraction will always equal the denominator. Now the probability of getting a head equals one. In general, **whenever an outcome is certain, its probability is one.**

The probability of any outcome is a number between and including zero and one. That is, the probability of any outcome ranges between impossibility and certainty. If we add up the probabilities of all the possible outcomes of an experiment, the total we get equals one (certainty). Thus if there are two possible outcomes and the probability of one of them is .37, then the probability of the other must be .63 (1.00 − .37 = .63).

9.2 The Word "Probability" in Everyday Language

Many theorists believe that our definition of probability covers the way the word is used in common parlance. The belief is that whenever we say "probably," we are applying knowledge of what occurs in the long run to a specific instance.

The statement "There is a 70% chance of measurable precipitation tomorrow" is a typical weather-forecast statement. It sounds as if it were about tomorrow and no day but tomorrow. However, in actuality the statement refers to a whole collection of days, and tomorrow is only one member of this set. The statement implies that on 70% of the occasions when there has been a weather situation like today's there has been measurable precipitation on the following day.

It is worthwhile illustrating the importance of clearly defining an experiment or process leading to an outcome before trying to determine the probability of that outcome. Two people may conjecture about the same outcome, but if they view the procedure that leads to it differently, their expectations are apt to be different.

For instance, suppose Person A knows only that a friend was hit by an automobile and brought to the hospital three months ago. Let us be diabolical and think of the accident as an experiment. Consider one particular outcome—that the victim was seriously injured. Suppose it is known that one-half of hospitalized auto-accident victims are seriously injured. Person A, who knows that this friend was hospitalized following an auto accident, would make a prediction. The friend would be classified as a member of the group of people hit by automobiles and brought to the hospital. This would lead to the conclusion that the probability is one-half that the friend was seriously injured.

However, suppose we know (but Person A doesn't) that the victim left the hospital one week after being admitted. Furthermore, we know that one-fifth of the individuals hit by automobiles and detained in hospitals for one week are seriously injured. On the basis of what we know, it is correct for us to state that the probability is one-fifth that the victim was seriously injured.

Person A has arrived at a probability statement correctly and so have we. The two statements differ because they are based on different information. The moral is that we must be careful to describe in detail our experiment or process when we discuss probability. If we omit information, we cannot expect another person to attach the same probability value to an outcome that we do.[1]

9.3 Sample Spaces and Events

Often the situation arises in which we are interested in determining the probability of an occurrence that is more easily viewed as a group of outcomes rather than a single outcome. In such cases we usually refer to the occurrence as an **event**. For instance, suppose our aim is to determine the probability of rolling a 3 or greater on one roll of a fair die. The set[2] of outcomes that immediately comes to mind is {1, 2, 3, 4, 5, 6}.[3] Note that this set satisfies the three criteria found in Definition 9.2.

> **Definition 9.2 A set of outcomes that satisfies the following three criteria is called a sample space for an experiment, designated by *S*.**
> **1. The outcomes are defined and the experiment must lead to one of them.**
> **2. No two outcomes can occur together.**
> **3. The sum of the probabilities of all the outcomes in the set is one.**

The event whose probability we are interested in determining, rolling a 3 or greater, occurs if a 3, 4, 5, or 6 appears when the die is rolled. Since each outcome has the same probability of occurring (1/6), our event will occur in the long run in four out of every six trials. Now this does not imply that exactly four of every six rolls will yield a result greater than or equal to 3; but rather, were the experiment to be repeated indefinitely, four-sixths of the time we could expect the event in question to occur.

Many probability problems can be greatly simplified if an attempt is made first to determine a sample space consisting of outcomes that have the same probabilities of occurring. To illustrate, let us determine the probability that a randomly selected family with three children will have two girls and a boy. Assuming that any child born is just as likely to be a boy as a girl, it follows that any arrangement of boys and girls is just as likely as any other. So we begin by listing the various possibilities. The first, second, and third children may all be girls; the first two may be girls and the third a boy; and so on. Using G to represent girl and B to represent boy, we can write the entire sample space as follows:

S = {GGG, *GGB*, *GBG*, *BGG*, GBB, BGB, BBG, BBB}

[1]For an interesting discussion of probability, see "It's More Probable Than You Think," by Martin Gardner in the November 1967 issue of *The Reader's Digest*.

[2]A set is any collection of objects. The usual notation is to enclose the objects in brackets { }.

[3]Other sets of outcomes are possible for the toss of a die, for example {even, odd}.

Our interest lies in the arrangements that contain two girls and a boy. There are three that satisfy this requirement; they have been italicized in the sample space. Since every arrangement is as likely as any other, one-eighth of all families with three children should fit into each arrangement category. Therefore, three-eighths of the families would be expected to have exactly two girls. It follows that the probability we are required to find is also 3/8.

It is worthwhile to consider another simple problem in which determining the sample space in advance is certainly advantageous. A tutor has agreed to give weekly help to four students enrolled in a psychology class. The tutor has time available for the service on Monday and Tuesday evenings only. If the students are all just as likely to select one evening as the other, what is the probability that the tutor will have all four students on the same night?

We could proceed as before by starting to list the possible outcomes. Letting M represent Monday and T Tuesday, our first outcome might be MMTT (first and second students on Monday and third and fourth students on Tuesday). We would soon discover, however, that it is very easy to overlook one or more outcomes. In such cases, drawings, graphs, pictures, or other aids may be of value. In the problem at hand, the following diagram, often referred to as a *probability tree*, proves to be of great assistance (Fig. 9.1).

Fig. 9.1

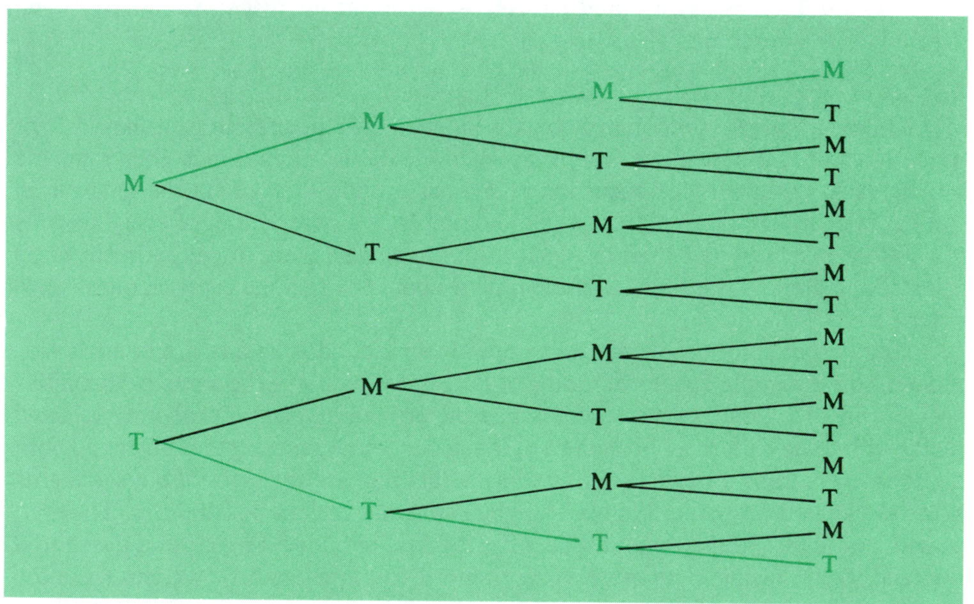

This probability tree makes it possible for us to write the sample space and be confident that all outcomes have been included.

S = {MMMM, MMMT, MMTM, MMTT, MTMM, MTMT, MTTM, MTTT,
TMMM, TMMT, TMTM, TMTT, TTMM, TTMT, TTTM, TTTT}

Only two of the 16 outcomes (the first and the last) show all four students meeting with the tutor on the same evening. So the probability that all four students will request help on the same evening is 2/16 or 1/8.

The procedures discussed in this section are intended to provide the reader with only a brief encounter with probability theory. At this level, concepts are most easily and clearly illustrated by using discrete distributions involving dice, cards, tossed coins, and the like. Advanced techniques in counting, however, make it possible to apply many of the things discussed here to nearly every facet of modern learning and technology.

9.4 Probability and the Sampling Experiment

One kind of experiment in which we make probability statements is of primary concern to researchers. It involves selecting a random sample consisting of one or more terms from a population. The variable of interest is one that may assume different values in different samples. For instance, the values of the variable may be the means of the samples. Its value in the particular sample chosen is considered the outcome of the experiment.

Each sample value or outcome has some probability of occurring. Once a particular value is specified, we can determine the probability that the sample drawn will have this value. Our focus now is on how to determine the probability of getting any particular outcome when an experiment is performed. In this section, we will assume that the sample drawn consists of one term. That is, the experiment involves choosing a single term at random from the population. The problem is to determine the probability that the term chosen has some specified value.

An illustrative experiment was conducted not long ago on a farm. The farmer, a man named Hiram, owned 100 chickens—70 of the Plymouth Rock variety and 30 Rhode Island Reds. Hiram was working in the barn and could not see the hawk that soared high overhead. Lazily, the hawk descended and circled until he was at an angle where the barn momentarily concealed him from the chicks. An instant later he bolted forward and was gone. Ninety-nine chicks were left scurrying about the yard.

Their chattering brought Hiram out of the barn, and he quickly realized what had happened. One thing made him curious though. Was it a Plymouth Rock chick or a Rhode Island Red that was at this moment sailing through the heavens? Probably it was a Plymouth Rock chick, the farmer thought. There were 70 of them and only 30 Rhode Island Reds. A hawk can be easily seen when he attacks and he has no time to choose his prey during his momentary plunge.

Hiram's problem may be formulated this way. We shall assume as he did that the hawk made a random selection of one chick from the population of 100. The problem is to determine the probability that the victim was a Plymouth Rock chick and the probability that it was a Rhode Island Red. Note that the variable in this experiment is the discrete variable, type of chicken, and the two possible outcomes are the two values this variable may assume.

The probability of an outcome of an experiment is a statement (in the form of a proportion) of what would happen if the experiment were to be repeated indefinitely. The experiment described cannot be repeated in actuality, but we can conceive of it being repeated, say once a day, and that is enough for us to reach a solution.

Suppose that every day for an indefinitely long period the *same* 100 chickens could be placed in the same yard. Each day the hawk swoops down to make his random selection of one of them. The fact that each single chick is an equally likely candidate to be chosen means that in the long run each chick is chosen on one out of every 100 days. Thus on 70 out of every 100 days, the abducted chicken is a Plymouth Rock chicken. We are considering as one possible outcome of the *sampling experiment, the abduction of a Plymouth Rock chicken*. The probability of this outcome is 70/100.

By the same reasoning, on 30 out of every 100 days in the long run the abducted chicken would be a Rhode Island Red. Thus the probability that a Rhode Island Red chicken gets taken in a single experiment is .30.

The point is that the members of the Plymouth Rock group comprise a certain proportion of the population (70/100) and this proportion tells us the probability that in a single experiment a member of the Plymouth Rock group will be selected. The group of Rhode Island Reds also comprises a proportion of the population (30/100) and that proportion tells the probability that a Rhode Island Red will be selected in a single experiment.

In general, suppose the experiment of selecting a random sample is performed. We wish to know the probability that the sample has some particular value or characteristic. The proportion of samples in the population with that value or characteristic tells us the probability that the single sample chosen will have that value.

For example, suppose a single United States inhabitant is chosen at random; consider the outcome that the person chosen lives in the State of New York. Suppose there are about 21 million New Yorkers and about 210 million U.S. inhabitants all together. Then New Yorkers would comprise about one-tenth of the population and the probability that the person chosen will be a New Yorker is about 1/10.

By the same reasoning, suppose there are roughly a quarter of a million medical doctors residing in America. Then about one person out of every 840 is a medical doctor. Therefore the probability that the person chosen at random will be a medical doctor is about 1/840.

So far we have discussed only the situation in which the sample consists of one term. The generalization to samples made up of more than one term will be made later on. It is not difficult, as we shall see.

9.5 Probability and the Normal Distribution

Our knowledge about the normal distribution enables us to make probability statements in connection with it. Consider the experiment of drawing a single term at random from a population of normally distributed terms. Suppose that a prediction about this term is made. For instance, it might be conjectured that the term that is picked will have a z score larger than 0 or that it will have a z score between 0 and 1. Our knowledge about the normal distribution enables us to determine the probability that a prediction of this kind will come true.

Remember how to determine the probability that a randomly drawn sample of one or more cases will have a particular characteristic. A certain proportion of samples in the population have the characteristic; this proportion tells us the probability that a single sample chosen will have the particular characteristic.

Now we are proposing to draw a sample of one case from a normal distribution. Consider the event that the term drawn will have a z score larger than 0. In a normal distribution the probability of this event equals the proportion of terms in the population that have z scores larger than 0. Therefore, the probability is 1/2 that the term drawn will have a z score larger than 0.

Instead we may specify that the term drawn will have a z score between -1 and $+1$. To determine the probability of this event, we must find the proportion of terms in a normal distribution which have z scores between -1 and $+1$. Appendix III shows that in a normal distribution about .68 of the terms have z scores between -1 and $+1$. Therefore the probability is roughly .68 that a randomly chosen term will have a z score in that interval. The area representing these z scores has been indicated in Fig. 9.2.

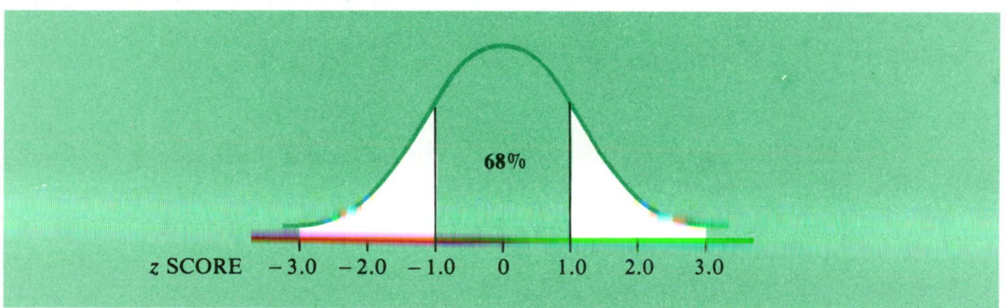

Fig. 9.2

Another event is that the term chosen will have a z score that is not between -1 and $+1$. That is, it may be specified that the term chosen will have a z score that is either less than -1 or greater than $+1$. For this event to occur, the term must come from one of the "tails" of the distribution depicted in Fig. 9.2. These two tails considered together include about 32% of the terms. Therefore, the probability is 32/100 that a randomly chosen term from a normal distribution will have a z score not in the interval -1 to $+1$.

About 95% of the terms in a normal distribution have z scores between -2 and $+2$. Therefore the probability is about 95/100 that a randomly chosen term will have a z score between -2 and $+2$.

Only 5% of the terms in a normal distribution have z scores that are either less than -2 or greater than $+2$. In other words, about 5% of the terms are two or more z score units from the balance point in either direction. These terms are included in the tails in Fig. 9.3. The probability that a randomly chosen term will have a z score that is either less than -2 or greater than $+2$ is about 5/100.

About 2.5% of the terms in a normal distribution have z scores larger than $+2$. That is, about 2.5% of the terms are in the right tail of the graph in Fig. 9.3. Therefore the probability is about .025 that a term chosen at random from a normally distributed population will have a z score greater than $+2$.

Fig. 9.3

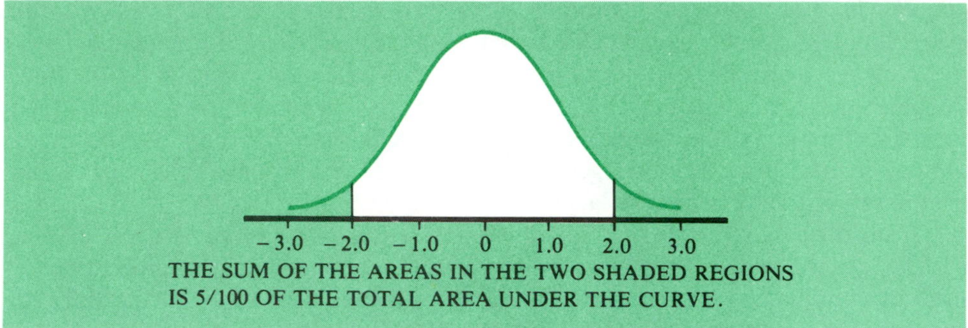

THE SUM OF THE AREAS IN THE TWO SHADED REGIONS IS 5/100 OF THE TOTAL AREA UNDER THE CURVE.

The material just discussed may now be applied to several problems. Suppose the heights of the men in a certain large city are normally distributed with a mean of 68 inches and a standard deviation of 3 inches. One of these men is to be randomly selected from the population. What is the probability that this man is between 68 inches and 71 inches?

The first step in solving this problem is to convert the specified interval into a z score interval. The value 68 inches corresponds to a z score of 0. The value 71 inches has a z score of 1.0. Thus the original problem leads us to ask what is the probability that a term randomly chosen from a normal distribution will have a z score value between 0 and 1.0. Appendix III tells us that the probability of this happening is 34/100. Thus the probability that the man who is chosen is between 68 inches and 71 inches is 34/100.

One might ask what is the probability that a randomly chosen man will be over 71 inches. After we translate into z score values, our problem leads us to determine the probability that a randomly chosen term has a z score value greater than 1. Appendix III tells us that this probability is 16/100.

One last problem is to determine the probability that the man who is chosen will be either shorter than 62 inches or taller than 74 inches. After translating into z score values, we are led to determine the probability that a randomly chosen term will have a z score that is either less than -2 or greater than $+2$. The probability of this happening is about 5/100. Thus the probability is 5/100 that the man who is chosen is either less than 62 inches or taller than 74 inches.

The turning point in solving problems like those given is to note that the interval specified in the problem defines a corresponding z score interval. The first step is to find this z score interval and the second step is to determine the proportion of terms included in it. This proportion tells us the probability that the term chosen will be a member of the interval specified in the problem.

9.6 Probability Statements about Sample Means

In scientific experiments, usually a sample consisting of many cases is drawn at random from a population. The mean of this sample is found. Suppose as before that a particular interval is specified when the experiment is performed. We can now ask what is the probability that the sample mean will be in the specified interval. The method of solving a problem of this kind is similar to the method used when a single term is drawn. The only difference is that we now refer to the distribution of sample means and not to the distribution of individual terms. The reason is that we are dealing with a sample mean instead of an individual term.

Consider the same population as in the last problem. The distribution of men's heights in this population has a mean of 68 inches and a standard deviation of 3 inches. This time a squadron of 100 men is randomly drawn from the population and the mean height of these men is found. When the experiment is performed, what is the probability that the mean height in the squadron is between 67 inches and 69 inches?

Our reasoning goes this way. We were asked the probability that the mean of a sample of 100 cases would lie in a specified interval. That is, if a vast number of similar samples were drawn, in the long run what proportion of them would have means inside the specified interval? To answer the question, we imagine that an infinitude of random samples of 100 cases each are drawn from the given population. We can determine what the distribution of the means of these samples looks like. For one thing, the distribution of sample means is normal.[4] Its standard deviation (that is, the standard deviation of means) is $3/\sqrt{100} = .3$ inches. Therefore the distribution of sample means, which we have conjured up to solve the problem, looks like Fig. 9.4.

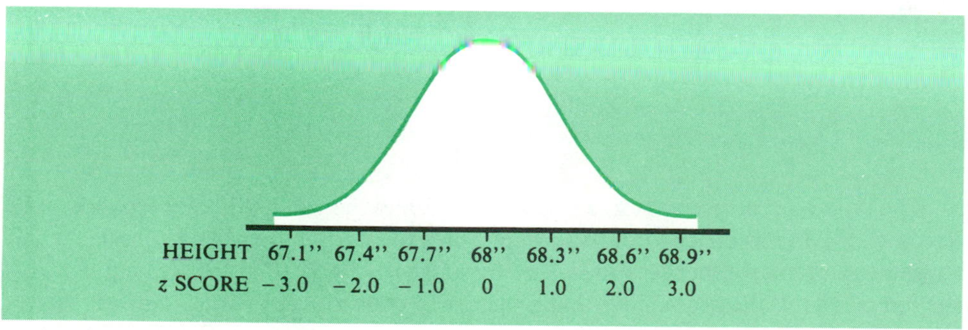

Fig. 9.4

Our job is to determine the probability that the mean of a particular randomly chosen sample will be between 67 inches and 69 inches. In the distribution of sample means the value 67 inches has a z score of -3.3. The value 69 inches has a z score of $+3.3$ inches. Thus the original problem leads us to ask what proportion of terms

[4]The normality is owing to Theorem 8.2. When considering the mean of a large-sized sample, it is not necessary to assume normality in the population.

(sample means) in the distribution of sample means have z scores between -3.3 and $+3.3$.

Appendix III provides an answer. One can see that virtually all the sample means are in the specified interval. Thus the probability that a randomly chosen sample will be in the specified interval is almost 1. The probability is roughly 999/1,000.

Note that a sample mean of 68.3 inches has a z score of 1.0. Appendix III tells us that approximately 34% of the sample means in the distribution are between 68 inches and 68.3 inches. The probability is about 34/100 that the mean height in a randomly chosen squadron of men will be between 68 inches and 68.3 inches.

Now consider this problem. Suppose that the mean length of films produced during the last decade is 100 minutes and that the standard deviation of film lengths is 24 minutes. If 36 different films are randomly chosen, what is the probability that the mean length of a film in this sample is either less than 92 minutes or greater than 108 minutes?

To solve this problem we imagine that a multitude of similar samples are drawn. That is, different samples of 36 films each are gathered and the mean length of each sample is found. The means of different samples are put into a distribution.

This distribution of sample means is normal. Its mean is 100 and its standard deviation turns out to be 4 minutes. It looks like Fig. 9.5.

Fig. 9.5

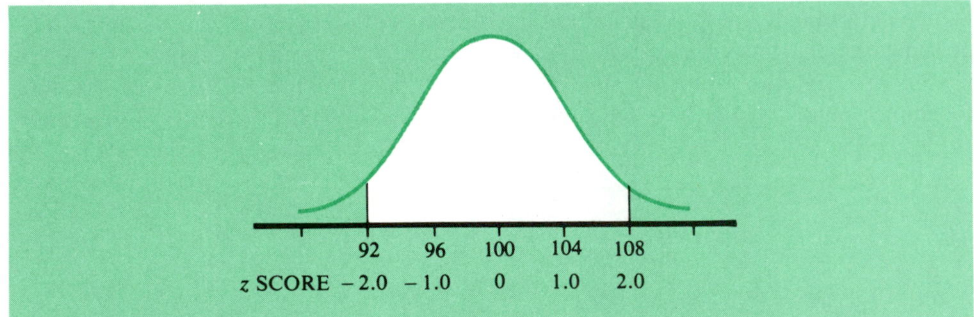

The value 92 corresponds to a z score of -2 and the value 108 corresponds to a z score of $+2$ in this distribution. We were asked the probability that a sample mean chosen at random is either less than 92 or greater than 108 minutes. An equivalent question is what is the probability that a randomly chosen sample mean has a z score either less than -2 or greater than $+2$. The answer is about 5/100. Thus the probability is about 5/100 that the mean length in a sample of 36 randomly chosen films is either less than 92 minutes or greater than 108 minutes.

The question might have been asked in a different form. One might have asked what is the probability that the mean length of films in the sample chosen is *at least 8 minutes away from 100 minutes* (the mean length of all the films produced). The answer is that the probability is about 5/100 that the mean length in a sample of 36 films will differ from the mean film length by as much as 8 minutes. In other words, only 5 samples out of 100 will have means that differ from the population mean by as much as 8 minutes.

9.7 The Normal Approximation to the Binomial Distribution

There is one other kind of experiment in which we will want to make probability statements. **Think of a variable like the flip of a coin that must show one of two outcomes in a single observation. Such a variable is called *dichotomous*.** The applications when using this variable are unlimited. Consider for a moment all the things that happen in one of two ways: test questions are often answered true or false; patients respond or they do not; legislators favor a bill or they do not; assembly-line products are defective or nondefective. The list is endless.

Now suppose that some dichotomous experiment, flipping a fair coin for example, is performed 100 times and the number of successes is recorded. We shall, arbitrarily, call a head a success and use the value X to refer to the number of successes. So for the first set of 100 trials we might have $X = 52$. If we perform 100 dichotomous trials a second time, $X = 47$ would be a reasonable result. **We do this repeatedly until we have innumerable X values each representing the number of successes on 100 trials of the dichotomous variable. These X values form a new distribution that is called *binomial*.** In our experiment we might expect the results to look something like this: 52 47 48 55 45 50 51 54 40.... Were we to organize these data into a graph, the result would look like Fig. 9.6. The mean would be 50 and the standard deviation 5. Take note of the fact that *the binomial is a discrete distribution*.

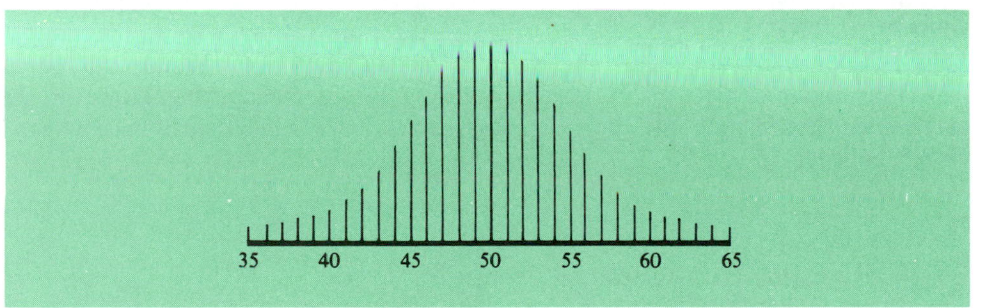

Fig. 9.6

Not all binomial distributions have a mean of 50 because two things vary from distribution to distribution: (1) the size of the samples used to build the distribution, and (2) the probability that on any one trial the dichotomous variable will be a success.

Consider a weighted coin that shows heads three out of five times; that is, $P = 3/5$, where P stands for the probability that a head will turn up on any one toss. In general, when we talk about a dichotomous variable we single out one of the two possible occurrences and use the letter P to stand for its probability. The distribution of outcomes or numbers of successes on successive experiments each of which is a sample of 50 flips has a mean of 30 and a standard deviation of $\sqrt{30\,(2/5)} = 3.46$. This was determined by references to the following theorem.

THEOREM 9.1 Suppose for a dichotomous variable an occurrence of interest has a probability of happening, designated by *P*. A vast number of equal-sized samples, each of which consists of *N* observations, is collected. In each sample the outcome (designated by *X*) is defined as the frequency of the occurrence of interest in the particular sample. The distribution of outcomes (or values of *X*) obtained from an infinite number of such samples approximates the normal distribution.[5] The mean and standard deviation are given by $\mu = NP$ and $\sigma = \sqrt{NP(1-P)}$.

Although we will not go into the proof of this theorem, we will point out why the binomial distribution approximates the normal distribution. Recall the Central Limit Theorem from Chapter 8. It states that *any* distribution formed by summing a multitude of equal-sized random samples from the same infinite population is approximately normal. The binomial distribution is formed in just such a manner. The infinite population from which the samples are taken is made up of values assumed by a dichotomous variable. By representing successes by 1 and their nonoccurrence by 0, we arrange for the sum of each sample to be the frequency of the occurrence of interest.

The probabilities of various events occurring in a binomial distribution can be determined by applying Theorem 9.1. Some difficulties are encountered because the *continuous* normal curve is used to approximate the *discrete* binomial graph. The discrepancy is traditionally minimized by using what is called a .5 correction factor (read "point five"). The need for such a factor can be illustrated by considering a drug that cures a specific disease 75% of the time ($P = .75$). If the drug is administered to 100 patients suffering from the malady, the probability that 75 or fewer or that 76 or more will be cured is 1. It is an assured happening. However, the area when shaded under the approximating normal curve is less than 1 since a "gap" is left between 75 and 76 (Fig. 9.7).

Fig. 9.7

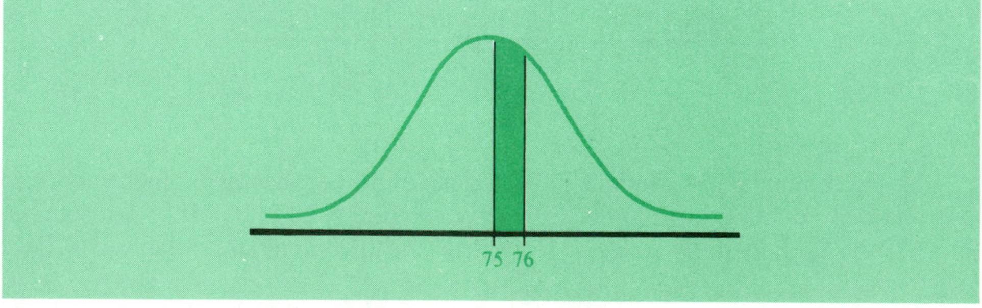

[5]The value of the mean and also the value of $N(1-P)$ should be at least 5 for the approximation to be satisfactory.

In fact, since the binomial only assumes whole number values, many gaps appear under the normal curve that approximates it. To correct this situation *we let the interval extending .5 unit either side of a number represent that number under the normal curve.*

With this correction in mind, let us find the probability that, using the previously discussed drug ($P = .75$), 160 or more of 200 patients will be cured. Since $NP = 200(.75) = 150 \geq 5$ and $N(1 - P) = 50 \geq 5$, we are justified in applying Theorem 9.1 and using the normal distribution to approximate the binomial. If X represents the number cured, the required binomial probability is that $X \geq 160$. This is approximated by the normal probability that $X \geq 159.5$. Note that this includes the interval representing 160 (159.5 − 160.5). Since the distribution in question has a mean of $NP = 200(.75) = 150$ and a standard deviation of $\sqrt{NP(1-P)} = \sqrt{150(.25)} = 6.12$, we are interested in the probability that the z score is greater than $\dfrac{159.5 - 150}{6.12} = 1.55$. Reference to Appendix III shows a probability of .0606 that 160 or more patients will be cured (Fig. 9.8).

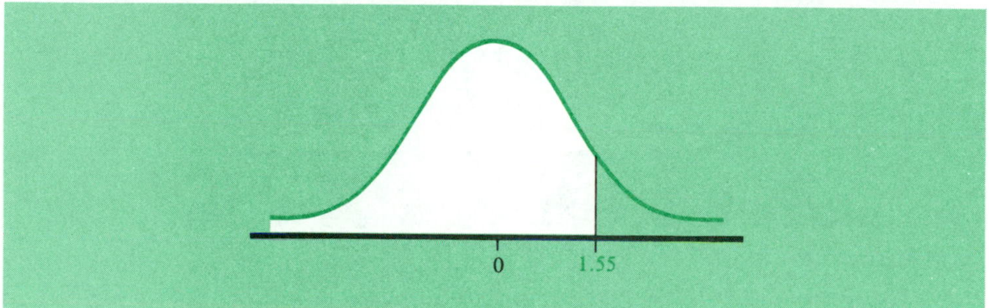

Fig. 9.8

Unlike the normal distribution in which every term comprises a negligible percent of the terms, the binomial distribution allows us to calculate the probability that a term selected at random will equal some constant. The probability that the drug being discussed will cure exactly 160 of 200 patients is the area under the normal curve between 159.5 and 160.5. By using $\mu = 150$ and $\sigma = 6.12$, we can convert to z scores of 1.55 and 1.72. The area between these scores (.0179) can be conveniently calculated using methods discussed in Chapter 7 (Fig. 9.9).

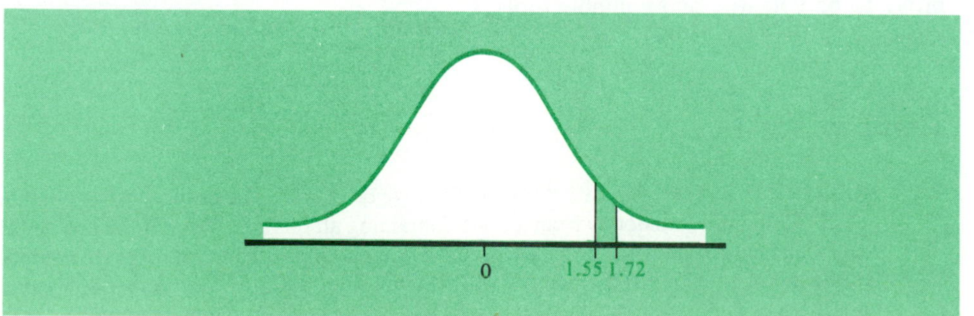

Fig. 9.9

The binomial distribution and its normal distribution approximation are so important that we shall consider another illustration. This time suppose it is known that 35% of high school students consider themselves shy. We shall consider a success to be that a student considers himself or herself to be shy ($P = .35$). Of 200 high school students selected at random, what is the probability that the number who would identify themselves as shy is between 65 and 80, including 65 and 80? Unlike cases involving the normal distribution, whether or not we include the endpoints of an interval affects the final answer. It should be noted at this point that the normal approximation may be used since $NP = 200(.35) = 70 \geq 5$ and $N(1-P) = 200(.65) = 130 \geq 5$.

Now the binomial probability we are seeking is the probability that $65 \leq X \leq 80$. The correct normal approximation is the probability that $64.5 \leq X \leq 80.5$, which is equal to the area shaded in Fig. 9.10.

Fig. 9.10

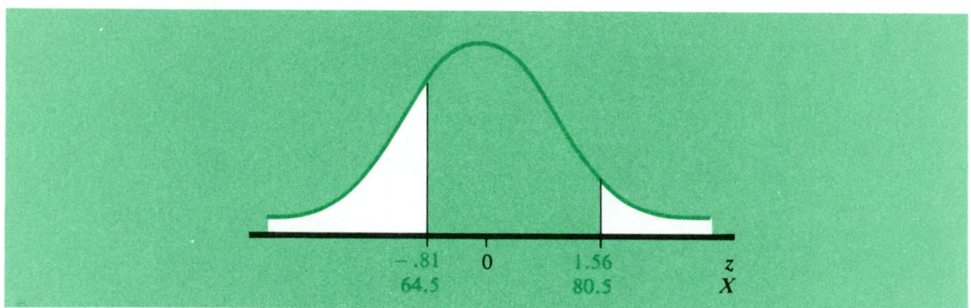

Using the mean $NP = 200(.35) = 70$ and standard deviation $\sqrt{NP(1-P)} = \sqrt{70(.65)} = 6.75$, we can transform the interval endpoints to z scores of $\dfrac{64.5 - 70}{6.75} = -.81$ and $\dfrac{80.5 - 70}{6.75} = 1.56$. Appendix III shows that .7316 [.2910 + .4406] of the area under the normal curve falls between the two z scores. So the required probability is also .7316.

Binomial distributions will greatly increase the number of situations about which we can make probability statements. They and distributions of sample means will prove to be versatile and valuable tools.

Problem Set A

9.1 Raffle tickets numbered from 1 through 728 have been sold. The draw consists of taking one of these tickets randomly from a crate containing all of them. What is the probability that (a) the number 628 will be chosen? (b) the number 545 will be chosen? (c) the number will be less than 225? (d) the number will be more than 600? (e) the number 839 will be chosen? (f) a number in the seven hundreds will be chosen? (g) the number will be either less than 125 or greater than 725? (h) the number will be either less than 50 or greater than 700?

9.2 Three tennis matches are arranged, each between American and Russian players of similar ability, training, and experience.
   a. Draw a probability tree and write the sample space for the experiment that will help answer the following questions.
   b. What is the probability that the Americans will win at least one match?
   c. What is the probability that the Russians will win two of the three matches?

9.3 According to the American Cancer Society, a 25-year-old male who smokes a pack of cigarettes per day gives up a mean of 5.5 years of life. Assuming the number of years given up to be normally distributed with a standard deviation of 1.25 years, what is the probability that a male who begins smoking a pack a day at age 25 will decrease his life span by (a) more than six years? (b) less than two years? (c) between three and seven years?

9.4 Under normal conditions, 85% of a certain strain of rosebush flower the first year. A landscape artist plants 150 bushes. Find the probability that more than 90% of them will flower before a year has passed.

9.5 The times that adults over the age of 21 spend daily reading a newspaper are distributed with a mean of 17.5 minutes and a standard deviation of 3 minutes.
   a. What is the probability that the mean time for 49 randomly selected adults is more than 18 minutes per day?
   b. Suppose the mean for 64 randomly selected adults is 20 minutes. What might you conclude?

9.6 The Internal Revenue Service reports that approximately 55 of 100 returns filed have no errors. What is the probability that of the 300 returns Agent Olson is reviewing, at least 170 of them will be error-free?

9.7 The burning times of candles produced by the Flasho Candle Company are found to be normally distributed with a mean time of 92 minutes and a standard deviation of 23 minutes. Find the probability that a single candle chosen at random from this population will burn (a) over 120 minutes, (b) between 60 and 120 minutes, (c) under 40 minutes, (d) neither under 50 minutes nor over 100 minutes.

9.8 The operators of the Ding-A-Ling Telephone Company receive a mean of 28 requests for information per day. The number of such calls is approximately normally distributed with a standard deviation of 7. Find the probability that 15 operators selected at random will receive a mean of 30 or more requests per operator in a single day.

9.9 Harry Househusband wants to have a cake ready when his wife returns from the saltmines. The recipe calls for two eggs. Being a relatively inexperienced cook, Harry cracks the eggs right into the cake batter rather than using a side bowl to be sure the eggs are fresh. Unfortunately, three of the five eggs in Harry's refrigerator are rotten. He selects two at random and adds them to the other ingredients.
   a. Write the sample space for this experiment using R_1, R_2, R_3 for rotten eggs and F_1 and F_2 for fresh eggs.
   b. What is the probability that Harry will ruin his cake by adding at least one rotten egg?
   c. Find the probability that no rotten eggs are selected.
   d. Compare the answers to b and c. How are they related?

9.10 A drug company reports that the miscarriage rate for pregnant women using a new drug to fight morning sickness is 5%. Dr. Rich prescribes the drug for 150 pregnant women. What is the probability that none of the women will miscarry?

9.11 Ethel Eagle-Eye has been a proofreader for her high school newspaper, *Central Chatter*, for the past three and a half years. The number of errors she finds per issue is approximately normally distributed with a mean of 35 and a standard deviation of 8.

a. Why is the distribution in question at best "approximately" normal rather than perfectly normal?
b. Find the probability that for 25 issues selected at random, the number of errors found by Ethel will have a mean of 39 or more per issue.

9.12 According to Tricity Police Major A. R. Rester, the number of tickets given to pedestrians in a one-month period for jaywalking and other traffic-related violations is normally distributed with a mean of 75 and a standard deviation of 12. For a month selected at random, what is the probability that the number of tickets written will be (a) between 60 and 90? (b) less than 55? (c) equal to 75? (d) greater than 80 but less than 85? If the monthly totals are used to compute yearly means per month, what is the probability that (e) this year's mean will exceed last year's mean of 77 tickets per month?

9.13 A car with four sparkplugs has one that is defective.

a. If one of the plugs is replaced at random, what is the probability that the right one will be replaced?
b. If two plugs are replaced at random, what is the probability that the defective one will be replaced?
c. If two plugs are defective and two are replaced at random, what is the probability that both defective plugs will be replaced?

9.14 The mean cost for a college text is $15.85. Assuming that the costs are normally distributed, with a standard deviation of $3.85, find the probability that the eight texts required in Joe College's classes will cost less than or equal to the $120 he has budgeted. (Hint: This means the mean cost per book must be less than $120/8.)

9.15 Historically, 65% of Senator Pete Low's constituents have approved of his decisions. In a recent random sample of voters in his district, more than 65 of 90 people approved of his action on the state banking amendment. What was the probability of this occurring (based on his historical approval rate)?

Problem Set B

9.16 Professor Scrooge gives his classes a surprise five-question true-false quiz that is so difficult that nobody has the slightest idea what any of the correct answers are. Find the probabilities that Ty Agin's quiz will have the following numbers of correct responses.

a. five   b. zero   c. three or more   d. two or three

9.17 The probability of a convicted felon committing a crime within five years after participating in a state rehabilitation program is .2. Of 300 inmates recently released from the program and prison, what is the probability that more than 50 will not be law-abiding during the next five-year period?

9.18 A rat maze has one entrance and three equally accessible exits, A, B, and C.

a. What is the probability that Ralph Rat will exit the maze at C?
b. Find the probability that Ralph and Rita Rat will both exit at C (in separate experiments).
c. What is the probability that at least one of the two rats will exit at A?
d. What is the probability that at least one of the two rats will exit at A or B?

9.19 The weights of women's handbags are normally distributed with a mean of 4.2 pounds and a standard deviation of .7 pound. Find the probability that a random sample of handbags has a mean in excess of 4 pounds if the sample contains (a) 9 handbags, (b) 49 handbags, (c) 144 handbags.

9.20 Scores on a dental aptitude test are normally distributed with a mean of 65 and a standard deviation of 15. One of the vast number of subjects who took the test is to be chosen at random. What is the probability that his or her score will be (a) above 88? (b) below 88? (c) between 58 and 88? (d) either below 58 or above 88? (e) between 57.5 and 72.5? (f) over 95?

9.21 Librarian Marion Clarion receives and processes a mean of 850 pamphlets from numerous sources each week. The number received varies with a standard deviation of 150. What is the probability that the weekly totals for an entire year would have a mean between 825 and 870?

9.22 The first-year drop-out rate at State U. is 35%. A random sample of 50 freshman students is monitored and the number of dropouts is recorded at the end of the year. Find the probability that 20 or more will not be on the school rolls.

9.23 Slicker Sales, Inc., sells a mean of 975 lots of swampland per year with a standard deviation of 125. Assuming normality, what is the probability that this year's sales will surpass the goal set by the corporation president, I.M. Slicker, of 950 lots? What is the probability that the company will sell exactly the 950 lots demanded by President Slicker?

9.24 CB radio license applicants in a certain state must wait a mean of 55 days after application to receive their licenses. If the waiting time is distributed with a standard deviation of five days, what is the probability that 36 randomly selected applicants will have a mean waiting period of less than 53 days?

9.25 A cafeteria offers a choice of any one of five entrees and any one of four vegetables. If every combination plate must appear in the serving line, how many types of plates must be prepared? (Use a probability tree.) What is the probability that the first two people served will choose the chicken entree?

9.26 Women make up approximately 30% of law school enrollments in public institutions throughout the region. On the basis of this figure, what is the probability that the university's next class of 75 students will be between (and including) one-third and two-fifths female?

9.27 One of the plants of Slippery Oil, Inc., refines a mean of 200,000 barrels of crude oil per day. Its daily production is normally distributed with a standard deviation of 25,000 barrels.

a. To meet its projected quota, this particular refinery is expected to produce a mean of 195,000 barrels per day for the next 90 days. What is the probability that the refinery will fail to meet its production quota?

b. If the probability that the plant meets its production quota for the time period is .75, approximately what quota was set?

9.28 A leading psychologist has devised a short test designed to measure a person's ability to concentrate deeply for a short period of time. The test consists of locating a sequence of letters in a large grid. The time required for individuals over 18 years of age to complete the test varies normally with a mean of 4 minutes and a standard deviation of 51 seconds. Find the probability that a person taking the test will complete it in (a) less than 3 minutes, (b) less than 3 minutes but more than 2.5 minutes, (c) not more than 5 minutes. (d) Find the probability that for 100 people taking the test, the mean completion time will be between 4 minutes 9 seconds and 4 minutes 51 seconds.

9.29 The city commission is accepting bids on three street projects. Three companies submit proposals but the sealed bids will not be opened until the regular commission meeting in two weeks. Assume that no additional bids will be submitted and no information is available concerning the bids already submitted.

a. What is the probability that at least one of company A's bids will be accepted?
b. What is the probability that the same company will be awarded all three contracts?
c. If a company can be awarded at most two contracts, find the probability that company A will receive at least one.

9.30 Assuming that finance companies require approximately four out of five loans granted to be cosigned, what is the probability that of the last 50 loans granted by Frugal Finance (a) more than six were granted without a cosignature? (b) exactly 45 required a cosignature?

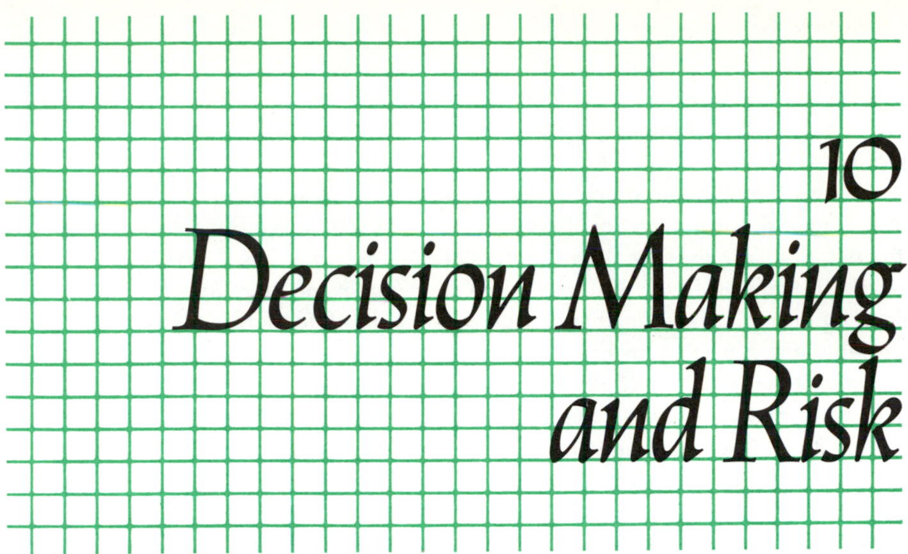

10.1 Introduction

We are constantly faced with the necessity of making decisions, large and small, and acting in accordance with them. Sometimes the inferences that lead to these decisions are made unconsciously and the decisions that spring from them resemble reflex actions. This is true, for instance, where a traveler steps to the side of the road upon hearing the noise of a vehicle behind. It is often true where one decides which tie or sweater to wear on a given day or which elective courses to take in a given semester. However, sometimes the evidence considered is quite well defined, and prolonged consideration is needed before a conclusion can be reached and action taken.

Very important decisions are made every day by the management of large corporations and by other persons in positions of responsibility. In many cases a knowledge of probability and sampling procedures plays an important role in the making of these decisions. In such cases inferences are made based on certain available evidence. The purpose of this chapter is to explain some of the theory behind the making of such inferences and in particular some of the risks taken when they are made.

Two prime considerations are present whenever an action is to be made. First, one must consider the probability that the inference that prompted the action is correct. Second, one must weigh the possible gains and losses. Statistics enables us to find the probability values attached to different kinds of inferences. However, it usually tells us nothing regarding how costly a mistake will be if a decision turns out to be incorrect.

When there is a choice of alternatives, it can sometimes be surmised that a mistake in one direction will be roughly as costly as a mistake in the other. Then it is reasonable to choose the action with the higher probability of being correct. However, sometimes when there are alternative actions, a mistake in one direction threatens to be much more costly than a mistake in the other. Then the potentially costly action should be taken only when it has a considerably higher probability of being correct.

It follows that the same probability value might indicate a particular action in one situation and not in another. For instance, a stock market investor who already has diversified holdings is apt to purchase a commodity that has a 7/10 probability of increasing its value. However, suppose that a man is on trial for murder and the evidence against him is such that 7 out of every 10 defendents in his position are guilty in the long run. The probability is 7/10 that this defendant is guilty, but under these circumstances a jury would most likely vote for acquittal.

The reason has to do with the stakes that are involved. The penalty for a faulty conviction, the life of an innocent person, threatens to be much greater than the penalty for a mistaken acquittal. Because of this imbalance, the decision would most likely be to exonerate the defendant despite the 7/10 probability that he is guilty. The jury, it is hoped, would reach the verdict with the lesser probability of being correct.

A make-believe incident dramatizes the importance of the stakes of the game in decision making. Consider the plight of Tom, a boy stranded in a strange town. The bus back home costs a dollar and Tom has only 99 cents. He is too proud to beg or borrow and now he is both angry and sad as he ambles toward the edge of the city.

Tom has always liked to flip coins for small sums. Soon he approaches a boy about his own age. "I'm gonna flip a coin and you take your choice," he says. "Guess either heads or tails when the coin is in the air. The winner gets a penny."

The boy, however, refuses to play and so do several others whom Tom approaches.

At this point our hero is about to hurl his 99 cents into a nearby lake when a daring plan enters his mind. The last boy he challenged looked hesitant before refusing his offer. Tom accosts him once more.

"Same deal as before," Tom says. "But this time if you win you get 99 cents and if you lose you pay me just one cent."

The deal is made. Tom's 99 cents and the boy's penny are put under a rock. Now the coin is in the air.

"Heads," the boy cries, but the coin falls tails. Tom scrambles to pick up the money—99 cents of his own and the penny he has won.

"Let's try it once more," the boy asks. But by now Tom has already scampered down the road on his way to the bus terminal.

The decision that Tom made was to risk 99 cents in order to possibly win a penny if the other boy did not guess the outcome of the coin correctly. The venture would ordinarily be a foolish one, but the stakes of the game made it worthwhile. The probability was 1/2 that Tom would win the bet, but from his point of view Tom was actually getting incredible odds in his favor. The loss of 99 cents would not have changed his stranded status. The conquest of one penny meant that he was no longer stranded.

Note that the gamble was wise for the other boy too. For him the stakes of the game were favorable. He could win 99 cents and stood to lose only one cent and he had the same chance to win as to lose.

It often happens that a gambling or business venture is favorable to both parties, since their stakes are likely to be different. To elaborate would bring us into the worlds of gambling and finance. The purpose here is merely to sensitize the reader to the importance of stakes. Their influence is ever present, even in the realm of science.

In experimental work, there are stakes in avoiding different kinds of errors. For instance, there may be a heavy penalty for accepting the thesis that a drug is beneficial when it is not. Thus there is a stake in not making this error. On the other hand, the risk of failing to discover a useful drug because of exaggerated skepticism is also to be avoided. We shall see in the pages to follow how an experimenter runs various risks of error which he or she cannot avoid entirely. In fact, the more one reduces the chance of one kind of error, the more one increases the chance of another. The experimenter must consider carefully the penalties that may accrue from making each kind of error and then choose a procedure. Often the approach that reduces the risk of making one kind of error entails increasing the risk of making another kind—but one that would be less costly.

10.2 Hypothesis Testing

In any branch of science, it becomes necessary to evalute hypotheses that arise. Sometimes these hypotheses, or conjectures about the nature of phenomena, are suggested by experimental findings. Sometimes they grow out of practical work in a field.

A hypothesis has the status of a bill presented to the legislature. We are the decision-making body and we must evaluate the hypothesis to determine whether to accept or reject it.[1] **Our procedure is to carry out what is called a *hypothesis-testing experiment.* This experiment is a formalized procedure for gathering and interpreting evidence relevant to deciding whether to accept or reject a hypothesis.** We shall discuss this procedure and later we shall present different applications of it, designed to illustrate tests of different kinds of hypotheses.

To clarify some issues that arise when a hypothesis is tested, consider the simple hypothesis that a particular coin is fair. This hypothesis states that the probability is 1/2 that the coin will fall heads when flipped. Imagine that an experimenter takes the coin and walks to the front of a huge auditorium filled with scientists. It is decided that the experimenter will flip the coin into the air 100 times in plain view of all. The outcome of each flip will be announced, and when the experiment is completed the assemblage will decide either to accept or reject the hypothesis that the coin is fair.

The experiment begins and the audience is silent. The coin is flipped into the air over and over again. After each flip, the coin is shaken up in a little box from which it is thrown the next time. Thus each flip (or observation) is independent. As the experiment proceeds, a giant scoreboard records the number of heads and tails.

[1]As we shall discover in Chapter 11, there is some objection to the use of the word "accept" in this context. "Not reject" is actually preferred. Until the reason why becomes clear, we shall use "accept," as it allows a clearer explanation of the concepts we are now considering.

When the experiment is over, the scoreboard reads:

HEADS: 55 TAILS: 45

The question is now put to the learned audience: "Judging by what you have seen, do you accept or reject the hypothesis that the coin is fair?"

Dr. R, a young researcher, rises from her front-row seat. "Reject the hypothesis," she says. "The evidence contradicts the hypothesis that the coin is fair. The number of heads obtained exceeded the number of tails by enough margin to indicate that the coin is biased."

Unexpectedly, Dr. A, an elderly man, stands up at the back of the auditorium. "On the contrary, accept the hypothesis that the coin is fair. One cannot condemn a coin that turns up 55 heads and 45 tails. A fair coin may turn up in that ratio. Our evidence is consistent with the hypothesis that the coin is fair."

Dr. R is becoming angry. "What would it take to get you to reject the hypothesis?" she asks. "I mean what kind of disparity between heads and tails would convince you that a coin is unfair?"

Dr. A replies, "At least 90 heads as opposed to 10 tails or 90 tails as opposed to 10 heads." He flicks on a cigarette lighter and holds it to his pipe. "Remember that I am much older than you and have seen much more. A fair coin when tested may frequently show a great disparity between heads and tails." (Dr. R is becoming quite angry, but dares not interrupt the speaker, who is often referred to as the dean of American scientists.) Dr. A continues: "If our policy is to reject a coin that shows a 55 to 45 ratio or more, this policy will lead us to condemn many fair coins that by chance happen to show up that way."

Dr. A is choosing his words slowly, because he is poised and because he senses that they are infuriating Dr. R. "I am sorry. You make it too easy to condemn a fair coin. Your plan will too often lead us to reject the hypothesis simply because some chance phenomenon occurs during an experiment."

Young Dr. R is now at the bursting point. "Why, that's ridiculous. You demand that a coin show a 90 to 10 ratio before you'll say that it is biased. Your overcredulity will make it almost impossible ever to reject a hypothesis about a coin. Granted, we will seldom reject a fair coin. But we will find it too hard to reject even a lopsided coin. Even lopsided coins do not usually show a disparity as great as 90 to 10. Following your plan, we will repeatedly make the mistake of accepting the hypothesis that a coin is fair when in actuality it is lopsided."

The arguments of Dr. A and Dr. R are both inherently sound, but the stakes of the two scientists differ. To focus on the difference, suppose that the experiment is repeated with a different coin each day over a long time period. Naturally it is not known whether the coin used on a particular day is fair or biased.

Dr. A's rule is to accept the hypothesis that a coin is fair unless the ratio in an experiment is as unequal as 90 to 10. Presumably, Dr. A abhors the error of condemning a fair coin. He has gone out of his way to minimize the chance of making this kind of error. He runs little risk of condemning a fair coin, since a fair coin will almost never show a ratio as unequal as 90 to 10. However, the price he pays is to lower greatly the power of detection of the experiment. As statisticians say, he has made it extremely difficult to reject the hypothesis. A large number of biased coins will not show a disparity as great as 90 to 10 and consequently these coins will go undetected. When Dr. A happens to be testing a biased coin, he runs considerable risk of mistakenly accepting the hypothesis that the coin is fair. Dr. A may be said to have a predilection to accept hypotheses. In fact, A is short for "Accept." As a consequence, when the hypothesis is true, he will seldom make the error of rejecting it. However, when the hypothesis is false, Dr. A will often make the error of accepting it. Dr. A is like the person who says "You can do anything you want but I have faith in your innocence." He refrains from condemning. He has the virtue of accepting true hypotheses even when coincidental factors throw them in a bad light. But he has the failing of being slow to revise an opinion even when considerable evidence contradicts it.

On the other hand, let us say that *Dr. R's rule is to accept the hypothesis that a coin is fair unless the ratio is as unequal as 55 to 45*. That is, she will accept the hypothesis only if the coin performs within narrow limits. Otherwise, she rejects the hypothesis. Dr. R. views the experiment like a government security check conducted during the threat of national calamity. She deems it essential to have high power of detection of biased coins. Her plan provides for minimum risk of accepting a false hypothesis. The price is to increase the number of times that fair coins will be rejected.

Dr. R may be described as having a readiness to reject hypotheses. In fact, R is short for "Reject." When the hypothesis happens to be false, she seldom makes the error of accepting it. When the hypothesis happens to be true, she will often make the error of rejecting it. Dr. A and Dr. R evaluate the context of the coin experiment differently. They disagree on the costs of making different kinds of errors. Thus they specify different cut-off points to indicate when to accept or reject the hypothesis in the experiment.

The ideal hypothesis-testing experiment obviously would always lead an experimenter to the correct decision. When a hypothesis happened to be true, the decision would be to accept it. When a hypothesis happened to be false, the decision would be to reject it. Knowing that an ideal experiment was being conducted, the experimenter could always bank on his or her decision. It would be known that when a hypothesis was accepted, it was in reality true, and that when rejected, it was in reality false. Of course this ideal experiment can never actually be conducted, because in any practical experiment there are inevitable risks of error.

In practice, two possibilities must be considered when a decision is made. Suppose the decision is to reject the hypothesis:

a. One possibility is that this decision is correct. In reality the hypothesis is false and deserved to be rejected.
b. The other is that the decision to reject is in error. The hypothesis is true but the finding from the sample gathered deluded the experimenter into rejecting the hypothesis. An error of this kind is called a *type-one error*.

Definition 10.1 When a hypothesis is true but the experimenter mistakenly rejects it, the experimenter is said to be committing a type-one error.

Suppose, on the other hand, the decision is to accept the hypothesis:

a. One possibility is that this decision is correct. In reality the hypothesis is true and deserved to be accepted.
b. The other is that the decision to accept is in error. The hypothesis is false, but the experimenter made the error of accepting it. An error of this kind is called a *type-two error*.

Definition 10.2 When a hypothesis is false but the experimenter mistakenly accepts it, the experimenter is said to be making a type-two error.

Table 10.1 Types of Errors

Decision	Hypothesis: True	Hypothesis: False
Reject	**Type I Error**	No Error
Accept	No Error	**Type II Error**

10.3 The Long-Run Outcomes with a Fair and a Biased Coin

Quite naturally, when any given experiment is carried out, the experimenter who has decided about accepting or rejecting the hypothesis does not know for certain whether his or her decision is correct. One cannot properly compare for accuracy Dr. A's decision to accept the coin hypothesis with Dr. R's decision to reject it, making reference only to what happened in the single experiment described. The coin, so far as we know, may have been either fair or biased, so that either one of them may have been correct.

As usual, the way to gain full insight is to imagine that the experiment described is carried out each day over a long time period. To begin with, let us suppose that a fair coin ($P = .5$) is used in carrying out the same experiment each day for a great many days. That is, each day a fair coin is flipped into the air 100 times and the number of heads is recorded. The fact that the coin is fair is being arbitrarily assumed by us, although of course this fact is not known to either Dr. A or Dr. R. In this way we can see what would happen were Dr. R to continue to use her rule and were Dr. A to continue to use his.

On any given day, whether it be the first or any other, the coin is flipped into the air exactly 100 times. (Each experiment consists of 100 flips and we shall consider an experiment as a sample for which $N = 100$.) On each given day the fair coin yields some number of heads, which is then recorded, and this total number is perceived by both scientists. The obtained number of heads is the only fact that becomes known to the two scientists, each of whom is left to make his or her own decision about whether the coin is fair.

Suppose that on the first day the coin shows 52 heads. This outcome may be indicated by placing a marker on a horizontal axis (see Fig. 10.1). Incidentally, on this particular day the score of 52 heads would lead Dr. A and Dr. R each independently to the correct conclusion that the coin is fair.

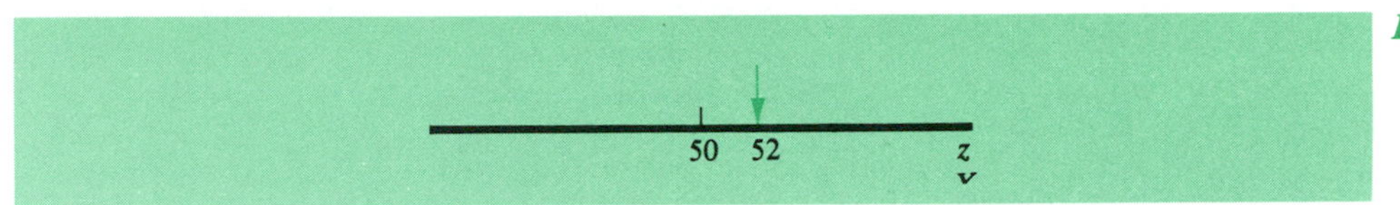

Fig. 10.1

On the next day the outcome might be 47 heads and the diagram would look like Fig. 10.2.

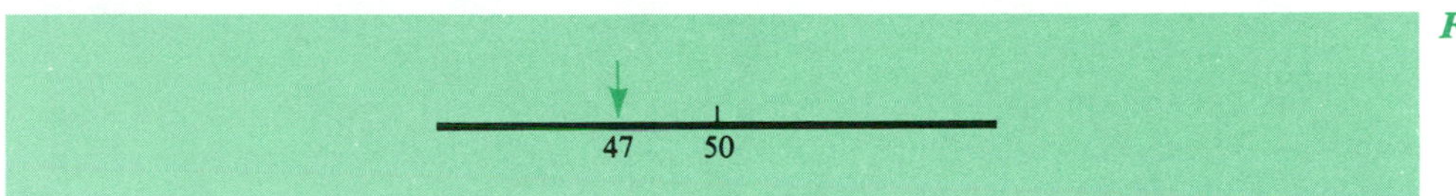

Fig. 10.2

Now suppose that the same experiment is performed on a vast number of successive days. The number of heads obtained on a given day would as a rule be close to 50, although on occasion it would be far removed. (We shall use the value X to refer to the number of heads obtained on a given day, so that for the first day $X = 52$ and for the second day $X = 47$.) **In the long run, the different numbers of heads (or different values of X) would form a binomial distribution, which, according to Theorem 9.1, may be approximated by a normal distribution with a mean of $NP = 100(.5) = 50$ and a standard deviation of $\sqrt{NP(1 - P)} = \sqrt{50(.5)} = 5$.** Our use of the approximation is justified by the fact that both NP and $N(1 - P)$ equal 50, which is greater than 5. The approximating normal curve is shown in Fig. 10.3.

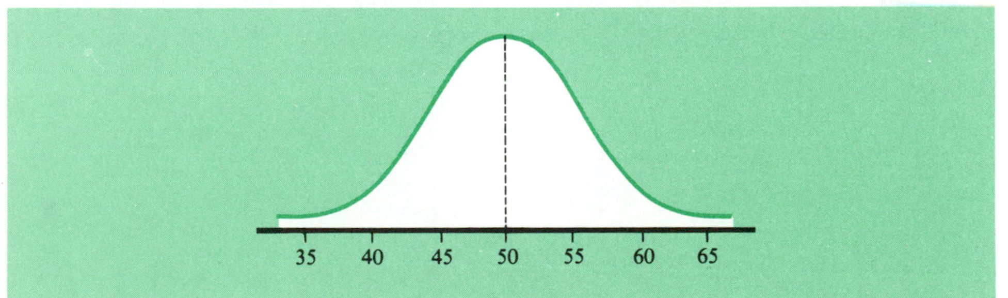

Fig. 10.3

We are interested in seeing what would happen with the rules of Dr. A and Dr. R if a fair coin were used repeatedly for all experiments. It must be emphasized that both are testing the hypothesis that the coin is fair. The only distinction is in their readiness to reject that hypothesis.

First consider Dr. A's rule, which is to accept the hypothesis if the number of heads obtained in the experiment is between 10 and 90. (Actually, the endpoints are 10.5 and 89.5 since we are approximating the binomial.) According to this rule, only if the number of heads is 10 or less, or 90 or greater, will Dr. A reject the hypothesis that the coin is fair. Dr. A's rule is shown in Fig. 10.4.

Fig. 10.4

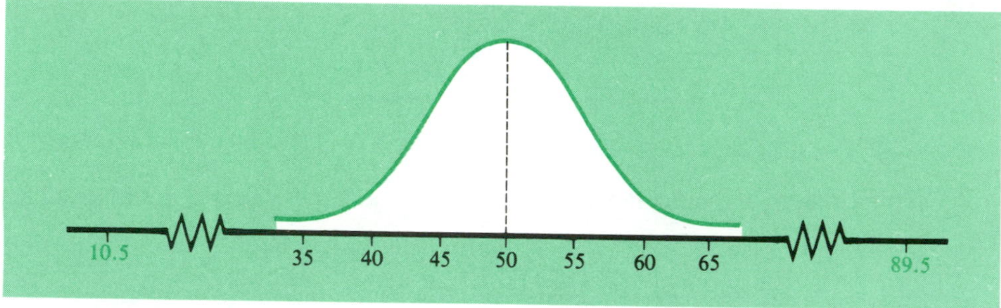

This illustration shows that when a coin is fair it will virtually always show a number of heads between 10 and 90 in an experiment. Since the acceptance region in Dr. A's rule covers 8 standard deviations on each side of the mean, a fair coin will almost never deviate from 50 so far as to lead Dr. A to reject his hypothesis mistakenly. In other words, he will hardly ever commit a type-one error.

Now let us consider the decision rule that Dr. R has advocated, which is to accept the hypothesis only if a coin shows heads more than 45 but less than 55 times. Fig. 10.5 shows what would happen if a fair coin were used repeatedly, this time in connection with Dr. R's decision rule.

We can compute the probability that a fair coin will show a finding in the acceptance region using Theorem 9.1. The endpoints of the acceptance region are 45.5 and 54.5, and each of these numbers differs from the mean of 50 by less than one standard deviation of 5. Theorem 9.1 allows us to change z scores: $(45.5 - 50)/5 = -.9$ and $(54.5 - 50)/5 = .9$. Appendix III tells us that the acceptance region contains about 63% of the total area ($.3159 + .3159 = .6318$) and that the shaded areas of Fig. 10.5 contain together about 37% of the total area. In

Fig. 10.5

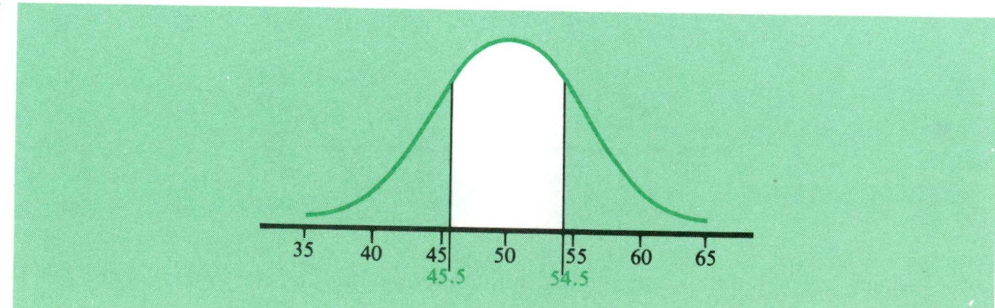

other words, the probability is approximately 63/100 that a fair coin will produce a sample in which the number of heads is in the acceptance region.

Of importance is the fact that the probability is about 37/100 that a fair coin will produce a sample in which the number of heads is in one of the rejection areas designated by Dr. R. This is the probability that even though the hypothesis is true, the finding will be such that Dr. R will mistakenly reject the hypothesis. In other words, Dr. R has set up a rule such that when a coin is fair, the probability of her making a type-one error is about 37/100.

So far we have considered what happens in the long run when the coin is fair. We have shown that Dr. A would over the long run virtually never reject the hypothesis mistakenly or commit a type-one error. Dr. R would over the long run mistakenly reject the hypothesis about 37% of the time, so that when a coin is fair the probability of her committing a type-one error is about 37/100.

Now suppose that the coin is biased—that is, its probability of turning up heads on a given flip is some fraction other than ½. Once again we are going to suppose that the coin experiment is repeated on a vast number of days, this time with the biased coin. But if the coin is not fair, then it may be biased to any degree. For instance, its probability of turning up heads on a flip might be 1/5 or 6/10 or 9/10 or any fraction. We have seen that the statement that the coin is fair was the specification of a particular value for P. However, the statement that the coin is biased is not a specification of a value for P, because a coin may be biased in any of an infinitude of ways. As we shall see, it typically occurs that a hypothesis may be true in only one way and false in many.

We want to evalute both Dr. A's and Dr. R's rules to determine how successfully each will manage to reject the hypothesis, now that it is proper to do so. However, the assumption that the coin is biased is insufficient to generate a picture of what would happen, so we cannot proceed yet. We must first specify a particular bias for the coin so that we can then use Theorem 9.1 to see what would happen.

Suppose for instance that the coin's probability of falling heads is 6/10 ($P = .6$). Again we are justified in using the normal approximation since $NP = 60 \geq 5$ and $N(1 - P) = 40 \geq 5$. The experiment with this biased coin is carried out each day and the number of heads obtained each time is considered the outcome for the experiment for the day. According to Theorem 9.1 the distribution of outcomes has a mean of 60 [$NP = (100)(.6) = 60$]. It has a standard deviation of about 4.9 [$\sqrt{NP(1 - P)} = \sqrt{100(.6)(.4)} = 4.9$].

Fig. 10.6 shows the long-run distribution of outcomes with the particular biased coin.

In considering a coin with a specific bias, we can now determine how successful Drs. A and R would be in rejecting the fair-coin hypothesis. To begin with Dr. A, his decision rule was to accept the hypothesis so long as the number of heads obtained is between 10 and 90. One can see in Fig. 10.6 or compute using Theorem 9.1 that even a biased coin would, in virtually every experiment, produce a number of heads somewhere in Dr. A's acceptance region. In other words, when the coin's probability of falling heads is 6/10, Dr. A would virtually always mistakenly accept his hypothesis and thus make a type-two error.

Fig. 10.6

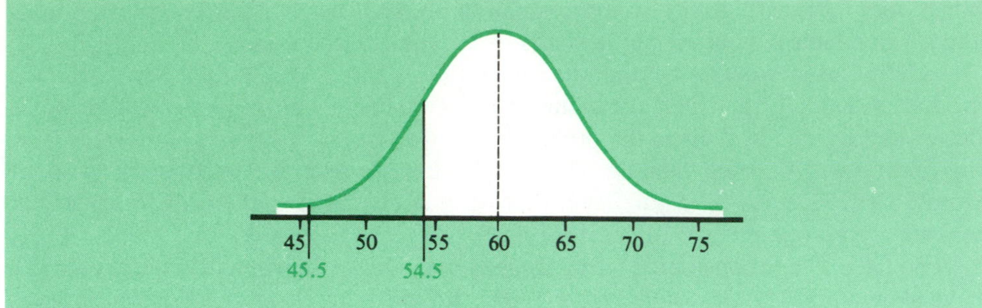

As for Dr. R, her decision is to accept the hypothesis only upon obtaining a finding somewhere between 45 and 55 heads. The area representing these findings is between 45.5 and 54.5 under the normal curve. She will mistakenly accept her hypothesis, making a type-two error, on those occasions when the biased coin produces a number of heads falling in this range. The shaded area in Fig. 10.6 shows the proportion of times that the biased coin will show a finding leading Dr. R to make a type-two error. Computing z values for each of the numbers 45.5 and 54.5 and using Appendix III, we find that the shaded area contains about 13% of the total area. In other words, on those occasions when a coin with a 6/10 bias is used, the probability is about 13/100 that Dr. R will mistakenly accept her hypothesis or commit a type-two error.

There is still one final angle. When contending with a coin of the bias described, the probability is about 87/100 that Dr. R will properly reject her hypothesis. This fact is important. With the same coin Dr. A had virtually no chance of properly rejecting the hypothesis, whereas Dr. R has a probability of 87/100 of doing so. The statistician says that *under the alternative described* (that is, a coin with a 6/10 bias) Dr. A has virtually no power to detect that the hypothesis is wrong, whereas Dr. R has considerable power to do so. Dr. R's test gives her a probability of 87/100 of properly rejecting the hypothesis in the case of the particular bias described. To put it another way, Dr. R's test is said to have a *power of 87/100 against the alternative described*. **The power of a test against a particular alternative is defined as the probability that the proper decision to reject will be made when the particular alternative occurs.**

To summarize, the test used by Dr. A would make his rejection of the hypothesis extremely rare. When the hypothesis is true, Dr. A will scarcely ever mistakenly reject it or commit a type-one error, but Dr. R will commit a type-one error 37/100 of the time. When the hypothesis is false and the coin happens to have a 6/10 bias, Dr. A will virtually always make a type-two error, which means that he will accept the hypothesis. His test gives him much less power of detection against the particular alternative than the test of Dr. R. The test of Dr. R leaves her a probability of 13/100 of making a type-two error and a power of 87/100 when the coin used has the particular bias.

We must remember that this discussion depends on our decision to consider what occurs with a coin of a particular bias. We have been considering only what would happen in the long run with a coin that in actuality has a probability of 6/10 of falling heads. Had we considered what would happen with some other kind of biased coin, we would have been led to consider a completely different distribution of outcomes, and for either Dr. A or Dr. R the chance of mistakenly accepting the hypothesis would have been different.

As one might expect, a coin with a more drastic bias would be even less likely to produce an outcome in the acceptance region of either person. For instance, let us consider an extremely biased coin, one that has a probability of 95/100 of falling heads. The distribution of outcomes of a vast number of experiments with such a coin would look like Fig. 10.7.

Fig. 10.7

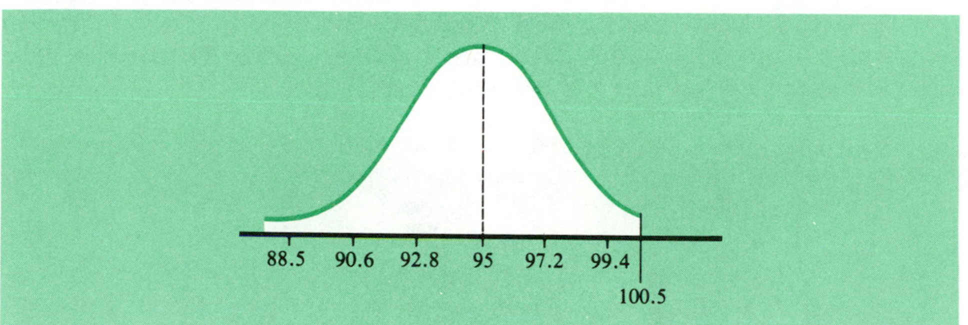

Both Dr. R and Dr. A are now virtually certain to reject the hypothesis. It should be seen that *the probability of either making a type-two error is dependent upon the particular alternative*—that is, the actual bias that happens to be the case. When the reality is particularly divergent from what is hypothesized, both are more apt to reject the hypothesis than when the reality only slightly belies the hypothesis. Another way to put it is to say that when the coin was only slightly biased both were more likely to be fooled into accepting the hypothesis. Now that we are assuming that the hypothesis is a ludicrous misstatement, both are more likely to reject it. It will be true for every decision rule we discuss that when the hypothesis is only slightly wrong the chance of making a type-two error is relatively great, but when it is very wrong the chance of making a type-two error will be relatively small. This property is favorable since it is usually less harmful to overlook a small inaccuracy than a large one.

10.4 Computing Errors Using Distributions of Sample Means

Many hypotheses are supported or refuted using evidence that comes from distributions other than the binomial. Consider a packing plant supervisor who suspects that one of his packing machines is malfunctioning. When operating properly, the machine processes 10 pounds of material per hour with a standard deviation of 1 pound. The supervisor decides to test the condition of his machine by randomly selecting 64 containers and weighing their contents. He decides ahead of time that if the mean per container is less than 9.75 pounds or more than 10.25 pounds he will call in a repairman. So the situation is as follows:

> Hypothesis: The machine is working properly and packing 10 pounds per container ($\mu = 10$).
>
> Acceptance Rule: Accept if mean weight for 64 containers is between 9.75 and 10.25 pounds.

Assuming the machine is working properly, what is the probability of a type-one error being made? That is, what is the probability the hypothesis is true ($\mu = 10$), but the sample of 64 gives results that do not support that conclusion? This happens if the mean of the sample is less than 9.75 or greater than 10.25 pounds. The proper area is shaded in Fig. 10.8.

Fig. 10.8

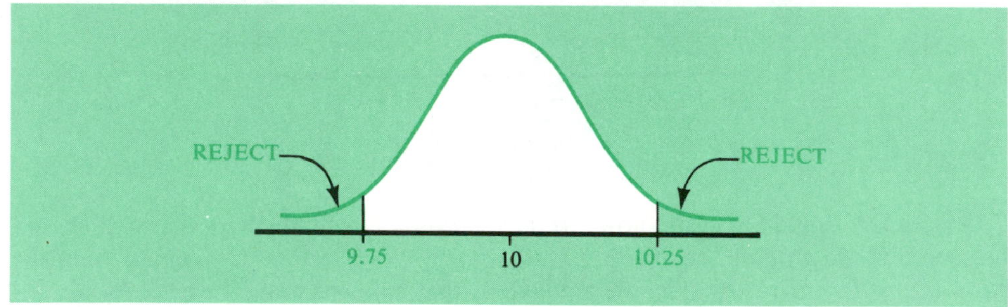

According to Theorem 8.2, the mean of the sample of 64 containers is a member of a normal distribution of sample means whose mean is 10 and whose standard deviation is $1/\sqrt{64} = .125$ ($\sigma_{\bar{X}} = \sigma/\sqrt{N}$). Changing to z scores $[(9.75 - 10)/.125 = -2$ and $(10.25 - 10)/.125 = 2]$ and consulting Appendix III, we find that the shaded area, and consequently the probability the supervisor will commit a type-one error, is .0456. Therefore, only about 4.6% of the time this test is applied will a repairman be called in unnecessarily.

On the other hand, suppose the machine is actually in need of repair and is only filling each container with a mean of 9.7 pounds with a standard deviation of .7 pound. What is the probability that the supervisor will commit a type-two error? Now the actual state of affairs is $\mu = 9.7$ and $\sigma = .7$. We must determine under these circumstances the chances of getting results from the sample that will lead to erroneous acceptance of the hypothesis. Putting it another way, what is the probability that a random sample of 64 terms from such a distribution will yield a mean between 9.75 and 10.25, the previously established acceptance interval? This time, according to Theorem 8.2, the distribution of sample means is normal with a mean of 9.7 and a standard deviation of $.7/\sqrt{64} = .0875$. The proper area is shaded in Fig. 10.9.

Changing to z scores $[(9.75 - 9.7)/.0875 = .57$ and $(10.25 - 9.7)/.0875 = 6.29]$ and consulting Appendix III reveals that .2843 of the time the supervisor will fail to have a defective machine repaired and thereby commit a type-two error.

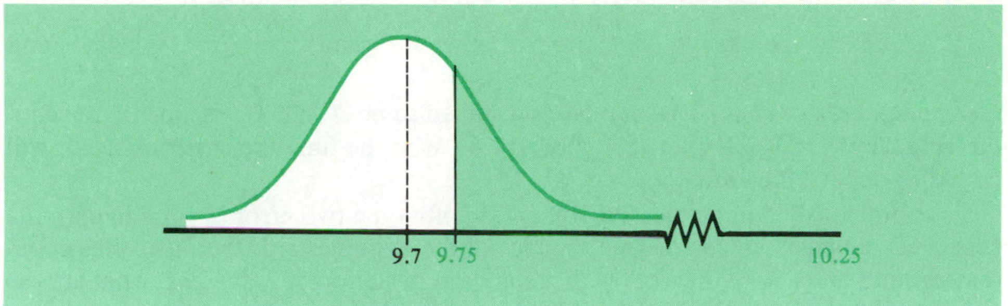

Fig. 10.9

The power of the test against the alternative $\mu = 9.7$ is .7157 (one minus the probability of a type-two error being committed). This indicates that the chance of detecting a malfunctioning machine with the supervisor's plan is .7157 if the machine is actually filling the containers with a mean of 9.7 pounds.

Let us consider one more situation in which a distribution of sample means is helpful in determining probability of error. This time the superintendent of a very large school system, in the interests of conservation, has announced that this winter all classrooms are to be kept at 67 degrees.

Because of insulation differences and varying types of classrooms, a standard deviation of 4 degrees will be allowed. Unless a random sample of 36 classrooms provides evidence to the contrary (a mean temperature below 66 degrees or above 68 degrees), the superintendent will be satisfied that the new temperature guidelines are being adhered to. The situation is as follows:

Hypothesis: The mean classroom temperature is 67 degrees ($\mu = 67$).

Acceptance Rule: Accept if mean temperature for 36 classrooms is between 66 degrees and 68 degrees.

To determine the probability of a type-one error, the superintendent must assume the hypothesis is true ($\mu = 67$) and determine the probability that the 36 classrooms will nonetheless have a mean temperature in the rejection region (shaded in Fig. 10.10). This distribution of means has a mean of 67 and a standard deviation of $\dfrac{4}{\sqrt{36}} = .67$.

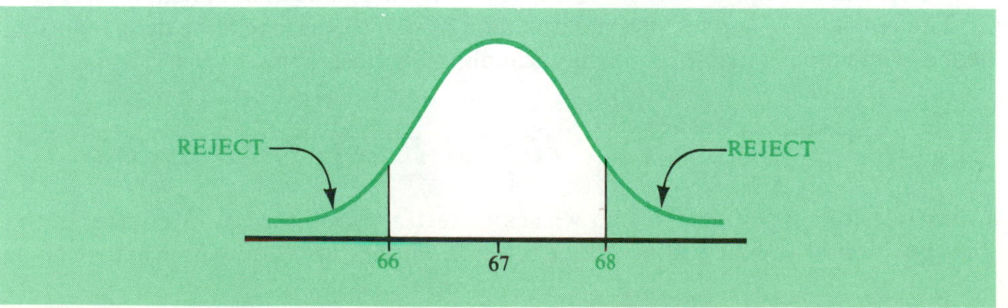

Fig. 10.10

z scores can be calculated $\left(\frac{66 - 67}{.67} = -1.49 \text{ and } \frac{68 - 67}{.67} = 1.49\right)$, and the shaded area in Fig. 10.10 can be determined to be .1362. Consequently, even if the guidelines are being followed, about 13.62% of the time the superintendent will be led to believe they are not.

As previously pointed out, determination of a type-two error (a false hypothesis accepted) depends on what is true in place of the hypothesis. If the mean classroom temperature is really 69 degrees with a standard deviation of 3 degrees, what are the chances that the mean of the 36 randomly selected classrooms will be in the 66- to 68-degree range required for acceptance of the hypothesis? Now the distribution of sample means the superintendent must consider is normally distributed with a mean of 69 degrees and a standard deviation of $\frac{3}{\sqrt{36}} = .5$. The acceptance interval is shaded in Fig. 10.11.

Fig. 10.11

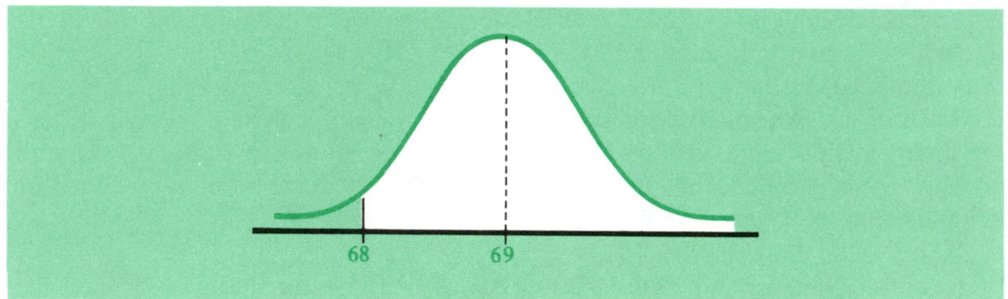

z score transformations yield $\frac{66 - 69}{.5} = -6$ and $\frac{68 - 69}{.5} = -2$. According to Appendix III the shaded area represents .0228 of the area under the curve. Thus, if the mean temperature is only 2 degrees above the guideline (with a standard deviation of 3 degrees), the superintendent will fail to find the hypothesis false only 2% of the time. That is, the probability of a type-two error being made is .0228.

Subtracting .0228 from 1 gives the power of the test against the alternative $\mu = 69$. The result, .9772, is the probability that a false hypothesis will be detected using the superintendent's plan of randomly sampling 36 classrooms.

10.5 A Final Word on Risk

In practice, the experimenter who wishes to test a hypothesis is, of course, in the blind position of deciding to accept or reject it without ever knowing whether he or she is right. One would naturally like to accept only what is true and reject only what is false. The trouble is that whatever decision rule is chosen leaves one open to risks. Some acceptance region with endpoints must be specified, and as soon as this is done the risk is run that even if the hypothesis is true one will come upon some stray but random finding that is not in one's acceptance region. Furthermore, even if the hypothesis is wrong there exists some chance that one's sample finding will be in one's acceptance region.

The question of how the experimenter is to determine the width of this acceptance region now arises. To use a narrow acceptance region means to run a high risk of making a type-one error. Increasing the width of the acceptance region decreases the chance of making a type-one error but increases the risk of making a type-two error. The point is that we cannot reduce one kind of risk without at the same time increasing the other.

As one might guess, the decision about how wide to make the acceptance region depends upon the readiness of the experimenter to make one kind of error as opposed to the other. In the case of Dr. R and Dr. A, the two differed essentially in terms of the risks they were willing to tolerate. Dr. A seems to have been motivated by a dread of condemning a fair coin. His expansive acceptance region suggests that he was relatively unconcerned with failing to detect a bad one. Dr. R seems to have picked her narrow acceptance region in dread of being fooled into accepting a bad coin. She seems scarcely to have cared that she would condemn many good coins in the long run.

In general, when one chooses an acceptance region one is motivated by the fear of mistakenly rejecting a true hypothesis and to some degree also by the fear of mistakenly accepting a hypothesis when it is false. The width of the acceptance region actually chosen represents a kind of compromise position.

This chapter has described what it is to test a hypothesis and has pointed up the risks always involved in reaching a decision. The next chapter will show specifically how one considers one's own willingness to tolerate risk in actually testing a hypothesis.

Unlike Dr. A and Dr. R, one makes an explicit decision about tolerance of risk and then follows a formalized procedure for actually testing a hypothesis. We have discussed the considerations underlying this formal procedure; next we shall discuss the hypothesis-testing procedure itself.

Problem Set A

10.1 Consider the hypothesis H: The defense posture of the United States is adequate to deter an attack (no change is needed). What risk is involved in a type-one error? In a type-two error?

a. Present strategies are scrapped and the country is attacked.
b. New strategies are adopted and the country is not attacked.
c. Present strategies are maintained and the country is attacked.
d. Present strategies are maintained and the country is not attacked.

10.2 Consider the hypothesis H: Chemotherapy methods for treating cancer are the most effective methods developed to date. What risk is involved in a type-one error? In a type-two error?

a. The chemotherapy method is used, but the recovery rate is not as high as if some other treatment had been used.
b. The chemotherapy method is not used, and the recovery rate is as high as possible with any known treatment.
c. The chemotherapy method is not used, and the recovery rate is lower than if it had been used.
d. The chemotherapy method is not used, but the recovery rate is higher than if it had been used.

10.3 Consider the hypothesis H: Universal University's admission policies are not discriminatory. What risk is involved in a type-one error? In a type-two error?

a. The policies are changed, but were not discriminatory.
b. The policies are retained, and discrimination continues.
c. The policies are changed, and discrimination is reduced.
d. The policies are retained, and there continues to be no discrimination.

10.4 Find the interval of outcomes that would allow acceptance of the hypothesis that 40% of children between the ages of 8 and 10 attend a movie on any given Saturday. Ninety students are surveyed and the decision rule is to accept any outcome that deviates less than five units from the mean.

10.5 The consulting psychologist for a large seaboard city believes that 45% of the city's gang members are White. The city crime commission decides to accept this hypothesis if a random sample of 100 records of gang members on file with the police department has from 40 to 50 White individuals (including these endpoints). What is the probability of a type-one error?

10.6 A union official hypothesizes that five-sixths of all employees in a certain industry favor a 35-hour work week. He surveys 84 workers selected at random. If the number of favorable responses is less than five units from the mean he will accept the hypothesis. (a) What is the probability of a type-one error? (b) What risk is involved in a type-one error?

10.7 An advertising agency promises that 65% of the market can be captured if a certain campaign is adopted (H: $P = .65$). After six months 900 people are asked if they use the product. The decision rule is to accept the hypothesis if the outcome is within 20 units of the mean (inclusive). Suppose, in fact, the campaign is adopted, but only 60% of the market is captured. Find the probability of a type-two error and the power of the test.

10.8 Baxter Business School's superintendent believes its graduates type a mean of 60 words per minute with $\sigma = 6$. To substantiate this, she tests 81 randomly selected recent graduates. She decides to accept the hypothesis if the mean for the group is between 59 and 61 words per minute. If the mean is really 58 words per minute with $\sigma = 5$, what is the probability of a type-two error and the power of the test?

10.9 The developers of an achievement test for children between the ages of 9 and 11 believe they have devised a new test whose scores will be distributed with a mean of 65 and a standard deviation of 15. If the test results of 150 students selected at random have a mean that deviates less than 1.5 standard deviations from its mean, the hypothesis will be accepted and published in promotional literature. (a) What interval of outcomes would allow for acceptance of the hypothesis? (b) Find the probability of a type-one error.

10.10 To test the hypothesis that the mean gasoline mileage for American-made cars is 18.5 miles per gallon on the highway with a standard deviation of 2 mpg, 121 such automobiles are driven 1,000 miles. Assuming the decision rule is to accept the hypothesis if the sample mean is within 2 standard deviations of the mean, find (a) the interval that will permit acceptance, and (b) the probability of a type-one error.

10.11 A construction company standard sets the breaking point of reinforced cylinders at 500 lb/in^2 with $\sigma = 4$ lb/in^2. To see if its cylinders meet the requirements, 100 cylinders are tested. If the mean breaking point has a z score between ± 1.5, the cylinders will be certified as meeting the standard. Find the probability of a type-one error.

10.12 Last year in his book *Executive on The Move*, E. X. Pert submitted that the American executive spends a mean of 75 minutes a day commuting to and from work with a standard deviation of 25 minutes. Mr. Pert plans to retest his hypothesis this year by randomly sampling 350 executives across the nation. If the sample mean deviates more than two units from its mean, he will reject last year's claim. If in reality the mean commuting time is 70 minutes with a standard deviation of 20 minutes, what is the probability that Mr. Pert will make a type-two error? Find the power of the test against the alternative $\mu = 70$.

10.13 The placement office of Hallowed Hall College contends that each year it places a mean of 175 graduating seniors in social work positions ($\sigma = 30$). The sophomore statistics class has been assigned to test this contention by contacting all social work graduates for the past five years. The decision rule to be used is as follows: Accept the hypothesis if the sample mean deviates less than 20 units from the mean.

a. What assumption(s) must the students make?
b. What is the probability of a type-one error if the hypothesis is true? What risk is involved?
c. If the true mean and standard deviation are 150 and 35 respectively, what is the probability of a type-two error? Find the power of the test.

10.14 The head of a state wildlife commission believes that only 35% of the pheasant population survived the winter. How many birds of a sample of 400 would have to survive for us to accept the hypothesis if the probability of a type-one error is .13 and the acceptance interval is centered at the mean?

10.15 The manager of the produce department of a large supermarket suspects that the mean diameter of the "2-inch apples" is really 1 3/4 inches with a standard deviation of a half inch. Into what interval would the mean of 49 randomly selected apples have to fall to support the manager's belief, if the probability of a type-one error is .05 and the acceptance interval is centered at the mean?

Problem Set B

10.16 Consider the hypothesis H: Sam is innocent. What risk is involved in a type-one error? In a type-two error?

a. Innocent Sam goes free.
b. Innocent Sam goes to jail.
c. Guilty Sam goes free.
d. Guilty Sam goes to jail.

10.17 Consider the hypothesis H: The picture tube does not need replacing. What risk is involved in a type-one error? In a type-two error?

a. A good tube is replaced.
b. A good tube is not replaced.
c. A bad tube is replaced.
d. A bad tube is not replaced.

10.18 Consider the hypothesis H: Mean class sizes will be the same as in the past. What risk is involved in a type-one error? In a type-two error?

a. Mean class size changes, but is accounted for by proper budget, faculty, and scheduling alterations.

b. Mean class size does not change, but plans anticipate a change.

c. Mean class size does not change, and planning continues as usual.

d. Mean class size changes, but planning fails to take the change into consideration.

10.19 What interval of outcomes would permit acceptance of the hypothesis that 45% of Creek City residents oppose the new flood control project? One hundred residents are surveyed and the decision rule is to accept outcomes that deviate 10 or fewer units from the mean.

10.20 Superintendent Biegler will file a report with the health department stating that 55% of the district's students have the Mongolian flu if this hypothesis is supported by the findings from a random sample of 80. He decides that if 50% to 60% (inclusive) of the sample is home with the flu he will accept this hypothesis. Find the probability that he will make a type-one error.

10.21 It has been claimed that 85% of Protestant ministers would choose their vocation again if given the chance. Find the probability of a type-one error if this hypothesis is tested by sampling 200 ministers with an acceptance interval of length 9 centered at the mean.

10.22 Find the interval of outcomes that would permit acceptance of the hypothesis that neighborhood safety patrols reduce burglaries by a mean of 75% with a standard deviation of 10%. One hundred neighborhoods with such patrols are to be studied, and the probability of a type-one error is to be held to .08. The acceptance area is to center about .75.

10.23 The president of Grey Goose Airlines maintains that, because of delays on the ground, the mean time that must be made up in the air is 12.5 minutes ($\sigma = 3$ minutes). A regulatory agency is willing to accept this contention if the mean of the times made up in the air of 64 randomly selected flights is between 12 and 13 minutes. What are the chances of a type-one error being made?

10.24 The manufacturer claims that St. Bernard puppies will increase their weight by 10 pounds during their third month of life if fed Puppy-Gro ($\sigma = 1$ pound). A veterinarian decides to accept this if the mean weight gained by the 36 St. Bernard puppies she is raising to sell is within .5 pound of the mean. Find the probability of a type-one error.

10.25 Add-On Corporation believes that only 3% of its calculators are defective. To substantiate this it randomly tests 600 models in question. The hypothesis will be accepted if between 15 and 25 (inclusive) calculators malfunction. If, in fact, the fraction of defective items is .05, what is the probability of a type-two error? Find the power of the test.

10.26 A statewide study of veterinarians asserts that half of the DVMs obtained their formal higher education outside the state. If the study is in error, and actually 65% received out-of-state degrees, what is the probability that the responses of a sample of 36 veterinarians will fail to point to the error? Assume the acceptance region includes 17, 18, and 19. What type of error is being considered?

10.27 The inventor of a new device purported to increase mileage 15% to 20% claims that 70% of all automobile owners would purchase the device despite the high price it would cost. A potential manufacturer decides to accept this claim if between 65% and 75% of a random sample of 600 car owners respond positively. If, in fact, only 63% of all owners would actually purchase the product, what is the probability of his making a type-two error? What risk(s) is involved in a type-two error?

10.28 The owner of a local tourist attraction claims that the typical viewer travels 300 miles to see his sight ($\sigma = 25$ miles). A reporter decides she will accept this if the mean distance traveled by the first 81 patrons she interviews is within one standard deviation of the mean. If the mean distance is actually 275 miles ($\sigma = 50$), what is the probability that the reporter will make a type-two error? Find the power of the test.

10.29 Find the acceptance interval that the mean speed (in miles per hour) of 36 hurricanes must fall into to support the hypothesis that the mean speed is 15.7 miles per hour ($\sigma = 3.9$ mph). Limit the probability of a type-one error to .10 and center the acceptance interval about the mean.

10.30 The president of the High Risk Insurance Conglomerate believes that 60% of the homeowners in a certain district are underinsured. How many homeowners in a sample of 900 would have to be underinsured to support this hypothesis, if the probability of a type-one error is .07 and the acceptance interval is centered at the mean?

Review II

Chapters 7–10 have added a whole new world of situations in which we may apply the statistical tools we have acquired. Here we use the phrase "a whole new world" as more than just a figure of speech, for we have discovered that nature obeys the same laws that are used to form the infinite distributions discussed in the preceding chapters.

The normal distribution is defined in Chapter 7 in terms of the z score and the percent of terms exceeded for each term. While an infinitude of normal distributions exists, one for each selection of μ and σ, each possesses a number of common properties. These properties (clustering, symmetry, equality of the mean, median, and mode, and the lack of upper and lower bounds) all stem from the relationship expressed in the Appendix Tables II and III between z scores and the percents of terms exceeded. The tables allow us to determine the percent of terms in *any* normal distribution that is above, below, or between the specified z scores. This is possible because the area under the curve can be thought to represent proportion of terms. Since every normal distribution shares the properties reflected in Tables II and III with every other normal distribution, raw scores must be converted to z scores before the tables can be used. Once this is done, many useful inferences can be made concerning the terms of distributions known to be normal or at least approximately so.

The importance of the Central Limit Theorem presented in Chapter 8 cannot be stressed too strongly. The results of this theorem and its sister theorem, the Central Limit Theorem for Means, have shaped much of the discipline we call statistics. It is this theorem that explains why natural phenomena so often exhibit the properties of normal distributions.

There are three ways for us to know, at this point, if a distribution is normal, or at least approximately normal: (1) to be told it is normal; (2) to recognize it as a distribution of sample sums; or (3) to recognize it as a distribution of sample means. Furthermore, the mean and standard deviation of a distribution of sample means are known to be $\mu_{\bar{X}} = \mu$ and $\sigma_{\bar{X}} = \dfrac{\sigma}{\sqrt{N}}$ where μ and σ are the mean and standard deviation of the distribution being sampled from. Knowledge of these measures allows us to make inferences regarding the means of samples selected at random from virtually *any* population. The procedures for doing so are also found in Chapter 8.

An understanding of the basic concepts of probability (outcomes, events, sample spaces, probability trees) makes it possible to determine the chances of any sample mean falling in some range of interest. Chapter 9 combines these basic concepts of probability with the procedures for finding areas under the standard normal curve. The applications are almost limitless, since the mean is almost always a revealing term.

As important as distributions of sample means are the binomial distributions, which can be viewed as special distributions of sample sums. Binomial distributions—which, like distributions of sample means, really form a family of distributions with common properties—differ from other distributions of sample sums. The parent population used to form a binomial distribution must contain exactly two values. P, the probability that a term selected at random will assume one of the specified values, must be known, as must N, the size of the samples used to obtain the distribution of sums. Under proper and often easily met conditions, the normal distribution may serve as an approximation for the binomial with $\mu = NP$ and $\sigma = \sqrt{NP(1-P)}$. Again we have enlarged the set of situations we are capable of addressing; and we have done so with essentially the same normal distribution that has been serving us so well! Now we have at our disposal the tools to consider questions involving distributions in which something is known or believed to happen a certain percentage of the time.

In preparation for work to follow, it is essential to be aware of the risks one takes in making decisions concerning an entire population based on the results of a random sample. These risks have been classified as either type-one errors in which true beliefs are rejected or type two errors in which false ideas are accepted. Furthermore, given enough information we can find the probability of either of these errors being made. The procedures for doing so are outlined in Chapter 10 for situations involving either sample means or sample proportions (binomial situations). A clear understanding of the hypothesis-testing procedures that will follow cannot be attained unless these procedures are mastered.

The next set of exercises should not only serve as a quick review of Chapters 7–10, but should also help pinpoint areas of weakness that should be strengthened before attempting to tackle new topics.

Problem Set A

II.1 Find the proportion of area under a normal curve that lies

a. to the right of $z = -.39$.

b. to the right of $z = .78$.

c. between $z = -1.83$ and 2.91.

d. within one-third standard deviation of μ.

e. to the left of P_{18}.

f. to the right of D_7.

g. between the median and the term exceeded by 30% of the terms.

II.2 A distribution is known to be normally distributed with a mean of 125 and a standard deviation of 30.

a. What score is exceeded by 95% of the scores?
b. Find P_{18}.
c. Find the term whose z score is 1.73.
d. What score has a percentile rank of 65?
e. Find PR_{140}.
f. Find PR_{50}.
g. Find an interval centered at the mean that includes 45% of the terms.
h. Find the mean and standard deviation of a distribution of sample means formed from samples of size 16 randomly selected from the distribution.

II.3 The pulse rate for college-age women is distributed normally with a mean of 75 per minute and a variance of 91. The physical performance lab is recruiting college females to take part in an aerobic exercise experiment and study.
a. What percent of the applicants can be expected to have pulse rates between 70 and 80?
b. What pulse rate is exceeded by 80% of the pulse rates?
c. If there are 45 applicants, what is the probability that their mean pulse rate will exceed 76 per minute?

II.4 The manufacturers of Happy Peanut Butter are offering taste samples of their product and those of three major competitors to supermarket patrons. The first two samplers both select Happy as their favorite. Of course, this is what the manufacturer would have hoped for. What is the probability that, had the shoppers no real preference, these results would have appeared by chance?

II.5 Suppose that samples of size 35 are randomly selected from a population in which 35% of the members are classified as transient. Let X be the number of transients in a sample selected at random. Find the probabilities of the following events occurring.
a. $X > 10$
b. $X \geq 10$
c. $X = 10$
d. $X < 15$ and $X \geq 11$
e. $X \geq 15$ or $X < 11$

II.6 The mean lifetime of *Sure-Ride* brand steel-belted radial tires is 55,500 miles. The lifetimes vary normally with a standard deviation of 4,000 miles. The tires are under warranty for the first 40,000 miles. Let $\bar{X}$ represent the mean lifetimes of 2,000 tires that will be sold through the Aberdeen distributor. Find the following.
a. The probability that $\bar{X} \geq 55{,}400$
b. The probability that $55{,}450 \leq \bar{X} \leq 55{,}550$
c. The probability that a tire sold at random to the mayor of Aberdeen will have a lifetime shorter than the warranty period
d. The probability that $\bar{X} = 55{,}500$
e. The lifetime that will exceed 60% of the lifetimes of tires sold nationally

II.7 The fraction of graduates in a certain discipline who find jobs in their field within a year of graduation is only 28%.
a. What is the probability that only one of the 25 seniors graduating in the field from Middlesap College will land a job related to this field within a year?

b. Find the probability that 10 or more will be lucky enough to find employment in the field.

II.8 To be effective, a certain vaccine must contain at least 2.57 cubic centimeters of a killed virus. Each dosage sold in premeasured filled syringes contains a mean of 2.6 cubic centimeters. The standard deviation is .01 and the distribution is approximately normal.

a. What is the probability that the mean dosage in each lot of 100 contains more than 2.59 cubic centimeters?

b. What is the probability that any one given vaccine will contain enough of the killed virus to be effective?

c. Only 10% of the syringes contain as much as what number of cubic centimeters of the virus?

d. Only 10% of the 100-syringe lots contain a mean of how many cubic centimeters or more?

II.9 Forty-five percent of the applicants for a job opening at Barry Sales International are women. After the initial screening process, all but 50 applicants are rejected. Ten women are still under consideration.

a. What is the probability that a random sample of 50 applicants would contain 10 or fewer women?

b. What is the probability that a random sample of 50 applicants would contain 20 or more women?

II.10 In an attempt to diversify her portfolio, an investor considers four risky stocks, each of which has an equal chance of either increasing or decreasing in value. What is the probability that (a) all of the stocks, (b) two or more of the stocks, (c) none of the stocks, and (d) at least one stock will increase in value?

II.11 An auditor hypothesizes that only 8% of all negative confirmations are returned. If he sends out 800 requests and plans to accept his belief if between 60 and 70 (inclusive) are returned, what is the probability of a type-one error being made?

II.12 Recent controversies regarding mandatory retirement age prompted sociologist Jack E. Taylor to survey 60 adults between the ages of 40 and 45. He hoped their responses to the question "At what age would you prefer to retire?" would support his contention that the typical worker in this age group wants to retire at 55 years of age. He will allow for a standard deviation of six years. Assuming his decision rule is to accept his hypothesis if the mean of the sample is within 1.75 standard deviations of what he expects it to be, find (a) his acceptance interval and (b) the probability of a type-one error.

II.13 A canning factory machine is designed to put 485 grams of applesauce in each can ($\sigma = 25$ grams). Each day 25 cans are opened and the contents weighed. If the mean for the 25 cans is not between 475 and 495, the entire canning process is shut down and the machine is serviced.

a. What hypothesis is being tested?

b. What is the probability of a type-one error if the machine is working properly?

c. If in fact $\mu = 473$ and $\sigma^2 = 150$, what is the probability of a type-two error and the power of the test against this alternative?

d. What risks are involved in making a type-one and a type-two error?

II.14 Believing that at least 55% of all eligible voters in his district approve of his gun control legislation, Representative Trigger surveys a random sample of 100 of his constituents. He plans to accept his theory if 50 or more support him. If in fact only 45% favor his bill, what are the probability of a type-two error and the power of the test against the alternative $P = .45$?

II.15 Let X be the distribution of the daily attendance for the city public schools. It is distributed with a mean of 85,450 and a standard deviation of 2,010. Each quarter (45 days per quarter) the mean daily attendance is reported for financial aid purposes. Let Y be the distribution of daily means per quarter. Let W be the distribution of the total numbers of school days attended by district students per quarter. Each year the mean daily attendance is computed for projective planning purposes. There are 180 school days per year. Let Z be the distribution of all annual mean daily attendance figures.

a. Which of Distributions X, Y, W, and Z are normal or approximately normal? Why?

b. Which, if any, of Distributions X, Y, W, and Z are continuous?

c. Which of Distributions X, Y, W, and Z have the same mean?

For parts d through g answer true or false.

d. $\sigma_Y < \sigma_Z$

e. $\sigma_Y = \sqrt{45}\,\sigma_X$

f. $\sqrt{180}\,\sigma_Z = \sigma_X$

g. $\sqrt{45}\,\sigma_Y/\sqrt{180} = \sigma_Z$

Problem Set B

II.16 A distribution is known to be approximately normal with a mean of 15 and a standard deviation of 2.5.

a. What fraction of the terms are between 13 and 14?

b. Find P_{78} and Q_1.

c. Find PR_{12} and PR_{18}.

d. If there are 500 scores between 12 and 14.5, how many terms are in the distribution?

II.17 A binomial distribution is formed by summing the terms of infinitely many samples of size 35 from a population in which $P = .8$.

a. Find the mean and standard deviation of the binomial distribution.

b. What percent of the terms exceed 30?

c. Find the term that exceeds only 15% of the terms.

d. Find PR_{30}.

II.18 A biologist studying the effects of drought periods on a certain species of desert snake captures 40 of the creatures to study. She weighs each snake, supplies no water for 48 hours, then again weighs the reptiles. If the mean weight loss for this type of snake subjected to these conditions for 48 hours is

5.7 grams ($\sigma = 1.1$), what is the probability that the biologist's sample will have a mean weight loss between 5.65 and 5.8 grams?

II.19 An educator who is considered an authority on learning disabilities speculates that up to 78% of all criminals have at least one learning disability that remained undetected and untreated during his or her formative years and years of formal education. The educator randomly selects 125 inmates from the state's penal system. Each is administered a battery of tests to diagnose various learning disabilities. If the educator's speculation is true ($P = .78$), what is the probability that more than 75% of the inmates will have previously undiagnosed learning disabilities? What is the probability that the number will be less than 87?

II.20 The number of work hours lost due to accidents per employee per year in the state's largest industry is approximately normally distributed with a mean of 7.9 and a standard deviation of 2.7. Each foundry within the industry employs about 45 workers. In what proportion of the foundries is the mean loss of work hours per employee per year (a) more than a full work day (8 hours), (b) less than 7 hours, and (c) between 7.5 and 8.5 hours?

II.21 A new capsule to combat the symptoms of the common cold has just undergone intensive tests. The drug was effective in all cases, but the time for the ingredient in the drug to take effect had a mean of 18 minutes and varied normally with a standard deviation of 4.7. If 75 subjects gained relief in between 10 and 13 minutes, how many people tested the new medicine?

II.22 The fraction of college-bound high school senior girls seeking a career outside the home is .87. For those girls terminating their formal education with high school graduation, the proportion is .46.

a. Find the probability that all of the 40 new pledges in a sorority plan to seek careers.
b. Of the 58 graduating senior girls in one class who are not planning to go to college, the probability is only 2% that more than what number will seek careers?

II.23 The self-confidence scores earned by professional actors on a test designed by a university research division are normally distributed with a mean of 67 and a standard deviation of 4. If the mean of a sample of professional actors is at the 16th percentile (in the distribution of sample means) and has a value of 66, how large is the sample?

II.24 A drug prescribed by a small minority of physicians 25 years ago is now known to cause stomach ulcers in 15% of persons who take the medicine for a one-week period or more.

a. The drug's manufacturer and its insurance company are compiling lists of those who took the drug. To date, 1,374 people have been located. What is the probability that 200 or more have developed stomach ulcers caused by the drug?
b. The company is assigning as many employees as needed to notify and follow up on each case. Each worker will be responsible for 100 names. What proportion of the workers will find between 10 and 20 (inclusive) drug-caused ulcer victims on their lists?

II.25 The weekly publication *The Tattler* sells a mean of 8,300 copies per week. The standard deviation is 950 and the distribution is approximately normal. Because of the large variability in sales, much of the time many copies are not sold. To remedy this costly problem the management plans to meet demand only 85% of the time.

a. How many copies will be printed per week?
b. What percent of the time will there still be a surplus of copies at the end of the week?
c. What percent of the time will the surplus be more than 1,000 copies?

II.26 Consider the hypothesis H: The PCB (a toxic chemical) level in the fish found in Lake Alexander is acceptable.

a. Describe what occurs if a type-one error is made.
b. Describe what occurs if a type-two error is made.
c. As the power of the test against some alternative increases, which type of error also increases?

II.27 The president of the city's board of practicing veterinarians believes from reports that he has seen that 30% of the city's registered dogs have an undiagnosed heartworm condition. He presents his theory to the board, which votes to accept it if, in a random sample of 75 dogs, between 20 and 25 (inclusive) test positively. Find the probability of a type-one error being made. What risk does it involve?

II.28 The commanding officer of a missile battery is attempting to substantiate his claim that the mean attack preparation time is 6 minutes ($\sigma = .5$). His superior officers decide that if the last 25 drills had a mean within one-half standard deviation of the mean ($\mu_{\bar{X}}$), the 6-minute figure will be accepted.

a. What chance is there that a type-one error will be made?
b. How can this probability be reduced?
c. Is it necessary to assume that the times for all drills are normally distributed?
d. Is this assumption justifiable? Why or why not?

II.29 The administrator of a proposed hospital that will offer a revolutionary type of atmosphere for the practice of the healing arts submits that the typical patient would travel 250 miles for treatment. This distance is large enough that local hospitals need not fear an added number of empty beds, and small enough that the trip would be feasible for both patients and family members. The results of a random survey conducted by an independent firm within a 1,000-mile radius will be used either to substantiate or to refute the administrator's claim. The hospital council agrees to accept the claim if the mean distance from the site of 144 people responding that they would use the facility if the need arose is between 225 and 275 miles. If in fact the mean distance is only 200 miles ($\sigma = 75$), what is the probability of a type-two error being made? What is the probability of a type-two error being made if the real mean is 275 ($\sigma = 100$)?

II.30 A sociologist believes that 55% of the inhabitants of an inner city are under 18 years of age. She is unwilling to accept figures from the last census, as she believes that economic changes have substantially affected the make-up of the community. She plans to select 60 people at random. She will consider results of between 30 and 36 (inclusive) as substantiation for her theory. If the true percentage is 50, what is the probability of a type-two error? Find the power of the test against the alternative.

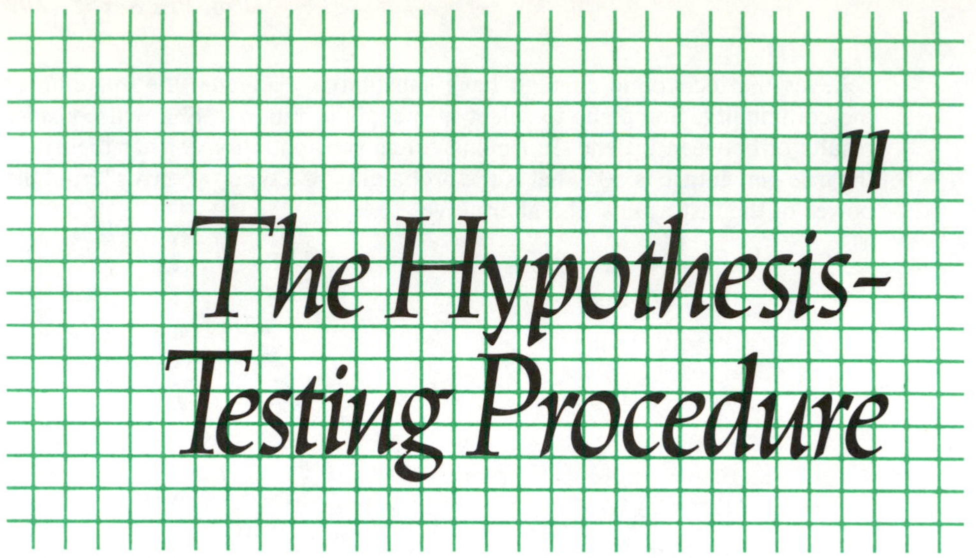

11 The Hypothesis-Testing Procedure

11.1 Introduction

There is a standard procedure for testing a statistical hypothesis, and it is used in psychology, education, the social sciences, and many other fields. In this chapter we shall discuss that procedure and illustrate its use in connection with several problems. We shall make practical use of the material in the previous chapter, including the discussion of a hypothesis and the relationships between the width of the rejection region and the various risks of error.

One major factor must be discussed at the outset—**the fact that evidence that is consistent with a hypothesis can almost never be taken as conclusive grounds for accepting it, whereas evidence that is inconsistent with a hypothesis does provide grounds for rejecting it.** Of course errors can always be made either way, but the point is important and needs elaboration. The reason for not necessarily accepting consistent evidence is that a finding that is consistent with a hypothesis would be consistent with other hypotheses too, and thus does not necessarily demonstrate the truth of the given hypothesis as opposed to these other alternatives. For instance, a finding of 51 heads out of 100 flips is consistent with the hypothesis that a coin is fair, but this finding would of course also be consistent with the assumption that the coin slightly favors heads. Therefore the finding that is consistent with the hypothesis does not demonstrate its truth. Even a finding of 50 heads would scarcely imply that the coin did not have some minute bias.

On the other hand, a finding of 80 heads out of 100 flips could virtually be taken to contradict the hypothesis that the coin is fair. Since a fair coin would have less than one chance in 10,000 of showing as many as 80 heads, such a finding might properly lead us to conclude with only a minute risk of error that the coin was not fair.

In the same way, suppose we hypothesized that the members of a particular group had a mean intelligence quotient of 100. A finding that the members of some sample taken from the group had a mean IQ of 102 might be consistent with the hypothesis. But the finding might also have been consistent with the hypothesis that the mean IQ was 101 or 99, and it certainly would have been consistent with the hypothesis that the mean IQ of the larger group was 102. Thus the finding would in no way demonstrate the truth of the hypothesis of 100 as opposed to these other possibilities. But now suppose the members of the sample had been found to have a mean IQ of 135. Assuming the sample size was large enough, we might show that, were the original hypothesis true, such a sample would virtually never be found. By inference, we might then properly construe the finding as contradictory to the hypothesis, and the risk of error in doing so might be infinitesimal.

What has been said amounts to the fact that the usual finding that is consistent with a hypothesis would have been consistent with other hypotheses too. Thus such a finding cannot be taken as grounds for accepting the particular hypothesis over its neighboring alternatives. However, we may always come upon a finding that would be so hard to reconcile with the hypothesis as to put it in considerable doubt. A hypothesis is like a precise story told by a defendant at a trial. He can never prove this precise account of what he did. Virtually any evidence he presents leaves open the possibility that he acted in a way somewhat different from that described. Yet a prosecutor can disprove his precise account by showing a single inconsistency.

11.2 The Null Hypothesis

It follows that when we fail to reject a hypothesis we are forced to conclude that it *may* be true. On the other hand, when we can reject it we conclude that it *definitely* is false.

This last statement is very important. When testing a hypothesis, we can reach a definite conclusion only if we can reject it. We therefore set up our experiment with the aim of *disproving* the hypothesis we test. That is, we also state a hypothesis that is an **alternative** to what we believe. By this approach, if we are in fact able to reject the hypothesis (to consider it disproven), we have then established its alternative, our own belief.

For example, if we wish to show that men differ in height from women, we test the hypothesis that there is *no difference* in their heights. We then hope to reject this hypothesis. To prove that members of one political party are different in some way from another, we test the hypothesis that there is *no difference* between the parties on whatever we are testing. Once again, our hope is that we can reject this "no difference" hypothesis and thereby establish our own thesis as the alternative.

Definition 11.1 The hypothesis stating that there is no bias or that there is no distinction between groups is called the null hypothesis.

We shall see many instances in which some form of a null hypothesis is tested, with the aim of rejecting it so as to "prove" its alternative.

Not every statement leads to a testable null hypothesis. As we have seen, the statement that a coin is biased is not specific enough to be translated into a specific hypothesis, and this is often the case. It follows from what has been said that *the only way experimenters can proceed is when they can translate the alternative to what they are trying to prove into a specific null hypothesis. Only then can they go ahead and test it, and hopefully reject it and so prove their theory*. It must be stressed that this is not always possible.

The first major step for an experimenter then is to put into statistical form the null hypothesis he or she hopes to *reject* in order to consider a theory, or alternative hypothesis, proven. He or she is now up to the point of carrying out the regular hypothesis-testing procedure.

11.3 When to Reject a Null Hypothesis; The Significance Level

To focus on the logic of his procedure, let us suppose that a particular experimenter has already set up some null hypothesis that he now hopes to reject in order to support his theory. (In our coin illustration let us say that he has decided to test the hypothesis that the coin is fair with the hope of rejecting it.[1]) Now he goes on to collect his sample and to calculate some value of interest from it (like the number of heads obtained in 100 flips).

His next step is to ask himself a crucial question: **"If my null hypothesis is true, what was the probability that I would draw a sample with a value as deviant from the expectancy as the one I obtained?"** (Suppose the coin is really fair and would fall heads half the time on the average. How likely is it that there would have been a number of heads in my sample as deviant from the expectancy of 50 as the number I actually found?) Note this question carefully. It is the key to virtually everything that is to follow.

The next step is to see whether, assuming the null hypothesis is true, the obtained sample finding would have been likely or not. Exactly what is meant by this statement and the computations involved in connection with several problems will be discussed later. To go on with the logic, the probability of getting a sample value like the one obtained from the hypothesized population may be relatively high or low, and this fact obviously determines whether the ultimate decision is to reject or not the null hypothesis.

In general terms, suppose that after calculation it turned out that, if the null hypothesis is true, the probability of having gotten a sample value like the one actually obtained would have been high. In such a case, the obtained sample value would have to be regarded as representative of the hypothesized population (as would be the case if 52 heads were found). That is, a value like the one obtained would have had too high a probability of turning up to allow us to reject the null hypothesis.

However, suppose instead that the obtained sample value is so remote from the

[1]The hypothesis that a coin is biased cannot serve as a null hypothesis. It is often the case that a hypothesis is not specific enough to generate a single distribution so that it can be tested. In other words, not every hypothesis satisfies Definition 11.1.

hypothesized expectancy that one like it would have had an extremely small probability of arising. (For instance, such would be the case if 93 heads had turned up.) To put it another way, suppose the sample value obtained is so deviant that calculations show that one like it would scarcely ever be found in a sample drawn at random, if the null hypothesis were true. Then the decision would be to reject the null hypothesis. Rather than conclude that an incredibly unusual event has occurred, we would consider it more likely that the null hypothesis was not true in the first place.

The process then is to assume the null hypothesis is true temporarily; and as will be seen, the next step is to calculate what the probability would have been of getting a value as deviant from the expectancy as the one actually obtained. The key question is how small should the probability of getting the sample value be (assuming for the time being that the null hypothesis is true) before the experimenter decides to reject the null hypothesis. Once again we get back to the issue of the stakes of the game, for the answer completely depends upon the experimenter's willingness to make a type-one error. The experimenter usually demands that the probability fraction be quite small, since the decision to reject the null hypothesis, when it is made, is in effect a decision to believe a new theory. Naturally it is important to stop incorrect theories from being accepted. For our experimenter it is important to minimize the number of times that coincidental factors fool him into rejecting a null hypothesis when it is true. Accordingly, before rejecting a null hypothesis he demands that a sample value be found that would have had an extremely small probability of arising if in actuality the null hypothesis is true.

In some scientific fields, the practice is to reject a null hypothesis only if the sample value obtained is one rare enough to come from the hypothesized population 5/100 of the time or less. In other fields, the practice is not to reject a null hypothesis unless the sample value obtained would be rare enough to have less than a 1/100 probability of coming from the hypothesized population. One thing is sure: an experimenter must demand that it be very unlikely that a sample would come from the hypothesized population before using the sample value as evidence that the null hypothesis is false.

The probability fraction that our experimenter does pick is called the significance level of the experiment. His choice of significance level must be made before he actually collects his data so that nothing that he comes upon during the experiment will affect him in his choice of it. Having chosen his significance level, the experimenter calculates some outcome of interest from his sample. He then assumes for the sake of argument that his null hypothesis is true, and under this assumption he determines whether a finding like his would be more or less probable than the significance level.

If, under the null hypothesis, a finding like his would be more probable than the significance level, then he fails to reject it, reasoning that his finding would not be sufficiently inconsistent with the null hypothesis to warrant his rejecting it and accepting the alternative in its stead. (For instance, suppose he chose the significance level of 5/100 and subsequently obtained 52 heads. If he calculates that a deviation of as many as two heads from the expectancy would occur more than 5/100 of the time by chance, then he cannot reject the null hypothesis.)

Suppose instead he calculates that, under the null hypothesis, a finding as deviant as his would be so rare as to occur with equal or smaller probability than the significance level. Now he can reject the null hypothesis. For instance, suppose that using the significance level of 5/100, he subsequently obtained a finding of 96 heads; and suppose he calculates that a deviation of this many heads from the expectancy would occur with a smaller probability than 5/100. The logic that leads the experimenter to reject the null hypothesis in such a case would be this: "Assuming the null hypothesis, a finding as deviant as mine would have been so improbable that I need not continue believing in the null hypothesis any longer." When this happens, the experimenter considers his theory, the alternative hypothesis, proven, and thus an experimenter always hopes to obtain a sample value that would have had *less* chance of arising under the null hypothesis than the significance level.

Once our experimenter has chosen his significance level, he has in effect made a definite rule for when he will fail to reject and when he will reject a null hypothesis. The significance level is a statement of when he will consider a finding to show a departure from the expectancy with small enough probability to induce him to reject the null hypothesis.

Now suppose the experimenter has picked his significance level, which for the sake of discussion we shall say is 5/100. He assumes that his null hypothesis is true, and according to protocol he gathers a random sample and computes some value of interest. We shall consider for a moment what is apt to happen when the null hypothesis is actually true (though of course the experimenter does not know this fact). Under the condition that the null hypothesis is true, the probability is 5/100 of obtaining a sample value so deviant from the expectancy as to induce the experimenter to reject this hypothesis mistakenly. **It follows that the significance level, which the experimenter chooses, tells precisely his probability of making a type-one error when the null hypothesis is actually true.**

As another illustration, suppose that the experimenter picks the significance level of 1/100 at the start of his experiment. Now when he rejects a null hypothesis he can say, "Well, if the null hypothesis is true, a finding like mine was one with less than one chance in 100 of arising; rather than believe that so great an unlikelihood occurred, I reject that hypothesis." But this means that, when the null hypothesis is actually true, by definition there will turn up one time in 100 (in the long run) the rarity that will fool the experimenter into mistakenly rejecting this hypothesis. By picking a significance level of 1/100, the experimenter is determining that if the null hypothesis is actually true, there is one chance in 100 that he will get a finding that will induce him to commit a type-one error.

In sum, as the experimenter picks his significance level, he is fixing the risk of a type-one error. The smaller his significance level, the less chance there is of his making a type-one error. However, the smaller his significance level, the harder it is for him to reject his null hypothesis and the more likely it is for him to make a type-two error if that hypothesis is false.

The choice of a significance level should be seen as the choice of a decision rule. For our problems there would be no analogous way to fix the risk of making a type-two error as a way of determining a decision rule, because the risk of making a type-two error varies with the alternative, as we have seen. The risks involved in a type-two error (accepting a false null hypothesis), are generally less serious than those involved in a type-one error. Since the null hypothesis is usually a statement about the status quo or the way things are now, when a type-two error is made, we continue to believe an erroneous idea. At worst we are in the same situation we were in before the hypothesis test. All we do is to fail to change. So it is the probability of a type-one error, in most cases, that we strive most vigorously to control. Having understood this much, we are now ready to discuss the formal procedure for testing a hypothesis, which entails fixing the significance level for an experiment and then testing our hypothesis according to it.

The steps of the hypothesis-testing procedure are presented here in formal order. It is important not only to go over them but to comprehend the total approach, since we shall make references to this approach repeatedly.

1. Statement of Hypotheses. The experimenter must **state two hypotheses, the null hypothesis** that will be tested and **the alternative hypothesis** or theory he or she wants to confirm. H_0 is generally placed before the null hypothesis and H_1 before the alternative hypothesis to identify them. (In our example we have "H_1: The coin is biased." and "H_0: The coin is fair.") It is especially important that H_0 satisfy Definition 11.1.

2. Description of Test. The experimenter must arbitrarily **select a significance level** based on the considerations already discussed. This step should also contain a statement as to whether a **one- or two-tail test** is being performed. (This will be explained in Section 11.6). As new types of tests are added in later chapters this step can be used to **describe what kind of test is being made.**

3. Gathering Information. In practice the experimenter must **gather a random sample and compute the sample value of interest.**

4. Computation. In this step, three things must be done. First, the experimenter must **assume the null hypothesis is true.** Second, the experimenter must **change the value of interest to a z score** (or other calculated score in later chapters). Finally, the experimenter must **compare the z score with some critical value.** This comparison should be done using a distribution curve, often the normal. This procedure will be illustrated shortly. In essence, this step determines under the null hypothesis how small the probability would have been of getting a finding as deviant from the expectancy as the one obtained.

5. Decision. If, assuming the null hypothesis is true, the experimenter's calculations reveal that the probability of having gotten a finding as deviant as the one obtained would have been greater than the significance level, the experimenter must fail to reject the null hypothesis. Some experimenters say that they "accept" their null hypotheses when their findings do not result in their rejecting them. However, there are frequent objections to the use of the word "accept" in this context. To say that we accept a hypothesis suggests that we are accepting it as opposed to all others, whereas in most experiments the data allow acceptance of not one but many hypotheses. Moreover, findings are sometimes not deviant enough to allow rejection of the null hypothesis but are deviant enough to create doubts in an experimenter's mind about its validity. Especially where these doubts lead to new experimentation, the statement that the null hypothesis is accepted does not seem as accurate as the statement that on the basis of the findings one could not reject the null hypothesis. If, assuming the null hypothesis is true, an experimenter's calculations indicate that the probability of having gotten a finding as deviant as the one obtained would have been less than or equal to the significance level, then the experimenter must reject the null hypothesis.

Step 5 is actually very short. The experimenter must *always* state one of two things—that is, either **"reject H_0" or "do not (or fail to) reject H_0"** depending on the results in Step 4.

6. Conclusion. This step is very important. As Step 5 is a statement about the rejection of the null hypothesis, Step 6 is a statement about the acceptance of the alternative hypothesis. The experimenter must *always* state one of two things—that is, either **"accept H_1" or "do not (or fail to) accept H_1"** depending on what Step 5 forces the experimenter to do. It is a good idea to translate Step 6 into words so it is known exactly what conclusion the test yields.

Step 5		Step 6
Reject H_0	→	Accept H_1
Do not reject H_0	→	Do not accept H_1

To conclude our coin example, suppose the finding is that of 55 heads. The null hypothesis that the coin is fair is being tested at the .05 significance level.

We are up to Step 4. Our experimenter temporarily assumes that the null hypothesis is true. He supposes that in actuality the coin is fair, and he now conceives of the sample he drew as one of a vast number of random samples of 100 flips each. This is the standard logic. The experimenter assumes that, under the null hypothesis, his particular experiment was one of a vast number of identical experiments that were carried out, and pictures the distribution of sample values that would be obtained.

The distribution of outcomes of a vast number of experiments, were they to be carried out with a fair coin, would be normal. This distribution, which is depicted in Fig. 11.1, would have a mean of 50 heads and a standard deviation of 5 heads (Theorem 9.1). The obtained value is 55 heads. However, since the continuous normal distribution is being used to approximate the discrete binomial distribution, it must be noted that 55 is represented under the curve as the area from 54.5 to 55.5.

Fig. 11.1

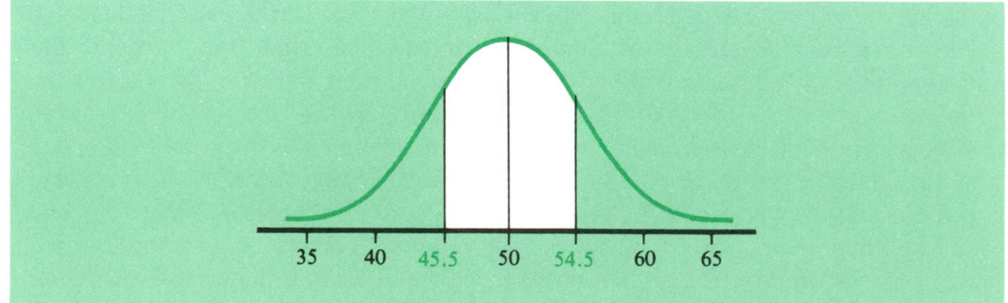

Assuming that the null hypothesis is true, the deviation from the exact expectancy was 4.5 heads. The question is: Assuming the null hypothesis to be true, what was the probability that the number of heads in a random sample of 100 flips would be as many as 4.5 from the expectancy? For the distribution of outcomes depicted in Fig. 11.1, the value 54.5 has a distance score of 4.5 from the expectancy. To get the z score of the value 54.5, we divide its distance score by the standard deviation of the distribution, which is 5. Thus the z score of the finding of 54.5 heads is .9.

To make things more straightforward for other similar problems, we shall use Theorem 9.1 to write a single formula to explain what was done. In general, the formula for the z score of an outcome when the sample consists of N observations of a dichotomous variable is as follows:

$$z = \frac{\text{distance score}}{\text{standard deviation}} = \frac{(O \pm .5) - NP}{\sqrt{NP(1 - P)}} \tag{11.1}$$

where O is the obtained outcome (or number of times the event occurred in the sample), N is the sample size, and P is the probability of the event occurring. Recall that the normal distribution may be used to approximate the binomial distribution only if $NP \geq 5$ and $N(1 - P) \geq 5$. For scores above the mean subtract .5; for scores below the mean add .5.

Note that NP is the value of the hypothetical expectancy, which is the theoretical guess indicated by the null hypothesis. One naturally anticipates finding some departure from this guess even when the null hypothesis is true, but it is the degree of the departure as measured in z score units that becomes crucial.

In our case, $O = 55$, $N = 100$, $P = .5$, and 55 is above the mean.[2] Thus

$$z = \frac{(O - .5) - NP}{\sqrt{NP(1 - P)}} = \frac{54.5 - (100)(.5)}{\sqrt{(100)(.5)(.5)}} = .9$$

To verbalize the formula, because of our null hypothesis we were actually assuming that $P = .5$, and under this assumption we found that the lower boundary of our outcome would have had a z score of .9 in the distribution of outcomes if a vast number of identical experiments had been carried out.

Remember the experimenter's question: Assuming the null hypothesis is true, what would have been the probability of obtaining an outcome that deviated from the expectancy by as many as five heads? The question when translated becomes: What was the probability of randomly obtaining an outcome with a z score value of as much as .9 unit from zero? Reference to Appendix III indicates that as many as approximately 37% of the terms in the normal distribution have z scores at least .9 unit from the expectancy. That is, all those terms in the shaded regions of Fig. 11.1, or about 37% of the terms, are more deviant than the outcome of 55.

Thus the outcome of 55 heads is such that, even were the null hypothesis true, an

[2]Since $NP = 50 \geq 5$ and $N(1 - P) = 50 \geq 5$ we are justified in using the normal approximation.

outcome this deviant or more would have turned up by chance about 37% of the time. Since the probability of so deviant an outcome is higher than the significance level of .05 which we set, we cannot reject the null hypothesis that the coin is fair. It may be said that the finding would not have been unlikely enough to motivate us into rejecting this hypothesis.

Now suppose instead that the coin had turned up 35 heads. This time the deviation from the expectancy is 15 heads and we see at a glance that a fair coin would be much less likely to produce such a finding. Fig. 11.2 indicates that, assuming the null hypothesis is true, the obtained value with the .5 correction factor would have been 2.9 standard deviations from the hypothetical expectancy.

Fig. 11.2

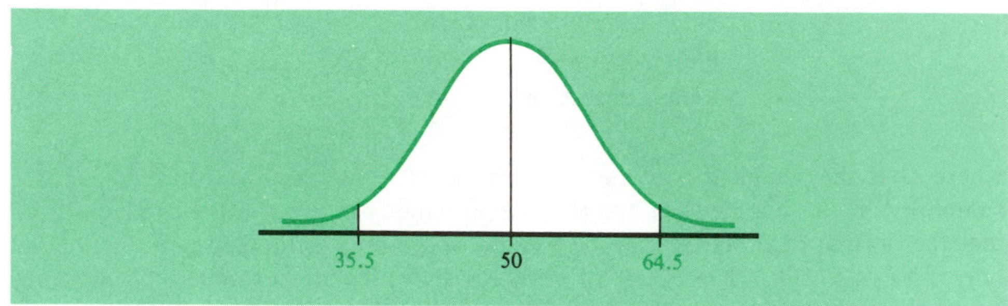

Using Formula (11.1), we see more formally that the z score of the value 35.5 is $z = -2.9$.[3]

$$z = \frac{(O \pm .5) - NP}{\sqrt{NP(1-P)}} = \frac{35.5 - (100)(.5)}{\sqrt{(100)(.5)(.5)}} = -2.9$$

The positive sign was used since 35 is below the mean.

Appendix III tells us that the probability of having gotten so deviant a z score would have been about 4 in 1,000. That is, the terms beyond the shaded tails of Fig. 11.2 comprise about 4/1,000 of all the terms in the distribution. Since this probability is less than the significance level, the decision must be this time to reject the null hypothesis that the coin is fair and to regard as confirmed the theory that the coin is biased.

The purpose thus far has been to present and illustrate the hypothesis-testing method and to discuss its logic. The basic steps and their rationales should be clear. First there must be an exact statement of the hypotheses, both null and alternative. Then the experimenters choose their significance level and calculate the outcome of interest (and sometimes other data) from their sample. Then, assuming that the null hypothesis is true, they determine how small a probability there would have been of obtaining a finding as deviant from the hypothetical expectancy as was their particular finding. Where they decide that a finding as deviant as theirs would have been at most as probable as the significance level, they reject the null hypothesis.

[3]Since $NP = 50 \geq 5$ and $N(1-P) = 50 \geq 5$ we are justified in using the normal approximation.

Unless their finding is this deviant, they cannot reject the null hypothesis and consider their original theory confirmed.

Experimenters who make regular references to the normal distribution when testing hypotheses get to know important relationships between z score distances and probabilities. For instance, they know that the probability of randomly coming upon a term from the normal distribution with a z score further than 1.96 units from the mean is 5/100. Thus when they test a hypothesis at the .05 level of significance, they automatically know not to reject it when z is between -1.96 and $+1.96$. If z is less than or equal to -1.96 or greater than or equal to $+1.96$, they reject the null hypothesis. Therefore 1.96 is called the **critical z value** for a two-tail test (as distinguished from one-tail, which is explained in the next section) at the .05 significance level. The tails of the curve can clearly be seen to be the rejection regions in Fig. 11.3A. At the .01 level of significance it can easily be shown using Appendix III that ± 2.58 are the critical z values for a two-tail test. They are shown delineating the rejection regions in Fig. 11.3B.

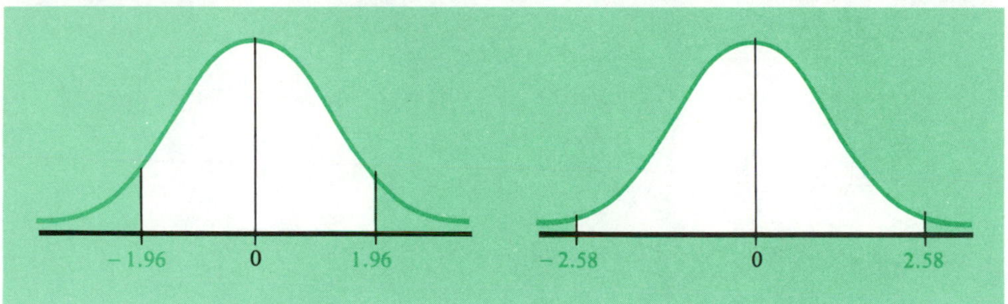

Fig. 11.3A

Fig. 11.3B

The basic methodology and logic we have been discussing is the same when applied to a whole variety of problems. Hypotheses concerning means, medians, standard deviations, and numerous other kinds of hypotheses are tested according to the identical procedure, though naturally the computational approach varies with the problem at hand.

11.4 Three Hypothesis-Testing Problems

We shall now look at three problems and their solutions. The first one will involve a dichotomous variable and be similar to the one already worked out. Thus we shall not go into as much detail, but will present it as an experimenter or researcher would be expected to do. The other two will be hypotheses concerning the unknown mean of a population. On the first problem we shall go into much explanation, but will present the second as a model to follow. It will be seen how the procedure already described is applied in each case.

Problem 1. Suppose a researcher believes that the voters in a particular state are not going to follow their usual trend in some particular year. Ordinarily two-thirds of them are Republicans and one-third of them are Democrats, but this time the belief is that some recent event in the state is going to have an unpredictable but strong impact on the voters' preference. That is, the researcher's theory is that the event will be meaningful, though she would not like to venture a guess as to which party will profit from it.

The researcher plans to test this null hypothesis at the .05 significance level with the hope of being able to reject it. Actually she is hoping to show merely that the usual 2/3 to 1/3 ratio does not remain. The researcher polls 450 voters in all and it turns out that 225 of them are Republicans.[4]

Hypothesis Test for Problem 1

1. H_1: The proportion of Republicans has changed and is no longer 2/3. ($P \neq 2/3$)

   H_0: The proportion of Republicans has not changed and remains 2/3. ($P = 2/3$)

2. .05 significance level

3. $O = 225$ $\qquad N = 450$

4. Assume H_0 is true. ($P = 2/3$)

   $\mu = NP = 450(2/3) = 300$

   $\sigma = \sqrt{NP(1-P)} = \sqrt{450(2/3)(1/3)} = 10$

   $$z = \frac{(O + .5) - NP}{\sqrt{NP(1-P)}} = \frac{(225 + .5) - 300}{10} = -7.45$$

   For a two-tail test at the .05 significance level, the critical z values are +1.96 and −1.96.

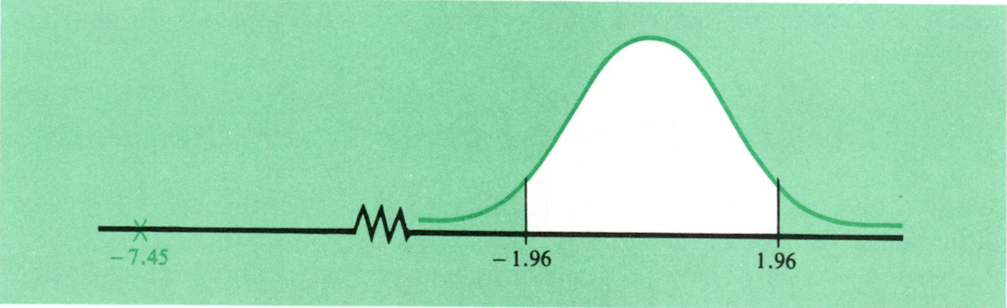

5. Reject H_0.

6. Accept H_1: The proportion of Republicans in the state is no longer 2/3.

Problem 2. The procedure is identical when we test a hypothesis about a population mean, except that now we use our *sample mean* as the evidence instead of

[4]Since $NP = 300 \geq 5$ and $N(1-P) = 150 \geq 5$ we are justified in using the normal approximation.

simply the number of times an event of interest occurred. We are now going to consider the simplest case of testing a hypothesis about a population mean—namely, when we know the standard deviation of the population.

As before, we begin by stating both a null and alternative hypothesis and setting a significance level. We then draw a random sample from the population, and this time the mean of our sample is measured for its discrepancy from the hypothetical expectancy. As before, the null hypothesis is assumed to be true and the process is to envision that a multitude of equal-sized random samples have been drawn from the population, and that the obtained sample is merely a member of this multitude. The *means* of these different samples would form a distribution, and, since we are assuming that the null hypothesis is true, we can determine what this distribution would look like using Theorem 8.3 and Theorem 8.4.

Once we have constructed the distribution, the next step is to determine what the z score of the obtained sample mean would have been. From this z score, we can determine whether the obtained sample mean would have been commonplace, or whether the discrepancy noted would have had a small probability of occurrence. Once again, if the discrepancy is such that it would have occurred with larger probability than the significance level, then the decision is to fail to reject the null hypothesis. If the discrepancy would have occurred with equal or smaller probability than the significance level, then the decision is to reject the null hypothesis.

To illustrate, the question recently arose of whether boys of age 8 who want to be firemen differ in IQ from the average. It was believed that as a group their IQs did differ and that this could be shown using the Stanford Binet IQ Test. A study was planned in a city where the mean IQ of boys of age 8 was 100 and the standard deviation of their IQ scores was 20 points. These facts had emerged from a city-wide study done previously.

To demonstrate that boys who wish to be firemen differ in IQ, it was decided to test the null hypothesis that they are an exact cross-sectional representation so far as IQ is concerned. The significance level of .05 was picked and naturally the hope was to be able to reject the null hypothesis. Sixty-four boys whose primary concern was to be firemen as determined from a questionnaire were selected randomly from the larger population. It turned out that the mean of their IQ scores was 108.

The null hypothesis that these boys were in no way special with respect to IQ was assumed to be true. According to this hypothesis, their 64 IQ scores could simply be viewed as a random sample of IQ scores chosen from the larger population of scores. The assumption that the null hypothesis was true implied that the particular sample of 64 IQ scores could be viewed as merely one of similar randomly drawn samples of 64 IQ scores each. Theorems 8.3 and 8.4 tell us that, under the null hypothesis, the distribution of means of these samples should itself have a mean of 100 and a standard deviation of 2.5 points. This distribution is illustrated in Fig. 11.4. The place of the obtained sample mean of 108 is indicated by an X.

Fig. 11.4 shows that, under the assumption that the null hypothesis is true, we have come upon a sample mean that is eight points above the balance point in a distribution that is normal and that has a standard deviation of 2.5 points. In other words, where we assume the null hypothesis is true, we must conclude that we have come upon a sample mean that is 3.2 standard deviations above its expectancy.

Fig. 11.4

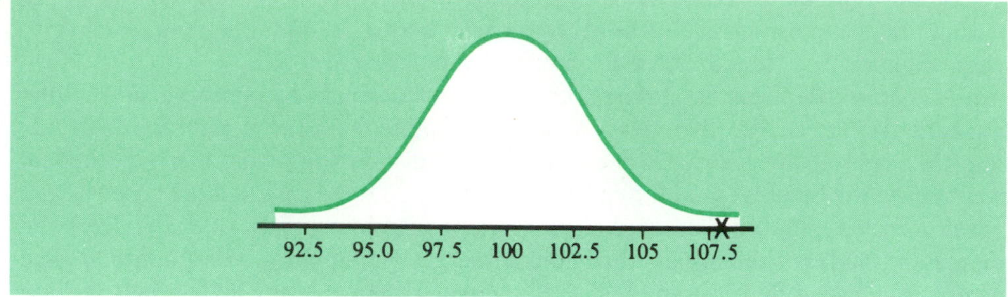

Appendix III tells us that, in a normal distribution, the probability is less than one in 1,000 of getting any one value by chance which has a z score more than three units on either side of the balance point. Thus, assuming that the null hypothesis is true, the obtained sample value was so deviant from the expectancy that its probability of occurrence was less than one in 1,000. Upon obtaining such a finding, our decision would naturally be to reject the null hypothesis, and to construe that boys who wish to be firemen differ as a group in IQ from others their age; that is, accept the alternative hypothesis.

It is instructive to go back over what we have done and to use a formula for the z score of a sample mean in the distribution of sample means. This formula will enable us to do similar problems without spelling out the steps in so much detail. The z score of a sample mean may be determined as follows:

$$z = \frac{\text{distance score for the mean}}{\text{standard deviation of means}} = \frac{\bar{X} - \mu}{\frac{\sigma}{\sqrt{N}}} \qquad (11.2)$$

Here $\bar{X}$ stands for the obtained value of the sample mean (which in our problem was 108); μ stands for the mean of the distribution (we were working under the assumption that $\mu = 100$); and σ stands for the standard deviation of the individual's scores (which in our case was known to be 20) for the entire population.

To substitute the appropriate values in Formula (11.2):

$$z = \frac{\bar{X} - \mu}{\frac{\sigma}{\sqrt{N}}} = \frac{108 - 100}{\frac{20}{\sqrt{64}}} = 3.2$$

It was after using the z score value of 3.2 and making reference to Appendix III that the decision was made to reject the null hypothesis at the .05 level of significance. We could also have rejected the null hypothesis on the basis that our calculated value, 3.2, is larger than the 1.96 needed to reject at the .05 level of significance. (See Fig. 11.5).

Fig. 11.5

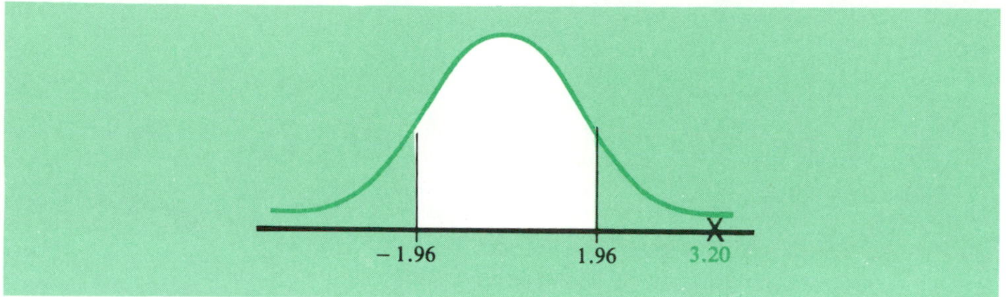

Problem 3. We shall now look at one more problem. It was decided to test the hypothesis that eighth-grade students in a particular school are a cross-sectional representation of students of their age in ability to spell. Actually these students had been taught spelling in a different way, which some contended was better and others contended was worse. At any rate, it was decided to test the null hypothesis at the .01 significance level. A city-wide spelling test was chosen for use, partly because a vast number of students like those to be used were known to have achieved a mean of 72 in this test. It was also known that the standard deviation of their scores was 12 points. There were 36 students in the eighth grade whose spelling scores were used as the sample for testing the hypothesis. The average spelling mark of these 36 students turned out to be 74.

Hypothesis Test for Problem 3

1. H_1: Eighth graders at the school differ from other eighth graders in spelling achievement. ($\mu \neq 72$)

   H_0: Eighth graders at the school do not differ from other eighth graders in spelling achievement. ($\mu = 72$)

2. .01 level of significance

3. $N = 36$ $\quad\quad \bar{X} = 74$

4. Assume H_0 is true. ($\mu = 72$)

$$z = \frac{\bar{X} - \mu}{\dfrac{\sigma}{\sqrt{N}}} = \frac{74 - 72}{\dfrac{12}{\sqrt{36}}} = 1$$

For a two-tail test at the .01 level of significance the critical z values are ± 2.58.

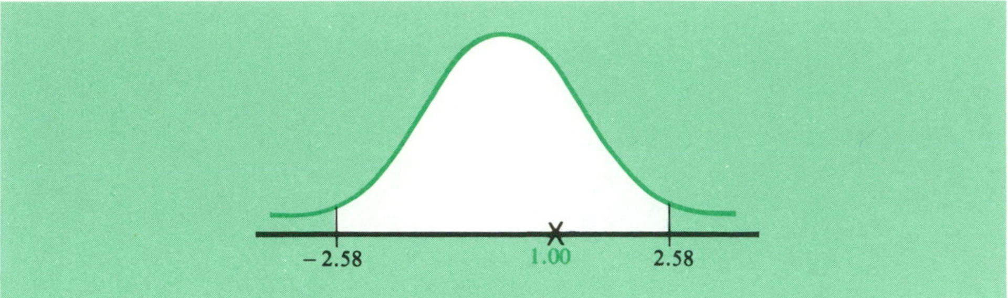

5. Do not reject H_0.
6. Do not accept H_1: We have failed to show that the eighth graders from the school perform any differently than other students.

11.5 Some New Language

A sharp distinction must be made between what is called a population value and one obtained from a sample. For instance, the mean IQ of boys in a given community might be 100, whereas the mean of a sample of IQs of boys taken from this community might be 93 or 110 or 135. A population value—for instance, the population mean of 100 in this case—is called a **parameter**. In contrast, a value obtained from a sample is called a **statistic**. The standard deviation of scores in a population would also be a parameter. Note in Problems 1 and 3 of the previous section that the hypotheses were stated first in words and then in terms of parameters ($P = 2/3$, $\mu = 72$).

Any statement of a hypothesis is one about a parameter. To test a hypothesis, we gather a random sample and compute from it some statistic. We then use the statistic to facilitate our reasoning about whether or not to reject the hypothesis. We have already used the phrase "to test a hypothesis at the .05 significance level." By this phrase, it should have been clear that we meant that we had chosen to test the hypothesis using the significance level of 5/100. In the future, we shall regularly refer to testing a hypothesis at a given significance level, by which it shall be meant merely that the given significance level was selected.

11.6 One-Tail Tests

So far our purpose has been to reject the hypothesis that a parameter—such as a population mean—was some hypothesized value. Our procedure was to test the null hypothesis that the mean had the particular value and to reject it upon finding a sample value sufficiently above or sufficiently below the parameter value. For instance, the theory (alternative hypothesis) was stated that boys of age 8 who want to be firemen differ from the larger group in IQ. The null hypothesis that these boys have a mean IQ of 100 was tested, and the hope was to be able to reject this null hypothesis upon finding a sample of would-be firemen whose mean IQ was either significantly higher or significantly lower than 100.

Now consider the situation in which a researcher wishes to show that members of some particular group *differ in some specified way* from those in a larger population. That is, we not only speculate that our sample value will differ from the parameter, but we also speculate about the direction in which it will differ. As an example, suppose we speculate that boys who wish to be firemen tend as a group to have higher IQs than members of the large group. Accordingly, we predict that the mean IQ in the sample of would-be firemen will be higher than 100. Now we are no longer concerned with merely finding a discrepancy, but with predicting what statisticians call the "direction" of the discrepancy as well. To find a sample mean even ten standard deviations lower than the parameter value will mean that we are wrong.

To proceed in this case, we begin by stating our null hypothesis—that the mean IQ of would-be firemen is identical to that of the larger population. We assume it is known that the distribution of IQ scores of the larger population has a mean of 100 and a standard deviation of 20 points. Let us say once more that we decide to test the null hypothesis at the .05 significance level and that we obtain the IQs of 64 boys who wish to be firemen. Once more we look at the sample mean, and under the null hypothesis we assume that this mean is that of a random sample drawn from the larger population of IQ scores. In other words, we assume that the fact that we chose would-be firemen does not constitute a selective factor, so that we can construe our sample mean as one in the distribution of sample means that would be obtained if a vast number of equal-sized samples had been drawn from the larger population. If the null hypothesis is true, the distribution of sample means that would turn up should look like Fig. 11.6.

The question now is, which sample means, if obtained, should lead us to reject the null hypothesis and consider our theory confirmed? To answer this question, remember that we have already set the .05 significance level for use in testing the null hypothesis. This means we have decided that, when the null hypothesis is true, we shall expose ourselves to the risk of making a type-one error exactly 5/100 of the time. We must pick some interval such that when the null hypothesis is true only 5/100 of the sample means which would be obtained in the long run would fall into that interval.

Our theory is that the mean of would-be firemen is more than 100. Our null hypothesis to be tested is that this mean is 100. Thus we certainly do not want to reject the null hypothesis when we obtain any mean which is less than 100. We want to pick the interval such that, when our theory is correct, we have the best chance to confirm it. Therefore we choose the interval containing the largest 5/100 of the sample means that would turn up.

According to Appendix III, the largest 5/100 of the terms in a normal distribution are the terms that are at least 1.65 standard deviations above the mean. In the distribution of sample means in Fig. 11.6, 1.65 standard deviations amount to 4.1 points. $[(1.65)(20/\sqrt{64}) = 4.1]$ Thus the largest 5/100 of sample means that would be obtained if the null hypothesis is true are those with values more than 4.1

points above the hypothesized population mean. In other words, the largest 5/100 of the sample means that would be obtained are those with values greater than 104.1 points. These are represented by the shaded region in Fig. 11.6.

Fig. 11.6

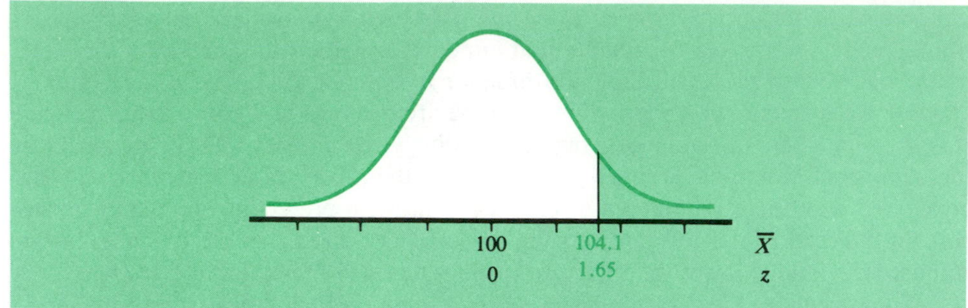

Our decision rule must be to reject the null hypothesis and to consider our theory confirmed when a sample mean larger than 104.1 turns up. Otherwise the decision is to fail to reject the null hypothesis. We have found the particular .05 level decision rule that makes the most sense in view of our theory. Another way of looking at this same rule would be to convert the sample mean to a z score and reject the null hypothesis when it exceeds or equals the critical 1.65 z score as is also indicated in Fig. 11.6.

The test of significance described is called a *one-tail test* because the rejection region is completely in one tail. Suppose we were trying to prove that the mean IQ of would-be firemen is less than 115. Then we would have wanted to reject the hypothesis that the mean is 115 only when we obtained a sample mean sufficiently deviant on the other side. We would have used analogous reasoning and the rejection region would have been the shaded area in Fig. 11.7. We have again assumed a standard deviation of 20 for the population and a sample size of 64. The reader should verify that the cut-off point is 110.9 for the .05 level.

Fig. 11.7

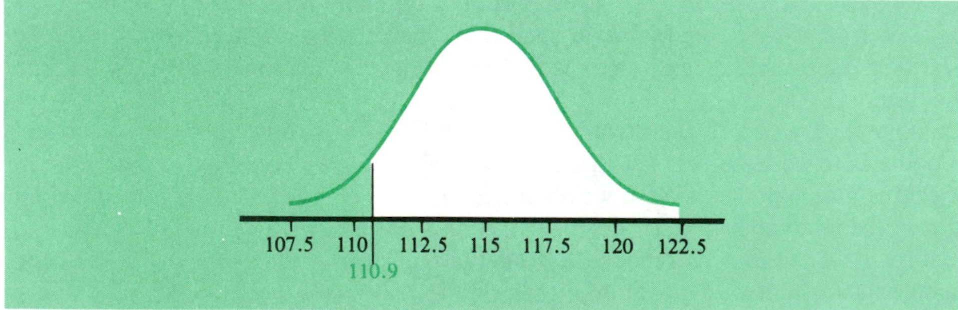

When we were merely trying to prove that a mean was different from some specified value, we used a *two-tail test*. We compared our sample mean with the hypothetical expectancy and we construed an extreme deviation from the expectancy in either direction as evidence contradicting the hypothesis. **But when we are trying**

to prove that the parameter differs in *some specified direction* from a hypothetical value, we use a *one-tail test*. The rejection region still is determined by the size of the significance level, but the logic of the problem tells us to put the entire rejection region in one tail. If we are trying to prove that the mean is larger than some value, we put the entire rejection region in the upper tail. If we are trying to prove that the population mean is smaller than some value, we put the entire rejection region in the lower tail.

The researchers whose theory predicts *how* the obtained sample mean differs from the expectancy profit by using a one-tail test. They give themselves a better chance to prove their theory using a one-tail test than by using a two-tail test. For instance, in our first illustration, if we were using a two-tail test our rejection region would be represented by the shaded area of Fig. 11.8.

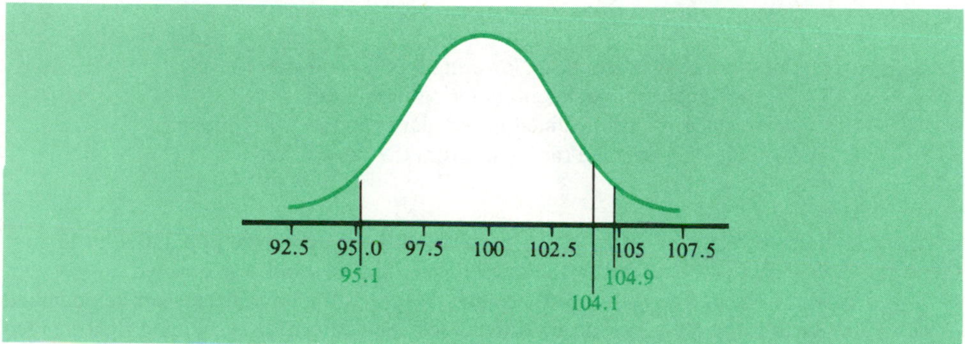

Fig. 11.8

Values between 104.1 and 104.9 would not have been sufficiently deviant from the expectancy to enable us to prove our theory using a two-tail test. But to obtain a value in this interval would allow rejection of the null hypothesis using a one-tail test.

In essence, in the case of the two-tail test, the theory merely predicts that a sample mean will depart from the expectancy. We consider the theory proven if the mean does in fact depart by a particular amount from the expectancy. In the case of the one-tail test, the theory predicts that the sample mean will be on a particular side of the expectancy. The fact that this prediction comes true is itself partial confirmation of this theory. In other words, it constitutes some evidence for the theory, and thus the magnitude of the departure needed to convince us fully of the theory is less than in the case of the two-tail test, since the prediction of direction has already been borne out. The one-tail test procedure is used to test hypotheses about other kinds of parameters. For instance, the procedure is sometimes used to prove that a population variance is larger than some specified number. We shall use other applications of a one-tail test in later examples.

Because they are used so often, the critical z values found in Table 11.1 should be memorized or kept handy for reference.

Table 11.1 Critical z Scores

	.05	.01
Two-tail	1.96	2.58
One-tail	1.65	2.33

Problem Set A

11.1 Classify the following hypotheses as to form, either null (H_0) or alternative (H_1).

a. $\mu = 50$
b. $P \neq .75$
c. The new drug is ineffective in lowering blood cholesterol levels.
d. The agency shows prejudice in its hiring practices.
e. Cigarette smoking has declined in popularity in the last 10 years.
f. The sample was selected randomly from the population.
g. $\mu \leq 100$
h. $p > .15$
i. Females perform better on the dexterity test than does the population at large.
j. Production of coal has not increased since the new bill was enacted.

11.2 Hook and Hack, Inc., is a major manufacturer of golf clubs. Because of low demand they historically have produced only one left-handed set for every nine right-handed sets. The company recently surveyed a number of pro shops across the country to determine if they should change the proportions of left- and right-handed sets produced. It was learned that of 350 sets sold, 40 were left-handed sets. Use Formula (11.1) to perform a two-tail hypothesis test at the .05 level of significance to see what decision the company should make.

11.3 A trash hauling company sets up its crews and schedules under the assumption that the typical family on one of its routes produces 22 pounds of garbage and trash per collection period with a standard deviation of 3 pounds, a result obtained from a five-year-old study. The union representing the workers claims that the mean has significantly increased and therefore the company should assign more men per crew. To substantiate its claim the union surveys 64 families at random and finds the mean output per family to be 24 pounds. Use Formula (11.2) to see if the union's claim is supported by a one-tail test at the .05 significance level.

11.4 A.W.L. broadcasting network has been severely criticized by civil rights groups who claim that women play major roles in only one-fourth of their productions. As a result, the network revises its policies. A study following the revisions indicates that of 45 major productions, women play major roles in 14.

a. Classify this problem as either a mean or proportion problem.
b. Should a one- or two-tail test be performed?
c. Have the revised policies been effective in increasing female participation in major roles at the .05 significance level?

11.5 Five years ago a large nursery sold hundreds of trees to the city through the Department of Parks and Recreation with the promise that within five years they would reach a mean height of 5½ feet with a standard deviation of 9 inches. This week the city manager randomly selected 81 trees and found that their mean height is 5 feet.

a. Classify this problem as either a mean or proportion problem.

b. Would the sample results support a city request for a partial refund from the nursery due to a breach of contract at the .01 significance level?

11.6 In the past it has been found that annually 18% of the members of "Dessert Raters," an international gourmet club, get *fed up* and resign their membership in the group. This year a random sample of 225 people who were members at the beginning of the year shows that only 30 have resigned. At the .05 significance level does this indicate a decrease in the desertion rate of "Dessert Raters"?

11.7 Safety standards dictate that the Ph level (measure of acidity and alkalinity) for a public swimming pool be 7.5 with a standard deviation of .12. Because of numerous complaints regarding the level at Lazy Brook Pool, an investigator is sent to take water samples on 36 randomly selected occasions. The mean of the 36 cases is computed to be 7.39. Do the results call for corrective action at the .01 significance level?

11.8 a. Find the critical z score for a two-tail test at the .07 significance level.

b. Find the critical z score for a one-tail test at the .04 significance level.

c. Find the critical raw score values for a two-tail test at the .03 significance level if the distribution of sample means is normal with a mean of 110 and a standard deviation of 20.

11.9 The supervisors at Smythe Industries are responsible for seeing that their drilling machines function properly. This means that the size of the holes drilled must have a mean of .75 inch in diameter with a standard deviation of .08 inch. Each day 49 holes are selected at random from each machine and measured to determine if the machine is working properly.

a. What mean diameters should cause the supervisor to call in a repairman, if the probability of a type-one error must be held to .01?

b. Suppose that the mean diameter for the holes measured from machine A is .74 inch. Should the machine be repaired?

c. Suppose that 100 holes are used in the sample rather than 49. Now what mean diameters in the sample should cause rejection of the null hypothesis?

d. What happens to the rejection area as sample sizes get larger?

11.10 A manufacturer of light bulbs claims a mean life of 1,000 hours for his product, with a standard deviation of 50 hours. The manufacturer tests his claims semiannually to make sure his advertising does not need to be changed.

a. Test at the .05 significance level if a random sample of 25 bulbs has a mean life of 875 hours.

b. Find the largest value (less than 1,000) needed for rejection of H_0: $\mu = 1{,}000$.

c. Suppose the sample contained 100 rather than 25 bulbs. Should the null hypothesis be rejected?

11.11 Sam Suave has been approached to run for the city council. He agrees to accept only if it can be shown that more than 45% of the voters currently support him. He samples 300 voters and finds 145 supporters. If he is willing to accept only a 1% chance of running without more than 45% support, what decision should he make?

11.12 Rising utility rates have been a major concern of Bullock City's Citizen Coalition. They have been monitoring a random sample of monthly bills to see if the city-owned Natural Gas Company has kept its promise not to allow the mean charge per residence to exceed last year's mean of \$32.85 (with a standard deviation of \$8.75). Of 65 bills monitored, the mean charge was \$36.79. Should the coalition confront the company with the evidence and claim that it has failed to honor its commitment to hold a lid on increases? Use the .05 level of significance.

11.13 The manager of a new restaurant, Eland's Eatery, estimates that the typical group of patrons will take a mean of 45 minutes to eat with a standard deviation of 10 minutes.
  a. He times the first 30 groups in his establishment to see if he needs to revise his estimate. If the mean for the 30 groups is 42 minutes, what conclusion can be drawn at the .05 significance level?
  b. If the mean for a certain number of groups is 49 minutes and this is the smallest mean greater than 45 needed to reject the null hypothesis at the .05 level, how many groups were sampled and timed?

Problem Set B

11.14 Classify the following alternative hypotheses as calling for either a one- or two-tail test.
  a. The proportion of users has significantly increased.
  b. Students of today react differently.
  c. $\mu > 100$
  d. $P \neq .7$
  e. The sample is not random.

11.15 A certain network maintains that switching its highest rated situation comedy, "Fanny and Pheenie," from Monday to Friday night will have no effect on the show's popularity, which is currently 53% of the viewing audience. The sponsor is not so sure, but has no opinion as to whether the change will be beneficial or harmful. A trial Friday night showing is agreed to and 500 homes in which someone is watching television are randomly surveyed. Of these viewers, 48% are watching "Fanny and Pheenie." Use Formula (11.1) to determine if the sponsor's belief is supported at the .05 significance level.

11.16 A large corporation, Swindle and Sons, wants to move a portion of its operation to a new locality. As an inducement for employees to follow the move, it promises a bonus to anyone who relocates. Furthermore, it promises that the mean bonus will be 10% of the employee's annual salary. (The standard deviation is 2%.) Leigh Z. Mann accepts the move but is unhappy with her bonus, which is only 3%. To see if the company was lying, she questions 25 of the other employees as to their percents. The mean is 9½%. Use Formula (11.2) to perform a one-tail test at the .05 significance level to see if the company was lying.

11.17 A manufacturer of steel cable formerly claimed a mean breaking strength of 55,000 pounds, with a standard deviation of 500 pounds. Because of an improved manufacturing process, the manufacturer now claims an increased breaking-strength value. A sample of 50 cables is tested and the mean breaking strength is found to be 55,250 pounds.
  a. Classify this problem as either proportion or mean.
  b. Does this situation call for a one- or a two-tail test?
  c. Is the manufacturer's claim supported at the .01 level of significance?

11.18 The television set was once considered a luxury. Sociologist Sally Social believes that this is no longer the case and sets out to establish that the proportion of homes below the government-established poverty level that have a TV set is higher than the 59% of such homes that had sets five years ago. Of 36 qualifying homes surveyed, 24 had sets.

a. Classify this problem as either proportion or mean.

b. Do the sample results support Ms. Social's contention at the .05 significance level?

11.19 Last year the state legislature passed a bill aimed at decreasing the mean expenditure of college-age students on package liquor. At the time, studies indicated the mean to be $48 per year per student, with a standard deviation of $8.60. This year a sample of 85 college students yields a mean of $45. At the .01 significance level, does this indicate that the bill was successful?

11.20 The Department of Recreation and Tourism, in an attempt to bolster the state's depressed tourist industry, has begun a campaign to encourage vacationers to remain in the state for longer periods of time. Studies for the past five years indicate that only one-fourth of all tourists stay three or more days. A recent survey of 200 tourists reveals that 55 of them stayed at least three days. At the .05 level of significance, does this support the campaign's effectiveness?

11.21 a. If a two-tail test is performed and the probability of a type-one error is held to .09, what are the boundaries of the rejection region in terms of z scores?

b. If the critical z score for a one-tail hypothesis test is 1.28, at what level of significance is the test being performed?

11.22 The manufacturer of a popular preservative for beef franks contends that this product will preserve franks kept at 38 degrees for a mean of 35 days with a standard deviation of 4.5 days.

a. What would be the critical number of days needed to reject this contention in a two-tail test at the .01 significance level if 144 franks are tested?

b. Suppose the mean time it takes the 144 franks to spoil is 36 days. What conclusion should be drawn? What conclusions should be drawn for 34.5 and 34 days, respectively?

c. Answer part b based on a sample of 25 rather than 144 franks.

11.23 The mean weight of lobster tails sold by Davey Jones Locker is 8 ounces with a standard deviation of 1.5 ounces. Chef Liz Luscious orders 75 of them to serve a New Year's Eve crowd. The 75 tails weigh 35.625 pounds (570 ounces). Did the chef receive 75 tails selected at random by the locker? Use the .05 significance level.

11.24 Lush Lawn, Inc., a monthly lawn care service, promises to retreat a yard free of charge if a customer is dissatisfied with weed, insect, or crabgrass control. If more than 30% of the customers consistently request a second treatment, the company will not make a profit. A random sample shows that of 300 treatments, 32% had to be repeated. At the .05 significance level, does this mean the company failed to make a profit?

11.25 Dr. Coughman, a leading allergy specialist, has developed what he believes to be a breakthrough in the treatment of allergic scratchitis. Only 65% of patients respond to conventional treatment. If Dr. Coughman tried his new treatment on 64 sufferers and was able to conclude at the .01 level of significance the superiority of his method over conventional ones, at least how many patients responded favorably?

11.26 During last year's telethon, the mean telephone pledge totaled $13.50, with a standard deviation of $3. The mean of a sample of the first pledges this year is $15. Since its z score is 3.0, it can be concluded that the mean pledge is running ahead of last year. How many pledges were in the sample that led to this conclusion?

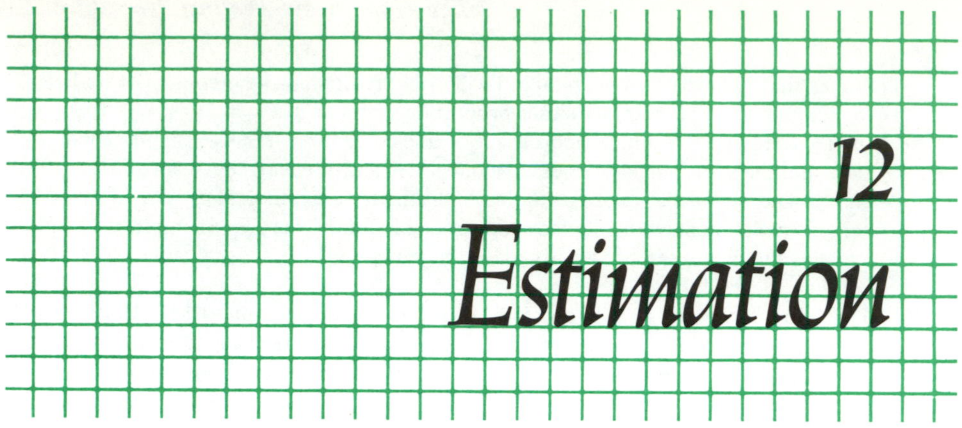

12 Estimation

12.1 Introduction

Suppose it is planned, as part of an experiment, to test the hypothesis that the mean IQ of eighth grade, private-school children in a large city is 110. This time, however, the standard deviation of the IQ scores of the members of the population is *not known*. Therefore, when the mean of the obtained sample is found, it is not possible to use the ordinary procedure for testing the hypothesis, as discussed in Chapter 11. To use that procedure it was necessary to know the value of the population's standard deviation.

The situation described is almost invariably the one that actually arises. A researcher who tests a hypothesis has no more knowledge than that contained in the data gathered. In particular, the researcher has no definite knowledge of the standard deviation of scores in the distribution from which the sample came. What must be done first in order to test a hypothesis about an unknown population mean is to make some kind of enlightened guess about the value of the population's standard deviation.

The problem of using data in a sample to make an enlightened guess of a population value arises in diverse contexts. The guess that one makes is called an **estimate**. In this chapter we shall discuss various considerations that arise when one makes an estimate, and it will soon be seen that there is more to the issue of estimation than meets the eye.

12.2 Unbiasedness of an Estimate

To begin with, we shall consider the simplest case—namely, the case when our purpose is to use the data comprising a sample in order to estimate the mean of the population from which it was drawn. To be concrete, let us suppose that we are researchers whose task it is to estimate the mean IQ of the private-school children already described. We obtain a sample of five IQ scores of children randomly selected from this population of interest. These scores turn out to be 90, 100, 105, 145, and 100. Our task is to derive from this set of scores the best possible estimate of the mean of the population from which they came. The totality of our knowledge consists of these five scores. The question is, what shall we do to arrive at the best possible estimate? In other words, what measure shall we use as our **estimator** of the mean IQ of the private-school population? At this point, note that the word *estimator denotes the kind of measure used for the purpose of estimation. An estimate is the numerical value of an estimator* derived from a particular sample.

The answer, as one might guess, is that the proper procedure would be for us to compute the mean of the sample of five scores. We would then use this obtained sample mean, whatever it is, as our estimator of the population mean. The mean of the five scores turns out to be 108 [$\bar{X} = 108$]. Note that this particular estimate of 108 is merely a guess made on the basis of the sample obtained. To have obtained a different sample would almost certainly have meant to have arrived at a different guess.

The next question is more difficult: Why was it proper for us to use the mean of our sample as the estimator of the population mean? What made it better to use this estimator than, say, the midpoint between the highest and lowest value in the sample? That estimator would yield the value 117.5 as an estimate. What made it better than using the largest sample value as an estimator? That estimator would yield an estimate of 145. To answer this question leads us to consider the more general issue of how to evaluate a good estimator. Actually there are many criteria for a good estimator, but here we shall consider only the most important one—namely, that of *unbiasedness*.

The way to evaluate an estimator is to evaluate the estimates that it would send forth from a multitude of different random samples drawn from the same population. (Of course in an actual experiment only one such sample is available.) Let us suppose this time that we have obtained five different IQ samples of five cases each. The purpose is to estimate the mean of the population of IQ scores from which these samples came, which in practice would be unknown to us.

In Table 12.1 the values of the five different random samples are listed, and the means of these samples are indicated.

Table 12.1 Samples

	1	2	3	4	5
	90	120	105	150	110
	100	110	120	113	95
	105	100	95	107	115
	145	115	130	90	95
	100	90	115	115	110
$\bar{X}$	108	107	113	115	105

Suppose now that we had access to the data comprising only the first sample. Using the sample mean as our estimator, the estimate we would obtain would be 108. Similarly, the estimate we would obtain from the second sample would be 107; from the third, 113, and so on. The mean of the five estimates listed in Table 12.1 is 109.6.

Consider now what would happen if we added more and more estimates from random samples of five cases each. As we added more estimates, the mean of these estimates would tend to get closer to the value of the unknown population mean. The reason is that, as more and more random samples are drawn from a population, the mean of their means approaches the mean of the population itself. For instance, by the time we had collected 5,000 samples, the mean of their means would be unlikely to differ from the population mean by as much as half a point.[1]

To sum up, suppose in general that a population mean is to be estimated and that the estimator to be used is the mean of a random sample drawn from the population. The estimate derived from one particular random sample may be relatively accurate or not. But were more and more estimates to be obtained from different samples, the mean of these estimates would approach the population mean. The sample mean, because it has this property, is said to be an unbiased estimator of the population mean. More generally, **an estimator of a parameter is called unbiased if the mean of estimates obtained from independent samples drawn from the population would approach the population value, were more and more samples to be included.** It is not implied that each new sample estimate must bring the mean closer to the unknown population value. The implication is only that, as new sample estimates are brought in, the *trend* is for the mean of these estimates to get closer to the population value.

Now we shall turn to the problem of computing from a sample an unbiased estimate of the variance of the population from which the sample was drawn. We shall contrast the case in which the mean is known with that in which it is not known. For convenience, we shall use for illustrative purposes the same data as before. That is, we are selecting our terms from a population of IQ scores of eighth grade private-school children. In theory we could start listing the scores as if we were able to look at each one in the vast population. We shall say that the first score we come upon is 90, the next is 100, the next is 105, and so on. The first 15 scores we might come upon are listed in Column (1) of Table 12.2.

To get the actual population variance, we would have to subtract the true population mean from each score. Let us now pretend that we know this true population mean and that it is 110 [$\mu = 110$]. The subtraction was done for the 15 scores in Table 12.2. The distance scores appear in Column (2). Were we actually to have obtained distance scores for all the terms in the population, their sum would be zero since they are distances from the mean (Theorem 2.1). However, our concern is with the squared distance scores in Column (3) of Table 12.2. It is the mean of all the squared distance scores that is the variance. It is this that we would like to estimate.

[1]To see why this is so, we may view the mean of the 5,000 sample means as an average of one gigantic sample of 25,000 cases. Even if the standard deviation of the IQ scores were as big as 20 points, there is little chance that the mean of a sample of 25,000 cases would differ from the population mean by as much as half a point. The reader should be able to verify this fact by using Theorem 8.4.

Table 12.2 Variance Using Population Mean

	(1)	(2) $X - \mu$*	(3) $(X - \mu)^2$	
1	90	−20	400	
2	100	−10	100	
3	105	−5	25	
4	145	35	1,225	
5	100	−10	100	1850
6	120	10	100	
7	110	0	0	
8	100	−10	100	
9	115	5	25	
10	90	−20	400	625
11	105	−5	25	
12	120	10	100	
13	95	−15	225	
14	130	20	400	
15	115	5	25	775

*μ stands for the population mean.

Suppose we had access to only the first five scores, and our purpose was to estimate the population variance. If we knew the population mean was 110, we might accordingly subtract 110 from each of these scores. We would obtain the five squared distance scores in Column (3), and their mean would give us an *unbiased estimate of the variance*. In other words, if many estimates were made in the same way, each time using a different sample of size five, the mean of the estimates would approach the population variance. After all, we would simply be performing on a small sample what should be performed on the entire population in order to get the population variance. The sum of the first five squared distance scores is $400 + 100 + 25 + 1{,}225 + 100 = 1{,}850$. Our estimate of the population variance then would be $1{,}850/5 = 370$, and that estimate would be unbiased.

However, the usual situation is quite different. *We usually have access to a sample of scores from a population but do not have knowledge of the value of the population mean*. Let us imagine ourselves in this position, once more with access to only the first five scores. The most obvious thing that may come to mind may be to estimate the population mean first, by determining the sample mean. Then using our estimate of the population mean in place of the real one, it appears that we might be able to proceed as before. Table 12.3 shows what would happen if we were to proceed in this way.

To begin with, when we compute the value of the mean from our sample of five cases, it turns out to be 108. In Table 12.3, Column (2), we have subtracted this value from each of the terms in the sample, and in Column (3) we have squared the distance scores thus obtained.

Table 12.3 Variance Using Sample Means

(1) X	(2) $X - \bar{X}$	(3) $(X - \bar{X})^2$
90	−18	324
100	− 8	64
105	− 3	9
145	37	1,369
100	− 8	64
		1,830

$$\frac{\Sigma(X - \bar{X})^2}{N} = \frac{1,830}{5} = 366$$

The sum of the squared deviations from our sample mean turned out to be smaller than the sum of squared deviations of the same five terms from the population mean when we used it. Using the population mean, the sum of the squared deviations was 1,850. Here it is 1,830. Thus our estimate of the variance using the sample mean is smaller than the estimate we just made using the population mean.

What happened here was not coincidence. It is always the case that a sum of squared deviations from a sample mean will be as small or smaller than what it would have been had we used the population mean. The reason is given by Theorem 2.2. For any set of terms, the sum of the squared differences of the terms from their (own) mean is less than the sum of the squared differences from any other point. We have been considering terms that comprise a sample, and their mean is the sample mean. With regard to these five terms, the population mean is an alien point. Thus it turned out necessarily that the mean of the squared deviations of these terms from their sample mean was as small as or smaller than the mean of their squared deviations from the population mean.

Now look at the second sample of five terms listed in Table 12.2. There the mean of the sample is 107. Once again the sum of the squared deviations from this sample mean turns out to be less than what it is when the population mean is used as a reference point. The reader should verify that this sum of squared deviations is less than 625.

What has been said boils down to the fact that one cannot make an unbiased estimate of a population variance from a sample by simply getting the mean of the squared deviations from the sample mean. Such an estimate would *not be unbiased*; or, more specifically, in the long run the mean of such estimates would be too small. So far then, three things have been established. First, the mean of a random sample is an unbiased estimator for the population mean. Second, the mean of the squared differences of sample values from μ, the population mean, is an unbiased estimator for the population variance. Its use is predicated on a knowledge of μ, however. Third, the mean of the squared differences from the sample mean, $\bar{X}$, is not an unbiased estimator for the population variance. In the long run, it yields estimates that are too small.

We must now consider an important concept in statistics—namely, that of **degrees of freedom.** Once this concept becomes clear, we can go on to consider how to derive from a sample an unbiased estimate of the variance of the population from which it came even when the population mean is unknown. In fact, the concept of degrees of freedom is quite important and we shall refer to it in a later context.

12.3 The Concept of Degrees of Freedom

A meaningful distinction must be made between the number of independent facts that have been ascertained and the total number of facts ascertained in the data gathering. We shall look at this distinction in a general way first and then we shall consider it specifically later, because it is highly relevant to the topic of estimation.

In the simplest case, suppose we are told that John wrote an essay on which he got a score of 87. We are told also that he lost 13 points on the essay and, furthermore, that the top possible mark on the essay was 100.

It is clear that from any two of the given facts we might have deduced the third. Thus we have actually been given only two independent facts, though we have been given three facts altogether. A fact is said to be *independent* of one or more other facts when it is new in the sense of not having been implied by them. A fact is said to be *dependent* of one or more others when it might have been deduced from them.

It would not be advantageous for our purposes to single out two particular facts from the three concerning John's performance and to say that they are independent and that the third one is dependent upon them. It is sufficient to say that only two independent facts are contained in the set, and that the facts that comprise the total set are dependent. We use the phrase *degrees of freedom* to indicate the number of pieces of information that are free of each other in the sense that they cannot be deduced from each other. Thus two degrees of freedom comprise our information concerning John's performance on the essay question.

Note that we can communicate the total information concerning John's performance by presenting two facts. These might be John's score of 87 and the total possible score of 100. Or they might be John's score of 87 and the number of points he lost, which was 13. Or finally, we might have presented the number of points John lost, 13, and the total possible score on the test. The minimum number of independent facts that convey the total information is always the same no matter which facts we choose. Therefore we can always demonstrate the number of degrees of freedom involved by presenting independent facts that account for the totality of information. The number of facts entailed comprises the number of degrees of freedom involved. Incidentally, in the case described, the total possible test score would undoubtedly be known to be 100, so that for all practical purposes there would be one degree of freedom of interest. That would be the degree of freedom that provides information concerning John's performance.

The statistician must often make a careful count of the number of degrees of freedom that go into his making of an estimate. Actually we are discussing elementary techniques in this book, and there are simple rules for figuring the number of degrees of freedom involved in making an estimate whenever the need to do so arises. As might be guessed, one can put relatively little stock in an estimate when it is based on relatively few degrees of freedom. **As the number of degrees of freedom increases, the dependability of the estimate increases.**

The main concern is with making the distinction between the number of degrees of freedom involved and the total number of facts reported, which may be considerably larger. A serious error is apt to occur when an estimate based on relatively few degrees of freedom is mistakenly thought to have been based on many more. The error is in the direction of overrating the soundness of the estimate.

The following medieval folk tale, treated by several German poets, may help the reader to distinguish between degrees of freedom and observations in general. The city states had not yet welded themselves together as the Prussian empire. The enlightened monarch of Wierstrasse, a mythical northern state, had died leaving his twin sons, Hans and Fritz, as heirs to his throne. The king had left written instructions with the council to choose as his successor the son with the superior talent of prophecy.

In accordance with the will of the king, a contest before the courtiers was arranged so that each boy could demonstrate his skill. Hans was the first to step before the crowd in the marketplace. He handed the prime minister a golden coin with an engraving of his deceased father on one side (heads) and one of the royal gardens on the other (tails). The minister leaned over a wooden table and then tossed the coin high in the air.

Hans shouted, "The picture of my father will face the sky when the coin has fallen." He was correct. A second time the minister threw the coin in the air and Hans shouted, "The picture of my father will be on the underside of the coin this time." Again he was correct, for the coin turned up tails and the picture of his father was on the underside as he had predicted.

The coin was thrown for a third and final time. Hans shouted, "Now the coin will have the tail side facing the sky." Again he was correct. The crowd applauded. Hans had made three correct predictions and his performance was over.

His brother Fritz now asked for the golden coin and handed it to the minister. The minister threw the coin in the air and Fritz shouted: "I make two predictions. First, I predict that the picture of my father will face the sky. Also I predict that the underside of the coin will have imprinted on it the picture of the royal gardens." Both predictions were correct. The heads side did face the sky and the tails did rest against the table on which it fell. Fritz smiled.

The minister threw the coin in the air a second time. "Again I make two predictions," Fritz shouted. "I predict that the tails side of the coin will face the sky and that the heads side of the coin will be the underside and rest against the table." Again both predictions were correct. Fritz now turned to the assemblage, for his performance was over.

He said, "I have made four correct predictions, whereas my brother has made only three. I have proved myself superior as a visionary. Humbly and gratefully do I accept the crown of Wierstrasse."

The courtiers were perplexed and began talking among themselves. The murmur grew and then one of them shouted, "Long live Fritz, King of Wierstrasse." Soon the others picked up the chorus. "Long live Fritz, King of Wierstrasse," they shouted three times, when Clarence, the king's philosopher, stepped to the platform and showed his palms to the assemblage asking for silence.

"You have just witnessed a contest," he began. "You have witnessed three correct predictions made by Hans and four correct predictions made by his brother Fritz. Consequently, you esteem Fritz a better prophet than Hans." He paused. Then he said, "Ladies and gentlemen of the court of Wierstrasse—you have been duped!"

Loud talking began in the throng, but Clarence brought silence as he began to speak again. "Let us examine the evidence closely. The evidence in favor of Hans consists of three correct independent observations in the form of predictions made by him." He emphasized the word *independent*. "I say that the observations are independent since knowledge of the outcomes of any two of them would give us no information about the outcome of the third. Our purpose is to estimate the ability of Hans as a prophet. The estimate we make is based upon three independent observations, or *degrees of freedom*, as they are called.

"As for Fritz, our four observations in the form of his correct predictions are not independent. Certain of these observations necessitated others. Fritz made two predictions at each flip of the coin. He predicted what its top side would be and he predicted what its underside would be. At each flip, we made two observations. The two observations made at each flip are dependent in the sense that either observation necessitated the other. Knowledge of the outcome of either one of these predictions would tell us the outcome of the other. The two observations made at each flip are based only upon one degree of freedom, and therefore our complete estimate of Fritz's skill is based upon two degrees of freedom.

"The soundness of an estimate depends upon the number of degrees of freedom used in making it, rather than the number of observations counted blindly. Our estimate of Hans's prophetic skill is based upon three degrees of freedom. Our estimate of Fritz's prophetic skill is based upon only two degrees of freedom."

The legend is that Hans made an excellent king and that his reign was a happy one. The legend emphasizes the importance of counting the number of degrees of freedom that go into an estimate. *The number of degrees of freedom relates to the probability that the estimate is accurate,* or, as we might say loosely, the dependability of the estimate.

12.4 Estimating the Variance

In our example of the population of IQ scores of eighth grade private-school children, we discussed two situations in which we might make an estimate of a population variance. The first occurs when we have the data of the sample and we know the population mean. We shall refer to it as Situation A. The second occurs when we have the data of the sample but have no outside information. We shall refer to the context when this occurs as Situation B.

Remember that degrees of freedom refer to the number of independent observations used to make an estimate, for we are now going to see what happens when we estimate the variance in Situations A and B. Let us return to the population of IQ scores of the eighth grade private-school children, which we are pretending has a mean of 110. In particular, let us again contrast the problem of estimating the population variance, knowing the value of the mean and not knowing it.

To begin with, suppose our sample consists of only one score. We shall say that the IQ of the first child randomly selected happens to be 90. We shall compare what we would do to estimate the population variance in each situation.

Situation A: Since we know that the population mean is 110, we compute the difference between our single score of 90 and this reference point. The difference is − 20 points. The fact of this difference is one piece of information on which to base our estimate. Our procedure would be to square the difference and divide by one to obtain the estimate of the population variance. Note that we are making an estimate based on one degree of freedom:

$X - \mu$	$(X - \mu)^2$
(−20)	400

Our estimate of the variance is

$$\frac{400}{1} = 400$$

Situation B: Now let us see what happens when we proceed in ignorance of the population mean. The mean of our sample of one term naturally has the same value as the term itself. Thus one term gives us no information about the variability of the terms in the population. We cannot estimate the population variance, since one term is not enough to give us one difference from a reference point.

Suppose that we now add a second IQ score to our sample, which we shall say happens to be 100.

Situation A: This second score gives us a new independent difference from the population mean. The first score led to a different score of −20 points and now this second score leads to a difference score of −10. The first difference score would not permit us to infer the other. Note that these two independent differences are what go into the variance estimate. To put it in more technical language, our estimate of the variance is based on two degrees of freedom.

$X - \mu$	$(X - \mu)^2$
(−20)	400
(−10)	100

Our estimate of the variance is

$$\frac{500}{2} = 250$$

Situation B: The two scores give us a new location for the sample mean. Where there are two scores, their mean is the point midway between them. Since the two scores are 90 and 100, their mean is 95. Thus the first difference score with reference to the sample mean is -5 and the second difference score is 5. Remember that the differences of a set of terms from their own mean must add up to zero (Theorem 2.1). Therefore the two differences from the sample mean are mutually dependent. Were we to know either one, we could figure out the other. The two terms thus provide us with only one independent piece of information concerning the population variance. Even though we have a sample of two terms, our estimate is based upon only one degree of freedom.

Now to be general, let us suppose that our sample consists of some larger unspecified number of terms. We shall say it consists of N terms.

Situation A: Each IQ score in our sample gives us an independent difference score. That is, were we to know the value of all but one of these differences, we could still not possibly compute the value of the last one. Thus, when there are N terms in the sample, our estimate of the population variance becomes based on N independent difference scores. Technically, **were we to know the value of the population mean, our estimate of the population variance would be based on *N* degrees of freedom.**

Situation B: This time we have chosen the reference point ourselves. Now were we to know all the differences from the reference point except one, we could deduce the value of the last difference. Because the reference point is the sample mean, the sum of all the differences from this reference point must be just large enough to make the sum of all the differences total zero. For instance, if all but one of the differences total -4, then the last difference must be $+4$ in order to make the sum of all the differences exactly zero. In essence, there are N terms in our sample, which give us N differences. However, we have only $N-1$ degrees of freedom included in our N differences. Technically, **there are *N* − 1 degrees of freedom that go into our estimate of the population variance when the population mean is unknown.**

It may be shown mathematically that to make an unbiased estimate of a population variance from a sample, the proper procedure is to take the sum of the squared differences from the sample mean and to divide it by the number of degrees of freedom involved in making the estimate. In other words, the denominator must not be the number of terms in the sample in Situation B, but it must be one less than that number. For instance, if there are 30 terms in a sample, the procedure must be to compute the sum of the squared differences of these terms from the sample mean. This sum must then be divided by 29 in order to get an unbiased estimate of the population variance.

More generally, we can put what has been said in the form of a theorem.

THEOREM 12.1 Suppose a random sample of N terms is drawn from a population that has an unknown mean and variance. Then the sum of the squared deviations of the N terms from the sample mean divided by $N - 1$ provides an unbiased estimate of the unknown population variance. The square root of this value is an unbiased estimate of the population standard deviation.

The usual letter used to designate a sample estimate of a population variance is s^2. That used to designate a sample estimate of population standard deviation is s. In other words, the value of s^2 provides an unbiased estimate of σ^2 for the population from which the sample was drawn. The value of s is our best estimate of σ for the population from which the sample was drawn.

The formulas for s^2 and s are as follows:

$$s^2 = \frac{\Sigma(X - \bar{X})^2}{N - 1}$$
$$s = \sqrt{\frac{\Sigma(X - \bar{X})^2}{N - 1}} \tag{12.1}$$

Here the random sample consists of N terms. Its mean is $\bar{X}$. The values of X are those of the terms of the sample.

Remember that it is only correct to use the denominator $N - 1$ when the purpose is to obtain a value that is an estimate. That is, when the data are being construed as a sample drawn from some larger population that is of interest, it is proper to use $N - 1$ when estimating the variance of that larger population. When the data themselves are thought of as comprising some total population, then the value computed is σ^2 and not s^2. Under this condition, the denominator must be N. Unfortunately, it has become commonplace in the field to refer to the value of s^2 as a *sample variance*, and to refer to the value s as a *sample standard deviation*. Technically, however, they do not represent the variance and standard deviation of the sample, but rather *unbiased estimators for the variance and standard deviation* of the population. It is especially important for the reader to have the distinction clear because this practice is so prevalent.

Finally, to complete a kind of cycle, let us get back to our initial problem. We are researchers whose task it is to estimate the variance of the scores of eighth grade private-school children in a vast community. We obtain a sample of five IQ scores of children from this community. These scores are 90, 100, 105, 145, and 100. We quickly compute that the mean of this sample, $\bar{X}$, is 108. Now our purpose is to estimate the variance of the scores in the larger population. Following Formula (12.1), we find the difference between each score and the sample mean, square them, and then add up the squares. We then divide this sum of squares by 4 to get what is loosely called a sample variance. This value, which turns out to be 457.5, becomes our unbiased estimate of the population variance. Its square root, 21.4, is our estimate of the standard deviation of the larger population of scores of eighth grade private-school children. The computations appear in Table 12.4.

Table 12.4 Estimate of Standard Deviation

90	−18	324
100	−8	64
105	−3	9
145	37	1,369
100	−8	64
		1,830

$$\bar{X} = 108, N = 5, N - 1 = 4$$

$$s^2 = \frac{\Sigma(X - \bar{X})^2}{N - 1} = \frac{1{,}830}{4} = 457.5$$

$$s = \sqrt{457.5} = 21.4$$

12.5 Computational Formulas for s^2

The formula for s^2 (12.1) is not the simplest one to work with when the mean is not a whole number as is usually the case. In earlier chapters we used what were called computational formulas in addition to the defining formulas, as for instance when we discussed σ^2 and σ. The computational formulas, which make the actual arithmetic work much easier where one wishes to obtain the value of s^2, are as follows:

$$s^2 = \frac{\sum X^2 - \frac{(\sum X)^2}{N}}{N - 1} \quad \text{or} \quad \frac{N\sum X^2 - (\sum X)^2}{N(N - 1)} \qquad (12.2)$$

When the sample values are given in grouped frequency form the applicable formula is

$$s^2 = \frac{\sum FX^2 - \frac{(\sum FX)^2}{N}}{N - 1} \qquad (12.3)$$

In Formula (12.3) X stands for interval midpoint.

12.6 Interval Estimation

So far we have considered how to derive from a random sample the single best guess of the mean and variance of the population from which the sample was drawn. However, as mentioned, even the best guess we make is virtually certain to deviate at least somewhat from the exact parameter value. What is not only desirable but necessary in practical situations is to make some determination of how close the guess is likely to be. That is, even the best guess may be dependable or it may be made as a shot in the dark, and, along with the guess itself, what is needed is some communication of the likelihood of its accuracy.

The language used in this section has been the subject of much controversy among statisticians, and the word "probability" is used here in a sense that is not acceptable to some of them. Therefore especially close attention is required. Once the conceptualization is clear, the use of the word "probability" may be seen as appropriate and need not be a source of confusion.

To begin with, when we make a best guess about a parameter value—like the mean of a population, for instance—we cannot talk about the probability that this guess is correct. For example, suppose that on the basis of a sample we make the guess that the mean height of adult American males is 5 feet 9 inches. It is virtually certain that this guess would prove to be at least somewhat off, were we actually able to measure the heights of all adult American males and average them. Therefore we cannot talk about the probability that such a guess is accurate, if for no other reason than that this guess is virtually certain to be at least somewhat off.

Instead of specifying a single point, the practice we are now going to consider is that of specifying an interval and giving the probability that it includes some unknown parameter between its endpoints. We call this kind of estimate an **interval estimate.** The practice is to specify some desired probability fraction (like 95/100 or 99/100) and then to find two points such that the probability that the unknown parameter is between them is the specified fraction. A typical statement would be that the probability is 95/100 that the mean height of American adult males is between 5 feet 8 inches and 5 feet 10 inches.

Many statisticians argue that to use the word "probability" even here is incorrect on the grounds that a parameter value is fixed and thus one cannot talk about the probability that it is contained in a given interval. For them, the notion of probability is only applicable when one is talking about where a random-sample value will appear.[2]

Before proceeding, let us look more closely at what we mean when we say that the probability is 95/100 that the height of the average American male is between 5 feet 8 inches and 5 feet 10 inches. To begin with, remember how we defined probability. We defined what we called an outcome of an experiment. We then said that the probability of the outcome occurring when the experiment takes place is defined as the proportion of times that the outcome would occur were the experiment to be repeated over the long run. In other words, the probability of a particular coin falling heads on a single flip is the proportion of times that the coin would fall heads were it to be flipped over and over again in the long run.

Our question now is, what do we mean when we say that the probability is 95/100 that the height of the average adult male is between 5 feet 8 inches and 5 feet 10 inches? In particular, what is the outcome of interest, and what is the experiment which we conceive of as being repeated over and over again? Suppose that on Monday we have gathered a single random sample and have set up an interval, and

[2]The interested reader may look further into the topic of logical probability and its use with set theory.

we are predicting that this interval contains the unknown parameter value. We must conceive of ourselves going through the same procedure of gathering a random sample of heights from the same population and each day setting up a particular interval that we hope contains the unknown population mean. On Tuesday, following the same procedure, we might set up a different interval that happens to include the height of the average American adult male. On Wednesday, our sample might be such that the interval we set up does not include this unknown parameter value, and so on. When we say that the probability is 95/100 that the interval we have set up today does contain the unknown parameter value, what we mean is the following: We are using a procedure which, if we repeated it each day, would in the long run lead us to establish intervals 95% of which would include the unknown parameter value.

An analogy should make clear exactly what a probability fraction means when it is given in connection with an interval estimate. When we make an interval estimate, we can conceive of ourselves as operating like a bomber pilot hurtling over an enemy city. The pilot, who is out to destroy a munitions factory, has only one bomb, which he drops through the clouds. The range of the blast of the bomb is, let us say, 10 miles. Thus the pilot knows that if the bomb strikes within 10 miles of the factory, it will destroy it. Otherwise the blast will not extend to the factory and will thus leave it unharmed.

The bomb actually falls on a particular spot that corresponds to the bomber's estimate of the location of the factory. The blast of the bomb has a range of 10 miles and this range corresponds to an interval estimate of a parameter. By now you may have guessed that the exact location of the factory is the unknown parameter. A bomb that catches the factory in its blast is like an interval estimate, which includes the actual value of the parameter between its endpoints.

Now remember the definition of probability. The probability of the outcome of an experiment is the proportion of times that the outcome would occur were the experiment to be repeated indefinitely. Thus, to determine the probability that the bomb blast destroys the factory, we must conceive of the identical mission being carried out over and over again.

Suppose we could show that, during 95 out of every 100 missions in the long run, the pilot would destroy the factory with the blast of his bomb. Then it would be correct to say that the probability is 95/100 that the pilot will destroy the factory with his single bomb blast. Analogously, our purpose is to set up an interval such that were we to repeat our procedure identically, 95/100 of the interval estimates we would make in the long run would include the unknown parameter value. Referring to this fact we shall say that the probability is 95/100 that our interval estimate includes the parameter.

12.7 Interval Estimation of the Population Mean

We are now ready to consider specifically the procedure for making an interval estimate of a population mean. The procedure involves making a best guess of the population mean as before. This time, however, we are going to construct an interval around our best guess such that the probability that the population mean is inside of the interval is some specified fraction. For instance, instead of guessing that the mean IQ of a population is 117, we might now end up saying that the probability is 95/100 that the population mean is between 113 and 121.

It is of course desirable to have an interval estimate as narrow as possible. The narrowness of an interval indicates the specificity of the estimate. For example, compare the following two interval-estimate statements:

1. The probability is 95/100 that the average weight of people with disease K is between 100 and 180 pounds.
2. The probability is 95/100 that the average weight of people with disease K is between 139 and 141 pounds.

The same probability fraction, 95/100, is attached to each statement. However, the narrowness of the interval in the second statement pinpoints the stipulation and makes it much more meaningful.

Note also that **the higher the probability fraction attached to a statement, the stronger is the interval-estimate statement.** For example, compare the following two statements:

1. The probability is 90/100 that the mean life span of fox terriers is between 10 and 14 years.
2. The probability is 99/100 that the mean life span of fox terriers is between 10 and 14 years.

We might say that the second statement is stronger than the first in the probability sense. Consequently, a researcher puts more stock in the second statement than in the first.

We are now ready to attack the problem of making an interval estimate of a population mean. To think in terms of a specific problem, suppose we are interested in setting up an interval estimate of the mean IQ of American medical students. Specifically, we wish the endpoints of our interval to have the probability of 95/100 of including between them the mean IQ of these students. We shall suppose that we know that the standard deviation of the population of medical students' IQs is 16 points. Let us say that we gather a random sample of the IQs of medical students and that it turns out that the mean IQ of these scores is 118.

As mentioned, we are going to conceive of a multitude of researchers, each of whom follows the same procedure as we do. That is, each of these researchers must be conceived of as gathering a random sample of 64 cases from the same population, computing its mean, and setting up an interval estimate as we are about to do. Our purpose now is to discuss a procedure, which, if followed, would lead 95/100 of all the researchers to set up intervals that actually include the unknown population mean.

To proceed, we must reason in a somewhat general way. The population of IQ scores of medical students, from which our sample emanated, has a mean that is of course unknown. Its standard deviation is 16 points. Now conceive of each researcher, including our research team, as computing the mean of an obtained sample of 64 cases. According to Theorem 8.2, the distribution of obtained sample means is normal. Its mean is unknown, but its standard deviation is 2 points.

$$\sigma_{\bar{X}} = \frac{\sigma}{\sqrt{N}} = \frac{16}{\sqrt{64}} = 2$$

$\sigma_{\bar{X}}$ is called the "standard error of the mean" or "the standard deviation of means." In the distribution of sample means each term is the mean of a sample of 64 IQ scores. More specifically, each term in it is the best guess of the mean IQ of medical students that a particular researcher would make, having drawn a particular sample. One sometimes refers to such a best guess as a *point estimate* to distinguish it from the *interval estimate*, which we are now considering.

Since this distribution of sample means is normal and since it has a standard deviation of two points, it follows (according to Appendix III) that roughly 68/100 of the sample means included will be less than one standard deviation away from the unknown population mean. This means that roughly 68/100 of the obtained sample means will be within two points of the population mean.

According to Appendix III, in a normal distribution approximately 95/100 of the terms are less than 1.96 standard deviations from the mean. Therefore in Fig. 12.1 roughly 95/100 of the sample means are within 1.96 standard deviations of the unknown population mean. *More specifically, about 95/100 of the point estimates that would be made by researchers would be within 3.92 points of the unknown population mean.* This is true because 1.96 standard deviations amount to 3.92 points.

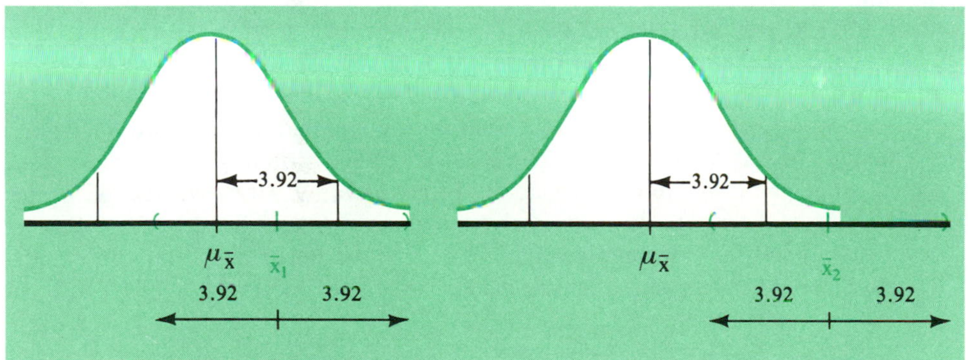

Fig. 12.1

Now let us look at what must be the point of view of any researcher who has computed the mean of this particular sample. That researcher may say, "Of the vast number of my colleagues, 95/100 have computed sample means that are within 3.92 points of the unknown population mean. I may be among this 95/100 or I may be among the 5/100 of researchers whose sample means are more than 3.92 points from the population mean. Since 95/100 of researchers who operated like myself are in the first category, the probability is 95/100 that I am too. In other words, the probability is 95/100 that my particular sample mean is within 3.92 points of the unknown population mean."[3]

Suppose in particular that as a research team we have computed the mean of our sample to be 118 as specified. We are able to say that the probability is 95/100 that our particular sample mean is within 3.92 points of the population mean (which is the average IQ score of medical students). This means that the probability is about 95/100 that the population mean is within 3.92 points of 118. To put it another way, the probability is about 95/100 that the mean IQ of medical students is between 114.08 and 121.92. We have constructed the interval containing all values that are within 3.92 points of our obtained sample mean. This is called a **95% confidence interval** for the mean.

Note that were we to repeat our procedure each day we would get a different sample mean with each new sample. Therefore, each day the center of our interval estimate would be different. We can only say that, on about 95 out of every 100 days in the long run, the interval that we managed to construct would include the value of the unknown mean IQ of medical students. The interval constructed about X_1 in Fig. 12.1 is one such interval. The interval about X_2 does not include μ since X_2 deviates more than 3.92 units from μ.

We might have wanted to construct an interval such that the probability was 99/100 that it included the value of the unknown population mean. In this case, our initial logic would have told us that about 99/100 of the terms in a normal distribution are within 2.58 standard deviations of its mean. Therefore, the probability is about 99/100 that our obtained sample mean is within 2.58 standard deviations of the unknown population mean. And it would follow that the probability is about 99/100 that our obtained sample mean of 118 is within 5.16 [2.58 × 2] points of the unknown population mean. Finally we would have said that the probability is about 99/100 that the unknown mean IQ of medical students is between 112.8 and 123.2. This is called a **99% confidence interval** for the mean.

To sum up, these are the steps we followed in this process: We built our interval estimate around our obtained sample mean. We computed the standard deviation of the mean (which is what the standard deviation of the distribution of a myriad of such means would be). Where our probability fraction was 95/100, we arrived at the z score value of 1.96. The lower end of our interval turned out to be 1.96 standard deviations below our obtained sample mean, and the upper end was 1.96 standard deviations above our obtained sample mean. In terms of a formula, a 95% confidence interval for the mean is

$$\left[\bar{X} - (1.96)\frac{\sigma}{\sqrt{N}}\right] < \mu < \left[\bar{X} + (1.96)\frac{\sigma}{\sqrt{N}}\right] \qquad (12.4)$$

Similarly, a 99% confidence interval for the mean is

$$\left[\bar{X} - (2.58)\frac{\sigma}{\sqrt{N}}\right] < \mu < \left[\bar{X} + (2.58)\frac{\sigma}{\sqrt{N}}\right] \qquad (12.5)$$

[3]To satisfy the mathematician, the researcher might say, "Before I gathered my sample the probability was 95/100 that I would obtain a mean within 3.92 points of the population mean. I have no new information on the subject. Therefore the probability is still 95/100 that my sample mean is within 3.92 points of the population mean."

In Formulas (12.4) and (12.5) μ stands for the unknown population mean. $\bar{X}$ stands for the obtained sample mean, σ stands for the standard deviation of the original terms, N stands for the sample size, and $<$ stands for "less than." Incidentally, $\frac{\sigma}{\sqrt{N}}$ stands for the standard error of the mean, $\sigma_{\bar{X}}$.

In the problem that we just discussed, $\bar{X} = 118$, $\sigma = 16$, and $N = 64$. Therefore, using Formula (12.4), our interval turned out to be

$$\left[118 - (1.96)\ \frac{16}{\sqrt{64}}\right] < \mu < \left[118 + (1.96)\ \frac{16}{\sqrt{64}}\right]$$

$$114.1 < \mu < 121.9$$

Using Formula (12.5), our interval was

$$\left[118 - (2.58)\ \frac{16}{\sqrt{64}}\right] < \mu < \left[118 + (2.58)\ \frac{16}{\sqrt{64}}\right]$$

$$112.8 < \mu < 123.2$$

Note that the 95% confidence interval in Formula (12.4) will always be smaller than the 99% confidence interval in Formula (12.5), as a result of the fact that the value 1.96 is less than 2.58. This makes sense because, where less probability of being correct is demanded, one can issue a statement that is relatively precise. As one demands a higher probability that the interval constructed will actually contain the unknown parameter value, it becomes necessary to construct an accordingly wider interval.

Problem Set A

12.1 Six jars of commercial spaghetti sauce were randomly selected from a grocer's shelf. Their salt contents per one cup serving in milligrams were 912, 1070, 864, 632, 1285, and 450. Now consider the population of all jars of spaghetti sauce and determine

a. the unbiased estimate, $\bar{X}$, of the mean, μ, of their salt contents.
b. the unbiased estimate, s^2, of the variance, σ^2.
c. the unbiased estimate, s, of the standard deviation, σ.
d. the degrees of freedom the estimate in part b is based upon.

12.2 Water samples were taken from 15 North American cities and carcinogen levels measured. Following are the results in parts per billion: 1.5, 7.9, 88, 35, 97.3, 105, 124, 150, 110, 79, 300, 407, 135, 120, and 110.

a. Estimate the mean, variance, and standard deviation (using unbiased estimators) of the distribution of carcinogen levels of all water samples in North America.
b. How many degrees of freedom are the estimates for σ and σ^2 based upon?

12.3 Orchestra conductors as a group appear to possess longevity as far as life spans are concerned. The ages at which 10 conductors selected at random from various countries died are given below.

87.5	88	85
93	95.25	97.75
72.75	78.5	
60.25	83	

a. Determine unbiased estimates for the mean age at death of all orchestra conductors.
b. Find unbiased estimates for the variance and standard deviation and the number of degrees of freedom these estimates are based upon.
c. Suppose it is known that the mean life span for all conductors is 82.5 years. Find an unbiased estimate for the variance that is more dependable than the estimate obtained in part b.

12.4 The lengths of the engagement periods of 50 randomly selected couples are given here in months.

Interval	*X*	*F*
30–36	33	2
24–30	27	5
18–24	21	7
12–18	15	9
6–12	9	15
0–6	3	12

a. Find an unbiased estimate for the mean length of all engagements.
b. Compute s^2 and s.
c. Are the estimates in part b less, more, or just as dependable as the estimates in part b of Problem 12.3?

12.5 The heights of 64 hybrid corn plants are presented in the group frequency table below.

Height Interval (in inches)	*F*
48.5–49.5	3
47.5–48.5	10
46.5–47.5	18
45.5–46.5	17
44.5–45.5	12
43.5–44.5	4

a. Find X, s^2, and s.
b. What do the statistics found in part a represent?
c. How many degrees of freedom were the calculations of s^2 and s based upon?

12.6 The mean time per bill necessary for the bookkeeper at Helen's Health Haven to process a stack of 144 bills is 2.5 minutes. Assuming the population of times to process bills to have a standard deviation of .75 minute, find the 95% confidence interval for the population mean.

12.7 The mean disposable income for 49 randomly selected teenagers between the ages of 16 and 18 is found to be \$23.75 a week. If the population standard deviation is \$2.25 find the 99% confidence interval for the mean of all such incomes.

12.8 A random sample of 75 tranquilizers sold under the name "Slow Downers," which were obtained from a number of different pharmacies, is found to contain a mean of 33 milligrams of diazepam per pill. Assuming the population is known to have a standard deviation of 2.4 milligrams, find both the 95% and 99% confidence intervals for the population mean.

12.9 The dollar amounts provided foster parents by state agencies for each foster child per day are provided below for 30 randomly selected foster families.

4.80	3.75	5.15	4.30	4.70
3.25	3.80	5.25	3.20	4.65
5.75	6.10	7.00	3.80	5.25
6.00	5.00	3.00	3.75	5.35
5.10	4.75	6.15	6.50	4.50
4.50	4.80	7.50	9.35	3.85

The population standard deviation is known to be \$1.40. Find the 95% confidence interval for the mean of the population.

12.10 The IQs of 49 students selected at random to participate in a three-year reading program are measured after completion of the program. The mean is found to be 107.

a. Assuming the standard deviation is 15, find the 95% confidence interval for the mean IQ of numerous graduates of the program.

b. Again assume the standard deviation is 15 and find the 95% confidence interval; but this time assume the sample consists of 100 rather than 49 students.

c. What relationship exists between the sample size N and the length of the confidence interval?

12.11 The lengths of 256 boats involved in accidents have a mean of 22.25 feet. Assume that the population of lengths of all boats involved in accidents has a standard deviation of 3 feet.

a. Find the 95% and 99% confidence intervals about the sample mean.

b. Find the 98% confidence interval.

c. What happens to interval length as confidence level increases?

12.12 The mean monthly income of students at a certain college is to be estimated from a random sample. If the mean income is to be estimated within \$20.00 with a probability of 95%, what is the smallest size sample needed for this estimate? Assume that the standard deviation is \$100.

12.13 Determine the size of the smallest sample needed to estimate the mean cost of dental insurance for a four-member family within one standard deviation of the sample mean. Use a 95% confidence level.

Problem Set B

12.14 The amounts (in dollars) of money spent on a maternity wardrobe by seven women who were included in a clothing industry survey were 250, 375, 200, 480, 385, 270, and 275.

a. Find the unbiased estimate for the amount that the typical pregnant woman spends on her wardrobe.
b. Find unbiased estimates for the variance and standard deviation of the distribution.
c. How many degrees of freedom are the estimates in part b based upon?

12.15 The lifetimes of eight transistors delivered to Lucy's Light-up Lab, a fix-it shop for amateur electricians, are found to be (in hours) 1,200, 1,170, 1,260, 1,085, 1,305, 1,275, 1,208, and 1,194.
a. Estimate the mean, variance, and standard deviation of the distribution of all lifetimes of the transistors in question.
b. How many degrees of freedom are involved?
c. Would knowing the population mean, μ, allow a more dependable estimate of the standard deviation? Why or why not?

12.16 The workers at Amalgamated Associates recently took part in a program to increase motivation and output. One week after the program's completion, the outputs of 100 workers who had taken part were measured and recorded as below.

Output Intervals (units per hour)	*F*
70–80	1
60–70	10
50–60	17
40–50	43
30–40	18
20–30	10
10–20	1
0–10	0

a. Find the unbiased estimate for the mean number of units per hour produced by workers who have completed the training.
b. Find s^2, s, and the number of degrees of freedom they are based on.

12.17 Anna Lee, this year's PTA president, plans to propose a major money-making project for this year's board. She contacts 36 other organizations that have completed a similar project. Their profits are presented below.

Profit (thousands of dollars)	*X*	*F*
4–5	4.5	4
3–4	3.5	8
2–3	2.5	7
1–2	1.5	10
0–1	.5	6

a. Estimate the mean amount of profit for all groups that pursue this project.
b. Estimate the variance and standard deviation of the profits.
c. State the degrees of freedom involved.

12.18 Studies for the past three years (36 months) indicate the mean number of runaways between the ages of 9 and 15 to be 75,550 per month. Assuming the standard deviation to be 4,000 youngsters, find the 95% confidence interval for the mean number of runaways per month.

12.19 The records of Universal Hospital show the mean weight of 250 infants born to mothers whose weights had increased during pregnancy by 40 pounds or more to be 8 pounds.

Assuming the standard deviation to be .75 pound, find the 99% confidence interval that estimates the mean weights of all babies born to such mothers.

12.20 The lengths of 81 "yardsticks" are found to have a mean of 36.25 inches.
   a. Assuming the standard deviation of the lengths of all such sticks produced to be .5 inch, find both a 95% and a 99% confidence interval for the mean.
   b. How does changing the confidence level affect the width of the interval?

12.21 According to a sample of 40 parents, the mean expense incurred while raising a child from the time the child leaves the hospital until the first birthday is $1,600.
   a. Assuming the standard deviation of the population to be $350, find the 95% confidence interval for μ.
   b. Find the 95% confidence interval for μ, assuming the standard deviation to be $450.
   c. What effect does a larger standard deviation have upon the size of the confidence interval if all other variables are kept constant?

12.22 Fifty showers with identical standard shower heads were run full strength for one minute. The heads were then replaced by various water-saving heads and the water was turned on full strength again for one minute. A total of 65 fewer gallons of water was used with the devices. Find the 99% confidence interval for the mean amount of water in gallons saved per minute due to the installation of a water-saving head. Assume the standard deviation to be .3 gallon per minute.

12.23 In Mr. Greg's ward, figures from how many past elections would be needed to estimate the mean number of voters per election within 150 voters with a confidence level of 95% if the standard deviation is known to be 98?

12.24 Find the sample size necessary to estimate the mean number of traffic accidents per intersection within .15 standard deviation of the sample mean at a 99% confidence level.

12.25 Determine the 90% confidence interval for the mean number of hours that successful CPA candidates claim to study per day for the national exam. A survey of 200 successful applicants taken by Commissioner Bob Otto shows a mean of 5 hours per day during the three-week period preceding the exam. Assume the standard deviation to be 1½ hours.

12.26 a. Explain why the mode of a sample is not an unbiased estimator for the population mean.
   b. Verify that the mean of a sample of 25,000 from a population whose standard deviation is 20 will rarely differ from the population mean by more than .5.

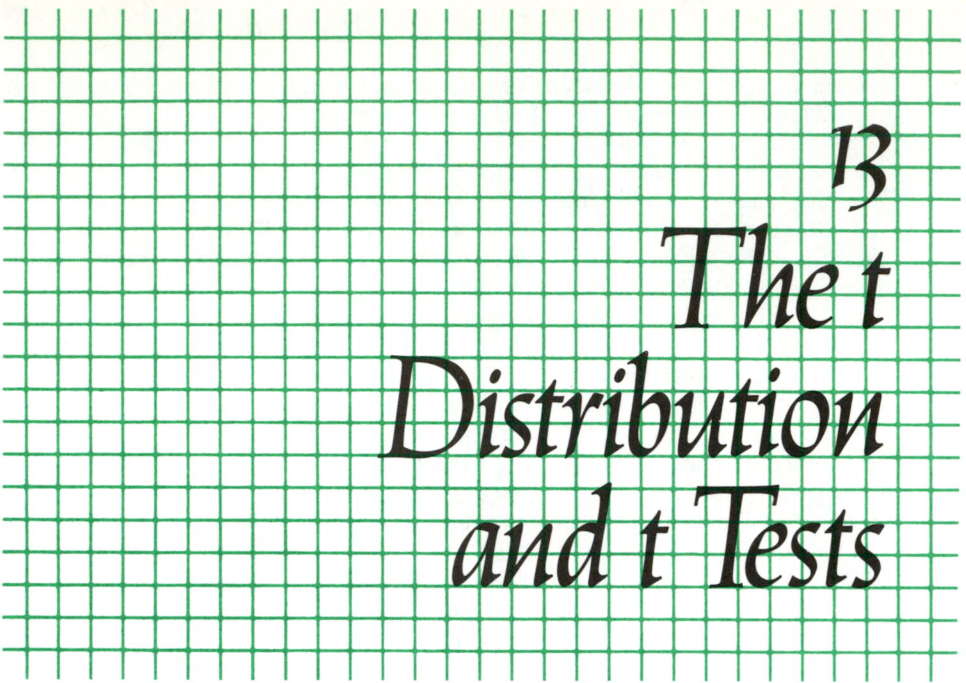

13 The t Distribution and t Tests

13.1 Introduction

In this chapter we are going to discuss procedures that embody much of what has already been said. However, up to now we have presupposed that the experimenter testing a hypothesis about a population mean had direct knowledge of the standard deviation of the population from which the sample was drawn. This is seldom the case. Nearly always the situation is such that a researcher has as the only information available the data comprising the sample drawn. On the basis of the data comprising the sample and virtually no other information, he or she must contrive to test whatever hypothesis is stated concerning a population mean or some other parameter.

We shall now consider a method of testing a hypothesis concerning an unknown population mean when the experimenter has no information other than that contained in the sample. *If N is small, the one new assumption that must be made is that the terms in the population from which the sample is drawn are normally distributed, or at least there must not be a drastic departure from normality in their distribution. If N is moderately large, this assumption is virtually always unnecessary.*[1] Thus, the hypothesis tests to be discussed have very wide applicability. Given the assumption,

[1]Nearly always N is considered moderately large if it is at least 30.

it is possible for an experimenter to use a sample mean to test a hypothesis about a population mean, even when there is no actual knowledge of the standard deviation of the population from which the sample was drawn.

The best way to make the logic and procedure clear is to think in terms of a specific experiment, and we shall begin with one that is arbitrarily simple. Suppose an anthropologist wants to demonstrate that the mean height of natives on a particular island is smaller than 5 feet 2 inches (62 inches). She decides to test the null hypothesis that the mean height of these natives is exactly 5 feet 2 inches. She might have set up a confidence interval, as described in the last chapter. She might then have asked whether the value 5 feet 2 inches was inside that interval. Instead, let us say, she chooses to test the null hypothesis that $\mu = 62$ inches against the alternative that $\mu < 62$ inches. She chooses the .01 significance level and hopes to be able to reject the null hypothesis by getting a sample mean that is too small.

We shall assume that the distribution of heights of natives on the island is normal, or approximately so. The anthropologist randomly selects five natives and determines their heights (though of course in practice she would not be content with so small a sample). The heights of these five natives turn out to be 54 inches, 56 inches, 57 inches, 60 inches, and 63 inches. The mean of these five heights is 58 inches. The anthropologist now wishes to base upon her sample a statistical test that will tell her whether to reject or not reject her null hypothesis.

To set the stage, let us go back for a moment and suppose she knew the exact value of the population standard deviation. For instance, suppose she knew that the standard deviation of natives' heights was 5 inches. She could then test the hypothesis in the manner described in Chapter 11. She would conceive of her random sample as one of a vast number gathered by similar anthropologists. Under the null hypothesis, the distribution of the means of such samples would have a mean of 62 inches. Knowing the value of the standard deviation of the natives' heights to be 5 inches, she could compute the standard deviation of the theoretical distribution of sample means. That is, given that $\sigma = 5$, she would compute

$$\sigma_{\bar{X}} = \frac{\sigma}{\sqrt{N}} = \frac{5}{\sqrt{5}} = 2.24 \text{ inches}$$

In other words, assuming the null hypothesis was true, she would ascertain that her obtained sample mean, 58 inches, would have been 4 inches below the expectancy in a distribution of sample means with a standard deviation of 2.24 inches. Assuming that the null hypothesis was true, by dividing she would see that the sample mean she had obtained would have had a z value of -1.79. We may sum up what her calculations would have been:

$$z = \frac{\bar{X} - \mu}{\sigma_{\bar{X}}} = \frac{58'' - 62''}{2.24''} = \frac{-4''}{2.24''} = -1.79 \qquad (13.1)$$

On the basis of the obtained z score, she would have decided whether to reject or not reject the null hypothesis.

But we shall not pursue her calculations further. The point is that now our anthropologist does *not* know the actual value of the standard deviation of the natives' heights. She cannot proceed in the same way. This time she is going to test her hypothesis in ignorance of σ. What she must do first of all is to make an enlightened guess of the population standard deviation using the data of her sample. She must compute the value of s, which she knows will be an estimate of σ though not identical with it. The value of s that she would compute, using Formula (12.1), is 3.54. Note that this value of s is an estimate of the actual value of σ. Next our anthropologist must make an estimate of the value of the standard error of the mean. We shall call this estimate $s_{\bar{X}}$ and remember that it is an estimate of $\sigma_{\bar{X}}$. As one might guess, our anthropologist makes use of Formula (13.2), which is analogous to our Formula (8.1) for $\sigma_{\bar{X}}$. The essential distinction is that here we are using s, an estimate of σ, whereas in (8.1) it was assumed the experimenter knew σ. Here we write:

$$s_{\bar{X}} = \frac{s}{\sqrt{N}} \tag{13.2}$$

In the case described, $s = 3.54$ and $N = 5$. Therefore,

$$s_{\bar{X}} = \frac{s}{\sqrt{N}} = \frac{3.54}{\sqrt{5}} = 1.58$$

If our anthropologist could be certain that her estimate, $s_{\bar{X}}$, was perfectly accurate and was the actual value of $\sigma_{\bar{X}}$, she could proceed smoothly. She could compute the z value of her obtained sample mean using Formula (13.1) with $s_{\bar{X}}$ in place of $\sigma_{\bar{X}}$. That is, she could determine this value by solving for the value of

$\dfrac{\bar{X} - \mu}{s_{\bar{X}}}$ instead of having to solve for the value $\dfrac{\bar{X} - \mu}{\sigma_{\bar{X}}}$.

The trouble is that the value $s_{\bar{X}}$ is only an enlightened estimate of $s_{\bar{X}}$ and is not identical to it. Therefore, the best our anthropologist could do would be to compute what she might view as an estimate of the z score value of her sample mean. We shall call the value she computes a t value and write

$$t = \frac{\bar{X} - \mu}{s_{\bar{X}}} \tag{13.3}$$

In particular, our anthropologist would compute the t value of -2.53 for her sample mean.

$$t = \frac{\bar{X} - \mu}{s_{\bar{X}}} = \frac{58'' - 62''}{1.58''} = -2.53$$

Note that in the case described, the t value obtained was numerically greater than the z value would have been. The reason is that the estimate, s, was smaller than the hypothetical value we used for σ. Had the value of s, computed from the sample, been larger than the value of σ, then the t value computed would have been smaller than the actual z value. The point is that the value of t in any particular case depends upon not only the obtained sample mean but also the value of s that happens to be computed from the particular sample.

13.2 The t Distribution

Let us assume that the mean height of natives on the island is actually 62 inches, so that the null hypothesis is actually true. Under this assumption, let us suppose further that our anthropologist is one of a vast number of anthropologists, each of whom gathers a random sample of the heights of five natives. We shall assume that each anthropologist now computes the mean and s value of the sample, and then the t value of the particular sample mean. In other words, assuming that the mean height on the island is actually 62 inches, there would come into existence a distribution of t values obtained by the different anthropologists. In each case, the particular anthropologist would be obtaining a t value as a result of using Formula (13.3). This distribution of t values, which would turn up in the long run, is crucial to know. With reference to it, a particular anthropologist like the one described can determine whether a particular t value would be commonplace or rare.

The mathematician has considered the problem and has devised what is called the t distribution. This is the distribution of t values that would be obtained from independent random samples of equal size drawn from a normal population. For now we shall stick to the case where each sample consists of five cases. The graph of the distribution of t values appears in Fig. 13.1. Note that the distribution of t values, based on samples of five cases each, is similar in appearance though not identical to the normal distribution. The proportions of cases in various sectors are indicated. Observe that the t distribution depicted here resembles the normal distribution. For different degrees of freedom, the t distribution looks different, and thus the t distribution is a whole *family of curves*. The more the degrees of freedom determining the particular shape, the more closely the t distribution resembles the normal distribution. Fig. 13.2 shows the t distributions for 5, 10, and 20 degrees of freedom.

The researcher who has obtained a t value for the mean of a sample of five cases

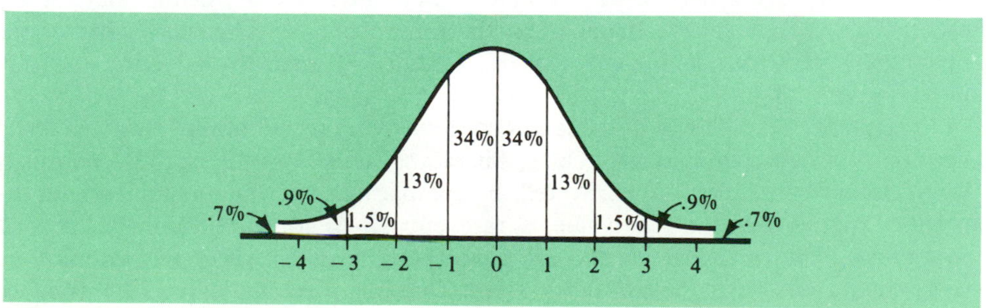

Fig. 13.1

Fig. 13.2

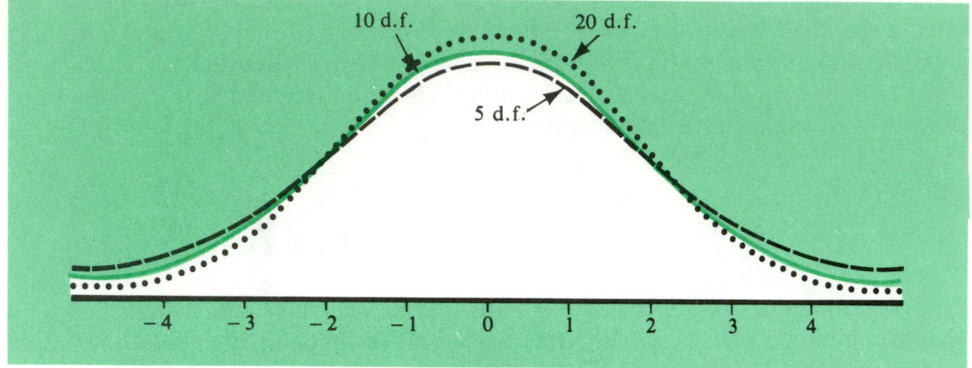

may refer to the graph in Fig. 13.1. Remember that one assumes that a null hypothesis is true before computing a t value. Our anthropologist was prepared to reject her null hypothesis only if she obtained a t value so small as to have a probability of less than 1/100 of appearing by chance. According to Fig. 13.1, this means that only if her obtained t value was less than -3.75 would she reject her null hypothesis. Actually her obtained t value was -2.53, which means that it was not sufficiently small to indicate rejecting her null hypothesis. Accordingly, her decision must be not to reject her null hypothesis that the mean height of natives on the island is 5 feet 2 inches (62 inches).

Remember that the number of degrees of freedom used to compute s, which was the estimate of σ, is one less than the sample size. In the case described, the anthropologist based her estimate on four degrees of freedom because her sample consisted of five cases. This number of degrees of freedom used to compute s also becomes described as the number of degrees of freedom used to compute t. In other words, we say that the anthropologist based her t value on four degrees of freedom. Were it the case that her sample consisted of 50 terms, then it would be the case that 49 degrees of freedom went into her estimate of t.

The number of degrees of freedom determines the shape of the t distribution. Or to put it another way, for any given number of degrees of freedom there is a particular t distribution with its unique set of critical values for significance. Therefore the worker whose sample consists of five cases must think of the obtained t value as a member of the t distribution based on four degrees of freedom and not any other. This fact becomes crucial when the time comes to determine whether the obtained t value is significant (that is, whether it indicates rejecting the null hypothesis).

Appendix IV makes it possible to determine what proportion of t values would be smaller than certain critical t values, if in fact the experimenter's null hypothesis is true. This allows the researcher to tell, by comparing the t value obtained with the proper critical value, if that value is rare enough to force rejection of the null hypothesis. The entries at the top are proportions of cases with t values less than those indicated. Right below these entries are numbers indicating the level of significance the hypothesis is being considered at and whether it is one- or two-tail.

Each row relates to a t distribution with a unique number of degrees of freedom. For instance, our anthropologist would have to go to the fourth row because her t value must be considered a member of a t distribution based on four degrees of freedom. Since *the t curve, like the normal curve, is perfectly symmetric*, the left and right halves of the curve are mirror images. Therefore, only positive critical values need appear in the table. The anthropologist would be interested in Column (3) since she is doing a one-tail test at the .01 significance level. The critical value in Column (3), line four, is 3.75. We are interested in the negative of that value, of course (a value we have already discovered by examining Fig. 13.1). Fig. 13.3 shows clearly that the anthropologist's obtained t value, -2.53, fails to fall in the rejection region.

Fig. 13.3

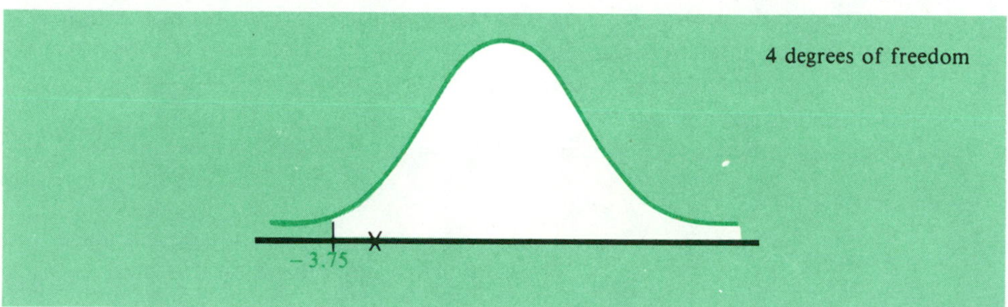

One more fact of interest is that, **as the samples yielding their means become larger, the *t* distribution approaches the normal distribution**. When the size of the samples is large, the values of s computed from the different samples become relatively close to σ as a group. As a result, the t values computed from the different samples become virtually identical to what the z values would have been. Thus in the 33rd row of Appendix IV, where the number of degrees of freedom for the distribution is 33, the cut-off points or critical values for the t distribution are approximately the same as those for the normal distribution. As a matter of fact, it is generally thought adequate to use the normal distribution table so long as one's sample is made up of at least 30 cases.

To return finally to our anthropologist, her sample mean was computed to have a t value of -2.53. The critical question for her to ask is whether the probability of having gotten a sample mean as small as hers was less than 1/100. According to Appendix IV, the t value of -3.75 is larger than exactly 1/100 of the t values in the distribution. Thus our anthropologist's obtained t value of -2.53 must have been higher than more than 1/100 of the terms. (To put it another way, Fig. 13.3 shows that the sample mean with a t value of -2.53 is too large to get into the .01 rejection tail.) The anthropologist has not found a sample mean small enough and is unable to reject her null hypothesis that the mean height of natives on the island is exactly 5 feet 2 inches. Therefore, she has failed to obtain statistical evidence to support her theory that the mean height of the natives on the islands is less than 5 feet 2 inches.

We shall consider one more example. Suppose a researcher knows that the mean score obtained by 12-year-old boys taking a mechanical aptitude test is 72. He is willing to assume that scores on this test are roughly normally distributed. He is interested in showing that boys who apply for the boy scouts at age 12 are higher as a group than average in mechanical ability. Our experimenter plans to obtain the mechanical aptitude scores of 20 would-be boy scouts. He wishes to test the null hypothesis that their scores comprise a random sample taken from the larger population. He plans to test this null hypothesis at the .05 significance level with the hope of rejecting it by obtaining a sample mean that is significantly high. The mean of his sample of scores, $\bar{X}$, turns out to be 86, and s turns out to be 8. The researcher's formal test of his hypothesis would read as follows:

Test of Hypothesis (t Test)

1. H_1: Boys who apply for the boy scouts at age 12 are higher than average as a group in mechanical ability. OR The sample obtained is not random because its mean is too high. OR $\mu > 72$.

   H_0: Boys who apply for the boy scouts at age 12 do not differ significantly in mechanical ability from other 12-year-old boys. OR The sample obtained is random. OR $\mu = 72$.
2. .05 significance level/one-tail
3. $N = 20$ $\bar{X} = 86$ $s = 8$ (The computations used to obtain these statistics may appear in this step.)
4. Assume H_0 true. ($\mu = 72$)

$$t = \frac{\bar{X} - \mu}{\frac{s}{\sqrt{N}}} = \frac{86 - 72}{\frac{8}{\sqrt{20}}} = 7.83$$

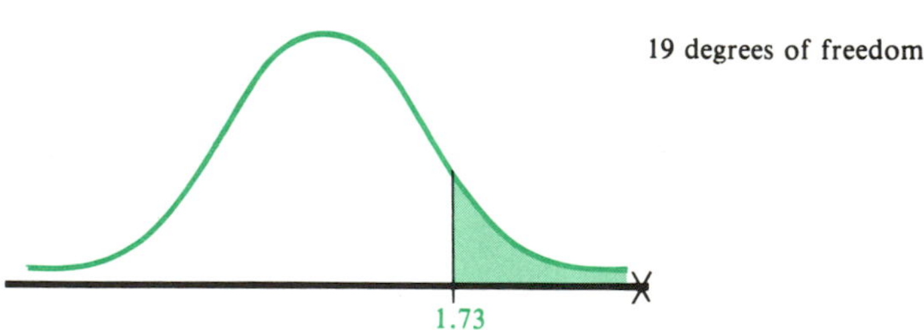

5. Reject H_0.
6. Accept H_1: The boys in question do have more mechanical ability as a group than the average.

Our experimenter may reject the null hypothesis as a consequence of his finding. He may conclude that his sample of boy scouts' scores is not merely a random sample taken from the population of mechanical aptitude tests. Having rejected the null hypothesis, he may conclude that there is some relation between having mechanical aptitude and choosing to be a boy scout (at least for the particular age group studied). Note that we used a one-tail test in each of the foregoing examples, since in each case we were concerned with a theory that involved difference in a specific direction from the hypothetical value. Otherwise we could have used a two-tail test, proceeding in the same way as when we used such a test with z values.

The test of significance, which has been described in this section, is often called a *t test*. The ratio given in Formula (13.3) is often called a *t ratio*. As one may imagine, there is much more practical use made of the t distribution than even the normal distribution when one tests a hypothesis. Even though one must conceive of a sample mean as a member of a distribution that is normal, the fact that one can usually only estimate the standard deviation of this distribution causes the crucial statistic computed to be a t value and not a z value.

We shall now consider another hypothesis test for which we make reference to the t distribution. This test is also known as a t test, but it is more complex than the one we have considered. We shall use virtually all of our previous logic in considering this test.

13.3 The t Test for the Difference between Means

The context that we shall now consider is the one that arises most often for researchers. They have no information concerning the mean or standard deviation of a population. They come to a field with no validated facts, but only with the desire to demonstrate some phenomenon that they think they have either observed or inferred. For instance, a teacher may have designed a method that he believes is more effective than a usual one. He has no validated facts concerning the old method, much less the new one, but he would like to demonstrate that his new method is superior. Or a drug company may wish to compare two drugs for efficiency, and because both drugs are new there may be no information concerning either.

As the reader may have noticed, several chapters back we started with the simplest kind of hypothesis test, for which we assumed considerable previous knowledge on the part of the researcher. We have subsequently considered tests where the researcher has had successively *less* information, and these tests have been successively more subtle to grasp. At the same time we have been moving toward considering more usual contexts. Now we are considering the one in which the researcher has no prior knowledge about any parameter and yet wishes to prove a theory. The test that we are about to consider requires that the experimenter deal with two sample means and compare them.

This time it will be simplest to begin with a theoretical discussion. Suppose that there is a normally distributed population of terms to which we have access. We gather a random sample of, let us say, 15 terms and we compute its mean and variance. We shall call this our first sample, since we are going to gather another independent one. Accordingly we shall use the symbol N_1 to stand for the number of cases in this sample, so that here $N_1 = 15$. We shall use the symbol $\bar{X}_1$ to stand for the mean of the first sample and the symbol s_1^2, to stand for its variance.

We do exactly the same for a second sample which, let us say, consists of 20 cases. We designate the number of cases in this sample by N_2, so that here $N_2 = 20$. We designate its mean by $\bar{X}_2$ and its variance by s_2^2.

We now compute the difference between the means of the two independent samples. Specifically we shall compute $(\bar{X}_1 - \bar{X}_2)$. Since both samples came from the same population, the expectancy is that the value of $(\bar{X}_1 - \bar{X}_2)$ will be zero. (This is true because the mean of each sample has an expectancy of being the same as the population mean.) However, in practice where the samples are random we can obviously anticipate that the mean of one sample will be higher than that of another, though of course we cannot say which will be higher. Thus in practice the value of $(\bar{X}_1 - \bar{X}_2)$ will almost certainly be some value either less than zero or more than zero.

If it turns out that $\bar{X}_1$ is smaller than $\bar{X}_2$, then $(\bar{X}_1 - \bar{X}_2)$ will be a negative number. If it turns out that $\bar{X}_1$ is larger than $\bar{X}_2$ then $(\bar{X}_1 - \bar{X}_2)$ will be a positive number. Now suppose that each one of a vast number of researchers was to do as we have done. That is, each researcher goes to our same normally distributed population and draws two random samples, the first comprising 15 terms and the second comprising 20 terms. Each individual researcher computes the value of $(\bar{X}_1 - \bar{X}_2)$ from a particular pair of samples. For instance, suppose for one pair the values $\bar{X}_1 = 50$ and $\bar{X}_2 = 60$; then $(\bar{X}_1 - \bar{X}_2) = -10$. For the next researcher the value of $(\bar{X}_1 - \bar{X}_2)$ might be 2. For the next one it might be $-.5$, and so on.

Table 13.1 shows what might happen for the first few researchers from among the vast multitude. For each researcher there is an obtained value of $\bar{X}_1$ and one of X_2, and also a difference score, $(\bar{X}_1 - \bar{X}_2)$. To simplify we are going to use the letter d in place of $(\bar{X}_1 - \bar{X}_2)$ from now on.

It is the column of d values, each of which was obtained by a different researcher, that will be of concern to us from now on. When we conceive of these differences as terms in a distribution, we find that **in the long run the mean of these differences, μ_d, is zero and that they are normally distributed.**

Table 13.1 Sample Differences of Means

	$\bar{X}_1$	$\bar{X}_2$	$d = (\bar{X}_1 - \bar{X}_2)$
First researcher	50	60	−10
Second researcher	48	46	2
Third researcher	41	41.5	− 0.5

The standard deviation of these d values may be designated by the symbol σ_d. σ_d is sometimes called the "standard error of the difference."

$$\sigma_d = \sqrt{\frac{\sigma^2}{N_1} + \frac{\sigma^2}{N_2}} = \sigma \sqrt{\frac{1}{N_1} + \frac{1}{N_2}}$$

where σ is the standard deviation of the population from which each pair of samples was drawn.

But remember that we are talking from the position of a single researcher, in particular the first one, who has no knowledge of the value of σ. The sum total of his information is contained in the two samples he has drawn. Suppose this researcher starts to think of his obtained d value (of -10) as a member of the distribution of d values similarly obtained. He knows that the mean of this distribution is zero and that it is normal, but he can only estimate the standard deviation of this distribution. That is, he must use the information he has gathered to estimate σ_d. He computes the value of s_d, which is the symbol we shall use for his estimate of σ_d. As might be suspected (from our previous use of s in place of σ), it can be shown that

$$s_d = s\sqrt{\frac{1}{N_1} + \frac{1}{N_2}}$$

Now, however, s^2 is a variance based on *both* of the samples the researcher has drawn rather than on just one sample as in Section 13.1. It can be shown mathematically that

$$s^2 = \frac{(N_1 - 1)s_1^2 + (N_2 - 1)s_2^2}{N_1 + N_2 - 2}$$

(This may be described as a "weighted mean" of s_1^2 and s_2^2, each weighted by the number of degrees of freedom on which it is based.) When the square root of this is taken to obtain s and it is then substituted in the expression for s_d, the result is the rather formidable-looking Formula 13.4.

$$s_d = \sqrt{\frac{(N_1 - 1)s_1^2 + (N_2 - 1)s_2^2}{N_1 + N_2 - 2}} \cdot \sqrt{\frac{1}{N_1} + \frac{1}{N_2}} \qquad (13.4)$$

In other words, the first researcher may think of his obtained d value as a member of a normal distribution whose mean is zero and whose standard deviation he estimates by Formula (13.4). He cannot obtain the z score of his obtained difference but he can obtain a t value for this difference. To get this t value he must compute how far his obtained difference is from the balance point of this theoretical distribution of differences. But we have already seen that the balance point of this distribution of differences is zero. Therefore our researcher whose obtained d value was -10 has found a d value that is 10 units to the left of the balance point in the theoretical distribution of differences.

To get the t value of his obtained difference, our researcher would first have to compute the value of s_d. That is, he would have to make his estimate of the standard deviation of the theoretical distribution of differences depicted in Fig. 13.4. Suppose he computed that $s_1^2 = 210$ and $s_2^2 = 220$.

He would then use Formula (13.4) and get

$$s_d = \sqrt{\frac{(N_1 - 1)s_1^2 + (N_2 - 1)s_2^2}{N_1 + N_2 - 2}} \sqrt{\frac{1}{N_1} + \frac{1}{N_2}}$$

$$= \sqrt{\frac{14(210) + (19)(220)}{33}} \sqrt{\frac{1}{15} + \frac{1}{20}} = \sqrt{\frac{2940 + 4180}{33}} \sqrt{\frac{7}{60}}$$

$$= (\sqrt{215.8})(\sqrt{.1167}) = 5$$

Now to find the t value of his obtained difference of -10, he would have to divide the distance score by his estimate, s_d.

$$t = \frac{d - \mu_d}{s_d} = \frac{d - 0}{s_d} = \frac{\bar{X}_1 - \bar{X}_2}{s_d} \qquad (13.5)$$

$$= \frac{-10}{5} = -2$$

Fig. 13.4

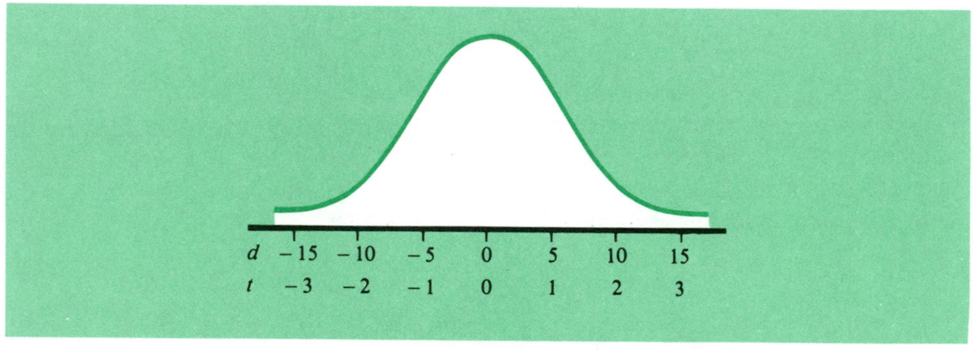

In the theoretical distribution of differences, our researcher has found a difference with a t value of -2. What he has done in effect is to construe his obtained difference, d, as the finding of interest. He has viewed this finding, d, as a member of a normal distribution with a mean of zero. The members of this distribution are other d values that in theory would be found by other researchers doing exactly the same thing. Our researcher has made an estimate of the standard deviation of the distribution of d values—that is, he has computed s_d. Finally, since he knows the distance score of his particular finding, he divides this distance score by the value s_d to get a t value for his obtained difference.

We can imagine each of our multitude of researchers doing the same thing. That is, we can imagine each researcher computing the value of d and also computing s_d. Thus each researcher divides d by s_d to get the t value for the obtained difference. Table 13.2 shows what the computations of a few such researchers might look like. The table has been left for the reader to complete.

Table 13.2 Computation of t Values

	$\bar{X}_1$	$\bar{X}_2$	d	s_1^2	s_2^2	s_d	$t = \dfrac{d-0}{s_d}$
First researcher	50	60	−10	210	220	5	−2
Second researcher	48	46	2	200	230		
Third researcher	41	41.5	−0.5	170	190		

It should be clear that each researcher has used the assumption that a particular pair of independent samples came from the same normally distributed population. Having had no further knowledge, a t value for the difference, d, was computed. That is, each researcher has obtained a t value for the difference $(\bar{X}_1 - \bar{X}_2)$ computed from a pair of samples.

We must now specify the number of degrees of freedom that went into the t distribution just described. Note that each researcher computed a value of s_1^2 based on $N_1 - 1$ degrees of freedom and a value of s_2^2 based on $N_2 - 1$ degrees of freedom. Both estimates went into the estimate of s_d so that this latter statistic was based on $(N_1 - 1) + (N_2 - 1)$ degrees of freedom inasmuch as the two samples are independent. In other words, the t distribution described is based on $N_1 - 1 + N_2 - 1$ degrees of freedom. To simplify, it is based on $N_1 + N_2 - 2$ degrees of freedom. Specifically for the instance described, since $N_1 = 15$ and $N_2 = 20$, the distribution of differences—some of which were listed in Column (7) of Table 13.2—is a t distribution based on 33 degrees of freedom. The reader may have noticed that the expression $N_1 + N_2 - 2$ appears in Formula (13.4).

What has been said may be summarized in the form of a theorem, which shall soon be of special use to us.

THEOREM 13.1 Suppose a vast number of pairs of independent random samples of size N_1 and N_2 are drawn from the same normally distributed population. For each pair the value $(\bar{X}_1 - \bar{X}_2)/s_d$ is computed, where $\bar{X}_1$ is the mean of the first sample, $\bar{X}_2$ is the mean of the second, and s_d is as defined in Formula (13.4). Then the distribution of values thus obtained is the *t* distribution with $N_1 + N_2 - 2$ degrees of freedom.

Even if the original population is not normal, Theorem 13.1 holds provided that the samples are large enough.

13.4 Application of the t Test for the Difference between Means

We are now going to make use of Theorem 13.1 to describe the hypothesis-testing procedure used most often in such fields as education, psychology, and the social sciences. An illustration will be useful to present the technique. Suppose a French teacher has taught vocabulary by two methods, one of which involves using visual aids whereas the other does not. He decides to do an experiment to determine if one method is more effective than the other for teaching French vocabulary to his classes.

We shall say that this French teacher has two comparable classes, one of 15 students and the other of 20, and that he decides to use the visual aid method with one class and the other method with the second class. At the end of the semester, he tests both classes on vocabulary words learned during the year. Actually he has taught 300 new words in all, and he obtains for each student in each class a score indicating the total number of words the student has learned.

As a researcher, the instructor is essentially interested in learning which of the methods was more effective, and he assumes that he is typical of an instructor using either. We shall say that he used the visual aid method on Class A, which consisted of 15 children, and that he used the other method on Class B, composed of 20 children. It turns out that the mean number of words learned by Class A was 220, whereas the mean number learned by members of Class B was only 200. The question our researcher asks is *whether the difference is meaningful*. In other words, may he infer from the finding that the advantage shown by Class A means that the visual aids method was really better? Or was the difference merely one that might have been expected to occur by chance so often that he can draw no conclusion from it whatsoever?

The first crucial step for our researcher is to set up the *null hypothesis that the two methods are identically effective*. Now he may conceive of a vast population of children with whom he has used a uniform method of teaching French vocabulary. Under this hypothesis, Class A would be merely a random sample of 15 children whom he has taught and Class B would be an independent random sample consisting of 20 children. The obtained difference between the mean of Class A, which we shall call $\bar{X}_1$, and that of Class B, which we shall call $\bar{X}_2$, is considered by him merely a difference between the means of two random samples from the same population. According to the null hypothesis, this difference, $(\bar{X}_1 - \bar{X}_2) = 220 - 200$, might as easily have been zero or have gone the other way.

The next step is for our researcher to conceive of himself as only one of a vast number of similar researchers who have performed the identical experiment. Each of these other researchers would also obtain a difference and a t value corresponding to it. That is, each researcher would, from a pair of samples, compute $(\bar{X}_1 - \bar{X}_2)$ and also s_d and then divide the former by the latter to get a t value. Our researcher must construe the particular t value that he obtains to be a member of this t distribution based on 33 degrees of freedom.

We may suppose for instance that he is testing his hypothesis at the .05 level of significance. The difference he has obtained between the mean vocabulary scores in the classes is 20, and we shall say that he computes the value of s_d to be 15. Therefore

$$t = \frac{\bar{X}_1 - \bar{X}_2}{s_d} = \frac{20}{15} = 1.33$$

The question he now asks is whether his obtained t value is extreme enough to have had a probability of less than 5/100 of occurring. The assumption that the null hypothesis is true led him to determine his t value. Now he cannot reject the null hypothesis unless it turns out that his obtained t value happens to be too extreme. According to Appendix IV, where t is based on 33 degrees of freedom, the smallest 2½% of t values are those less than -2.03, and the largest 2½% are those greater than 2.03. Therefore the most extreme 5/100 of t values are those less than -2.03 and those greater than 2.03, taken together. The t distribution based on 33 degrees of freedom is shown in Fig. 13.5.

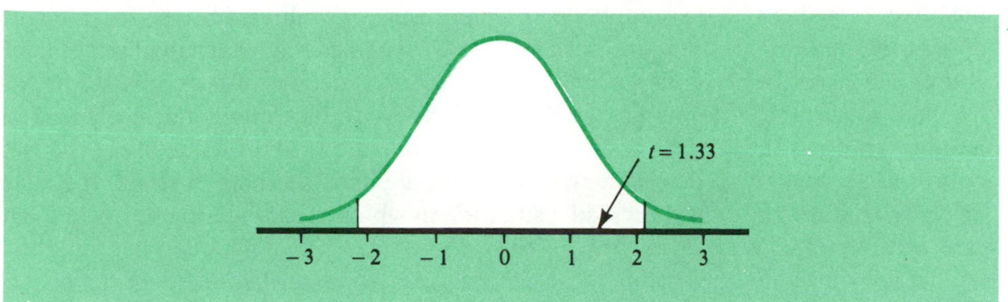

Fig. 13.5

The obtained t value of 1.33 is not in the rejection region of Fig. 13.5. Therefore the null hypothesis that the two classes represent random samples from the same population cannot be rejected. This is the hypothesis that there is a uniform effectiveness to the two teaching methods, so that the difference between mean achievement in the two classes cannot be considered meaningful.

Suppose instead that the obtained t value had been 4.00. This obtained t value would be in the rejection region of Fig. 13.5. The null hypothesis that Class A and Class B represent random samples from the same population of French vocabulary achievement scores would have to be rejected. Since the two classes may not be considered as samples from the same population, one must accept the alternative explanation that the actual teaching methods used in the classes must have made them different. The conclusion must be that the method used with Class A, the visual aid method, is different in effectiveness from the method used with Class B. Note that reference to the appropriate row in Appendix IV would have been sufficient to make the decision.

The procedure for doing a t test like that described is quite straightforward in practice. The experimenter picks his significance level. Then he determines the values of N_1, N_2, $\bar{X}_1$, $\bar{X}_2$, s_1^2, and s_2^2. He then solves for s_d using Formula (13.4). Then he calculates and looks up his obtained t value using the degrees of freedom given by $N_1 + N_2 - 2$ in Appendix IV. If the obtained t value indicates that he cannot reject the null hypothesis, he concludes that whatever difference was found between the two group means was apt to have been the result of chance. If he is able to reject the null hypothesis, he concludes that the two groups did not come from the same population in the first place. The conclusion is then that the difference between the two groups was the result of the different procedures used with them or to some other factor that distinguished them in the first place.

The point is that the experimenter knows which range of t values to expect when comparing means of two samples from the same population. He compares the means of the two groups treated differently by assuming that they were treated similarly and so are from the same population. If, using Appendix IV, he gets a t value not in the rejection region, he is unable to conclude that the differential treatment did have a differential effect. Otherwise he concludes that it did.

Perhaps the reader can already see the wide applicability of the t test for the difference between means. Whenever the effect of a method or treatment is to be studied, the t test for the comparison of group means is likely to be appropriate. Often the effect of a single process applied to human beings is evaluated by the researcher's comparing those on whom it has been applied with an equivalent group of subjects on whom it has not been applied. For instance, when the effect of a drug is studied, a frequent procedure is to administer it to one group and to administer a placebo to another group. Then a t test of the kind described is done and the means of the two groups are compared to determine whether the drug itself has had an effect. It should be noted that we have *not* discussed a situation in which the *same* group is used twice (see Sections 13.5 and 19.3).

It is worthwhile illustrating the procedure already described once more because of its applicability and importance. This time, however, we shall merely present the problem and give a formal presentation of the hypothesis test as it might be recorded to substantiate (or fail to substantiate) a theory.

Suppose that a certain industrial consultant who is an expert in employee motivation and performance theory is interested in testing her belief that, in a certain type of factory, humidity affects worker output. Therefore she selects two comparable factories and arranges for the humidity to be *high* in one and *low* in the other for a one-week period. Fifteen workers are selected randomly from each factory and their average daily outputs for the week (in number of articles) are recorded.

We shall say that the 15 values recorded for Group I were 35, 42, 50, 60, 68, 72, 74, 74, 76, 80, 81, 83, 85, 90, and 93.

The 15 values for Group II were 40, 44, 48, 65, 70, 73, 73, 73, 79, 84, 86, 87, 89, 90, and 91.

A significance level of .05 is selected and the test proceeds as follows:

Hypothesis Test (Difference Between Means t Test)

1. H_1: The output of workers exposed to *high* humidity differs from the output of workers exposed to *low* humidity. ($\mu_1 \neq \mu_2$)
   H_0: There is no difference in the outputs of workers exposed to *high* and *low* humidity levels. ($\mu_1 = \mu_2$)
2. .05 significance level/two-tail
3. $N_1 = 15$ $\quad \bar{X}_1 = 70.87$ $\quad s_1^2 = 294$

   $N_2 = 15$ $\quad \bar{X}_2 = 72.80$ $\quad s_2^2 = 287$

   (The actual calculations of these statistics may appear in this step. The reader should verify that they are correct.)
4. Assume H_0 true. ($\mu_1 = \mu_2$ or equivalently $\mu_d = 0$)

$$s_d = \sqrt{\frac{(N_1 - 1)s_1^2 + (N_2 - 1)s_2^2}{N_1 + N_2 - 2}} \sqrt{\frac{1}{N_1} + \frac{1}{N_2}}$$

$$= \sqrt{\frac{14(294) + 14(287)}{28}} \sqrt{\frac{1}{15} + \frac{1}{15}} = 6.22$$

$$t = \frac{\bar{X}_1 + \bar{X}_2}{s_d} = \frac{-1.93}{6.22} = -.31$$

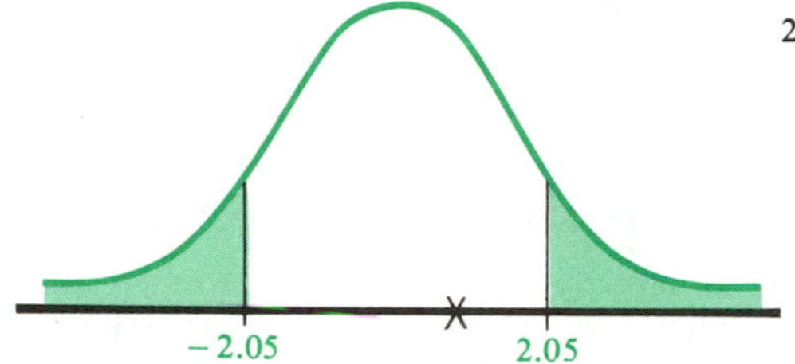

5. Do Not Reject H_0.
6. Do Not Accept H_1: We have failed to show any difference in the outputs of workers exposed to high and low humidity levels.

Only one other point might be mentioned in connection with the foregoing. We have used two-tail tests in our examples. A one-tail test could be used if the theory involved a prediction that a specific one of the two means was higher than the other. In this case, the test for the differences between means using a one-tail test would proceed in the same way as the one-tail tests presented earlier in the text.

13.5 The t Test for Matched Groups

Sometimes there is a natural *pairing* between scores in one group and those in another. This occurs, for instance, when subjects are each tested twice. The subject obtains two scores, and the *change* in the subjects' scores are of interest to us.

For instance, suppose 20 girls in a class are given a test of prejudice, are afterward shown a film calculated to reduce prejudice, and are then given the same (or an alternate form of the same) test. Of interest to us is whether the film changed their attitudes, as measured by the test.

We shall assume that the stability of the test is such that, unless there was some new learning, the scores would not change. This assumption would of course have to be demonstrated for any particular test, since practice sometimes produces change. We shall also assume that the changes in the subjects' scores do not deviate drastically from normality. If this assumption cannot be made, a test described in Section 19.3 must be used.

With natural pairings, we do not use the t test for means mentioned in the previous section because a more powerful test is available. The t test for the difference between means, mentioned in Section 13.4, is subject to a sort of error that we can eliminate when we use matched pairs of scores. Chance differences between subjects will not add to our error likelihood this time, because our approach is to deal wholly with the *differences* between performances by the same subject under the two conditions. The approach is also used when the experimenter has matched subjects so that corresponding to each subject in one group there is a particular subject in the other group serving as a counterpart. For instance, the matching may be done on the basis of IQ or on any other control variable.

Table 13.3 shows the prejudice scores of 20 girls both before and after seeing a particular educational film. Column (4) lists the difference scores for each subject. On this particular test, the higher the score the more the prejudice. It is desired to show that the film reduced prejudice—that the post-film scores are lower than the pre-film scores.

This time we operate entirely on the column of differences to do our significance test. We shall choose the .05 level of significance, and note that we are doing a one-tail test, since our theory or alternative hypothesis asserts that the scores after seeing the film will be lower than (and not merely "significantly different" from) the earlier scores.

Table 13.3 Scores on Prejudice Tests

Subject	Pre-Film Score	Post-Film Score	Difference Score
1	74	70	− 4
2	35	34	− 1
3	46	47	1
4	79	55	−24
5	55	30	−25
6	67	60	−7
7	44	41	−3
8	23	20	−3
9	78	67	−11
10	66	55	−11
11	43	32	−11
12	21	19	−2
13	90	45	−45
14	44	55	+11
15	35	40	+ 5
16	69	60	−9
17	88	82	−6
18	34	32	−2
19	25	12	−13
20	76	30	−46

$$N = 20 \qquad \sum d^2 = 6{,}230$$

$$\sum d = -206 \qquad N\sum d^2 = 124{,}600$$

$$\bar{d} = \frac{\sum d}{N} = \frac{-206}{20} = -10.3$$

The null hypothesis we are testing is that the film had no effect. From this it would follow that the difference scores average out to zero in the long run. Our finding of $\bar{d} = -10.3$ is, under this null hypothesis, a random departure from the expected value of $\bar{d}$, which is zero. Note that in this design N stands for the number of difference scores, which means that it is the number of matched pairs.

The question we are to answer by means of our significance test is whether our obtained $\bar{d}$ is significantly far away from zero. The t value used to test the null hypothesis, based on $N - 1$ degrees of freedom, is given by the following formula:

$$t = \frac{\bar{d} - \mu_{\bar{d}}}{s_{\bar{d}}} \tag{13.6}$$

Note that the same pattern is followed here as has been followed in the past. The distance score of any term (the score minus the mean of infinitely many such scores) divided by the estimate for the distribution's standard deviation yields the t score for the term. Mathematically it can be shown that the just-mentioned formula simplifies to

$$t = \frac{\Sigma d}{\sqrt{\dfrac{N\Sigma d^2 - (\Sigma d)^2}{N - 1}}} \tag{13.7}$$

$$t = \frac{-206}{\sqrt{\dfrac{124600 - (-206)^2}{19}}}$$

$$t = \frac{-206}{65.76}$$

$$t = -3.13$$

Here, $N - 1 = 20 - 1 = 19$. As the reader should verify, the obtained t value is significant at the .05 level of significance. Since we are forced to reject the hypothesis that the film made no difference in the expression of prejudice, we must conclude that the film reduced the degree of prejudice felt by the subjects or, at the very least, that it curtailed their expression of prejudice.

As long as the groups can be matched pairwise, the approach given in this section can be used.

A final example will be used to illustrate another matched pairs hypothesis test presented in formal form.

Suppose that 10 different metals are tested to see if a certain treatment will decrease corrosion (as measured on a continuous 100-point scale). Two specimens of each metal are used in the test, one treated, one not. After being exposed to the same conditions for equal amounts of time, the results are found to be those in Table 13.4.

High scores indicate more corrosive action has taken place than low scores do. Assuming the difference scores to be at least approximately normally distributed, a test at the .05 significance level would be presented this way:

Table 13.4 Results of Metal Corrosion Test

Metal	Not Treated	Treated
A	93	87
B	85	76
C	73	63
D	63	47
E	50	62
F	45	43
G	66	62
H	75	71
I	53	50
J	80	70

Hypothesis Test (Matched Groups t Test)

1. H_1: Treated metals show less corrosion than untreated metals. ($\mu_{\bar{d}} < 0$)

   H_0: Treated metals show no less corrosion than untreated metals. ($\mu_{\bar{d}} = 0$)

2. .05 significance level/one-tail

3.

d	d^2
−6	36
−9	81
−10	100
−16	256
+12	144
−2	4
−4	16
−4	16
−3	9
−10	100
−52	762

$N = 10$
$\Sigma d = -52$
$\Sigma d^2 = 762$

(All of the data may appear in this step.)

4. Assume H_0 true. ($\mu_{\bar{d}} = 0$)

$$t = \frac{-52}{\sqrt{\dfrac{10(762) - (-52)^2}{10 - 1}}} = \frac{-52}{23.37} = -2.23$$

5. Reject H_0.
6. Accept H_1: Treated metals exhibit less corrosion.

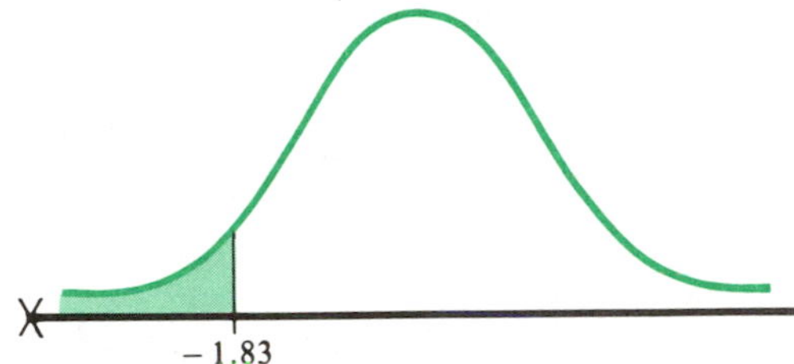

9 degrees of freedom

Problem Set A[2]

13.1 The annual report to the stockholders of J & J Sugar Refineries stated that the average American consumes 100 pounds of sugar per year ($\mu = 100$). In response to the report the North American Physicians Association conducted a year-long study to measure the sugar consumption of 25 randomly selected people. The sample mean was 93 pounds; the standard deviation was 4 pounds. Do the results refute the company's annual report at the .05 level of significance?

13.2 Efficient residential ovens should expend a mean of 4,500 kilocalories or less per hour of baking. A regional study by the Clint Cooperative Electric Company is aimed at showing that the mean usage per oven in the area is more than the allowable amount. Of 49 residential ovens tested, the mean amount used per oven per hour is 4,535 kilocalories with a standard deviation of 225. At the .05 level of significance, is the company's contention that residential ovens use too many calories per hour supported?

13.3 One of the strongest arguments in favor of halfway houses for nonviolent criminals is the cost factor. Government officials claim that the mean cost per house is only $1.50 per day per resident, since each resident is working and contributing to his or her own support. A neighborhood association opposed to the construction of a new halfway house in its area counters the argument by surveying 14 such houses. The sample results are $\bar{X} = \$1.65$ and $s = \$.35$. What conclusion would a hypothesis test yield at the .05 significance level?

13.4 The founder of Weight Off Clinic contends that men weigh more pounds over their ideal weights than do women. Is this supported by the following evidence at the .05 level of significance?

[2]Normality may be assumed in all distributions whenever necessary.

Men	*Women*
$N_1 = 9$	$N_2 = 6$
$\bar{X}_1 = 23$	$\bar{X}_2 = 18$
$s_1^2 = 16$	$s_2^2 = 36$

Each sample is from a distribution of pounds over ideal weights for individuals in the population.

13.5 The manufacturer of an iron supplement known as "Go Getters" (G) is conducting a test to show that its pills contain more iron than its major competitor, "Liver Discs" (L). A 150-size sample of each daily dose yields the following.

$$\bar{X}_G = 510 \text{ mg} \qquad \bar{X}_L = 504 \text{ mg}$$

$$s_G = 25 \text{ mg} \qquad s_L = 20 \text{ mg}$$

Can the manufacturer of "Go Getters" advertise that this product provides more iron per daily dose than its competitor at the .01 significance level?

13.6 Union representatives are concerned that in a certain factory benefits paid to management (population 1) are significantly higher than those paid to rank and file (population 2), even though both are supposedly covered by the same plan. Union leader Ray Mond gains access to insurance files through a court order and discovers the following as far as paid out benefits are concerned.

$$N_1 = 15 \qquad N_2 = 29$$

$$\bar{X}_1 = \$125 \qquad \bar{X}_2 = \$107$$

$$s_1 = \$10 \qquad s_2 = \$30$$

Is the union's position upheld at the .05 significance level?

13.7 Because so many women are returning to the workforce after raising a family, Ms. Margaret's Refresher School has been established to aid those in need of sharpening rusty skills. Among courses offered is "Getting Back Into Speed Typing." To demonstrate the effectiveness of the course, Ms. Margaret presents the pre- and post-course typing speeds (in words per minute) of eight graduates.

Before	*After*
45	47
53	59
50	63
65	60
73	79
68	75
52	60
49	80

Do these statistics vouch for the course's effectiveness at the .05 significance level?

13.8 Is it true that in spring a young man's fancy turns from studies? Ten of the brothers of a local fraternity have been comparing their fall and spring semester GPA's to see if spring grades really do plummet.

Brother	*Fall Average*	*Spring Average*
Bob	3.9	4.0
Dean	1.7	1.9

Brother	*Fall Average*	*Spring Average*
Doug	3.4	3.2
Bill	2.3	2.0
Greg	2.5	2.3
Leon	2.1	1.9
Pete	3.2	2.0
Jim	2.9	2.7
Don	3.1	2.0
Rick	1.8	1.9

Perform a test at the .01 significance level.

13.9 Red-X Insurance Company pays a flat rate of $495 for an appendectomy. The total cost charged for 11 such operations selected at random was $5,600. Does this suggest at the .025 significance level that the mean cost of the operation exceeds the amount the company will cover if the amounts in the sample varied with a variance of $300?

13.10 The makers of Slim, a new diet cola, advertise their product to contain 15 calories per bottle. Nutrition columnist and professional dieter Betsy Janes buys two cartons (12 bottles) of Slim and directs her agent to have them analyzed for calorie content. He presents her with these results (in calories per bottle): 14.9, 14.8, 15.7, 15.9, 16.3, 17, 17.5, 18.9, 15.7, 18.7, 20.9, 17.3. Should the columnist advise her readers that the company claim is too small at the .05 significance level?

13.11 Psychologist Al Wright believes that women working under male supervisors rate their supervisors differently than do men who are working under female supervisors. A comprehensive rating form is filled out by nine men and nine women. Each of the 18 subjects works for a supervisor of the opposite sex and no two work in the same industry or city. The women give their supervisors a mean score of 75 with a standard deviation of 15, whereas the men give a mean score of 70 with a standard deviation of 20. Should Mr. Wright consider his hypothesis supported at the .02 significance level?

13.12 Five years ago, Sue Knight's private air parcel post service, Fly-By-Knight, installed processing and sorting machines that were designed to handle packages with a mean length of 16 inches. After numerous malfunctions, Ms. Knight suspects the machines need to be replaced by others equipped with new specifications. Is she right at the .05 significance level in believing the mean package length is no longer 16 inches if the following is a random sample of 17 package lengths (in inches): 39, 16, 5, 17, 14, 12, 8, 28, 29, 30, 3, 37, 31, 30, 20, 23, 10?

13.13 A marketing consultant is studying two locations before advising a client where to locate his new fast food eating establishment. In one study 15 pairs of matched facilities are selected, one in each location. Each is from a national chain so each pair contains two units of similar size, with similar advertising budgets, menus, and procedures. The mean daily profit is determined for each test subject over a six-week period.

Establishment	*Location A*	*Location B*
Pearl's Pizza	395	295
McDingles' Burgers	232	250
Tom's Tacos	175	145
Sally's Stir Fry	250	200
Frank's Franks	135	150
Bologna Burgers by Barry	95	100
Phil's Fast Food	200	130
Louie's Luncheonette	205	154
Andy's	185	163
Imperial Eatery	130	125

Establishment	*Location A*	*Location B*
Gert's Greasy Grill	360	325
Snack Shack	215	200
Bingo Burger Barn	185	180
Pancake Heaven	150	90
Koffee House	135	85

Is there any reason to believe one location offers a better profit-making atmosphere than the other at the .10 level of significance?

13.14 Twenty-eight heart attack victims, 14 men and 14 women, were counseled at length as to the benefits of a daily regimen of walking. Each was instructed to keep a diary. After three months, the diaries were given to researchers at the Cardiac Clinic who compiled the following mean numbers of minutes walked per day for the 28 test participants.

Men		*Women*	
12.0	5.0	9.0	20.0
15.0	21.0	10.8	24.3
13.5	30.0	25.5	18.5
17.3	18.9	35.0	15.6
18.0	23.2	31.4	30.2
11.5	25.0	27.8	29.5
7.4	19.0	25.0	24.0

The researchers expected the women as a group to have a higher mean than the men, which they did. Can the higher mean be explained as a chance occurrence or does it indicate a trend at the .005 significance level?

13.15 Is there any significant difference at the .01 level in the growth of the following sample of the Zenus Adripedes jungle plant during the third and fourth years? The results (growth in inches) of 12 plants kept under identical controlled conditions for their first four years are presented here to aid in answering the question.

Plant	*Third Year*	*Fourth Year*
1	1.55	1.25
2	1.32	1.37
3	2.00	1.43
4	1.07	1.28
5	1.43	1.84
6	1.35	1.75
7	1.47	1.68
8	1.18	1.20
9	1.15	1.35
10	2.03	2.30
11	1.85	1.88
12	1.92	1.07

Problem Set B[3]

13.16 Three years ago the mean outstanding balance per charge customer at Hansen's Department Store was $28.39. A random sample of 16 accounts this month shows a

[3]Normality may be assumed in all distributions whenever necessary.

mean outstanding balance of \$33.18, with a standard deviation of \$5.00. At the .05 level of significance, does this indicate an increase over the last three years in the mean balance outstanding?

13.17 A realtor attempting to sell a bookshop advertises the daily profit to have a mean of \$55 a day. A prospective buyer, believing this figure to be too high, asks to test for herself. Selecting 15 days at random, she calculates a mean profit of \$49 and a standard deviation of \$13.75. If the store is purchased, should the buyer expect profits of less than \$55 a day? Use a .05 significance level.

13.18 The mean attendance at the Woodchucks' baseball games last season was 875 people. A random sample of 25 games this season has a mean attendance of 900 people with a standard deviation of 75. Has the hiring of a new coach affected game attendance at the .05 significance level?

13.19 A survey of 11 physicians in private practice showed that they felt that the pressures of their jobs necessitated taking a mean of 23.5 days of vacation per year. The sample standard deviation was 4.0 days. Eleven physicians in clinical practices with at least three other doctors responded that a mean of 19.0 days were necessary with a standard deviation of 3 days. At the .05 significance level do the two groups differ in their opinions as to vacation time necessary to avoid extreme stress?

13.20 The Toting Truck Moving Company has always given its drivers the option of taking the shorter route through Megapolis or the much longer route around the city. In an attempt to cut driving time, the company is interested in finding out if there is any time difference associated with the two routes. Ten drivers are recruited to drive each route with a full load, obeying all laws and speed limits. The results for the through-city sample are $\bar{X} =$ 75 minutes and $s = 18$ minutes. For the around-the-city sample $\bar{X} = 61$ minutes and $s =$ 8 minutes. At the .05 significance level is there any difference in the times it takes to traverse each route?

13.21 The effects of radiation on the time for mice to learn a specific behavior are being studied by the sophomore psychology students. Samples of exposed (E) and nonexposed (N) mice are taught the behavior. The sample statistics are summarized below.

$$N_E = 25 \qquad N_N = 25$$
$$\bar{X}_E = 15 \text{ minutes} \qquad \bar{X}_N = 13.5 \text{ minutes}$$
$$s_E^2 = 2.0 \text{ minutes} \qquad s_N^2 = 1.5 \text{ minutes}$$

Is there any evidence at the .01 significance level that exposed and nonexposed groups differ in the time required to learn the behavior?

13.22 A pharmaceutical company is studying the effectiveness of two kinds of *child-proof* lids for medicine bottles. Each of nine 5-year-old children is given one bottle with a type A lid. Then the same children are given bottles with type B lids. The times in minutes required for the children to open the bottles are recorded.

Child	*A*	*B*
1	5.0	4.3
2	7.5	8.0
3	10.0	9.2
4	3.4	4.0
5	11.0	3.5
6	4.3	2.0
7	3.9	2.1
8	8.0	5.0
9	7.5	4.0

Is there any indication at the .01 level of significance that the type A cap is more successful in keeping 5-year-olds out longer than the type B?

13.23 Kim Kegler offers a special series of 10 one-hour bowling lessons at a special discount. Ten people of various ages bowled a series of 10 games and then averaged their scores. Following 10 lessons from Kegler, they again bowled a 10-game series and averaged their scores.

Bowler	*Before*	*After*	*Bowler*	*Before*	*After*
1	97	103	6	50	65
2	105	110	7	75	79
3	150	140	8	110	100
4	88	105	9	65	85
5	93	93	10	100	130

At the .05 level, are the lessons effective in increasing scores?

13.24 The pork industry claims that the meat is declining in fat content because of new and better feeding programs for hogs. Ten years ago, 3-ounce servings contained a mean of 250 calories. A random sample taken by the Home Economics Department at Tyler Tech of 19 3-ounce servings from various supermarkets and restaurants has a mean calorie content of 230 calories with a standard deviation of 35 calories. Does this indicate a significant drop over the last 10 years at the .025 level?

13.25 Jeff and Jan received a new television set as a wedding present six years ago. Now the picture tube needs replacing. They are wondering if the mean life of all picture tubes differs significantly from the life of theirs. Eight of their acquaintances at the Racquet Club had picture tubes with lives (in years) of 8.5, 7, 4.5, 7.5, 6.3, 5.5, 8, and 7.3. Does the mean life of all picture tubes differ from six years at the .05 level of significance?

13.26 It has long been a mystery why a certain drug shows dramatic results in some people and has very little effect on others. Dr. Elizabeth Hughes has noticed that those who respond well show a strong dislike for milk. She conducts a test to measure the percent of the drug that has been absorbed into the bloodstream five hours after being administered to 22 milk drinkers (sample A) and 22 nonmilk drinkers (sample B). The results are

$$\bar{X}_A = 40 \qquad \bar{X}_B = 48$$
$$s_A^2 = 45 \qquad s_B^2 = 50$$

Can Dr. Hughes conclude at the .01 level of significance that those who abstain from drinking milk absorb more of the drug than those who do not?

13.27 According to EPA standards, waste materials dumped into the Ruby River may contain no more than three parts per million of toxic chemical no. 487. An environmental group concerned over the high levels of the chemical in the water is testing the waste materials of all the factories on the river's bank. The daily readings for one factory (in parts per million) are 3.1, 3.2, 3.3, 2.9, 3.5, 3.4, 2.5, 4.3, 2.9, 3.6, 3.2, 3.0, 3.7, 3.5, 3.9. Is this factory guilty at the .05 significance level of breaking the EPA guidelines?

13.28 To test a new formula to prevent fingernail splitting and breaking, 11 women are asked to file all their nails completely and apply the treatment. In 14 days, the mean growth per nail is recorded for each person. Then the nails are completely filed down again, but not treated. In 14 days the mean growth per nail is again recorded. The results are given in millimeters.

Subject	*Treated*	*Not Treated*
1	3.25	3.00
2	2.14	2.15
3	2.87	2.53
4	1.32	1.35
5	.41	0.00
6	1.59	1.44
7	1.33	1.30
8	3.01	2.50
9	2.87	2.43
10	2.53	2.21
11	1.77	.89

At the .05 significance level is the formula effective?

13.29 The mandatory attendance policy at Bennett College has long been a point of controversy. The administration agrees to a test. One hundred students are selected to form a stratified random sample to include all classifications of students. They are exempted from the attendance policy for one semester, but not told why. The mean of their GPAs for the semester is 2.57, with a standard deviation of .5. One hundred other students selected at random to form a stratified random sample have a mean GPA of 2.61, with a standard deviation of .4. At the .10 significance level, do the results support the value of a mandatory attendance policy?

13.30 As part of Handicapped Employee Awareness Week, 17 employers are given a test to measure their prejudice against hiring handicapped workers. They are then shown an animated movie aimed at attacking any prejudice they might have. Finally, a second test is administered to each employer. Do the results indicate that the movie was successful at the .005 significance level in reducing prejudice against hiring the handicapped? High scores indicate high degrees of prejudice.

Employer	*Before*	*After*
A	43	45
B	18	20
C	93	85
D	87	80
E	85	81
F	30	28
G	29	25
H	47	45
I	54	50
J	16	16
K	11	10
L	65	60
M	67	60
N	80	81
O	79	75
P	75	73
Q	70	61

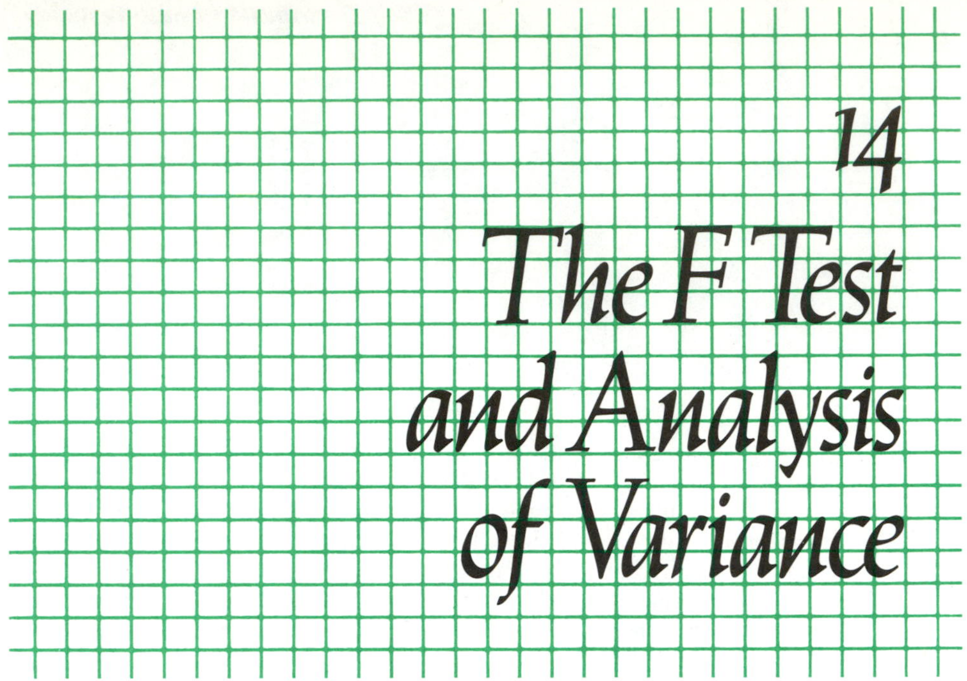

14.1 Introduction

We sometimes want to test the null hypothesis that two populations have the same, or equal, variances. For instance, two methods of teaching arithmetic are tried with sixth grade students, and samples of scores of students taught by each method are gathered. We shall assume that the distribution of arithmetic scores is known to be normal. We compute the variances of the two samples and designate them s_1^2 and s_2^2. *Our null hypothesis is that the two teaching methods do not produce different effects on the variability of students' scores.* Under this hypothesis, our two sample variances would be independent estimates of the same unknown σ^2, the variance of scores in the larger population.

Having obtained our values of s_1^2 and s_2^2, we wish to compare them in some way to test our null hypothesis. But how? If the two obtained values were identically equal, we could not possibly reject the null hypothesis. Almost certainly though, the obtained values would be different. The question is, "How are we to tell whether the obtained discrepancy is sizeable enough to allow us to reject our null hypothesis at whatever significance level we choose?" The F test answers this question for us.

We make reference to the F distribution whenever we test the hypothesis that two sample variances, s_1^2 and s_2^2, are independent estimates of a single population variance, σ^2. *We assume normality in the parent population*, though departures from normality tend not to invalidate the procedure unless they are drastic. After considering the F distribution, we shall do several problems testing the null hypothesis that a pair of sample variances are both estimates of a common σ^2.

14.2 The Theoretical Model of the F Distribution

Picture a giant wooden crate with an inexhaustible supply of slips of paper, each with a number written on it. Let us say that the distribution of numbers on the slips is normal, with unknown mean and variance, which we call μ and σ^2, respectively. We choose two numbers arbitrarily, which we shall use to determine the sizes of samples to be drawn. The numbers, let us say, are $N_1 = 21$ and $N_2 = 25$.

We now draw a pair of random samples from the crate, the first of 21 slips and the second of 25. The variance of the sample of 21 slips is, let us say, $s_1^2 = 35$, and that of the second sample is $s_2^2 = 20$. We divide the variance of our first sample by that of the second.

$$\frac{s_1^2}{s_2^2} = \frac{35}{20} = 1.75$$

We repeat the process, again drawing random samples of sizes 21 and 25, and again dividing the variance of the sample of 21 slips by that of the sample of 25. This time the variances are $s_3^2 = 10$ and $s_4^2 = 30$, where s_3^2 was computed from the sample of 21 slips and s_4^2 from that of 25 slips.

$$\frac{s_3^2}{s_4^2} = \frac{10}{30} = .333$$

Our next pair of samples gives us a ratio with a value of 1.25.

$$\frac{s_5^2}{s_6^2} = \frac{25}{20} = 1.25$$

Let us say we repeat the process of drawing a pair of random samples, the first of size 21 and the second of size 25, and each time divide the variance of our 21 slip sample by that of our 25 slip sample. Imagine, for instance, that the two hundredth time we do it, our ratio turns out to be .20.

$$\frac{s_{399}^2}{s_{400}^2} = \frac{4}{20} = .20$$

We record the computed ratios by piling up squares on a line. After we've obtained 500 ratios, our graph looks like that shown in Fig. 14.1.

Remember that the variance of our sample of 21 cases has $N_1 - 1$ degrees of freedom, which is 20; and that the variance computed from our sample of 25 cases

Fig. 14.1

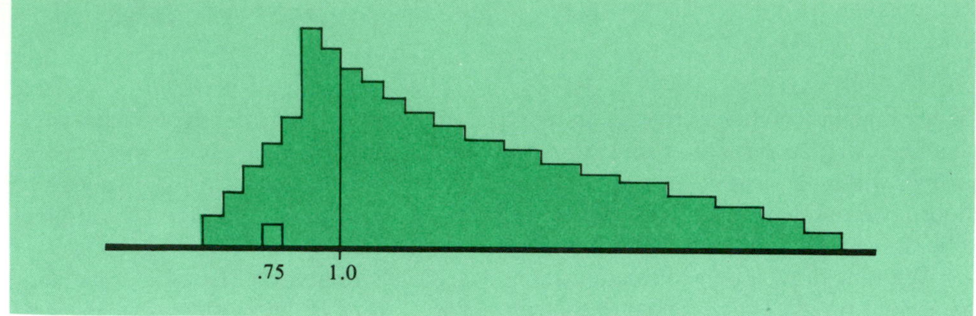

has $N_2 - 1$ degrees of freedom, which is 24. We have obtained and graphed a collection of ratios, each of which has a numerator based on 20 degrees of freedom and a denominator based on 24 degrees of freedom. In each fraction, both numerator and denominator are estimates of the unknown σ^2, and the numerator and the denominator of each fraction are independent.

It might be argued that the numerator of any given fraction is as likely to be bigger than the denominator of that fraction as it is to be smaller than the denominator; and thus as we increase the number of ratios in our collection, it should become increasingly likely that about half of the ratios would have values less than 1.000 and half of them more than 1.000. They would tend to cluster around 1.000. Actually, it can be shown mathematically that only when N_1 and N_2 are the same will the median be exactly 1.000. If N_1 and N_2 are nearly equal, as in our example, the median will be close to 1.000. The true median is .994 in our case.

Unlike the normal curve, the long run distribution we are approaching this time is **not symmetrical.** Since variances are never negative, **no ratio can be less than zero,** whereas **some ratios are very large.** As we continue adding new ratios, the graph of their frequencies approaches the smooth curve in Fig. 14.2.

Fig. 14.2

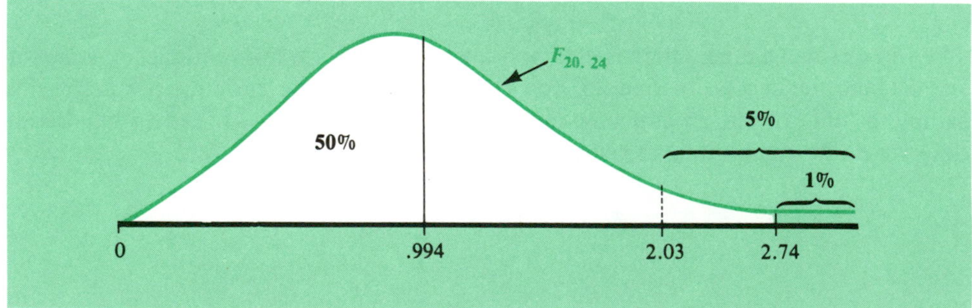

We have generated a new distribution, the F distribution, with degrees of freedom 20 and 24. The first number, 20, indicates the degrees of freedom (d.f.) that went into each numerator estimate of σ^2; and the second number, 24, reports the d.f. in each denominator estimate. Fig. 14.2 shows that as we continue our procedure, half of our obtained ratios are expected to be less than .994 and half more than .994.

Half the area is to the left of the vertical line at .994. Five percent of the ratios are expected to be bigger than 2.03 and 1% are expected to be bigger than 2.74 in the long run.

Suppose, instead, that we had arbitrarily chosen pairs of samples of 16 and 10 cases each, and had each time divided the variance of our 16 slip sample by that of our 10 slip sample. This time the degrees of freedom in each ratio would be 15 and 9. The F distribution generated by continued collection of such samples would appear as in Fig. 14.3.

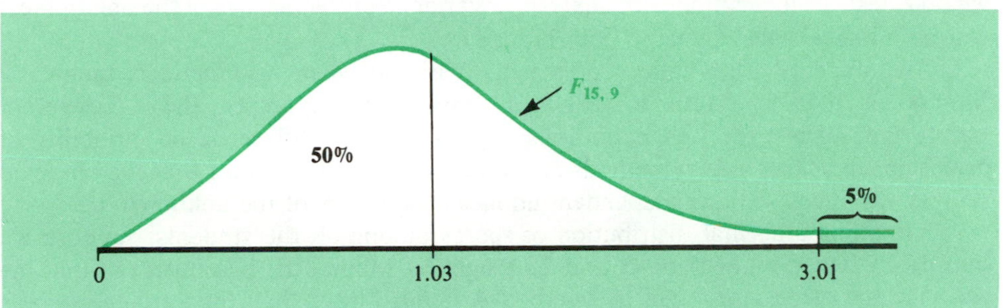

Fig. 14.3

This time it can be shown mathematically that the median is 1.03, so the axis at 1.03 divides the area in half, reflecting the fact that ratios are as likely to be less than 1.03 as more than 1.03. The curve is again nonsymmetrical, but slightly different in shape from the last one. Five percent of the area is to the right of 3.01.

When we talk about F curves, we are talking about a **family of distributions.** Two members of this family were graphed in Figs. 14.2 and 14.3. Each is a long run distribution of ratios, in which the numerator and denominator are independent estimates of a common σ^2. The d.f. in the numerator and that in the denominator must both be given to designate a member of the F distribution. We denote a member of the F family by writing $F_{r,s}$ where r is the d.f. in the numerator and s in the denominator. Thus we have discussed $F_{20,24}$ and $F_{15,9}$. We use our knowledge of the F distribution whenever we test the null hypothesis that two independent samples we have drawn from a population assumed to be normal are each providing unbiased estimates of the same σ^2.

Before exploring the usefulness of an F distribution in hypothesis-testing procedures, let us briefly enumerate the properties of the distribution discussed thus far.

1. An F distribution is continuous.
2. An F distribution is asymmetric.
3. All terms are nonnegative.
4. The median is one if $N_1 = N_2$. Otherwise it is close to one. (The closer N_1 and N_2 are, the closer the median is to one.)
5. An F distribution is one of a family of distributions. There is one for each pair of degrees of freedom.

When testing hypotheses and when we need to divide one variance by another, *we shall always be dividing the bigger of our two sample variances by the smaller one, and then using a form of the F table which is constructed for use with one-tail tests*. Although this is not essential, it is desirable. This practice, which assures that ratios which we look up in the F table when doing problems will always be bigger than one, simplifies computations and allows us to work with less elaborate tables than would be needed if we did two-tail tests. All that is necessary is that we state our alternative hypothesis in terms of "greater than" or "more variable than" as opposed to "less than" or "less variable than." This should create no problems. Later, when we do F tests in connection with analysis of variance problems, use of the one-tail test will be a logical consequence of our procedure.

Are students' arithmetic scores more variable when arithmetic is taught by Method A than by Method B? Let us choose, for example, the .05 level of significance to test the null hypothesis that there is *no* difference in variability of performance using the two methods. This null hypothesis assumes that the obtained sample variances will be independent unbiased estimates of the unknown σ^2.

Assuming a normal distribution of scores among all the students, suppose we sample 21 taught by Method A and 25 taught by Method B. If it turns out that the variance for the group taught by Method A is *not* bigger than the variance for the group taught by Method B, we need proceed no further, since we cannot possibly reject our null hypothesis in this situation. However, if it turns out that the group taught by Method A does have a bigger variance than the group taught by Method B, we must see whether the variance is enough bigger to permit us to reject our null hypothesis. Let us suppose that s_1^2, the variance of the group taught by Method A, is 46 and that s_2^2, the variance of the group taught by Method B, is 20. Since there were 21 people taught by Method A, s_1^2 is based on $N_1 - 1 = 21 - 1 = 20$ d.f. s_2^2 is based on $N_2 - 1 = 25 - 1 = 24$ d.f.

We divide the bigger sample variance by the smaller:

$$\frac{s_1^2}{s_2^2} = \frac{46}{20} = 2.30$$

Appendix Table V shows the 1% and 5% critical points for the F distribution where it is assumed that the bigger of two obtained estimates of a common σ^2 has been made the numerator. We must find the critical point relevant to our problem, where the numerator and denominator have 20 and 24 d.f. respectively. In Table V, the column values pertain to the d.f. in the numerator. Here $N_1 = 20$. The row value in the table, indicating the d.f. in the denominator, is $N_2 = 24$. In Table V, the number in Roman type in the box in row 24, and the column headed 20, is 2.03. This number gives us the beginning of the rejection region when the .05 level of significance is used. Since our obtained value of 2.30 is bigger than 2.03, we reject the null hypothesis that s_1^2 and s_2^2 are each providing independent estimates of the same unknown σ^2.

If the experiment was conducted properly (and this implies, among other things, that the students were assigned the different methods randomly), we may conclude that the alternative hypothesis has been established; that is, the teaching methods exert different effects on the students' variability of performance, with Method A producing the more variable results. Intuitively, it should be seen that since we are always putting the bigger variance on top when we need to compute a ratio, then the bigger the ratio we find, the less likely it is that the two sample variances were estimates of the same σ^2. Since, under the null hypothesis, a ratio as big as ours would be found less than 5% of the time by chance, we were able to reject the null hypothesis in our one-tail test. Note that if we had used the .01 level of significance, the ratio would not have been big enough to allow us to reject the null hypothesis. Our obtained value was less than the .01 value (of 2.74) given in boldface type. For our one-tail tests, a bigger ratio is always needed to reject at the .01 level than at the .05 level.

We should like to consider one more problem, beginning with raw scores this time, in order to emphasize notation that will be needed shortly. An experimenter wants to prove the alternative hypothesis that at age 21, men are more variable in mechanical aptitude than women are. She gives an aptitude test to 41 men and 31 women in this age group, planning to reject the null hypothesis only if s_1^2/s_2^2 is significantly bigger than one, where s_1^2 is the estimate of σ^2 based on the men's scores and s_2^2 is the estimate of σ^2 based on the women's scores. This means that if s_2^2 turns out to be bigger than s_1^2, the null hypothesis cannot be rejected. As before, the logic of our experiment dictates that the test be one-tailed. Our theory says to reject only if the numerator is sufficiently larger than the denominator. The two sets of aptitude scores are shown in Table 14.1. The squares of the individual scores have been computed and will be used in the final computations. Our values of s_1^2 and s_2^2 will be computed by Formulas (12.2).

$$\text{d.f.} = N_1 - 1 = 40$$

$$s_1^2 = \frac{\sum X^2 - \dfrac{(\sum X)^2}{N_1}}{N_1 - 1} = \frac{40{,}660 - \dfrac{(1{,}230)^2}{41}}{40} = \frac{40{,}660 - 36{,}900}{40} = \frac{3{,}760}{40} = 94$$

$$\text{d.f.} = N_2 - 1 = 30$$

$$s_2^2 = \frac{\sum Y^2 - \dfrac{(\sum Y)^2}{N_2}}{N_2 - 1} = \frac{21{,}775 - \dfrac{(775)^2}{31}}{30} = \frac{21{,}775 - 19{,}375}{30} = \frac{2{,}400}{30} = 80$$

If the null hypothesis is true, the ratio of these two estimates of σ^2 is a member of $F_{40,30}$.

Table 14.1

	Group One Men			Group Two Women	
	X	X^2		Y	Y^2
1.	14	196	1.	28	784
2.	29	841	2.	31	961
3.	37	1,369	3.	20	400
4.	20	400	4.	17	289
5.	38	1,444	5.	25	625
6.	29	841	6.	3	9
7.	52	2,704	7.	27	729
8.	32	1,024	8.	31	961
9.	34	1,156	9.	8	64
10.	31	961	10.	19	361
11.	23	529	11.	14	196
12.	31	961	12.	27	729
13.	34	1,156	13.	23	529
14.	29	841	14.	25	625
15.	8	64	15.	24	576
16.	16	256	16.	25	625
17.	37	1,369	17.	38	1,444
18.	32	1,024	18.	34	1,156
19.	36	1,296	19.	16	256
20.	25	625	20.	33	1,089
21.	29	841	21.	19	361
22.	38	1,444	22.	21	441
23.	30	900	23.	27	729
24.	51	2,601	24.	17	289
25.	58	3,364	25.	43	1,849
26.	29	841	26.	38	1,444
27.	25	625	27.	40	1,600
28.	35	1,225	28.	25	625
29.	29	841	29.	27	729
30.	30	900	30.	30	900
31.	32	1,024	31.	20	400
32.	28	784	$N_2 = 31$	$\Sigma Y = 775$	$\Sigma Y^2 = 21,775$
33.	27	729			
34.	28	784			
35.	9	81			
36.	27	729			
37.	21	441			
38.	28	784			
39.	33	1,089			
40.	30	900			
41.	26	676			
$N_1 = 41$	$\Sigma X = 1,230$	$\Sigma X^2 = 40,660$			

$$\frac{s_1^2}{s_2^2} = \frac{94}{80} = 1.175$$

We cannot reject our null hypothesis at the .05 level of significance, since Appendix V shows that a value as big as 1.79 or bigger is needed; and thus we cannot conclude that men are more variable than women in mechanical aptitude, at least not in the population sampled by our experimenter.

The formal presentation of a hypothesis test involving an F distribution follows Steps 1–6 presented in Chapter 11. The example just completed would look much like this.

Hypothesis Test (Difference Between Variances)

1. H_1: $\sigma_1^2 > \sigma_2^2$

   H_0: $\sigma_1^2 = \sigma_2^2$
2. .05 significance level/one-tail/F test
3. $N_1 = 41$ $\quad s_1^2 = 94$ $\quad$ (All computations may be presented in this step.)

   $N_2 = 31$ $\quad s_2^2 = 80$
4. $F = \dfrac{s_1^2}{s_2^2} = \dfrac{94}{80} = 1.175$
5. Do Not Reject H_0.
6. Do Not Accept H_1: We failed to show that men are more variable than women.

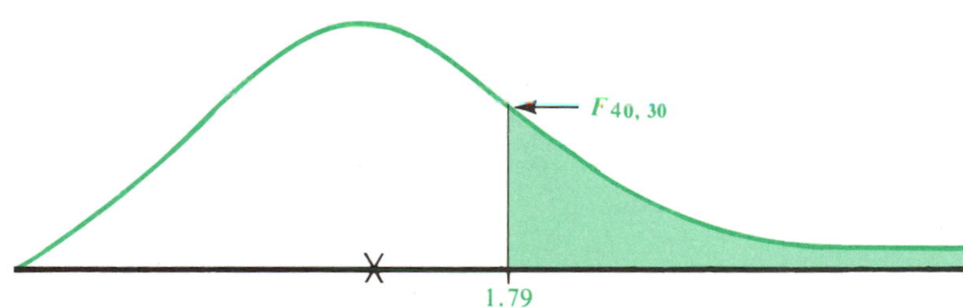

The F distribution was named after the late Sir Ronald Fisher, already mentioned, a genius at mathematical statistics and in particular at solving problems by intuitive geometrical devices when algebraic methods were opaque. We refer to the F distribution whenever we test the null hypothesis that the variances of two samples are estimates of the same σ^2 (or of σ^2's in different populations that, under our null hypothesis, are identical in value).

Note that as a formula we have

$$F = \frac{s_1^2}{s_2^2} \qquad (14.1)$$

14.3 Comparing More Than Two Sample Means

Suppose that children in the fourth grade of a school are randomly assigned to six groups, each to be taught spelling by a different method. At the end of the year, these children take a city-wide spelling test. The question we want to answer is "May we conclude that the methods differ in effectiveness, as measured by scores on the city-wide spelling test?" We shall assume that, ordinarily, fourth grade spelling scores are distributed normally, and that 30 children have been taught by each of the six methods. At the end of the school year, when the test is given to the six groups, the mean scores of these groups are

	Group					
	Method A	Method B	Method C	Method D	Method E	Method F
Mean Score	$\bar{X}_1 = 75$	$\bar{X}_2 = 72$	$\bar{X}_3 = 76$	$\bar{X}_4 = 79$	$\bar{X}_5 = 82$	$\bar{X}_6 = 72$

Are the differences wide enough for us to conclude that the methods really differ in effectiveness? Consider first how many *t* tests might be done involving pairs of sample means. Table 14.2 shows that there are 15.

Table 14.2

$\bar{X}_1$ with $\bar{X}_2$	$\bar{X}_2$ with $\bar{X}_3$	$\bar{X}_3$ with $\bar{X}_4$	$\bar{X}_4$ with $\bar{X}_5$	$\bar{X}_5$ with $\bar{X}_6$
$\bar{X}_1$ with $\bar{X}_3$	$\bar{X}_2$ with $\bar{X}_4$	$\bar{X}_3$ with $\bar{X}_5$	$\bar{X}_4$ with $\bar{X}_6$	
$\bar{X}_1$ with $\bar{X}_4$	$\bar{X}_2$ with $\bar{X}_5$	$\bar{X}_3$ with $\bar{X}_6$		
$\bar{X}_1$ with $\bar{X}_5$	$\bar{X}_2$ with $\bar{X}_6$			
$\bar{X}_1$ with $\bar{X}_6$				

With all those *t* tests to do, the experimenter is given many more opportunities of finding a significant *t* value than would be the case if only two samples had been collected and therefore only one comparison made. If instead of six, we had collected 50 samples and computed their means, and then performed *t* tests between all pairs of means, then even if the samples really came from exactly the same population, it is almost certain that *some* differences between pairs of means would be significant. For the very reason that we have given ourselves too much opportunity for a significant finding, the t test becomes inappropriate as soon as it is clear that more than two sample means are going to be compared with each other. When a comparison is to be made involving more than two sample means, another technique is needed to test the null hypothesis that these sample means, the $\bar{X}$'s, are *all* estimates of a common population mean, μ.

We are working toward a test that takes into account three or more group means *simultaneously* and tests the null hypothesis that all of them are estimates of the same population mean, μ. In the case we are considering, *our null hypothesis is that the different teaching methods were equally effective*. It would follow from this hypothesis that the spread among our sample means was a chance phenomenon. Our test must answer the question, "How unlikely would it be to encounter as wide a spread among sample means as this one, assuming that the null hypothesis is true?" So we will, in essence, be using the concept of variance to test for difference between means. It is the scatter or variance between the sample means that will be of interest to us. If it is significant, we will eventually conclude that the six methods are different and all six means are not equal.

14.4 Introduction to Analysis of Variance

Table 14.3 shows the means, variances, and sample sizes of the six groups. Note that all samples are the same size. In the next section we shall develop a technique that will eliminate this requirement. Our null hypothesis is that all six means have come from the same population, with mean $\mu_{\bar{X}}$ and variance $\sigma^2_{\bar{X}}$ both unknown. (The distribution is a distribution of sample means with $\mu_{\bar{X}} = \mu$ and $\sigma_{\bar{X}} = \dfrac{\sigma}{\sqrt{N}}$.)

Is the scatter among our six means big enough for us to reject the null hypothesis, say at the .05 level of significance? If, under the null hypothesis, our means would show at least as much scatter more frequently than five times in 100, we cannot reject the hypothesis; if, under the null hypothesis, scatter this big would be found as or less frequently than five times in 100, we shall reject the null hypothesis, concluding that the sample means were *pushed* apart by differences in the teaching methods.

Table 14.3

$\bar{X}$	$\bar{X}_1 = 75$	$\bar{X}_2 = 72$	$\bar{X}_3 = 76$	$\bar{X}_4 = 79$	$\bar{X}_5 = 82$	$\bar{X}_6 = 72$
s^2	$s_1^2 = 173.2$	$s_2^2 = 168.7$	$s_3^2 = 170.1$	$s_4^2 = 169.8$	$s_5^2 = 172.0$	$s_6^2 = 167.6$
N	$N_1 = 30$	$N_2 = 30$	$N_3 = 30$	$N_4 = 30$	$N_5 = 30$	$N_6 = 30$
d.f.	29	29	29	29	29	29

Before looking at the specifics of the test, we need to see the overall picture, the battle plan. Once the basic strategy is understood, we can turn our attention to the little skirmishes we will encounter along the way.

First we must consider the population of all fourth graders being taught spelling, by whatever method. The distribution of spelling scores has a mean μ and a standard deviation of σ^2. It is this σ^2 we are interested in. Our goal is to find two estimates for σ^2. One of them, which we shall call s_B^2, *will reflect the scatter* between the six sample means. The other, which we shall call s_W^2, *will not reflect this scatter*. Now, if the scatter between the means is signficant, s_B^2 and s_W^2 will be close to one another since they estimate the same thing and the scatter is exerting little influence. That is, their ratio will be close to one. On the other hand, if the scatter between the means is significant, then s_B^2, the estimate reflecting that scatter, will be affected (since now the scatter is exerting significant influence). In this case, the ratio of s_W^2 and s_B^2 will not be close to one. It is imperative that the reader understand everything said so far in this section. If not, it should be studied again before proceeding. Now, on to the specifics.

To begin with, we need a way of representing the amount of scatter we actually found among our sample means. We compute our measure of scatter, the variance of the means, which we denote $s_{\bar{X}}^2$. That is, using Formula (12.2) we calculate this variance of our six means, considering them as six members of a theoretical distribution of means.[1] Since means instead of individual cases are elements, $\bar{X}$ replaces X, and K, the number of means being considered, replaces N. Formula (14.2) is then obtained from Formula (12.2).

$$s_{\bar{X}}^2 = \frac{\Sigma \bar{X}^2 - \dfrac{(\Sigma \bar{X})^2}{K}}{K - 1} \tag{14.2}$$

The estimate $s_{\bar{X}}^2$ is based upon $K - 1$ degrees of freedom. Computations done with the data in Table 14.3 are as follows:

$$\begin{aligned} \Sigma \bar{X} &= 75 + 72 + 76 + 79 + 82 + 72 = 456 \\ (\Sigma \bar{X})^2 &= (456)^2 \\ \Sigma \bar{X}^2 &= 75^2 + 72^2 + 76^2 + 79^2 + 82^2 + 72^2 = 34{,}734 \\ K &= 6 \end{aligned}$$

To find the variance of the sample means, $s_{\bar{X}}^2$, we write

$$s_{\bar{X}}^2 = \frac{34{,}734 - \dfrac{(456)^2}{6}}{5} = 15.6$$

[1] $s_{\bar{X}}^2$ can be obtained in this problem, as always, by subtracting each of the terms from their mean, squaring the differences, and then dividing by one less than the number of terms.

Now if the null hypothesis is true and the scatter among our means is random, then in Formula (14.3), which is obtained by squaring both sides of Formula (13.2), we have

$$s_{\bar{X}}^2 = \frac{s^2}{N} \tag{14.3}$$

Here N is the number of cases in each of our samples, and s^2 is an estimate of σ^2, the variance of scores in the population of fourth graders. Rearranging Formula (14.3) yields

$$Ns_{\bar{X}}^2 = s^2 \tag{14.4}$$

In our problem, where $N = 30$,

$$Ns_{\bar{X}}^2 = (30)(15.6) = 468 = s^2$$

Now we have exactly what we were searching for—an estimate for the variance of the scores of all fourth graders, σ^2. Furthermore, this estimate does reflect the scatter between the sample means since it traces back to $s_{\bar{X}}^2$ (a measure of the scatter between the six means). So we can designate this estimate s_B^2 instead of s^2 to remind us that this estimate of σ^2 is based entirely on the scatter between the sample means. Now let us review the progression of equations.

$$s_B^2 = Ns_{\bar{X}}^2$$

[substituting s_B^2 for s^2 in Formula (14.4)]

$$s_B^2 = N\left[\frac{\sum \bar{X}^2 - \dfrac{(\sum \bar{X})^2}{K}}{K-1}\right]$$

[substitution using Formula (14.2)]

s_B^2 is based on $K - 1$ degrees of freedom.

In our example we have

$$s_B^2 = 468$$

remembering that s_B^2 is based on $K - 1 = 5$ d.f.

Suppose that our null hypothesis were false—that is, the methods really exerted different effects. As experimenters, we would not know this fact, but it would already have manifested itself by increasing the value of $s_{\bar{X}}^2$ and therefore the value of s_B^2, which is an estimate of σ^2. If the null hypothesis were false, s_B^2 would tend to be too big.

Having discovered Formula (14.5) for s_B^2, we now need to find a formula for s_W^2, the estimate that will *not* reflect the scatter between the sample means.

To discover this second estimate, reconsider Table 14.3. We want an unbiased estimate of σ^2, which would not be affected if there were real differences between groups. $s_1^2 = 173.2$ is such an estimate. It provides us with an estimate of σ^2, which, since it was based wholly on scores from Group One, cannot possibly have been influenced by differences between groups; and thus $s_1^2 = 173.2$ is the sort of value we want to compare with s_B^2. $s_2^2 = 168.7$ would also provide an unbiased estimate, and so would the sample variance of each of the groups. To find our best estimate, the one based on the most degrees of freedom, we shall use the data in all six sample variances. When samples are all the same size, we can add up the s^2's and divide by the number of groups, in this case six; when the samples differ in size, we must give proportional representation when computing our average variance. We weight each sample variance by its degrees of freedom, add, and then divide their sum by the total degrees of freedom. If there are N_1 scores in the first group, N_2 in the second, and so on, then s_1^2 is based on $N_1 - 1$ d.f., s_2^2 on $N_2 - 1$ d.f., etc. The weighted mean of these sample variances is the estimate of σ^2 we are looking for. We designate it s_W^2.

$$s_W^2 = \frac{(N_1 - 1)s_1^2 + (N_2 - 1)s_2^2 + \ldots + (N_K - 1)s_K^2}{(N_1 - 1) + (N_2 - 1) + (N_3 - 1) + \ldots + (N_K - 1)}$$

where K is the number of groups. [It may be recognized that the expression under the first square root sign in Formula (13.4) is the special case of this based on two samples.]

It is convenient to let $N_T = N_1 + N_2 + N_3 + \ldots + N_K$, so that we may write equivalently,[2]

$$s_W^2 = \frac{(N_1 - 1)s_1^2 + \ldots + (N_K - 1)s_K^2}{N_T - K} \tag{14.6}$$

This estimate is based on $N_T - K$ d.f.

From the values in Table 14.3, we can compute

[2] $(N_1 - 1) + (N_2 - 1) + \ldots + (N_K - 1) = N_T - K$.

$$s_W^2 = \frac{(30-1)(173.2)+(30-1)(168.7)+(30-1)(170.1)+(30-1)(169.8)+(30-1)(172.0)+(30-1)(167.6)}{180-6}$$

$$= \frac{(29)(1{,}021.4)}{174}$$

$$= \frac{1{,}021.4}{6}$$

$$= 170.2$$

This estimate is based on $N_T - K = (180 - 6) = 174$ d.f.

s_W^2 is the estimate of σ^2 we have been looking for. It is based entirely on the scatter among scores *within* the groups. At no time did a distance *between* mean scores in different groups enter the calculations, and thus even if the sample means had been pushed very far apart by the teaching methods exerting different effects on the student's performance, this estimate, s_W^2, would not have been in the slightest altered on that account. s_W^2 is independent of s_B^2.

If the null hypothesis is true, both s_B^2 and s_W^2 are unbiased estimates of σ^2. If it is false, we may expect that s_B^2 will be inflated, whereas even under that condition s_W^2 will remain an unbiased estimate of σ^2.

We may now test the null hypothesis by computing our ratio $F = s_B^2/s_W^2$ using Formula (14.1). Under the null hypothesis, this ratio would have an F distribution with $K - 1$ d.f. in the numerator and $N_T - K$ d.f. in the denominator. We shall test the hypothesis at the .05 level of significance. We have

$$F = \frac{s_B^2}{s_W^2} = \frac{468}{170.2} = 2.75$$

Here the d.f. are $K - 1 = 5$ and $N_T - K = 174$.

Although the exact entry does not appear in Appendix Table V, we see that if the d.f. had been 5 and 150, a ratio as big as 2.27 would have been significant. With the same d.f. in the numerator and even more d.f. in the denominator, a ratio of 2.27 must surely be significant. Any discrepancy becomes more meaningful as the number of cases from which it is computed increases. Thus we can safely conclude that our obtained value of 2.75 represents significance at the .05 level. We reject the null hypothesis that the six sample means were unbiased estimates of μ, and we conclude the alternative that something other than sampling error must have inflated the numerator estimate of σ^2. In general, if a finding would be significant with fewer d.f. in the denominator than we have collected, then it must be significant in the problem at hand, where there are more degrees of freedom.

To summarize and review the test just discussed, we shall present it in formal form. Recall that it applies only when all samples are the same size.

Hypothesis Test (Difference Between Many Means)

1. H_1: The six sample means do not all estimate the same population mean.

   H_0: The six sample means all estimate the same population mean μ. ($\mu_1 = \mu_2 = \mu_3 = \mu_4 = \mu_5 = \mu_6$)

2. .05 significance level/one-tail/F test.

3.

$\bar{X}$	$\bar{X}_1 = 75$	$\bar{X}_2 = 72$	$\bar{X}_3 = 76$	$\bar{X}_4 = 79$	$\bar{X}_5 = 82$	$\bar{X}_6 = 72$
s^2	$s_1^2 = 173.2$	$s_2^2 = 168.7$	$s_3^2 = 170.1$	$s_4^2 = 169.8$	$s_5^2 = 172.0$	$s_6^2 = 167.6$
N	$N_1 = 30$	$N_2 = 30$	$N_3 = 30$	$N_4 = 30$	$N_5 = 30$	$N_6 = 30$
d.f.	29	29	29	29	29	29

4.
$$s_B^2 = N\left[\frac{\sum \bar{X}^2 - \frac{(\sum \bar{X})^2}{K}}{K-1}\right] = 468 \qquad K - 1 = 5 \text{ d.f.}$$

$$s_W^2 = \frac{(N_1 - 1)s_1^2 + \ldots + (N_K - 1)s_K^2}{N_T - K} = 170.2 \qquad N_T - K = 174 \text{ d.f.}$$

$$\frac{s_B^2}{s_W^2} = \frac{468}{170.2} = 2.75$$

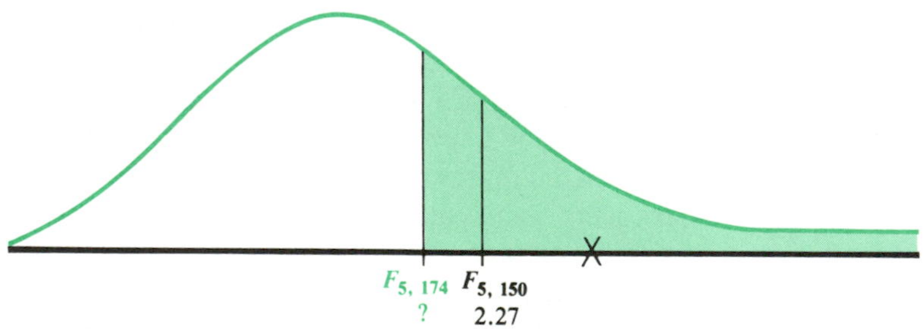

5. Reject H_0.

6. Accept H_1: The six methods differ in effectiveness.

Note that we cannot conclude that all differences between pairs of means are significant. So far, we can draw no conclusions regarding any particular pair of means. Strictly speaking, a significant finding does not even allow us to say that the biggest of our K means is significantly bigger than the smallest! Other tests of significance, discussed in more advanced texts, show us how to make individual comparisons. The F test we have done allows us to conclude only that *some* factor is scattering the means; and if our experiment was properly controlled, we may conclude that this factor was inherent in the teaching methods used.

The method just described is not commonly used in practice. It does, however, give us a good insight into the logic behind the analysis of variance. A more adaptable and widely used technique will now be explored.

14.5 The Technique of Analysis of Variance

The aim so far has been to convey the logic behind the technique of analysis of variance, using information already given to the reader in earlier discussions. In practice, when we must work from raw data, not having already computed the parts that go into our F ratio, it serves us well to put them into a table designed to give us particular parts when we need them and to help us interpret our findings. The procedure to be presented is widely used; it has the advantage of being easily extended for use with more complex analysis of variance problems than are to be presented in this book.

To make sure this procedure is conveyed clearly, we shall use numbers that make the operations easier to perform than they would be in practice, and we shall keep the sample sizes smaller than they would be ordinarily.

Suppose we wish to show that there are differences among three groups in "personal defensiveness." Our null hypothesis is that the three groups are the same in personal defensiveness and that whatever scatter we encounter among the sample means will be due to chance. This is the hypothesis that the means of the samples of the three groups are all unbiased estimates of a common unknown μ. We shall use the .05 significance level. The personal defensiveness scores and their squares appear in Table 14.4. Study the calculations closely. This is the time to make sure you are familiar with the notation. Though notation itself can be messy and should be kept to a minimum, a good notational system may be all we have to save us when computations become numerous and difficult to handle.

The computations made at the bottom of Table 14.4 have each been labeled. In the first row, the sum of each group is given; the subscript indicates the group. In the second row appears the sum of squares of each group. For instance, in Group B the sum of the squared terms, ΣX_2^2, is 70. The third row shows the number of terms in

Table 14.4 Table of Computations for Personal Defensiveness Scores

	Group A		Group B		Group C		
	X_1	X_1^2	X_2	X_2^2	X_3	X_3^2	
	2	4	5	25	5	25	
	4	16	4	16	2	4	
	6	36	0	0	1	1	
	8	64	5	25	0	0	
			2	4			Row Totals

	Group A	Group B	Group C	Row Totals
1. ΣX	$\Sigma X_1 = 20$	$\Sigma X_2 = 16$	$\Sigma X_3 = 8$	$\Sigma X_T = 44$
2. ΣX^2	$\Sigma X_1^2 = 120$	$\Sigma X_2^2 = 70$	$\Sigma X_3^2 = 30$	$\Sigma X_T^2 = 220$
3. N	$N_1 = 4$	$N_2 = 5$	$N_3 = 4$	$N_T = 13$
4. $(\Sigma X)^2$	$(\Sigma X_1)^2 = 20^2$	$(\Sigma X_2)^2 = 16^2$	$(\Sigma X_3)^2 = 8^2$	
5. $\frac{(\Sigma X)^2}{N}$	$\frac{(\Sigma X_1)^2}{N_1} = \frac{20^2}{4}$	$\frac{(\Sigma X_2)^2}{N_2} = \frac{16^2}{5}$	$\frac{(\Sigma X_3)^2}{N_3} = \frac{8^2}{4}$	$G = \frac{(\Sigma X_1)^2}{N_1} + \frac{(\Sigma X_2)^2}{N_2} + \frac{(\Sigma X_3)^2}{N_3}$
6. $\frac{\Sigma X}{N} = \bar{X}$	$\bar{X}_1 = \frac{20}{4} = 5$	$\bar{X}_2 = 3.2$	$\bar{X}_3 = 2.0$	$G = \frac{20^2}{4} + \frac{16^2}{5} + \frac{8^2}{4}$ $G = 167.2$

each group. The fifth row gives the value in the fourth row divided by the value in the third row. The sixth row gives the mean of each group. Each time, the group number is indicated by the subscript. The subscript T, indicating *total*, refers to the entire collection of observations in the experiment. For instance, ΣX_T^2 stands for the sum of the squares of all the observations made. ΣX_T^2 will soon be important. N_T shall hereafter be used to indicate the total number of observations made; here 13. The terms in row four do not have immediate value for us but were needed to calculate the terms in row five. The total of the terms in row five is a very important quantity, which we shall call G.

$$G = \frac{(\Sigma X_1)^2}{N_1} + \frac{(\Sigma X_2)^2}{N_2} + \ldots + \frac{(\Sigma X_K)^2}{N_K} \tag{14.7}$$

In more complex problems, we need to compute several such quantities, each used in a different part of the analysis. It is always useful to examine the means of our different groups to get some sense of what we are asking. Here the means are 5, 3.2, and 2.0; we are testing the null hypothesis, which says that the scatter among these three estimates of μ is due to chance and not to influences specific to particular groups.

We can now compute our two estimates of σ^2 from the data in Table 14.4. We use the following computational formulas. Remember that K is the number of groups; here $K = 3$.

$$s_B^2 = \frac{G - \dfrac{(\sum X_T)^2}{N_T}}{K - 1}$$

so that

$$s_B^2 = \frac{167.2 - \dfrac{(44)^2}{13}}{2} = 9.14$$

And

$$s_W^2 = \frac{\sum X_T^2 - G}{N_T - K}$$

so that

$$s_W^2 = \frac{220 - 167.2}{13 - 3} = 5.28$$

In these fractions, the denominators represent the numbers of d.f. that went into the estimates. s_B^2 is based on $K - 1$ d.f. and s_W^2 on $N_T - K$ d.f. Here s_B^2 is based on 2 d.f. and s_W^2 on 10 d.f.

Next we divide s_B^2 by s_W^2.

$$\frac{s_B^2}{s_W^2} = \frac{9.14}{5.28} = 1.73$$

We must now make reference to Table V and in particular to what it tells us about the distribution of $F_{2,10}$. Our value of 1.73 is not significant, being less than 4.10, which marks the beginning of the .05 rejection region. We have failed to reject our null hypothesis. The evidence does not allow us to conclude that these groups differ in personal defensiveness.

The method described would enable us to do any "one-way analysis of variance" problem—that is, any problem in which scores on only one variable were studied and the question asked, "Did the means show significant scatter on that variable?" In the future, we shall compute the numerator and denominator of each of our variance estimates separately, and enter each as we find it on a table like Table 14.5.

The reader should study the form of Table 14.5 and learn the new names for the column headings and be sure where the numbers come from.

We have based our computations on the following formulas, which should be studied carefully. The essence of the analysis of variance work discussed in this chapter is given in Table 14.5. The entries have been made using the following formulas.

$$SS_B = G - \frac{(\sum X_T)^2}{N_T} \tag{14.8}$$

$$SS_W = \sum X_T^2 - G \tag{14.9}$$

$$s_B^2 = \frac{SS_B}{K-1} \tag{14.10}$$

$$s_W^2 = \frac{SS_W}{N_T - K} \tag{14.11}$$

In our problem we find $G = 167.2$, and the results of the other computations are given in Table 14.5, culminating in

$$F = \frac{s_B^2}{s_W^2} = 1.73$$

Table 14.5 Analysis of Variance Table

Source of Variation	Numerator—Sum of Squares	Denominator—d.f.	Estimate of σ^2	Obtained F Ratio	Table Value
Between Group Means	SS_B 18.3	$K-1$ 2	s_B^2 9.14	1.73	F_{K-1, N_T-K} 4.10
Within Groups	SS_W 52.8	$N_T - K$ 10	s_W^2 5.28		

We shall conclude our discussion of analysis of variance with one more example. Suppose our raw data consists of the results of four procedures that are to be compared for their effectiveness in teaching English as a second langauge. Twenty non-English-speaking students were randomly assigned to the four groups using the four methods. Three students did not finish the course for nonacademic reasons, so in the end the group sizes were unequal. The final scores at the end of the term were as follows:

Group I: 32, 33, 36, 34, 32
Group II: 23, 32, 20, 19, 32
Group III: 27, 34, 31, 22
Group IV: 26, 17, 30

The work is often made easier if all computations are recorded on a worksheet similar to the one found in Table 14.6.

Table 14.6 Analysis of Variance Worksheet

H_1: The procedures differ in effectiveness.
H_0: The procedures do not differ in effectiveness.

	X_1	X_1^2	X_2	X_2^2	X_3	X_3^2	X_4	X_4^2	
	32	1024	23	529	27	729	26	676	
	33	1089	32	1024	34	1156	17	289	
	36	1296	20	400	31	961	30	900	
	34	1156	19	361	22	484			
	32	1024	32	1024					
ΣX		167		126		114		73	480
ΣX^2		5,589		3,338		3,330		1,865	14,122
N		5		5		4		3	17
$(\Sigma X)^2$		27,889		15,876		12,996		5,329	
$\frac{(\Sigma X)^2}{N}$		5,577.8		3,175.2		3,249		1,776.3	13,778.3
$\bar{X}$		33.4		25.2		28.5		24.3	

$$SS_B = 13{,}778.3 - \frac{(480)^2}{17} = 225.4$$

$$SS_W = 14{,}122 - 13{,}778.3 = 343.7$$

Source of Variation	*SS*	d.f.	Estimate of σ^2	*F* Ratio	Table Value
Between Groups	225.4	3	75.13	2.84	$F_{3,13}$
Within Groups	343.7	13	26.44	2.84	3.41

Decision: Do Not Reject H_0.
Conclusion: Do Not Accept H_1: We failed to show any difference in effectiveness.

The technique of analysis of variance is designed for testing the null hypothesis that the means of a set of K samples are all estimates of the same population mean, μ.[3] We test the hypothesis by comparing an estimate of σ^2 derived from the scatter among means with an estimate of σ^2 stable enough not to undergo change if the hypothesis is false. The F ratio obtained is the statistic used to test the hypothesis that the variable being studied had no effect on the group means— that is, did not push them apart.

This technique may be extended to problems where there are many variables and hence different sets of means to be studied. It allows us to examine separately the influence of different variables operating in a single experiment, and even to study the interactions among the different variables. No matter how intricate the design, the form of each and every F test of a hypothesis is such that the numerator is subject to being increased, if the hypothesis is false, and the denominator is left uninfluenced. This means that every F test we do, in simple designs and in complex ones, is one-tailed and that we reject the null hypothesis if the numerator of our obtained F ratio is too big.

Problem Set A

14.1 Intercollegiate academic debate involves two opposing sides, the negative (N) and the affirmative (A). Coach Andrews is interested in establishing that the number of matches won per season by negative teams is more variable than the number won by affirmative teams. For 31 randomly selected teams

$$N_N = 17 \qquad N_A = 14$$

$$s_N^2 = 64 \qquad s_A^2 = 40$$

Is the coach's belief supported at the .01 significance level?

14.2 Samples of two types of cable insulation were tested by the manufacturer of type A insulation in hopes of showing its resistance less variable than that of type B. At the .05 level of significance, was the test successful?

$$N_A = 20 \qquad s_A^2 = 58$$

$$N_B = 15 \qquad s_B^2 = 136$$

14.3 Medical tests have established that the mean number of treatments necessary to eliminate warts is the same for two products, Wart-Away (W) and Toad Touchers Remedy (T). Tests are now underway to determine if the variance for Wart-Away is less than that for Toad Touchers Remedy as is claimed by the former's developer. For 102 patients testing the formulas

$$N_W = 61 \qquad N_T = 41$$

$$s_W = 6.4 \qquad s_T = 8.6$$

Can the null hypothesis be rejected if the probability of a type-one error is .05?

[3] Experimenters sometimes envision themselves drawing samples from different populations where $\mu_1 = \mu_2 = \mu_3 = \ldots = \mu$. The hypothesis that this is so is statistically identical to the one stated.

14.4 a. Is an F ratio of 2.15 significant at the .05 significance level if the applicable distribution is $F_{8,50}$?
b. Is an F ratio of 2.05 significant at the .05 significance level if the applicable distribution is $F_{8,110}$?
c. Can it be determined from Appendix V if 2.64 is significant at the .01 significance level using the $F_{10,62}$ distribution? using the $F_{10,53}$ distribution?

14.5 A state licensing agency is looking into the lengths of stays for patients in three mental care centers. The files of 26 patients are randomly selected from each facility. A summary of the findings (in days) is given here.

$\bar{X}_1 = 49$	$\bar{X}_2 = 53$	$\bar{X}_3 = 47$
$s_1^2 = 33$	$s_2^2 = 46$	$s_3^2 = 48$

Is a difference in the mean stay per patient indicated among the three centers at the .05 significance level?

14.6 A leading criminologist, Sara Sleuth, has observed a difference in the ages of first offenders convicted of drug-related crimes in four geographical areas of the country. Is this observation indicative of a national trend or merely chance occurrence according to the following samples? Use the .05 level of significance.

Area I	*Area II*	*Area III*	*Area IV*
$N_1 = 20$	$N_2 = 20$	$N_3 = 20$	$N_4 = 20$
$\bar{X}_1 = 23.4$	$\bar{X}_2 = 27$	$\bar{X}_3 = 28.5$	$\bar{X}_4 = 29$
$s_1^2 = 46.5$	$s_2^2 = 58$	$s_3^2 = 48$	$s_4^2 = 51$

14.7 The mayor's safety council has been observing activity on the three main inbound thoroughfares during the morning and evening rush periods. It notes the number of reported accidents on each road for five different days.

I	*II*	*III*
6	4	5
7	5	5
7	5	3
7	5	3
8	6	2

At the .05 significance level is there a difference in the mean numbers of reported accidents per highway?

14.8 Fashionable Snob Heights has six private clubs that maintain year-round swimming facilities. Fifteen random checks at each pool show the water temperature means and standard deviations to be as shown in the accompanying table.

			Club		
Wealth	*Elite*	*Million*	*Rich*	*Leisure*	*Zenith*
$\bar{X}_1 = 70$	$\bar{X}_2 = 72.3$	$\bar{X}_3 = 69$	$\bar{X}_4 = 80.5$	$\bar{X}_5 = 80$	$\bar{X}_6 = 75.4$
$s_1 = 5$	$s_2 = 6$	$s_3 = 5.6$	$s_4 = 3.8$	$s_5 = 4$	$s_6 = 3.7$

Can the hypothesis that there is a difference between the various mean temperatures be accepted at the .05 significance level?

14.9 Four different fertilizers are used to treat four different groups of a certain strain of foliage plant. After six months the heights of the tested plants are measured. Results are given in inches.

Fertilizer Type

A	*B*	*C*	*D*
8	9	16	9
12	11	9	12
8	7	12	8
10	11	14	10
	8		7
	10		14
			12

Can the null hypothesis be rejected at the .05 significance level?

14.10 Is the theory that outputs per worker vary among the five shifts at Filbert's Nut Cracker Factory supported at the .01 significance level by the following samples of outputs?

Shift

Morning	*Day*	*Evening*	*Night*	*Graveyard*
5	13	8	4	7
7	10	11	8	10
6	11	12	10	9
8	7	7	6	8
4	13		7	6
				4

14.11 In an attempt to interpret the value and meaning of earned GPAs at the University, a task force sets out to determine if the claim by the law school dean is true. He contends that the grades in his school vary more than those in the medical college. Seventeen law students (L) and 12 medical students (M) are selected at random. Their GPAs are recorded here.

L: 2.60, 3.30, 3.40, 3.45, 2.10, 1.90, 3.75, 3.75, 3.80, 2.83, 3.90, 4.00, 2.00, 2.10, 2.30, 3.50, 3.20

M: 3.57, 3.81, 2.90, 2.85, 2.25, 3.90, 4.00, 2.80, 2.53, 3.56, 1.79, 3.21

Is the dean's contention supported by a hypothesis test at the .05 level of significance?

14.12 The relationship between age and satisfaction in life is the broad topic selected for discussion at a conference sponsored by the Psychologist Association of North America. Preparing to lead a discussion group, Dr. O. N. Couch administers a test to 35 of her patients to determine satisfaction with progress in profession or job. Do the test results presented here indicate any difference in the satisfaction levels of the different age groups among persons who have sought professional counseling? Test at the .01 significance level.

Age Groups

18–20	*20–25*	*25–35*	*35–45*	*45–55*	*55–70*
8	20	20	18	13	12
11	12	18	20	13	16
9	21	21	21	18	13
13	12	29	17	9	13
10		23	20	14	
		15	16	8	
		27	22		
		16			
		20			

14.13 Observers at three different locations on the earth's surface make coded measurements of radio signals received from outer-space sources. The results are compared with similar signals received on a space vehicle orbiting the earth.

Earth Observers			*Space Observers*
A	*B*	*C*	*D*
4	3	3	7
2	4	4	6
3	4	5	4
4	3	3	6
3	4	4	4
4	3	3	7

a. Test the null hypothesis that the means of *A*, *B*, *C*, and *D* are equal at the .05 significance level.
b. Test the null hypothesis that the means of *A*, *B*, and *C* are equal at the .05 significance level.
c. What do you infer about space observations?

14.14 Is there any difference in the mean hourly pay of union employees registered with various political parties? Perform a hypothesis test at the .05 significance level using the rates of the 30 employees given below.

		Party		
Actionites	*Conservatives*	*Liberals*	*Elitists*	*Anarchists*
5.84	6.08	8.45	4.60	3.50
5.30	7.44	6.40	3.85	6.80
5.27	6.05	7.28	5.90	5.95
6.21	6.40	7.28	6.00	
5.09	6.30	6.06		
	5.28	8.27		
	6.27	8.05		
	6.24	7.05		
		7.25		
		7.35		

14.15 The alternative hypothesis that states there is a difference in *K* means calls for a one-tail test as opposed to a two-tail test. Explain why.

Problem Set B

14.16 The mean age of new military recruits has changed little during this century. Has there been more variability of recruit ages during years of voluntary service (V) than during years of compulsory service (C)? Use the sample results supplied here and the .05 significance level.

$$N_C = 25 \qquad N_V = 25$$

$$s_C^2 = 23 \qquad s_V^2 = 31$$

14.17 Two samples of laboratory mice were given identical diets for two months. In addition, the first group received mineral and vitamin supplements. The mice were then timed on a maze-completion test. The sample sizes and variances are given below. Using the .05 significance level, can it be concluded that the supplement decreases variability of completion times?

$$N_{\mathrm{I}} = 22 \qquad s_{\mathrm{I}}^2 = 41.35$$

$$N_{\mathrm{II}} = 8 \qquad s_{\mathrm{II}}^2 = 87.56$$

14.18 The president of McCabe Industries has instructed the marketing department to investigate allegations that the price of their major product varies more in the three regions where manufacturer's suggested retail prices are not printed on the product (area I) than in the six regions where it is (area II). A total of 230 random purchases of the product in both areas yields the following sample statistics.

$$N_{\mathrm{I}} = 51 \qquad s_{\mathrm{I}} = 8.50$$

$$N_{\mathrm{II}} = 179 \qquad s_{\mathrm{II}} = 6.75$$

Are the allegations supported at the .05 significance level?

14.19 a. Define the F distribution with 10 and 20 degrees of freedom.

b. Do the distributions in Figs. 14.2 and 14.3 suggest that $F_{10,20}$ has a median greater than or less than one?

c. Make a general statement about when the median is less than one and when it is greater than one.

d. What is the median of $F_{10,10}$? Is $F_{10,10}$ symmetric?

14.20 The research division of Kermit Can and Bottling Corporation is studying six new canning processes designed to destroy bacteria economically. Twenty selected foods are subjected to each process. The bacterial count findings are presented here.

	I	*II*	*III*	*IV*	*V*	*VI*
$\bar{X}$	75	70	54	92	89	78
s	10	12	12	9	8	11

Can it be concluded that the six processes differ in their effectiveness to destroy bacteria at the .01 significance level?

14.21 Five strains of oats are each planted in six 10-acre fields in various Midwest locations (30 fields in all). After harvest the yields in bushels per acre are compiled, giving the following results.

	A	*B*	*C*	*D*	*E*
$\bar{X}$	29	28.3	26.7	27.8	27.5
s^2	1.7	2.8	3.9	4.0	5.0

Can one conclude at the .05 level of significance that the strains differ in yield per acre?

14.22 Consumer advocate Alfred Rader examines the service records of five makes of washing machines. The data he collects consist of the numbers of service calls necessary during the first five years of ownership for 37 randomly selected machines.

Make

Kwik Wash	Sure Clean	Pinnacle	Auto Wash	Sani Temp
3	4	5	9	0
7	3	5	4	1
6	2	6	3	1
2	2	3	3	0
1	2	4	3	3
1	1	5	5	4
5	5	2	5	
	0		4	
	2			

Should Rader conclude at the .05 level of significance that the mean number of service calls differs between makes and conduct further studies?

14.23 Farnham Factories, Inc., has five machines that produce a certain "5-centimeter" bolt. Periodically 10 bolts are selected by the product control manager from each machine to determine if there is any difference in the mean lengths of the bolts produced by the five machines. At the .05 significance level, what conclusion should be drawn from the following sample results? (Values are given in centimeters.)

Machine

	A	B	C	D	E
$\bar{X}$	5.04	4.91	5.14	4.83	5.36
s^2	.05	.01	.07	.06	.10

14.24 The nation has been divided into four sections for purposes of a societal mobility study. Forty-year-old people are randomly selected from each region and asked how many years they have lived within 10 miles of their present homes.

Section

I	II	III	IV
15	2	14	3
17	31	3	8
19	8	28	16
20	4	5	25
3	14	7	4
11	15	16	7
15	26	9	11
8	31	14	9

Draw the logical conclusion suggested by the data at the .05 significance level.

14.25 A continuing interest in the relationship between occupation and stress has led Ph.D. candidate Anne Pace to test 20 randomly selected persons from the four occupations mentioned in the table below. Each subject was tested to obtain his or her job-stress quotient.

Accountant	Dentist	College Professor	Engineer
84	62	37	89
93	69	49	75
71	94	57	74
64	67	83	72
85	51		53
60			

Do the results support a difference in the mean stress quotients for the four occupations at the .05 significance level?

14.26 Before setting up a foreign-language program, a team of teachers from Laura Carlo Junior High wants to determine if the incorporation of a mechanical laboratory into the program will tend to increase the variability of the program's success from year to year. Consequently, 20 other junior high schools in the area are contacted for information, 10 of which have a lab and 10 of which do not. The means of their students on a national standardized test are found here.

With lab:	73	70	71	69	70	84	86	85	80	79
No lab:	68	75	77	73	79	78	77	80	79	80

At the .05 level of significance, can it be concluded that variability is increased with a laboratory program?

14.27 Coach Doug Bine believes that the overall success of hitters depends upon time spent in the minors. He randomly selects 17 players from the league. Their batting averages are given here.

Number of Seasons Spent in Minors

0–1	2–3	4 or more
285	260	175
273	273	210
184	200	245
200	180	170
340	170	200
		150
		280

Do the results confirm Coach Bine's belief at the .05 significance level?

14.28 From the records of the Megapolis Police Department the following numbers of burglaries have been reported during the indicated seasons for the past 10 years.

Spring	Summer	Fall	Winter
94	90	86	57
92	86	84	53
90	84	80	90
90	82	78	65
86	87	78	73
86	80	72	82
84	78	78	59
82	86	78	58
108	84	70	60
125	82	69	51

At the .01 significance level can we reject the hypothesis that the mean numbers of burglaries reported are the same for each of the four seasons?

14.29 The salaries of nine university presidents are given below. Each is identified as the salary of a person heading either a state-owned school, a church-related school, or a private school with no religious affiliation. At the .05 level of significance is there any difference in the mean salaries drawn at the three types of institutions? (Hint: Squares of raw data may not fit on your calculator, so use scientific notation.)

State-Owned	*Church-Related*	*Private*
23,000	35,000	21,000
32,000	39,000	27,000
47,000	42,000	29,000

14.30 F tests for difference between two variances may also be two-tail in nature. Such problems usually require more extensive tables than those found in Appendix V.

a. The right-hand critical points can be determined from Appendix V for two different significance levels. What are they?

b. Find the right-hand critical F ratio for the two levels determined in part a for the F distribution with 50 and 80 degrees of freedom.

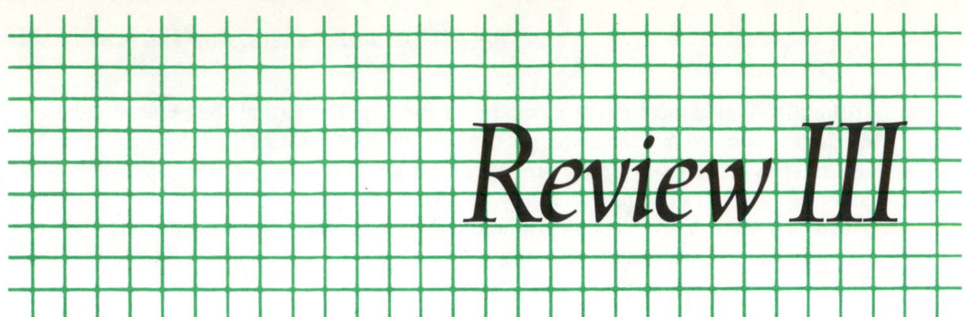

Review III

The big advantage of the methods discussed in Chapters 11 through 14 is that they enable us to make sound decisions even though those decisions might be based upon a given degree of uncertainty. The medical researcher must decide upon the effectiveness of a drug; the businessperson must appraise the economy. The political analyst must gauge public opinion, and the sociologist must assess changes. The common denominator is that all these decisions are made concerning entire groups or populations, including unobserved or unobservable members. They are based upon data obtained from a sample representative of the entire group. They require inferential methods.

At the heart of inferential statistics is the hypothesis test introduced in Chapter 11. A convenient six-step procedure that may be used for all tests was outlined early in the chapter. It distinguishes clearly between the hypothesis the researcher is hoping to substantiate (the alternative) and the hypothesis being directly tested (the null). The forms taken by these two types of hypotheses should be understood by the reader, as should the logic behind the test. It is this logic that permits acceptance of the alternative only through the rejection of the null hypothesis. Of course, a margin of error must be permitted, as there is always a risk involved. The setting of the significance level allows us to control the occurrence of the serious type-one error.

Even the most elementary types of tests have wide applicability. This was seen with the tests for population proportions and population means. Both tests draw upon our knowledge of the normal distribution, either in the form of the binomial distribution or the distribution of sample means. Calculations of z scores and their comparisons with critical z score values ultimately lead to the decision whether or not to reject the null hypothesis and accept its alternative.

Values obtained from random samples, commonly referred to as statistics, may be used as estimators of population values or parameters. Of course, some estimators are better than others. One of the essential properties of a good estimator is that of unbiasedness; that is, the mean of many estimates must approach the parameter being estimated. The sample mean, $\bar{X}$, has proven to be an unbiased estimator for the population mean μ, but unfortunately, the true sample variance, $\dfrac{\Sigma(X - \bar{X})^2}{N}$, is not an unbiased estimator of the population variance. Replacing $\bar{X}$ by the population mean will yield an unbiased estimator, but μ is often unknown. A discussion of the concept of degrees of freedom reveals that the sample variance $\frac{\Sigma(X - \bar{X})^2}{N}$ fails as an unbiased estimator because it gives estimates that are generally too small. If its indicator of dependability or degrees of freedom (the N that appears in the denominator) is decreased by one, the result is a useful unbiased estimator for the variance that is designated by $s^2 = \frac{\Sigma(X - \bar{X})^2}{N - 1}$. This statistic is commonly referred to as the sample variance.

Both $\bar{X}$ and s^2 (as well as s, which estimates σ) are point estimates. We don't expect these estimates to equal the parameters very often. Our hope is that they will be close. Also, there is an interval estimate for μ, called a confidence interval, that we expect actually to include the population mean most of the time. Depending on the confidence level selected, the failure rate can be controlled. Usually 1% or 5% of the time is allowed for error. The procedures for finding point estimates and confidence intervals are outlined in Chapter 12.

The initial discussion of hypothesis tests in Chapter 11 assumed a knowledge of the population standard deviation. Because σ is seldom known, it must often be estimated using techniques discussed in Chapter 12. z score calculations are then impossible as they also depend upon a knowledge of σ. The t distribution comes to our aid in Chapter 13. t scores, which estimate z scores using s in place of σ, form a distribution with many of the properties of the normal curve. However, the proportions of terms under various parts of the curve differ from the normal proportions, thus calling for slight variations in the hypothesis-testing procedure. For the reader well grounded in the techniques of testing hypotheses, these should pose no significant problems.

The t distribution allows us greatly to increase the number of hypotheses we are able to test. The means of two populations can be compared using the t test for difference between means or in special instances by the t test for matched groups. Both depend upon t distributions, as there is no knowledge of the standard deviations in either case.

In Chapter 14, we encounter our first theoretical distribution that is not normal, or even approximately so. The F distribution, however, is built using the ratios of variances of samples randomly selected from normal distributions. The result is a continuous, nonsymmetric distribution in which all terms are nonnegative. Two types of hypothesis tests are presented that depend upon this kind of distribution. The first is the comparison of the variances of two populations. The second, the comparison of a group of sample means, is in some ways an extension of the test for difference between two means. The mechanics of the new test, outlined in Chapter 14, are quite different from the t test, however, as they rely upon the comparison of two sample variances rather than two sample means. A worksheet for dealing with these situations organizes the data so as greatly to simplify computations.

The preceding four chapters have basically outlined the main techniques used to test hypotheses when the population from which samples come is known to be at least approximately normal.

Problem Set A

III.1 John blocks punts. Last year he set an all-city record by blocking 15% of all punts made by the opposing team while he was on the field. In the first six games of this season, he has blocked three of 40 punts. At the .05 level of significance has John's blocking record changed significantly from last year?

III.2 The World Hunger Committee is concerned over reports that the typical American throws away more than enough edible food per year to feed a family of four in a poverty-stricken country for a month. The committee orders a study in which the garbage of 50 randomly selected households (115 people) is examined for a three-month period. The conclusion is that the mean amount of food wasted by the 115 people is valued at $31.85 per person per year. Assuming that the population standard deviation is $5.85 and that a poverty-stricken family overseas can subsist for a month on $30 worth of food, are the reports substantiated at the .01 level of significance?

III.3 A promotion director of a large advertising firm is trying to convince the vice-president of his division that Sunday supplements to newspapers carry more than 15% of all cents-off coupons issued. The division has been planning its promotions around this figure. The director randomly selects 1,000 coupons redeemed in the target areas and has the origin of each coupon traced.

a. State the null and alternative hypotheses.

b. How many coupons in the sample would have to originate in Sunday supplements to accept the promotion director's belief at the .05 significance level?

c. How many coupons in samples of 500 and 2,000 would have to originate in Sunday supplements to force rejection of the null hypothesis at the .05 significance level?

III.4 Surveys of emergency rooms across the state reveal that accidents involving wet, ice-covered, or snow-packed pavement have been plentiful. The reported totals for the past 15 years are

987	599	739	888	890
873	694	862	523	667
622	735	791	911	529

a. Find an unbiased estimate for the population mean.

b. Find unbiased estimates for the population variance and standard deviation.

c. How many degrees of freedom are the estimates in part b based upon?

III.5 Annually, thousands of purses are left behind on subways, buses, trains, and planes. While an attempt is made to return the bags to the owners, often this is impossible because of lack of proper identification. The accompanying table gives the number of purses left by category (amount of money contained in purse) for one year on one airline.

Amount (in dollars)	*F*
100–500	43
50–100	130
25–50	189
10–25	287
0–10	395

Find unbiased estimates for the mean, variance, and standard deviation of the distribution of amounts found in purses left on the airline's carriers.

III.6 Find the critical values for the tests described.

a. A two-tail z test at the .06 significance level.

b. A one-tail t test at the .01 significance level involving seven degrees of freedom.

c. A one-tail F test using samples of size 25 (for the larger estimate) and size 35 at the .01 significance level.

III.7 Due to the burgeoning file of complaints dealing with lakefront subdivisions, the state regulatory agency has begun an investigation into the dealings of lake lot development companies. For the 36 companies being studied, the mean number of platted lots is 652. If the standard deviation (σ) is 175, find the 99% and 95% confidence intervals for the mean number of platted lots per company.

III.8 A physiologist who actively supports the participation of girls in competitive sports wants to determine if training and practice account for the popular theory that prepuberty boys are able to throw a small object farther than prepuberty girls can. Fifty 10-year-olds, 25 of each sex, are randomly selected from the public schools. Each tosses the same object with his or her nondominant arm to minimize the advantages of participation in organized sports. A summary of the results follows.

Girls	*Boys*
$\bar{X} = 37.5$	$\bar{X} = 40$
$s = 8$	$s = 11$
$N = 25$	$N = 25$

Do the results suggest a difference in the boys' favor at the .05 level of significance?

III.9 The mental health and social work division chief at a military medical command in Europe has noted the unusual number of stress-related complaints by American military personnel and dependents. In response, she decides to test a number of European-based Americans using a stress adjustment examination whose mean and standard deviation are known to be 80 and 13, respectively. Find the critical mean scores of 40 subjects that would indicate a significant deviation from the mean at both the .02 and .08 significance levels. (These are one-tail tests.)

III.10 Some apartment building owners want to find out if the installation of laundry facilities and added security devices significantly affects the vacancy rates. Twenty-six complexes of comparable quality located in comparable markets are divided into four groups: A (laundry facilities and security devices provided), B (laundry facilities only), C (security devices only), and D (neither laundry facilities nor security devices provided). Do the results below (vacancies per 50 units) indicate a difference among the four groups? Test at the .05 significance level.

A	*B*	*C*	*D*
3.50	1.00	2.00	4.00
7.00	4.00	1.00	8.00
6.00	6.30	.75	8.00
2.00	2.00	.50	7.00
.50	5.00	8.00	6.50
	9.00	3.00	3.00
		4.00	7.00
			9.00

III.11 Believing that the mean pay-off in sweepstakes conducted through the mail is less than $20, Lucky Larry sends for the complete list of prize winners from 20 contests. He randomly selects 35 winners and contacts them to verify the winnings. If the sample mean is $19 and the sample variance is $5, what, if anything, should Larry conclude at the .005 significance level?

III.12 The paralegal profession is relatively new, but is growing. It is becoming well established on the East Coast, though it has not yet caught on in the Midwest. There, apparently, not so many hire paralegal help. A young lawyer examines 35 East Coast firms and 29 Midwest firms. The aim is to determine if there is greater variability in the number of paralegals employed per lawyer in the Midwest. If the sample variances for the East and Midwest samples are $s_E^2 = 26.3$ and $s_W^2 = 41.3$, what conclusion can be reached at the .05 significance level?

III.13 A new aerodynamic shield is being tested that is purported to lower the drag coefficient, and thereby decrease the fuel consumption, of full- and mid-size automobiles. Eighteen cars are tested under identical controlled conditions, both with and without the device. The mpg figures follow.

Car	*Without Shield*	*With Shield*
1	18.2	19.0
2	17.5	17.6
3	16.9	20.3
4	9.0	10.1
5	25.4	25.8
6	13.7	14.6
7	14.0	14.5
8	16.5	18.3
9	18.0	16.0

Car	*Without Shield*	*With Shield*
10	12.3	14.4
11	23.0	21.7
12	17.2	18.5
13	16.1	19.3
14	20.9	24.0
15	16.3	18.0
16	14.1	15.0
17	15.3	15.0
18	19.0	24.0

At the .05 significance level what conclusion may be drawn as to the effectiveness of the shield?

III.14 Volunteer time from registered party members has been solicited by four methods in one major political group: (1) personal mail appeals, (2) general mail appeals, (3) telephone appeals, and (4) personal contact appeals. The national committee needs to determine if there is any significant difference in the volunteer time obtained by the four methods. A group of 160 people to be contacted, 40 by each method, are randomly selected from the party rolls. The time commitments obtained (in hours) are summarized here.

Personal Mail	*General Mail*	*Telephone*	*Personal Contact*
$\bar{X}_1 = 15$	$\bar{X}_2 = 16$	$\bar{X}_3 = 18$	$\bar{X}_4 = 23$
$s_1 = 6$	$s_2 = 5$	$s_3 = 6$	$s_4 = 4$
$N_1 = 40$	$N_2 = 40$	$N_3 = 40$	$N_4 = 40$

Can the hypothesis that there is no significant difference be rejected at the .05 significance level?

III.15 A new filing system has been established in an attempt to decrease the 2.3-hour mean that each office worker spends per day filing orders. The supervisor who devised the plan reports to the board of directors that it has been successful. He has found that 48 employees had a mean filing time of 2.15 hours per day per employee. If the standard deviation (σ) was .75 hour before the change, what is the smallest probability of a type-one error (as indicated by the significance level) the supervisor could be allowing for?

Problem Set B

III.16 The jogging craze may mean longer, healthier lives for many people. However, all too often novices sustain injuries due to lack of preparation, failure to obtain medical clearance, and improper shoes and clothing. Suppose the general population of runners sustains approximately 13.5 injuries per 100 joggers each week with a standard deviation of 4. "The Track," a running club with all-weather facilities and professional guidance, claims to have a superior injury record. The injuries to 100 jogging members picked at random were recorded each week for 14 weeks. The mean number of injuries per week proved to be 11.3. Are there significantly fewer injuries at "The Track" at the .05 significance level? at the .01 significance level?

III.17 A published report states that 8% of all girls give birth before the age of 18. Of the graduating senior girls at Central High, 140 have reached the age of 18. Of these, only two are mothers. What conclusions can be drawn concerning the report and/or the graduating senior girls at Central High?

III.18 The chief of the cardiac division of a major southern California hospital contends that heart-ailment complaints generally increase by more than 35% ($\sigma = 5\%$) during periods of second-stage smog alerts. The records of 49 such periods are studied, and the significance level is set at .01.

a. What is the smallest mean percent of increase in the sample needed to reject H_0?
b. If an additional 51 periods are studied, what is the smallest mean increase that would substantiate the claim?

III.19 A county's economy is heavily dependent upon fruit production. The annual yields for pears during the past 10 years (in numbers of crates) have been 8,750, 6,840, 7,850, 6,100, 4,300, 6,220, 4,570, 8,860, 4,930, and 7,110.

a. Find an unbiased estimate for the population mean.
b. Find unbiased estimates for the population variance and standard deviation.
c. Assuming the population of yields has a standard deviation of 1600, find a 95% confidence interval for the mean.

III.20 To estimate the absorbency of a new brand of paper towel, 125 sheets are tested and give the following results (in ounces of water absorbed).

Absorbency	*F*
2.4–2.5	20
2.3–2.4	23
2.2–2.3	50
2.1–2.2	27
2.0–2.1	5

a. Find an unbiased estimate for the population mean.
b. Find unbiased estimates for the population variance and standard deviation.

III.21 A comprehensive national health care plan has long been supported by liberal members of society who are generally thought to be younger members. A political survey turns up 18 individuals who favor such a plan. Their mean age in years is 23.7, and the standard deviation (s) is 4.5. Does this indicate at the .025 level of significance that the mean age of proponents of the plan is less than 25 years?

III.22 Scientists are studying a new disease spreading in the southeast section of the nation. Sixty people were reported coming down with the ailment, which has symptoms similar to those of Legionnaires' disease, a mean of 8.5 days after being exposed to a person known to have contracted it. Assuming the standard deviation (σ) to be two days, find a 99% confidence interval for the mean length of the incubation period.

III.23 Two commercial meat lockers maintain the same mean temperature. Fifteen random spot checks at each site reveal a standard deviation of 4.3 at locker A and 6.9 at locker B. At the .05 significance level, does the variance at locker B exceed the variance at locker A?

III.24 An area dam project was to provide a mean daily surplus of 200,000 gallons of water per day for Megapolis. The water commissioner doubts that the project has been successful in doing so. To investigate, he has the surpluses measured for 14 randomly selected days. The mean is only 195,000, and the standard deviation is 25,000. Can the commissioner conclude at the .05 significance level that the project is failing to provide the amount of water promised to the city?

III.25 A new no-frills supermarket has opened in the area. It provides low prices, but no services such as bagging, check cashing, or carry out. Since its profits depend upon volume sales, checkout stations must have a mean of more than 10 customers per register per hour. Thirty hour-long spot checks show the following numbers of patrons served per hour at various registers.

12	7	7	16	18
8	6	9	12	12
11	7	11	9	11
5	10	14	9	10
15	12	14	7	10
7	13	4	15	11

Do these statistics indicate that the supermarket is profitable at the .01 significance level?

III.26 Do high school students spend more time watching television or studying? To answer this question, the student council selected a student at random from each home room at Central High. The parents of half the test subjects were asked to record the study habits of their son or daughter without his or her knowledge. The parents of the other half of the students were asked to record TV habits, again without the students' knowledge. Do the results below support the student council's hope that more time is spent studying than watching television? Use the .05 significance level. (Times are given in hours.)

Study time:	$\bar{X} = 2.5$	$s = .5$	$N = 15$
Television time:	$\bar{X} = 2.1$	$s = .75$	$N = 15$

III.27 The pH of rain and snow varies from locality to locality. A meteorologist measures the pH levels of 19 rain samples known to originate in either coastal, desert, agricultural, or volcanic areas. Do the results below support

the accepted theory that these environmental factors affect acidity–alkalinity levels of rain water differently? Use the .05 significance level.

Coastal	*Volcanic*	*Desert*	*Farm Land*
4.6	5.1	6.2	6.1
5.9	5.3	6.0	5.9
5.8	5.6	5.9	6.2
6.2		5.7	6.0
		6.1	6.1
		6.3	6.3

III.28 The burning of coal may be the best avenue for many oil-importing nations to follow to achieve energy independence. The problem of sulfur dioxide (SO_2) emissions, however, is still a serious obstacle. At least five methods are being tried to reduce the emissions: (1) changing coal to liquid fuel, (2) changing coal to gaseous fuel, (3) improving combustion efficiency, (4) cleaning coal before burning, and (5) removing the SO_2 in the exhaust. Twenty controlled experiments involving each of these methods gave the results in the accompanying table. The figures refer to the percent of sulfur dioxide emissions eliminated.

Method	1	2	3	4	5
$\bar{X}$	30.5	28.7	35.2	29.0	29.5
s	4.5	4.3	6.8	2.0	4.1

At the .05 significance level do the methods differ in their effectiveness in removing the SO_2 emissions?

III.29 In response to the charge that young adults lack consumer skills in the areas of life insurance, savings, and investment, a new course is being offered at the senior high level. The 17 students enrolled in the course are pre-tested and then tested again upon completing the course. Two forms of a standardized test are used. The instructor is not allowed access to the exams prior to the time they are administered. Do the matched scores below indicate that the new course has been successful in teaching consumer skills? Use the .05 level of significance.

Student	*Pre-test*	*Post-test*
1	127	150
2	115	117
3	100	108
4	153	173
5	107	112
6	119	115
7	125	129
8	128	129
9	133	138
10	152	162

Student	*Pre-test*	*Post-test*
11	97	107
12	85	100
13	111	100
14	115	123
15	127	125
16	135	144
17	142	150

III.30 Cushing and Sons is planning to purchase a new computer for the payroll department. Because they have had so many maintenance problems with their present computer, they are planning to change models. A preliminary survey of other companies using models they are considering shows the following results, which indicate how many hours each company's computer was "down" last month.

Model A	*Model B*	*Model C*
15.3	23.4	33.4
16.7	26.9	19.7
12.4	14.2	12.2
29.8	18.9	25.3
17.3		14.6
14.9		17.0
15.0		

At the .05 significance level is there any difference between the three models in mean "down" time?

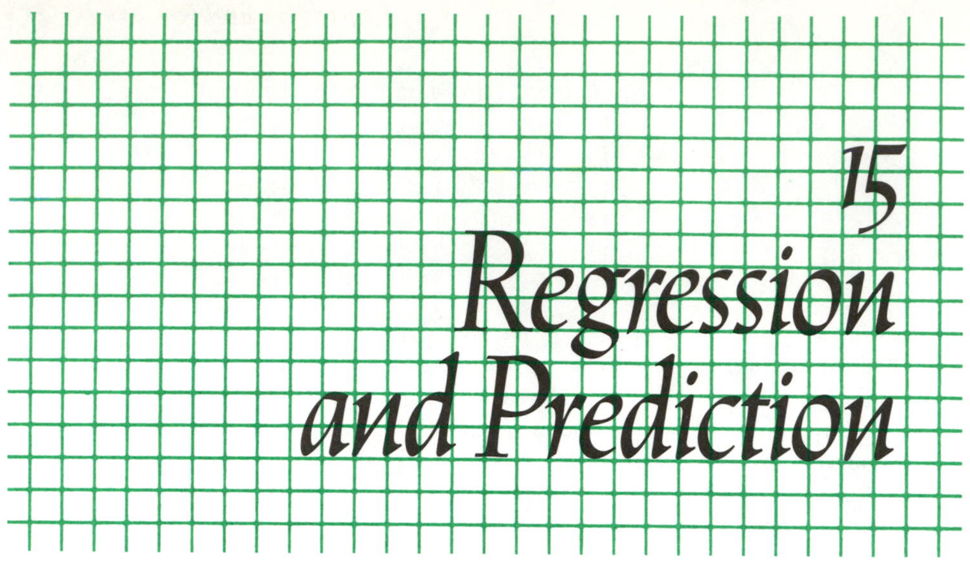

15.1 Introduction

The preceding chapters have dealt with problems encountered in describing distributions of one variable and in making predictions based on samples involving one variable. Many practical problems, however, involve more than one variable. For example, consider such questions as "To what extent is it possible to predict college success from entrance examination scores?" and "Is it possible to predict scores on the Graduate Record Examination with any success from a knowledge of college grade-point averages?" In each of these situations it is possible to identify two variables. In fact, much of science may be described as the discovering of new relationships between variables. The power to predict and control nature arises largely from a knowledge of existing relationships in which two or more variables are involved.

In the remainder of this book we shall be concerned primarily with problems involving two variables. The number of possible relationships between two continuous variables is infinite. Of course there may be no relationships at all, but, in the simplest case where one does exist, it may be that high scores on one variable tend to accompany high scores on the second. For instance, such is the case with the variables age and vocabulary. The younger one is, the fewer words one is likely to know, and the older one gets the more one is likely to have learned. The relationship described is sometimes called a positive one.

A second kind of relationship is one in which successively higher values of one variable tend to accompany lower values on the other. For instance, such is the relationship between degree of education and crime rate. Those who are poorly educated are the most likely to commit crimes, and those who tend to have successively more education are to that extent less likely to commit crimes. This type of relationship is often called a *negative* or *inverse relationship*.

More subtle kinds of relationships also exist quite commonly. For instance, successive increases in one variable may first accompany increases in the other. Then further increases in the first variable may go along with decreases in the second. The relationship between age and physical strength is of this type. As age increases, so does physical strength, up to a point; but beyond this point further increases in age are accompanied by decreases in physical strength. Thus, at the low age levels, to know that one person is 5 years younger than another leads to a best guess that the younger person is weaker; but to know the same thing for two people at advanced ages would be to make a best guess that the younger person is stronger.

Keep in mind as we proceed that we are talking merely about relationships between two variables and not necessarily *causal relationships*. For instance, one must not infer that a causal relationship exists between education and abstinence from crime merely from what was said. It may be that a third variable, like degree of economic security, is crucial in accounting for the relationship that was described. That is, the degree of one's economic security may account both for one's abstinence from crime and also for one's decision to increase one's education. It is important to remember that relationships may be causal or they may be *consequential relations* that exist because of some other variable that was not of interest to us. Our concern shall be only with the relationships as they exist and with their implications, not with how they came into existence. This is a problem for the theorist and the researcher.

15.2 Blind Prediction

One of the primary advantages of knowing about a relationship between two variables is that one can use the knowledge to facilitate making predictions. Specifically, with exact knowledge of an individual's score on one of two variables, one can use knowledge of the relationship to increase the accuracy of a prediction of the individual's score on the other variable. For instance, suppose there is a known relationship between IQ and students' freshman averages in a particular state university. Then an admissions office worker can use his or her knowledge of an individual's IQ score as an aid in making a prediction of what the student's freshman average will be. The fact that a relationship exists between two continuous variables is often relevant to the whole issue of prediction, though sometimes the relationship is not strong enough to produce sizable predictive advantages, and at other times there is no practical concern with making predictions. At any rate, some comments must be made about the issue of prediction before going ahead, since we shall make reference to this issue throughout the discussion.

To begin with, by a prediction we mean a best guess of what a single value or score will be. We often have occasion to make predictions defined in this way. For a college admissions officer, the value to be predicted may be the average that a person will get as an engineering student. For the student, the value predicted may be the income that a graduate from the same college will earn 10 years after graduating. For an insurance company, the value to be predicted may be the number of years that a woman of 62 with high blood pressure will live. Or the value to be predicted may be the number of years that an industrial machine will run before breaking down.

Definition 15.1 A prediction is a guess about the value of a term to be drawn from a specified population.

To make any kind of meaningful prediction, we must have at least some knowledge about the population that is to yield the single case. We are going to assume throughout that we know at least the mean and variance of this population. For instance, if we are going to make a prediction about the height of an adult American male, we are going to assume that we know at least the mean and variance of the heights of adult American males. We shall say, for the sake of discussion, that we know the mean height of American adult males to be 5 feet 8 inches and that the variance of these heights is 9 inches.

The simplest kind of prediction to consider is that of the value of a term when we know nothing more than the mean and variance of the population of which the term is a member. Such a prediction shall be called a *blind prediction*. The use of the blind prediction is worth considering so that we may make comparisons with it later on. Suppose, for instance, that we are going to go through a telephone book and pick out the name of an adult American male in order to predict his height. The name turns out to be Joe Brown, and now we must make our prediction. We are making a blind prediction because it is to be based on no information other than that concerning the population from which the single case has been derived.

Before going any further, note that there is a world of difference between the topic of estimation that we have already discussed and that of prediction, which is being introduced. *An estimate is a guess of a parameter value*, like a population mean, and is typically based upon information derived from a sample. When we make a prediction we know all we are going to know concerning the population, and *the prediction itself concerns a single case drawn from the population.*

Estimate → Parameter

Prediction → Term

Now we return to our problem of making our blind guess of the height of Joe Brown. It turns out that the best we can do is to take the population mean of 5 feet 8 inches and use this value as our guess. In other words, our best blind guess is that 5 feet 8 inches is the height of Joe Brown. In general, **the population mean is the best blind guess of the value of a single case drawn from a population.**

The reason for using the population mean as the blind guess may be seen by considering the situation when one makes many blind guesses, one after the other. For instance, suppose Joe Brown's actual height is 5 feet 10 inches and one guesses that his height is 5 feet 8 inches. Then the error attached to this guess, defined as the actual value of the term minus the guess, is 2 inches (5 feet 10 inches minus 5 feet 8 inches equals 2 inches). We shall say that once more a name is randomly chosen from the telephone book and once more the guess of 5 feet 8 inches is made. This time the actual height of the person turns out to be 5 feet 7 inches, so the error turns out to be negative 1 inch. The process is repeated *ad infinitum.*

To begin with, note that, when the mean is used as a blind guess, the error each time is really the deviation of the term from the mean of its population. The sum of these errors to be made in the long run is zero (Theorem 2.1). Also the sum of the squares of these errors is really the sum of the squared deviations of the terms from the mean of their own population. Suppose some value other than the population mean was repeated as a blind guess each time. Then for one thing the errors would not balance out to zero. Also the sum of the squares of the errors made in the long run would be larger than when the mean itself was used. The reason is that, for any population, the sum of the squared deviations about the mean is less than the sum of the squared deviations about any other value (Theorem 2.2).

It would not help to vary one's guess from one time to another. Only when the population mean is used as the guess of the unknown case each time is it true that the errors average out to zero in the long run and that the squared deviations are a minimum. Note that when the errors as a group are large, whether positive or negative, then the squares of these errors are large. The squares of the errors as a group may be small or large, but when they are small they indicate that the predictions as a group are relatively accurate. When the squares of the errors tend to be large, they indicate that the guesses as a group tend to be inaccurate, because the errors themselves must be running large in order for their squares to be large.

The classic measure of how accurate the guesses are as a group is the mean of the squares of the errors. For instance, if the errors in inches are − 3, 2, 0, 1, 2, then the squares of these five errors are 9, 4, 0, 1, 4. The mean of these five squares is 18/5. That is, the mean of the squared errors for the five guesses mentioned is 18/5. But in the long run, when one guesses the mean each time, each squared error is a squared deviation from the mean. As a vast number of guesses are made, the mean of the squared errors becomes the population variance. *In sum, when one makes blind guesses, the mean of the squared errors over the long run is the population variance, and thus the measure of accuracy is the population variance*. Later we shall see how a system using other information is apt to lead to more accurate predictions over the long run. *A system is considered more accurate than the blind predictive method when the mean of the squared errors made according to this system over the long run would be less than the population variance.*

It should be very clear how each particular error has been found—by subtracting the population mean from the actual value of the term of interest. Where the term was 5 feet 10 inches, the error was 2 inches. Where the term was 5 feet 7 inches, the error was negative 1 inch. Table 15.1 shows some terms and the errors accompanying them when the population mean was used as a guess.

Note that to find the error in each case the process was to take the original term and subtract the population mean from it. In other words, from each term a constant was subtracted to find the corresponding error. This means that the variance of the original terms is identical to the variance of the errors made in predicting them. (Theorem 3.1). With reference to Table 15.1, the variance of the values in the third

Table 15.1 Errors of Guesses

Guess	Term	Error
5′8″	5′10″	+2
5′8″	5′7″	−1
5′8″	5′8″	0
5′8″	5′11″	+3
5′8″	5′2″	−6
5′8″	6′1″	+5

column is identical to the variance of those in the second column, a fact that the reader should verify. We may talk about either the population variance or the variance of the errors made when the mean is used repeatedly as the predictive guess. To facilitate comparisons later on, we shall talk about the latter and call it the ***error variance***. For instance, we are supposing that when the mean is used repeatedly to predict the heights of adult American males, the variance of the errors made is 9. We shall call this value, 9, the error variance. The error variance in the case described is the measure of the accuracy of predictions made when we have no information aside from knowing the mean of the variable of interest. It should be noted that *the error variance is always the mean of the squared errors*. When guessing blindly, the error variance is also the population variance (in the long run). Keep in mind that our goal is to find a method of prediction whose error variance (or mean of the squared errors) is less than the population variance.

15.3 Predictor and Predicted Variables

Consider once more the variable, freshman average in some large state university. We shall say that any student who graduates from an accredited high school in the particular state is allowed to enroll as a matriculated student in the state university. Furthermore, we shall say that freshman averages, defined as averages obtained by students at the end of their first year, have a mean of 75 and a variance of 100. Note that we are not specifying anything about the distribution of these averages, so it may or may not be normal.

It follows that if a particular student were to come along, and we knew only that he was to begin at the university next September, our best guess would be that his average after one year would be 75. For each student in the same situation, our best procedure would be to make this prediction. Obviously we would make errors nearly every time, but in the long run at least they would balance out. The sum total of our errors in one direction would equal the sum total of our errors in the other. The variance of our errors would be 100.

We are going to consider now the simplest context in which we have information relevant to making predictions about individual students. Suppose we know the following facts to be true as a result of an extensive survey:

Those whose high school averages were below 80, considered as a group, received the poorest college freshman grades. We shall call them Group A. This collection of students got freshman grades that had a mean of only 70. Considered together, they got freshman grades that had a variance of 60.

Those whose high school grades were from 80 to 90, considered as a group, got freshman grades that averaged 78. We shall say that these students compose Group B. Considered together, they got freshman grades with a variance of 60 also.

Those whose high school grades were over 90, considered as a group, got college freshman grades that averaged 85. We shall say that these brightest students compose Group C. Considered together, their grades as college freshmen also had a variance of 60.

We are going to assume that the facts given have been roughly true over a long time period with only trivial variations from one year to the next. Therefore, there is reason to believe that they will continue to hold, at least roughly, for next year, too. For instance, students who fall into Group A and who are about to begin at the state university next year may be expected to get freshman averages with a mean of about 70 and a variance of 60.

Where we are concerned with making predictions, we must consider the variable, high school average, as the predictor variable. A **predictor variable** is one that provides relevant information for predicting what scores will be on some other variable. The variable, college freshman average, is called our predicted variable. A **predicted variable** is one about which predictions are made. We have not yet discussed the exact method of making predictions, but it should be clear that exact knowledge of the relationship between the predictor variable and the predicted variable is necessary if we are to make the most accurate predictions possible.

As might be expected, when we do have an interest in making predictions, we must designate as our predictor variable the one that yields scores to which we may have access before we know anything about scores on the predicted variable. The distinction is usually chronological in that values of the predictor variable come into existence first. But there are exceptions, such as when we wish to infer from later events what happened before them. For instance, one might wish to make inferences about a person's childhood from the person's behavior, in which case scores relating to the individual's current behavior constitute values of the predictor variable and those relating to childhood constitute values of the predicted variable.

Note that the grouping of students as described has been wholly according to their high school performances. *For grouping to be meaningful, it must always be in terms of the predictor variable*, as we shall see. Note also that those who compose any particular group in our example tend to perform more similarly as college freshmen than would a random sample of applicants. We know this from the fact that those composing any particular group, like Group A, get freshman grades with a variance of only 60, whereas applicants in general get grades with a variance of 100. This means, for instance, that members of Group A, considered exclusively, show less scatter in their freshman performances than do members of the larger population, considered altogether.

Note that in our illustration the students composing any one group on the predictor variable performed with the same variance of 60 on the predicted variable.

When changes in category on the predictor variable do not affect the variance of the predicted variable, the discussion becomes greatly simplified, although the main points remain the same. This phenomenon of unchanging variance on the predicted variable is in fact often the case, and is called homogeneity of variance or *homoscedasticity*.

Now suppose that we are members of the board of admissions of the state university described and a particular applicant comes before us. For various purposes, such as advising her concerning her program for the following year, we would like to predict what her freshman average will be. If we know nothing of her high school average, our best guess—as we have seen—is that her freshman average will be 75, since this is the average grade achieved by freshmen considered as a group. Were we to make this same prediction for each applicant, our predictions would have an error variance of 100, which is the variance of freshman grades. In sum, our blind predictions that each student will have a 75 average have an error variance of 100. But suppose instead that we have the information given concerning students' high school averages and their relationship to freshman grades. We are going to consider how to make the best use of this information, and we are also going to measure to what extent it is of advantage to us. To sum up what we know, it is that students with high school averages below 80 perform as freshmen with a mean of 70 and a variance of 60. Those with high school averages from 80 to 90 perform as college freshmen with a mean of 78 and a variance of 60, and finally those with high school averages over 90 perform as college freshmen with a mean of 85 and a variance of 60. For convenience, we have described these students as composing Groups A, B, and C respectively.

Our procedure must be first of all to determine for each student his or her high school average and to use this average to place the student in one of the three groups. Then we will use the mean of the particular group as the prediction of what the student's freshman average will be. In other words, if he or she is a member of Group A, based on high school average, then our prediction must be that his or her freshman average will be 70. If a member of Group C, then our prediction must be that his or her college freshman average will be 85.

To illustrate: When the first high school graduate comes before us, we promptly look up his high school average and determine that it was 74. Clearly this applicant is a member of Group A. That is, he is a member of a group who perform as freshmen with a mean of 70, and therefore we use the value 70 as our prediction of what his freshman average will be. The next applicant, we shall say, has a high school average of 93 and thus is a member of Group C. She is in effect a member of a group who perform as college freshmen with an average of 85 and, therefore, we predict that her freshman average will be 85. One by one as the applicants come before us, our procedure in each case is first to place each applicant in his or her appropriate group and then to use the mean of that group as our prediction of what his or her freshman average will be.

Previously, when we used the mean of freshman averages as our guess for each student, it turned out that the variance of our errors was 100. Now that we are using knowledge of the subject's high school performances to facilitate making predictions, it turns out that we are more accurate in the long run; or, to put it another way, *the variance of our errors is less*. Specifically, suppose that we have made a vast number of predictions using the new method described. To determine the variance of our errors, we must consider the applicants as falling into the three groups individually, though naturally they did not come in any specific order.

We made the prediction that all those who composed Group A would obtain a freshman average of 70. The variance of the errors made concerning members of this group was 60. We predicted that each member of Group B would get a 78, and here too the variance of the errors was 60. Our prediction for those in Group C was that each member would get an 85, and for this group also the variance of the errors made was 60. In other words, the variance of the errors made in each group was 60. In sum, the blind guesses led to an error variance of 100, whereas when high school averages were used the variance of the errors made was only 60. One might say that the information made it possible to reduce the variance of the errors to 3/5 of its original size.

We shall now consider one way of presenting findings graphically when each individual has received scores on each of two variables. We designate the predictor variable as the *X variable* and represent it along the horizontal axis. In our case the horizontal axis will be used to stand for high school averages, which we shall say may range from 65 to 100. We call the predicted variable the *Y variable* and shall represent it along the vertical axis. This variable, freshman averages, shall be said to have a possible range of from 50 to 100. (Note, however, that even when we are not concerned with the problem of prediction we may use the same mode of representation by arbitrarily choosing either variable as the *X* variable.)

In illustration, suppose we have obtained for a multitude of subjects their high school averages and also their college freshman averages. The first student attained a high school average of 70 and a college freshman average of 80. We shall use a minute square to represent his pair of scores in Fig. 15.1

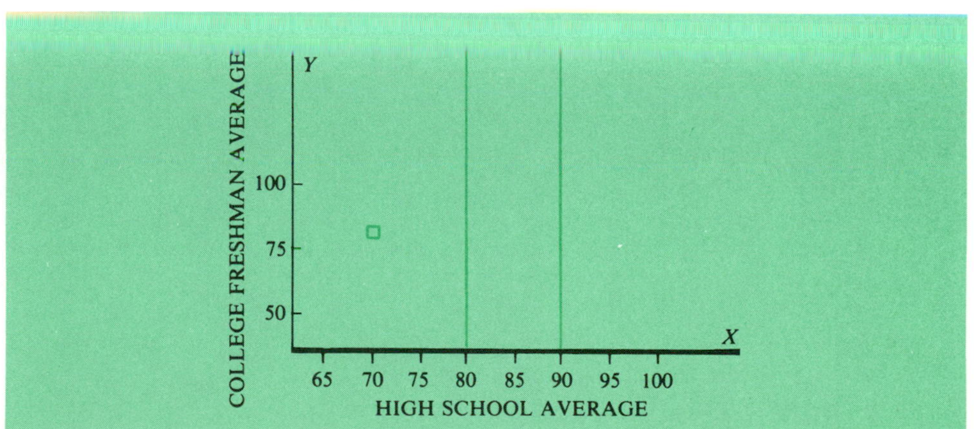

Fig 15.1

With respect to the horizontal axis, representing high school averages, this square rests on the value 70. With respect to the vertical axis, representing college freshman grades, it is opposite the value of 80. Note also that it is in Group A, which includes all those high school averages that were between 65 and 80.

Let us say that the next student attained a high school average of 92 and a college freshman average of 78. And a third student attained a high school average of 80 and a freshman average of 60. The squares representing the scores of these two students appear in Fig. 15.2 along with the square already shown representing the scores of the first student.

Fig. 15.2

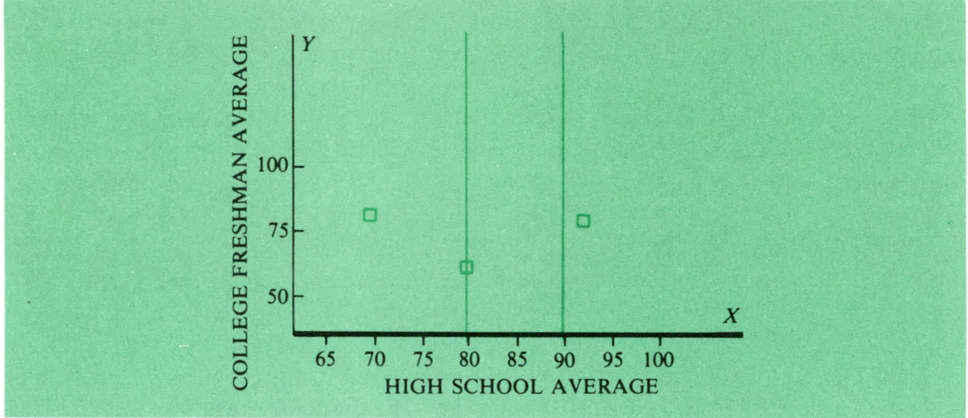

It might turn out that a fourth student would have gotten the same pair of scores as did the first one, in which case the square representing her pair of scores should go right on top of the square used for the first student. In other words, there should actually be a height dimension indicating where more than one student got the same pair of scores. The squares representing the scores of some of the multitude of students might assemble themselves as in Fig. 15.3.

Fig. 15.3

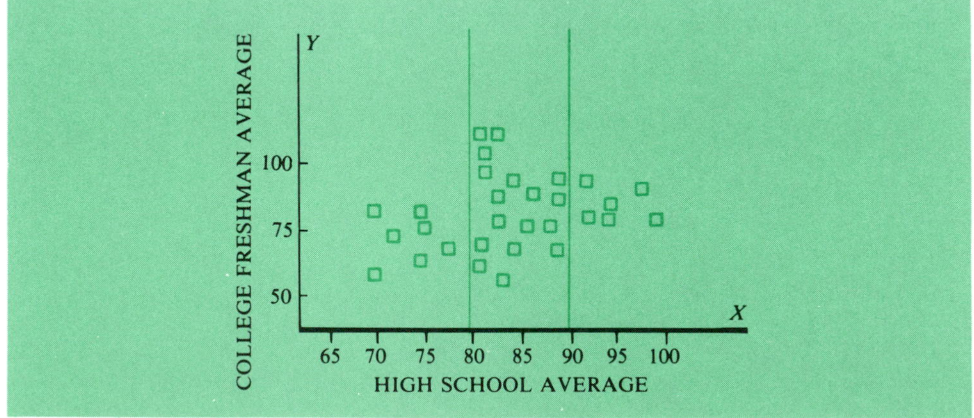

There is only one trouble with the mode of representation using squares as in Fig. 15.3. Where two students perform similarly but not identically, the squares representing their scores must be at a minute distance from each other. In fact, no matter how small we were to make our squares, sooner or later we would find, since our variables are continuous, that we had insufficient room to place some pair of squares side by side. Thus from now on as we proceed, instead of squares we shall use points, because they have no actual dimension. Note that the location of each point indicates how a single student fared both in high school and in college. That is, each point, like the square that it is replacing, has a value on the horizontal axis and also on the vertical one. It remains true for Fig. 15.4 that, where one student performed identically with another, one point should be precisely on top of the other, and the illustration does not actually make a perfect representation of this phenomenon. But this cannot be helped, since no height dimension is possible here. In any event, for our purposes it will be satisfactory to use points, and where they cluster most thickly we can imagine that they would be most likely to coincide. Fig. 15.4 is called a **scattergram**, which is a graph showing the scatter of points, each of which indicates values on a pair of variables. It shows the scores of only a few of the students.

Fig. 15.4

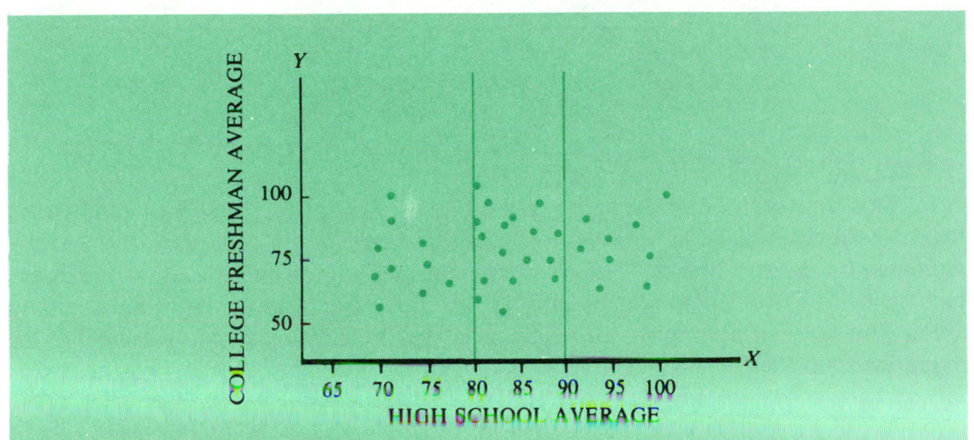

Our next step is to indicate the various means, or values, that we use as our predictions. To begin with, the mean of the freshman averages taken all together is 75. In Fig. 15.5, this value is indicated by the horizontal double bar. Note that this double bar intersects the value 75 on the vertical (or *Y*) axis and moves horizontally across the whole graph.

The single horizontal line in each group is at the level to indicate the value of the mean performance of students in that group. For instance, considering Group A alone, the mean freshman grade of the students composing this group was said to be 70. This mean is represented by the single horizontal line at the level 70, which appears only in Group A and not in the other groups. The mean of students in Group B so far as freshman average is concerned is 78, and this value is indicated by the bar at the appropriate level in Group B. Finally, members of Group C got freshman averages that had a mean of 85, and this value is indicated by the bar at the appropriate level in Group C.

Fig. 15.5

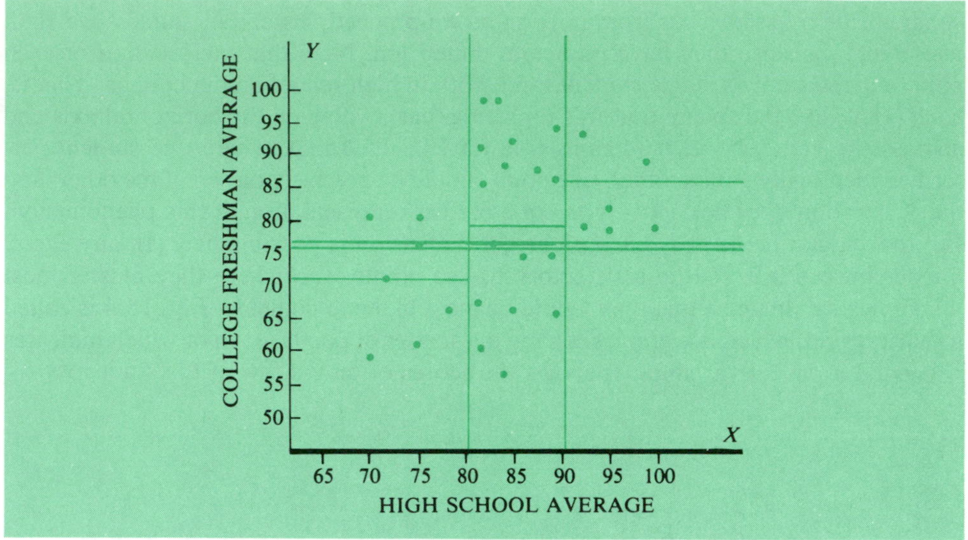

We are now ready to come back to the problem of prediction in the individual case. This time we shall make reference to Fig. 15.6. Consider first a student whose high school average is 85 and whose college average is 81. His pair of scores is indicated by the star in Fig. 15.6. We may consider what would have been the case had he merely walked into our admissions office, and without knowing anything about him we had attempted to predict what his freshman average was to be. As mentioned, our best guess would have been that his freshman average was to be 75. His actual freshman average of 81 would have been 6 points over our best guess. Our error is indicated by the broken line in Fig. 15.6, which is 6 units long.

Fig. 15.6

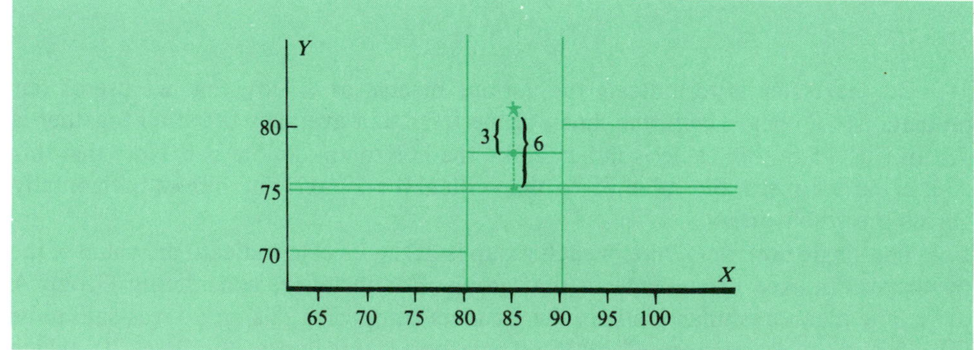

On the other hand, had we known the facts relating high school averages to freshman grades and had we known also that the particular student's high school average was 85, we would have proceeded otherwise. First we would have properly considered him as a member of Group B, and then we would have predicted his college freshman average to be 78. This time his actual average of 81 would have been only 3 points away from our predicted value.

In Fig. 15.6, the line connecting the actual value of 81 with our predicted value of 78 may be seen to be only 3 units long. Here the actual freshman average of the student was such that there resulted an error of only 3 points when we made our enlightened prediction.

Next, Fig. 15.7 shows what would have occurred in two other cases. The first one is that of a student whose high school average was 96 and whose college freshman average was 80. Her pair of scores is indicated by the star where the value on the horizontal axis is 96 and where the one on the vertical axis is 80. Once more our blind prediction would have been 75, and this time our error, indicated by the broken line, would have been 5 units. On the other hand, had we first thought of this student as a member of Group C, we would have predicted that her freshman average was to be 85. Here too our error, indicated by the solid line, would have been 5 units, but in the other direction.

Fig. 15.7

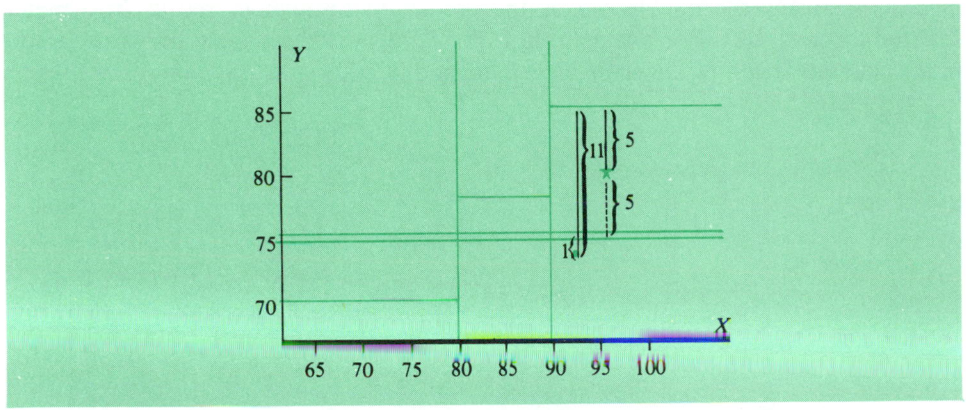

Finally, we may consider, also in Fig. 15.7, the case of a student whose high school average was 93 but whose college freshman average turned out to be only 74. With no information we would have used the value of the double line, 75, as our prediction of his college freshman average, and as indicated in Fig. 15.7 our error would have been only one point. On the other hand, the enlightened prediction we would have made, after properly locating the student in Group C, would have been the value at the level of the single line, the value 85. In this case our prediction would have turned out 11 points too high. In other words, this particular time our blind prediction would have been more accurate than the enlightened one. As one might guess, one does better by making enlightened predictions, and this last case cited goes against the trend.

From the illustration given, several facts should be clear. The first is that when we are without meaningful information, our prediction is that a student's freshman average is to be the value of the double line. Our error corresponds to the distance that the double line happens to be from the point representing the particular student's actual performance. When we have enough information to locate the student in one of the three groups, then our prediction is that his or her freshman average is to be the value of the single horizontal line representing the mean of that particular group. Our error in the particular case is the distance from that single line to the point representing the particular student's actual performance.

In sum, using the appropriate single bar for each subject in turn results in more accurate predictions in the long run than using the double bar (the population mean) repeatedly. Graphically, the best prediction of each student's freshman average is made by first placing the student in his or her proper group and then using the value of the single horizontal bar in that group as the prediction of his or her score. The single horizontal bar in any group comes closer to the points in that group than the double bar. The three single bars, taken together, come closer to the total collection of points than does the double bar.

Now let us go back to a particular statement made earlier—that use of the population mean leads to an error variance of 100, whereas use of the group mean leads to an error variance of only 60. Graphically, the first part of this statement means that using the double bar as the prediction throughout results in a mean squared error of 100. For example, in Fig. 15.8 where the double bar is used, the mean squared length of the error lines indicated is 100.[1]

Fig. 15.8

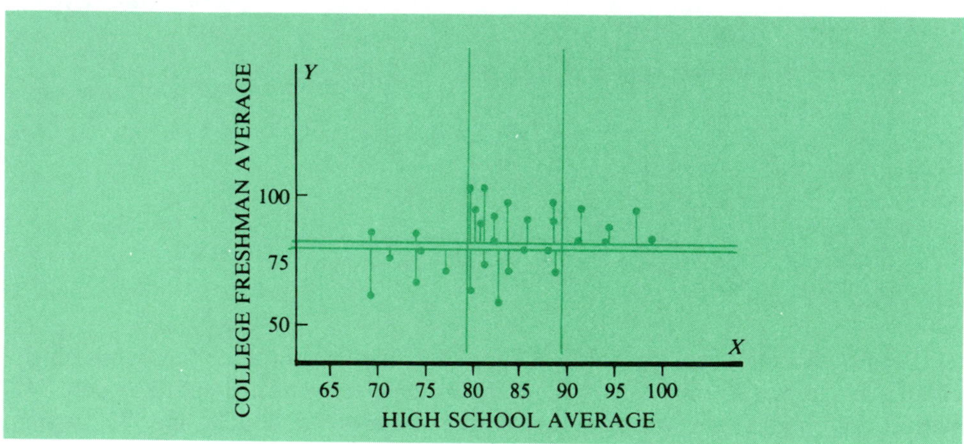

The fact that using the group means as predictions led to an error variance of 60 means that using the single bars led to a mean squared error of 60. The error lines in Fig. 15.9 are such that their mean squared length is 60.[1]

[1]Actually only some of the paired values are illustrated, in which case we expect the mean of the errors squared to be close to 100 in Fig. 15.8 and 60 in Fig. 15.9. However, if all paired values could be illustrated, the error variance would be exactly 100 in Fig. 15.8 and exactly 60 in Fig. 15.9.

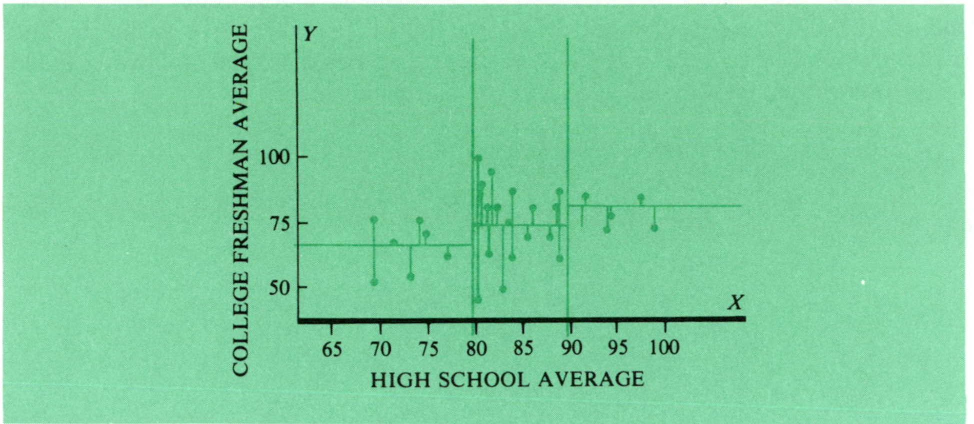

Fig. 15.9

15.4 Regression

The use of one variable to make predictions about another is called *regression*. More specifically, regression entails using known values of one variable to predict unknown values of the other. Here the known values actually used were the categories of the variable, high school average. Note also that using the mean of a group for predictive purposes meant that within any group the errors made averaged out to zero. Therefore, not only did the blind predictions lead to a mean error of zero, but also the regression predictions led to a mean error of zero. Note that in Fig. 15.5 as we move from left to right toward higher values of the variable, high school average, the mass of points taken together tends to rise. The fact that successive group means rise implies, intuitively speaking, that these lines are "going along" with the mass of points. Our information relating the two variables might have been more specific. Suppose for instance we had known the facts in Table 15.2. Since we are actually categorizing the continuous variable "high school average," we might have agreed to place borderline grades in the upper group. A high school score of 70, for instance, would be in Group B.

Table 15.2 Grouped Data

Group	High School Average	College Marks
A	65–70	62
B	70–75	68
C	75–80	73
D	80–85	75
E	85–90	80
F	90–95	82
G	95–100	90

If we had used Table 15.2, our predictions would have been better differentiated and, in the case described, would have resulted in a smaller error variance. The scattergram in Fig. 15.10 shows the predictions that would be made following Table 15.2. Remember that only a few of the scores are shown.

Fig. 15.10

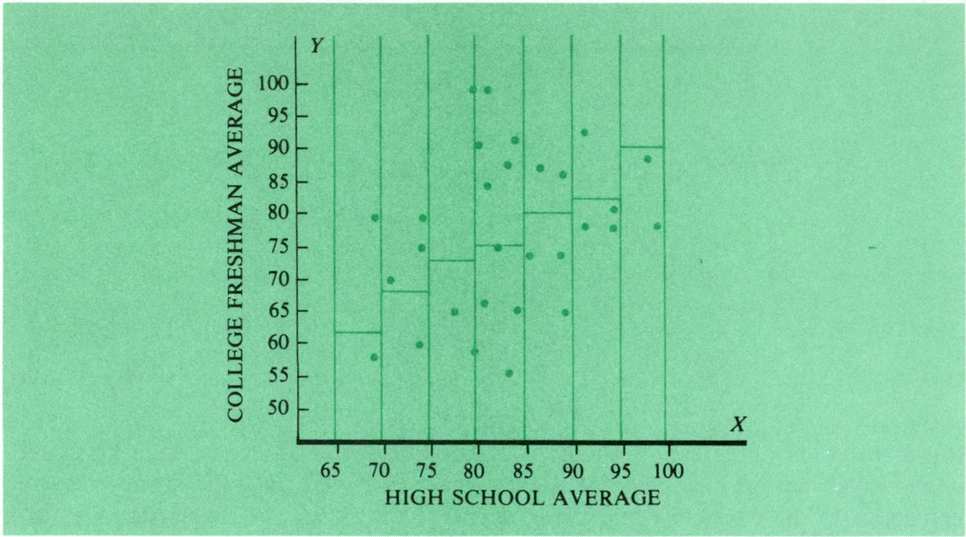

Finally, suppose we were able to consider our predictor variable, high school averages, as actually continuous rather than categorical for purposes of prediction. That is, suppose we were able to go from each single value of our predictor variable to a specific prediction. Now the situation would be that of having a continuous curve to indicate which predictions we were to make. This is shown in Fig. 15.11.

The prediction curve is called a **regression curve**. Admittedly we never have enough information literally to draw a curve like it. No matter how many cases we collect, it would obviously always be possible to find values of the predictor variable which have not appeared even once. Actually we supply the continuity of the curve ourselves. In practice, we either draw the regression curve after first plotting the scattergram or we use mathematical methods to determine its shape and position. In Chapter 17 we shall consider the construction of the regression curve in a particular case,[2] but for now we shall assume that we have drawn this curve precisely so that we can examine some of its meanings.

To look at Fig. 15.11, it should be clear that, for each value of the predictor variable, the regression curve tells what the prediction concerning the other variable must be. For instance, suppose a student arrives whose high school average was 82. The first step toward predicting what his college average as a freshman will be is to locate the value 82 on the horizontal axis in Fig. 15.11. Then we move straight up to the point where the regression curve is over the value 82. The height of the curve at

[2]The particular case will be where the regression curve is a straight-line curve.

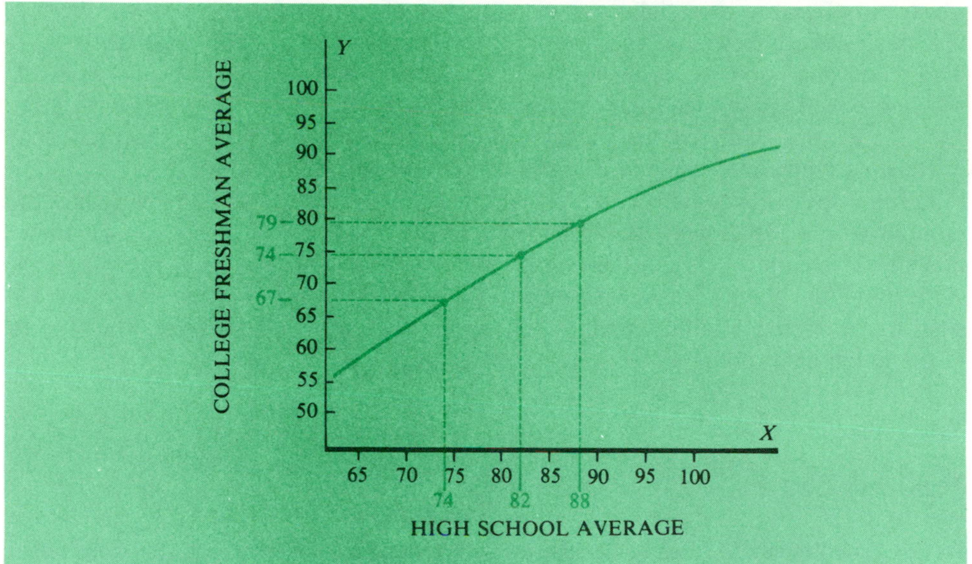

Fig. 15.11

this point, which is its value on the vertical axis, is 74. Thus, for the student whose high school average was 82, we predict that his freshman average at the state university will be 74. Our prediction is shown by the broken lines on Fig. 15.11.

The reader should verify using Fig. 15.11 that, for the student with a high school average of 74, the prediction is a college freshman average of 67. For the student whose high school average was 88, the prediction is a college freshman average of 79.

It should be clear that a regression curve relates every value of the predictor variable to some value of the predicted variable. In other words, for every single possible high school average, the curve identifies some prediction of a corresponding college freshman average. We might take any particular high school average and conceive of the multitude of students who attain that average over the long run. Then this particular collection of students would get some mean freshman grade at the state university. We may think of an infinitesimally narrow column directly over each point on the horizontal axis. The points in any given column represent performances of students with the particular high school average. As the curve passes through such a column, it takes on the value of the mean height of these points. In other words, the height of the curve as it passes through the particular column tells us what the students' mean college freshman grade would be.

In sum, the regression curve passes over the continuity of points on the horizontal axis. Its height at each point indicates what students with the particular high school average would get as their mean college freshman average. More generally, a regression curve indicates for each value of X what the mean Y value is. Thus *a prediction made using a regression curve is essentially the prediction of a mean.* One begins with a particular X value and predicts for the individual who received it that his or her Y value will be the mean of all Y values obtained by those who got the same X value. When we used the regression curve in a previous illustration we were essentially saying: "John's high school average was 82. What is the mean college freshman average of all those whose high school average is 82?" The regression curve by its height over the point 82 tells us that the answer is 74. Therefore we predict that John's freshman average will be 74.

We have considered the meaning of the regression curve and the procedure for making predictions using it. The subject of regression and its implications is quite broad and the purpose here is merely to introduce the reader to it. Having laid the groundwork, we shall next consider a very specific problem of regression—namely, that of correlation. Actually the correlation coefficient that we shall examine next is meaningful without making reference to the issue of regression. But as we use it we shall make reference to implications concerning regression, and they will be seen to throw much light on the whole issue of correlation.

Problem Set A

15.1 The Virginia Lee Realty Company has devised a table to help predict property values following annexation of homes in areas that are not presently within the city limits. The following averages are among those which have been determined.

Value Before Annexation (in thousands of dollars)	*Mean Value After Annexation (in thousands of dollars)*
55	66
65	82
75	97
85	105

The mean value for all homes after annexation is 75 (thousand dollars) with a variance of 10,000 (thousand dollars squared). The variance within each group given is 6,000 (thousand dollars squared). Use the above information to draw a regression curve and predict the value after annexation of the following homes in neighborhoods now outside the city limits:

a. a home selected at random
b. a home whose present value is $75,000
c. a home whose present value is $65,000
d. a home whose present value is $80,000
e. a home whose present value is $63,000

15.2 Draw a scattergram to represent the following actual appraised values of 16 homes before and after annexation.

Before	55	65	75	85
	60	76	82	93
After	64	78	83	99
	77	79	98	102
	79	83	110	106

Compute what the error variance would be for each group if you had used the means from Problem 15.1 for prediction.

15.3 a. What value would you have predicted for the homes in each of the groups in Problem 15.2?
b. What would your error have been for the fourth house in the first group had you predicted its value to be the mean value after annexation of homes in its group?
c. In the long run, what would the error variance be if repeated predictions were made using the values determined by the Virginia Lee Realty Company?

15.4 In which of the following pairs of variables would using regression probably be more accurate than using blind prediction?
a. position of the sun and time when the next solar eclipse will occur
b. price of tea in China and mean age of Tibetan monks
c. severity of disease and length of hospital stay
d. age of machine and frequency of repairs
e. mean GPA of Silo Tech students and the humidity in New York
f. student's examination scores and student's final grade
g. home state and weight of a member of Congress
h. rainfall and bushels of wheat per acre
i. advertising budget and sales
j. price of gold in New York and price of diamonds in Amsterdam

The following information applies to Problems 15.5–15.8. The mean yearly entry level (no experience) income of social workers in Megapolis is $10,845 per year with a standard deviation of $1,200. The mean entry level incomes of social workers for various levels of completed education are also available. Five of them are given below. A 12 indicates a high school graduate, a 13 completion of one year of college, a 14 completion of two years of college, etc. The standard deviation within each given group is $900.

Level of Education	*Mean Income (in thousands of dollars)*
12	6.8
14	8.3
16	12.5
17	13.4
19	15.1

15.5 a. What yearly income would you predict for an entry level social worker selected at random?
b. What yearly income would you predict for an entry level female social worker selected at random?
c. What yearly income would you predict for an entry level social worker at education level 16 selected at random?
d. In the long run what would the error variance be if you predicted the incomes of numerous entry level social workers selected at random to be $10,845 per year?
e. In the long run what would the error variance be if you predicted the incomes of numerous entry level social workers at level 14 to be $8,300 per year?

15.6 Use the mean incomes given preceding Problem 15.5 to draw a regression curve for predicting incomes of social workers, given educational background. Predict the income of a social worker who has completed (a) a bachelor's degree (four years of college),

(b) a master's degree (six years of college), and (c) four years of postgraduate study.

15.7 Below are the incomes (in thosands of dollars) of 20 social workers in five different categories who have just entered the work force in Megapolis.

Level of Education				
12	*14*	*16*	*17*	*19*
5.3	6.9	11.0	12.7	14.9
5.7	7.8	12.1	14.5	15.0
6.9	7.9	12.5	14.6	15.1
6.9	8.9	13.9	15.0	17.0

a. What prediction would you have made as to the incomes of the social workers in the above groups?

b. Compute the error variance for each group if you use the predictions in part a.

15.8 Draw a scattergram to represent the data in Problem 15.7. What error is associated with the first prediction of the first group? Represent that error on the scattergram. (Sketching in the regression curve from Problem 15.6 might help.)

15.9 a. Under what circumstances would using regression with two given variables not increase accuracy of prediction?

b. Which of the following scattergrams might describe sets of values from the variables in part a?

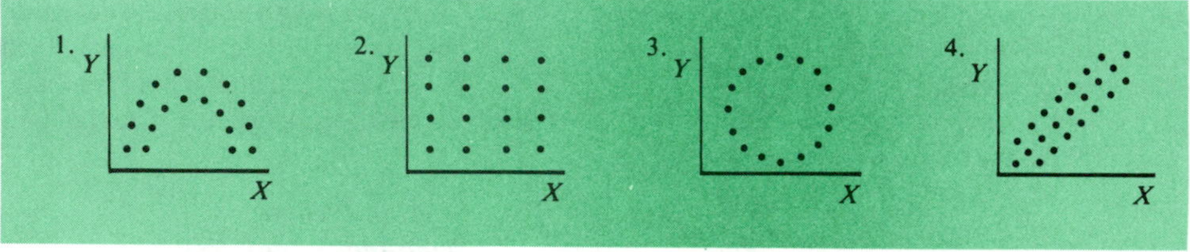

c. Draw the "regression curve" for the scattergram you selected in part b.

d. Give the equation of the regression curve in part c.

15.10 Explain why the variance of errors (error variance) is the same as the population variance in the case of blind prediction.

Problem Set B

15.11 The percentage of the workforce made up of women and the mean number of children per family both vary from state to state. The two have been found to be related and are useful as predictors of one another. Statistics gathered over the last 10 years yield the following means.

Percent of Workforce That is Female (X)	*Mean Number of Children Per Family (Y)*
29	4.3
35	3.1
47	2.9
58	1.7

a. Draw a regression curve using the above data and predict what the mean number of children per family is in a state in which 40% of the workforce is made up of women.
b. Are predictions always about future occurrences? Give three examples.
c. Five years ago only one-third of the workforce in a certain state was female. About how many children would you predict that each family in the state had?
d. Do women going to work cause family sizes to decline? Give some other "possible causes."

15.12 The following mean numbers of children per family have been determined for 12 states with the given percentages of workforce female.

Percentage	*29*	*35*	*47*	*58*
	3.9	4.8	2.5	2.1
Mean	3.2	2.0	2.8	2.5
	5.1	3.3	3.0	1.1

Compute the error variance for each group based on the use of the means in Problem 15.11 to predict the mean number of children per family.

15.13 A psychologist asked 500 grade school teachers across the country to rate their classes each day for one school year during recess as to compatibility as a whole on a scale from 1 to 25. Each day the outdoor temperature was also recorded. Only temperatures that occurred at least 1,000 times were used to obtain the results below.

Temperature	*Mean Compatibility Rating*	*Temperature*	*Mean Compatibility Rating*
15	4	54	19
23	11	68	21
27	13	75	15
39	18	83	12
46	20	91	8

Draw a regression curve and predict the rating of a class when the temperature is (a) 35 degrees, (b) 80 degrees, (c) 50 degrees, and (d) 54 degrees.

15.14 Suppose that after the means have been determined the study in Problem 15.13 continues. On a 75-degree day in Miami, Fla., five teachers turn in results of 14.8, 13, 16, 15, and 9. What would you have predicted the scores to be? What would your error variance have been?

For Problems 15.15–15.18 use the following information. Engineers often use winter snowfall in Graham County to predict the spring water level of Cracker Creek. The means below were determined from numerous past seasons. A zero indicates ideal creek level; negative numbers indicate how many feet the creek is below ideal level; and positive numbers indicate how far above the ideal level the creek is. The mean level is 1 foot below ideal (− 1) with a standard deviation of 4 feet. Within each prediction group the standard deviation is 2 feet.

Snowfall (in inches)	*Mean Creek Levels (in feet)*
30	+8
28	+7
19	0
13	−4
10	−9

15.15 a. What would you predict the creek level will be next spring?
b. What would your error variance be if you made the best blind prediction (the mean) each fall as to the creek level the following spring?
c. What prediction would an engineer make for the creek level after determining that the winter snowfall had been 13 inches?
d. If an engineer predicted the creek level each spring after determining the winter snowfall to fall into one of the categories in Problem 15.14, what would the error variance be?
e. Describe the relationship that appears to exist between the two variables.

15.16 Draw a regression curve using the mean creek levels. Predict the creek level when the snowfall is (a) 15 inches, (b) 17.8 inches, (c) 26.5 inches.

15.17 What does the prediction in part a of Problem 15.16 mean?

15.18 Suppose that the following levels actually occurred during 20 seasons.

	Snowfall				
	30 in.	*28 in.*	*19 in.*	*13 in.*	*10 in.*
	7.5	7.4	0.5	−2.0	−10.4
Levels	8.9	9.1	−2.0	−6.1	− 8.3
(in feet)	4.9	4.8	0.0	−3.7	− 7.1
	9.3	9.0	0.3	−4.2	−11.8

a. Draw a scattergram to represent these data.
b. Compute the error variance for each group if mean levels are used for prediction.

15.19 An actuary for a large national insurance company has devised a formula to determine the life expectancies of healthy White American males given their ages. The formula is used each time a policy is sold. The last 150 claims from South Dakota have not been consistent with formula predictions. What are some possible explanations for the large error variance in the 150 cases?

15.20 It has been stated that knowledge of a relationship between two variables aids in using one to predict the other. In what forms do you find relationships expressed?

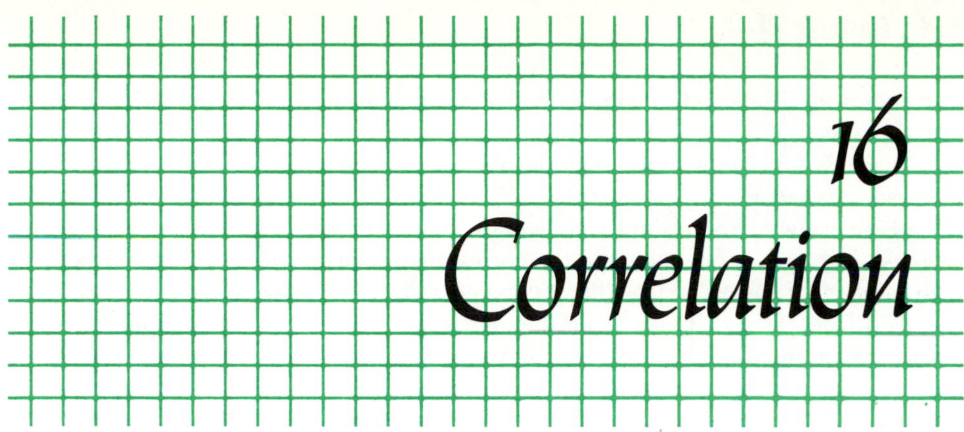

16.1 The Concept of Correlation

One particular type of relationship, called correlation, is of major interest to investigators. Two variables may show a direct or positive correlation, or no correlation, or finally a negative correlation, as we shall see. We shall illustrate correlation in its various forms before actually defining it; otherwise the formal definition will seem unnecessarily abstract and forbidding. As we have been doing, we shall use simplified illustrations to make clear the basic notion. In particular, we shall consider populations made up of only a few cases each. Since it is entire populations we are concerned with, our work will be descriptive in nature. In the final section of this chapter we shall address the idea from an inferential point of view; that is, we shall look at the correlation reflected in a sample rather than viewing the entire population as we shall do now.

Suppose the heights and weights of a population (A) of five men are as follows. The shortest man is 5 feet 5 inches tall and weighs 140 pounds. The next is 5 feet 6 inches tall and is 5 pounds heavier, and so on as indicated in Table 16.1.

Table 16.1 Population A

	Height	Weight
1	5′5″	140
2	5′6″	145
3	5′7″	150
4	5′8″	155
5	5′10″	165

The first thing we note is that for the five men there is a direct or positive relationship between their heights and weights. As the heights increase, the weights also increase. The shortest man is the lightest; the one following him is the next heavier, and so on.

But now let us look at the relationship more specifically. Note that, as we begin to move downward in Table 16.1, each inch in height corresponds to *exactly* 5 pounds in weight. For instance, as we go from the first person to the second, the increase in height is one inch and the corresponding increase in weight is 5 pounds. The same is true as we go from the second person, who is 5 feet 6 inches, to the third, who is 5 feet 7 inches. Once again the increase of a single inch corresponds to a 5-pound weight increase. Finally, even as we go from the fourth person to the fifth, the same correspondence exists. The increase of 2 inches corresponds to an increase of 10 pounds, which is the same as saying that each inch in height corresponds to 5 pounds in weight.

In sum, not only is there a positive relationship between height and weight in the population described, but the relationship is one of a very special kind. It is *perfect positive* correlation. It is characterized by the fact that the correspondence between 1 inch and 5 pounds is uniform throughout. **The notion of perfect correlation refers not merely to a relationship, but more particularly to one in which each unit change on one variable corresponds to a designated change in the other, and the correspondence is uniform.**

Fig. 16.1

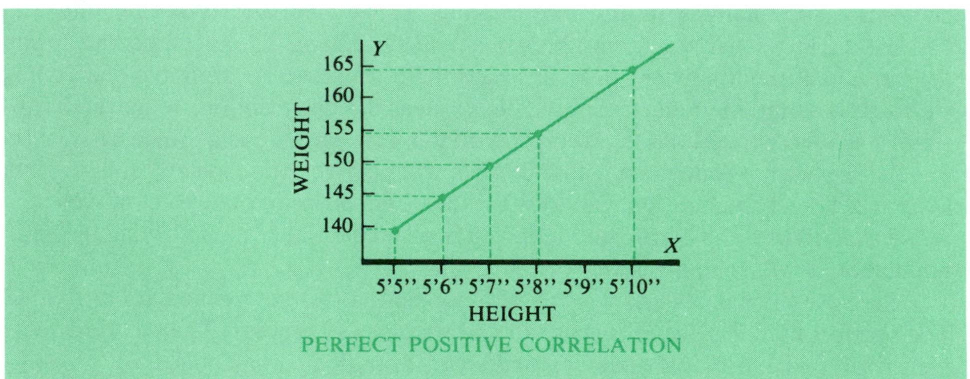

PERFECT POSITIVE CORRELATION

The scattergram for the five cases given in Table 16.1 appears in Fig. 16.1. All five points lie on the same straight line, and this line has actually been drawn in the illustration. Correlation is often described as linear relationship because it has the geometric interpretation that points representing cases lie on the same straight line. The degree to which two variables correlate is the degree to which a single straight line comes close to (or contains) the points representing cases in the population. (Only when there is perfect correlation does a straight line actually contain all the points.) To see why, let us consider the case of perfect positive correlation again. Here the increase of a given amount on one variable always implies some specified increase on the other. Graphically, for the variable on the horizontal axis, each unit increase means a shift of a particular distance to the right. Since there is perfect positive correlation, for each shift of this amount there must be a corresponding upward movement of some other particular amount.

We may start out with a line segment of any specified length (say length A) to represent a one-unit shift on the variable marked on the horizontal axis. We may pick any length (say length B) for our second line segment, which is to represent the corresponding change on the other variable. For instance, the two line segments might be those in Fig. 16.2.

Fig. 16.2

Now, beginning with our axes, suppose we make sure that for every horizontal shift of length A we make a vertical shift of length B. In other words, the only way we can move to the right a length A is to go up a length B. Following this rule, we would find that we always ended on the same straight line (see Fig. 16.3). *Perfect positive correlation implies that for every increase of length A on one variable, there must be a corresponding increase of length B on the other.* Note that an increase of length 2A implies a corresponding increase of 2B. Similarly, an increase of A/8 would imply a corresponding increase of B/8.

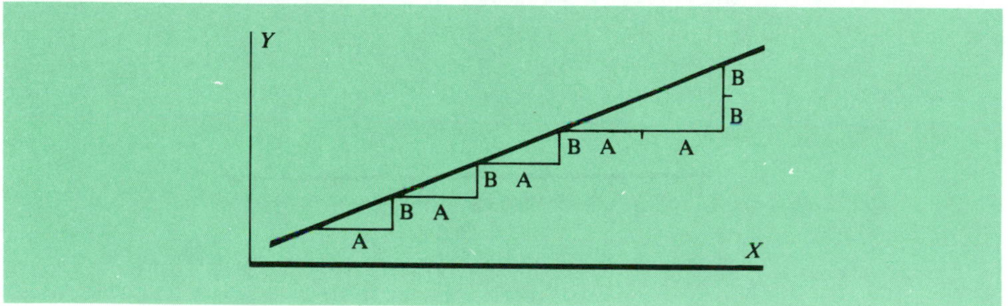

Fig. 16.3

So far we have been talking about the case of perfect or exact correlation. Naturally, one does not always expect to find an exact correspondence such as the one just described that will hold for all the cases in a population. More often we find ourselves considering situations where some relationship exists and where there is some degree of positive correlation as we shall define it, but the correlation is not perfect. To highlight further the notion of positive correlation, let us look at Population B, where the heights and weights of five men are as shown in Table 16.2. Here there is a positive relationship in the sense that the shortest man is the lightest, the next to shortest is the next to lightest, and so on. However, the correlation is not perfect since this time each increase in height does not have an identical meaning in terms of increase in weight. For the first three men, there exists a perfect positive correlation; or to put it another way, the relationship is positive and linear. As before, each increase of 1 inch corresponds to an increase of 5 pounds. But as we go from the third person to the fourth, the increase of 1 inch corresponds to an increase of 20 pounds. And as we go from the fourth person to the fifth, the increase of two inches corresponds to an increase of 50 pounds, so that now it is 1 inch per 25 pounds. The scattergram for the members of Population B appears in Fig. 16.4, and we can see that where the correspondence breaks down in the last two cases, the relationship between height and weight is no longer linear. Incidentally, the straight line coming

Table 16.2 Population B

	Height	Weight
1	5′5″	140
2	5′6″	145
3	5′7″	150
4	5′8″	170
5	5′10″	220

Fig. 16.4

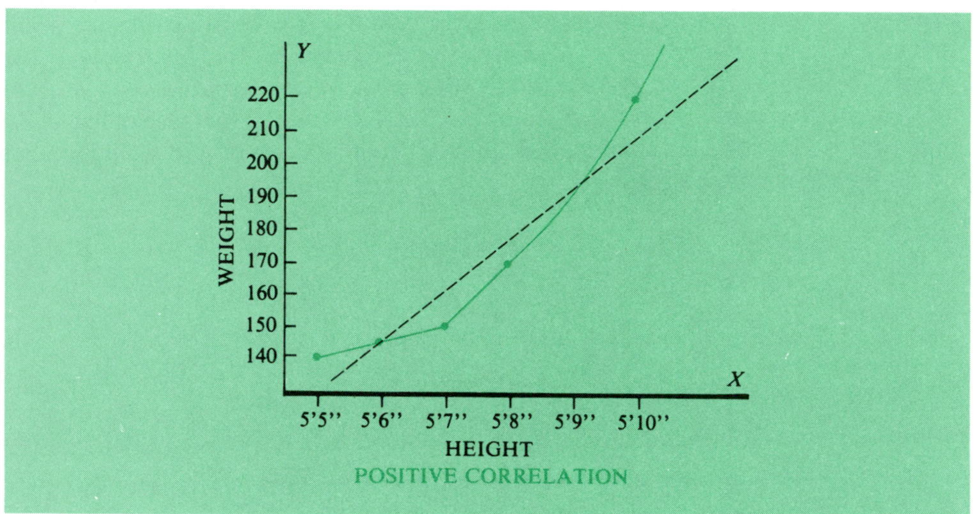

POSITIVE CORRELATION

closest to the five points has been drawn in Fig. 16.4. It no longer contains them, but still we shall have use for it later when we consider it as the regression line.

Next we shall consider the case of a *perfect negative correlation. This time each successive increase on one variable corresponds to some uniform decrease on the other*. We shall say that each of five male students was given first an adjustment scale and then a test to measure the degree of rage reaction that appeared when the student was thwarted in carrying out a defined task in front of the examiner. On the adjustment scale each student received a score indicating the adequacy of his personal adjustment. Each student's score on the frustration scale indicates his readiness to respond with an infantile rage reaction when thwarted. The data for Population C (Table 16.3) are contrived, since in practice a perfect correlation of any kind would scarcely ever appear, but the data do illustrate what a perfect negative correlation would look like.

We may note first that there is a negative relationship between the two variables, which means that the higher the individual's adjustment score is, the lower his rage-reaction score becomes. The first person, whose adjustment score is 38, has the highest rage-reaction score. Each succeeding person has a somewhat higher adjustment score than his predecessor and also a somewhat lower rage-reaction score.

Table 16.3 Population C

	Adjustment Score	Rage-Reaction Score
1	38	18
2	42	17
3	50	15
4	54	14
5	74	9

Finally the person with the highest adjustment score, the fifth person, is the one with the lowest rage-reaction score.

So far we have described what might be called a negative or inverse relationship. However, it may also be seen that this negative relationship is uniform and linear in the sense already described; therefore it is perfect. For each four-point increase in adjustment score, there is a one-point decrease in rage-reaction score. As we go from the first person to the second, the increase in adjustment score is exactly four points and the decrease on the rage-reaction variable is exactly one point. As we go from the second person to the third, the adjustment score increase is eight points and there is a corresponding drop of two points on the rage-reaction score. Next, as we go from the third person to the fourth, the increase of four points on adjustment once again corresponds to a one-point decrease on the rage-reaction scale. Finally as we go from the fourth person to the fifth, the increase of 20 points on the adjustment score corresponds to a decrease of five points on the rage-reaction scale. Once again each four-point increase on the adjustment scale corresponds precisely to a drop of a single point on the rage-reaction scale.

In sum, the data are such that for every unit increase on the adjustment variable there is a specific decrease on the rage-reaction variable. Two variables that enter a relationship of this kind are described as being *perfectly negatively correlated.* Two variables are perfectly negatively correlated when for each unit increase on one of them there is a corresponding decrease of a specified amount on the other.

The data of Table 16.3 appear in the form of the scattergram in Fig. 16.5. Since there is a perfect negative correlation between the two variables, the points that represent the five cases lie on the same straight line just as they did when there was a perfect positive correlation between the two variables. This time, each increase (or shift to the right) on the adjustment variable is coupled with a decrease or drop on the rage-reaction variable. Thus, as the straight line moves to the right, it moves downward. The reader should compare the scattergram of Fig. 16.1, where there was perfect positive correlation, with that of Fig. 16.5, which illustrates perfect negative correlation. In each instance the points all lie on the same straight line, but in Fig. 16.1 as the line moved to the right it moved upward, whereas now as the line moves to the right it moves consistently downward.

Of course not all negative or inverse relationships reflect perfect correlation. Consider the two variables, feet below surface level and temperature in Population D, which is composed of five specified locations in Grandee Lake. The values are listed in Table 16.4.

Fig. 16.5

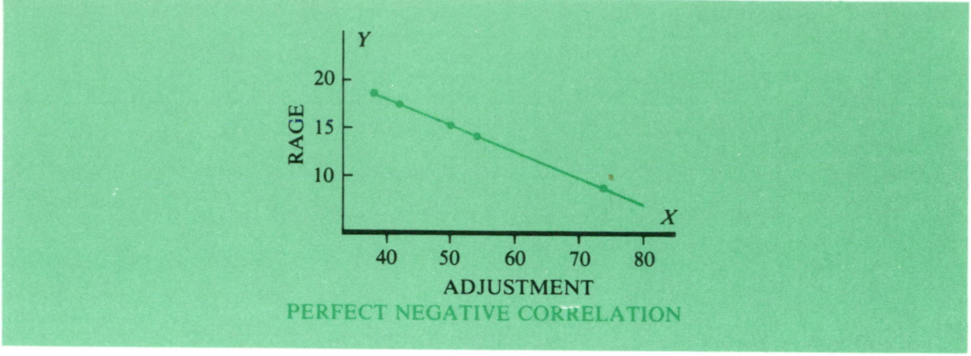

PERFECT NEGATIVE CORRELATION

Table 16.4 Population D

Location	Feet Below Surface	Temperature, Degrees
A	20	69
B	25	60
C	25	65
D	35	61
E	35	54

Population D differs from the populations already discussed in that the first variable takes on some values more than once. Two locations are 25 feet below the surface and two locations are 35 feet below the surface. The scattergram of the data in Population D (Fig. 16.6) reveals that no single increasing or decreasing curve can pass through all five of the points.

Fig. 16.6

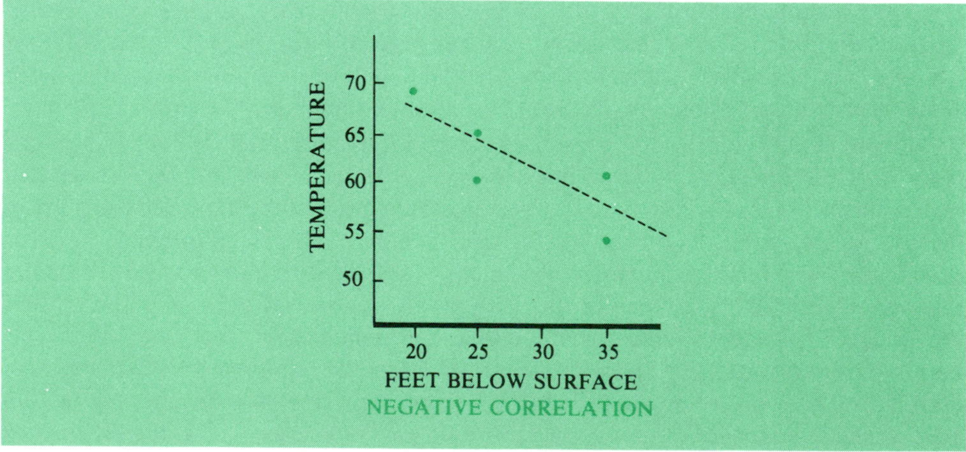

NEGATIVE CORRELATION

The tendency, nonetheless, is for temperature to drop as the distance below increases. The relationship then is negative. This is true even though one of the positions 25 feet below surface level is cooler (60 degrees) than one at 35 feet below

(61 degrees). *It is the overall tendency we are concerned with.* The scattergram in Fig. 16.6 clearly indicates that as distance below increases, temperature tends to decrease. Of course, other relationships may also exist between the two variables that may take into account other factors, such as the effects of currents, distance from shore, or numerous other influences. But at present it is the relationship of negative correlation that concerns us. Again, as in Fig. 16.4, the straight line that comes closest to the five points has been drawn in Fig. 16.6. Its usefulness as a regression line will become evident in the next chapter.

Finally we shall consider the case of no correlation between two variables. Specifically, suppose the heights and IQ scores of 10 people composing Population E are those listed in Table 16.5. In this table we see that, as the heights increase, the IQ scores do not show an apparent change in either direction. **There is in fact no correlation between the variables, which means that as one variable increases, the other variable does not show even the slightest overall tendency to increase or decrease uniformly.** The scattergram for this case of no correlation between the variables, height and IQ, is shown in Fig. 16.7.

Table 16.5 Population E

Height	IQ
5′5″	107
5′5″	99
5′6″	114
5′7″	94
5′8″	98
5′8″	106
5′9″	94
5′10″	114
5′11″	108
5′11″	98

This time there is no correlation. As we move to the right on the variable, height, there is no systematic upward or downward movement on the variable, IQ. We have drawn in the straight line that comes closest to all of the points considered together, and it turns out that this line is horizontal. We will see that this fact is important later on—that **where two variables do not correlate, the straight line that is "best-fitting" in the scattergram representing the case is perfectly horizontal.**

The two variables, height and IQ, as found in Population E in Table 16.5, show no relationship of any kind. We noted that in particular they do not correlate. However, of all possible relationships, the linear relationship, called correlation, is only one type. Thus it is possible for two variables to enter a complex relationship and for there still to be absolutely no correlation between them.

The data to be discussed are contrived once more for the sake of making the clearest possible illustration. Our purpose is to illustrate how two variables may enter a complex relationship that is of such a nature that the correlation between them remains nonexistent. Ten subjects who were members of a closely knit group were

Fig. 16.7

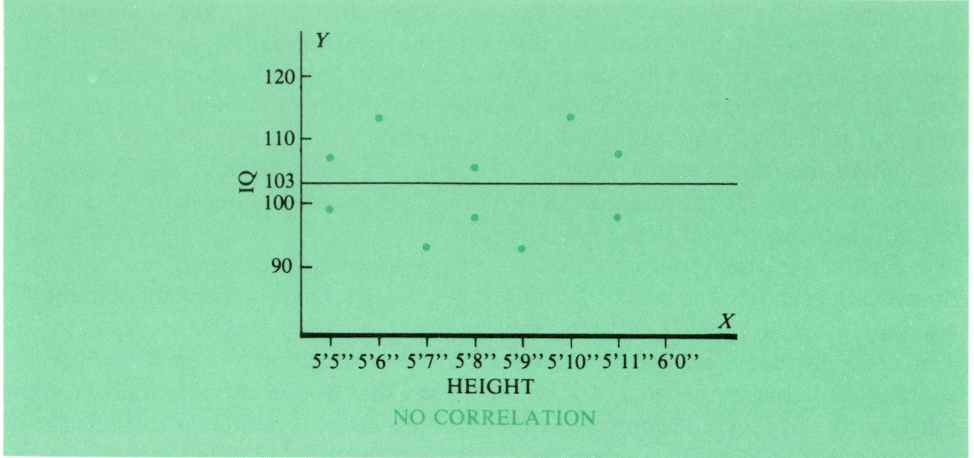

NO CORRELATION

given a well-known anxiety scale and each received a score indicating what is called his or her chronic anxiety level. A low score on this scale meant that the subject had a relatively low chronic anxiety level, and a high score meant a high one. The subjects were also each given a differentiation scale, which measured for each one of them the degree to which he or she was able to see personality differences among the group members. A relatively low score on this differentiation scale meant that a subject tended not to differentiate among the other group members; or, to put it another way, tended to see them all as relatively similar. A relatively high score meant that the subject tended to differentiate—that is, tended to perceive differences among the group members. The data for Population F appear in Table 16.6.

Note in Table 16.6 that the subjects are listed in order of their anxiety scores. The first subject is the one with the lowest anxiety score, the next subject is next lowest, and so on. Successive subjects after the first one show progressively increasing anxiety scores. As we go down the list, the differentiation scores start at their lowest and increase progressively up to the fifth subject. In other words, for the

Table 16.6 Population F

	Anxiety Scores	Differentiation Scores
1	20	1
2	24	2
3	26	4
4	30	6
5	35	8
6	40	8
7	45	6
8	49	4
9	51	2
10	55	1

Fig. 16.8

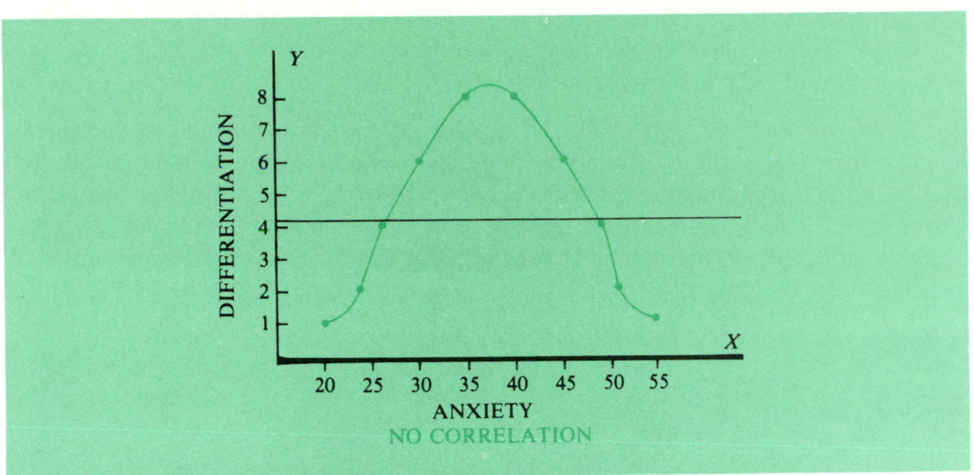

NO CORRELATION

first five subjects there is a positive relationship between anxiety and differentiation. But now as we go from the fifth subject to the sixth, we find that our subjects no longer continue to increase in their ability to differentiate. Now as the anxiety scores continue to increase, the corresponding differentiation scores cease to increase and instead they actually begin progressively to decrease. The scattergram showing the data of Table 16.6 appears in Fig. 16.8.

In Fig. 16.8 the points indicating the 10 cases have been joined by a smooth continuous curve. It may clearly be seen how, as we move to subjects with increasing anxiety, their differentiation scores first increase and then begin to decrease. Because the curve drawn in Fig. 16.8 depicts the relationship so well, this type of relationship is often called a *curvilinear* one. We have also drawn the straight line that comes closest to all the points; note that it is *horizontal*. An analysis indicates that for the first five cases there is a positive correlation and for the last five there is a negative correlation. The two variables are specifically related, but their overall relationship can by no means be defined as a correlation. We cannot say that as anxiety increases there is any kind of uniform change on the variable, differentiation. Loosely speaking, if we did we would necessarily be as wrong on some of the cases as we would be right on others.

We have presented two illustrations of how two continuous variables may be uncorrelated. In one case they may have no relationship whatsoever. In the other case, they may enter a very specific relationship, but we may not be able to conceptualize it when we restrict ourselves to thinking of a linear relationship. What has been said has been aimed at highlighting the fact that a correlation is a very specific kind of relationship between two variables—namely, a linear relationship.

We may now sum up what has been said with a formal definition.

Definition 16.1 Two variables are said to correlate positively when, as one variable increases, the other shows some trend to increase correspondingly in a uniform way. Two variables are said not to correlate at all when, as one of them changes, the other shows absolutely no overall trend to change in a

uniform way with respect to it. Two variables are said to correlate negatively when, as one variable increases, the other shows some trend to decrease correspondingly in a uniform way.

Whenever we have a population of individuals scored on each of two variables, the case of perfect positive correlation should be viewed as one extreme; the case of no correlation is a central point; and the case of perfect negative correlation should be viewed as the other extreme. If we think of a line representing possible correlations, as shown in Fig. 16.9, then we must conclude that the relationship between any two variables must be some point on that line.

Fig. 16.9

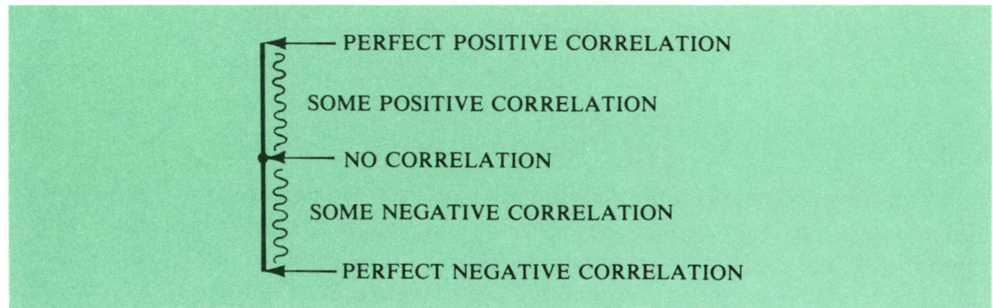

Remember that when we discuss correlation we are always talking about a trend that exists and not a necessary or invariable relationship. Also remember that the fact that two variables correlate does not at all imply that either variable *causes* the other to change. In any given case, the relationship may be coincidental or the consequence of some external cause that has not even been discovered.

Correlation refers to the degree to which two variables move uniformly with respect to each other, and it follows that two variables may show some correlation, even when their precise relationship is not perfectly linear. That is, the conceptualization of them as moving together in a linear way may be informative even where it is not the best possible conceptualization of how they relate. To put it another way, the straight line describing their relationship may be informative but still be only a meaningful approximation of some curve that would better describe it. Thus, where a correlation between two variables is found, we must remember that this correlation may not provide the ultimate statement of how the variables relate.

16.2 Correlation Construed as the Relationship between Two Sets of *z* Scores

The individuals composing a population are apt to be scored on each of two variables where the units of measurement are quite different. For instance, as we have seen, one variable may be height in inches and the other may be weight in pounds. *The units of measurement of the two variables have absolutely nothing to do with whether or not the two variables correlate*. In other words, the means and standard deviations of the variables may be entirely different, and still the variables may correlate moderately or perfectly in either direction. The point is that we may write down the scores of the individuals in ascending order on one variable and, once we have done so, the issue is whether the terms follow a linear order in one direction or the other on the other variable.

To illustrate, suppose each of nine students takes an IQ test and also another test, which we shall simply call Test A. On the IQ test the mean of their scores is 100 and the standard deviation is 5.16. On Test A the mean is 54 and the standard deviation is 12.9. For our first illustration, suppose that the IQ scores and those on Test A show a perfect positive correlation. The scores for the nine students are given in Table 16.7A.

Table 16.7A IQ Scores and Test A Scores Showing a Perfect Positive Correlation

Student	IQ Score	Test A Score
1	92	34
2	94	39
3	96	44
4	98	49
5	100	54
6	102	59
7	104	64
8	106	69
9	108	74

Examination of the data of Table 16.7A reveals that for each two-point increase in IQ there occurs among the students exactly a five-point increase in their scores on Test A. The means and standard deviations on the two variables are different, but their values have no bearing on the perfect positive correlation that exists between the two variables. There could have existed a perfect negative correlation between the variables, still assuming that both the IQ scores and the scores on the other variable had the same means and standard deviations as in Table 16.7A.

Table 16.7B IQ Scores and Test B Scores Showing a Perfect Negative Correlation

Student	IQ Score	Test B Score
1	92	74
2	94	69
3	96	64
4	98	59
5	100	54
6	102	49
7	104	44
8	106	39
9	108	34

Suppose that the IQ scores of the nine students had been exactly the same but the order of their scores on the other variable had been exactly reversed. Then the correlation between the scores on the two tests would have been a perfect negative one. In Table 16.7B the IQ scores of the same nine students are shown along with their scores on a fictitious Test B, which has the same mean and standard deviation as Test A. This time there is a perfect negative correlation as the reader may verify.

In Table 16.7C the same IQ scores are shown in correspondence with scores on a fictitious Test C. This time the IQ scores show no correlation with the test scores. Note that the mean and standard deviation of the scores on Test C are exactly the same as they were on Tests A and B since the scores are the same but appear in different order.

Table 16.7C IQ Scores and Test C Scores Showing No Correlation

Student	IQ Score	Test C Score
1	92	69
2	94	34
3	96	49
4	98	64
5	100	54
6	102	44
7	104	59
8	106	74
9	108	39

In our illustrations, IQ shows a perfect positive correlation with Test A, a perfect negative correlation with Test B, and no correlation with Test C. These facts are true even though the means and standard deviations of the three tests were identical. The point is that the fact of correlation relates specifically to how two variables go together and is independent of the values of the mean and standard deviation of the variables in question. It follows that **we preserve whatever correlation exists between two variables when we change the terms in each distribution to z scores.** The z scores of the terms in a distribution are in a sense simplified versions of the original terms and do not reflect the size of the mean or standard deviation of the original terms.

Table 16.8A shows the z score correspondence indicating the perfect positive correlation between IQ scores and those on Test A. Both the IQ scores and the Test A scores have been changed to z scores, and these appear in Columns (2) and (3) of Table 16.8A. In essence the scores that appeared in Table 16.7A have been replaced by z scores that appear in Table 16.8A. The same has been done in Table 16.8B for the IQ and Test B scores. The z score relationship, indicating the perfect negative correlation between IQ and Test B scores, may be seen in this table. Finally the z score correspondence between IQ and Test C scores may be seen in Table 16.8C. In each case the z scores have been rounded off. Thus the value 8.98, whose meaning will become clear shortly, is an approximation to 9.00, the value that would have been found were it not for rounding errors.

Table 16.8A z Scores of IQ and Test A Showing Perfect Positive Correlation

(1) Student	(2) IQ	(3) Test A	(4) Product
1	−1.5	−1.5	2.25
2	−1.2	−1.2	1.44
3	−.8	−.8	.64
4	−.4	−.4	.16
5	.0	.0	.00
6	+.4	+.4	.16
7	+.8	+.8	.64
8	+1.2	+1.2	1.44
9	+1.5	+1.5	2.25
			$\Sigma =$ 8.98

Table 16.8B z Scores of IQ and Test B Showing Perfect Negative Correlation

(1) Student	(2) (IQ)	(3) Test B	(4) Product
1	−1.5	+1.5	−2.25
2	−1.2	+1.2	−1.44
3	−.8	+.8	−.64
4	−.4	+.4	.16
5	.0	.0	.00
6	+.4	−.4	−.16
7	+.8	−.8	−.64
8	+1.2	−1.2	−1.44
9	+1.5	−1.5	−2.25
			$\Sigma =$ −8.98

Table 16.8C z Scores of IQ and Test C Showing No Correlation

(1) Student	(2) IQ	(3) Test C	(4) Product
1	−1.5	+1.2	−1.80
2	−1.2	−1.5	1.80
3	−.8	−.4	.32
4	−.4	+.8	−.32
5	.0	.0	.00
6	+.4	−.8	−.32
7	+.8	+.4	+.32
8	+1.2	+1.5	+1.80
9	+1.5	−1.2	−1.80
			$\Sigma =$ 0.00

The substitution of z score values for the original terms in a distribution actually moves us forward so far as the issue of correlation is concerned. In fact we shall make reference to the z score data of Tables 16.8A, 16.8B, and 16.8C now as we proceed to discuss the classic indicator of correlation, which is called the **correlation coefficient**.

16.3 The Correlation Coefficient

The correlation coefficient (often called the *Pearson product-moment correlation coefficient*), signified by the letter r, is a precise measure of the way in which two variables correlate. Its value is such as to indicate both the *direction* (positive or negative) and the *strength* of the correlation between two variables. We are now going to discuss the correlation coefficient in terms of z score data, since it may be best understood when considering such data. However, as we shall see later, we do not actually need to substitute z score values for original terms in order to compute a correlation coefficient.

To begin with, an examination of Table 16.8A shows that corresponding to each negative z score on IQ is a negative z score on Test A. Corresponding to each positive z score on IQ is a positive z score on Test A. Quite naturally, this type of correspondence reflects a positive correlation between two variables. It means that those who are below the mean on one variable are also below the mean on the other, and that those who are above the mean on the first variable are also above it on the second one.

Suppose now that each individual's pair of z scores are multiplied together, giving what we shall call a **cross product** in each case. These are shown in Column (4) of Table 16.8A. Remember that the product of two negative numbers is positive, as is the product of two positive ones. Thus, in each case, the perfect positive correlation reflects itself in positive or zero cross products. In Column (4) the nine cross products are indicated. They have been added and the mean found. The mean of these cross products is $+ 1.0$. (Were it not for rounding errors, the sum of the products would be 9.00 instead of 8.98. And $9.00/9 = 1.0$.)

It is the mean cross product of the z scores that is the correlation coefficient. Its sign, which was positive in the case just described, reflects the fact that the correlation was positive. Its size indicates the strength of the correlation in that the stronger the correlation is, the further away from zero is the value of the correlation coefficient. It may be shown mathematically that *the maximum value possible for the correlation coefficient is + 1.0, and that it takes this value only in the instance of a perfect positive correlation* as in the case described. But let us go on to compute the value of the correlation coefficient in other cases.

Table 16.8B presents the z scores for IQ and Test B where the two variables had a perfect negative correlation. There, negative z scores on IQ went with positive ones on Test B, and positive z scores on IQ went with negative ones on Test B. Since the product of a positive and negative number is negative, the cross products in this case all turn out to be negative or zero. Therefore the mean of these cross products—which is the correlation coefficient—is also negative, reflecting the negative correlation.

These cross products are shown in Column (4) of Table 16.8B, and their mean is −1.0. The negative sign of the correlation coefficient indicates that the correlation is negative. *The obtained correlation coefficient of −1.0 is the most extreme possible negative value that a correlation coefficient can take. It reflects the perfect negative correlation between the two variables.*

Finally, where there is no correspondence, as between IQ and Test C scores, the cross products are in some cases positive and in others negative, and they tend to balance each other out. The illustration in Table 16.8C was contrived to show the situation of exactly no correlation between two variables, and thus it turns out that the correlation coefficient is exactly zero. The computation of this correlation coefficient is shown in Column (4) of Table 16.8C. *A correlation coefficient of exactly zero reflects the fact that there is absolutely no correlation between two variables.* Incidentally, as one might guess, even when two variables would be absolutely uncorrelated in the long run, there typically turns out to be some small correlation between them in a given sample. This correlation is as likely to be positive as negative. Just as even a perfectly fair coin would be unlikely to show exactly 500 heads in 1,000 flips, a correlation of exactly zero as in this illustration would be extremely unlikely.

It is meaningful that any departure from perfect correlation tends to reduce the size of the correlation coefficient, where plus and minus signs are not involved. For instance, Table 16.8A shows the z scores and cross products where there is a perfect positive correlation. We may pretend that the z scores of the first two students on Test A are interchanged, in which case the table of z scores for IQ and Test A would appear as in Table 16.9. Note that the single interchange reduces the sum of the cross products. Hence this single lack of perfect correlation quite properly reflects itself by reducing the correlation coefficient. It is, in fact, true that any departure from perfect correlation, positive or negative, results in moving the correlation coefficient closer to zero, since only when the most extreme z scores on both variables are coupled does it turn out that the cross products add up to a maximum. Thus the correlation coefficient is precisely *sensitive* to the direction and degree of correlation that exists in each given case.

***Table 16.9** Altered z-Score Distribution For IQ and Test A*

Student	IQ	Test A	Product
1	−1.5	−1.2	1.80
2	−1.2	−1.5	1.80
3	−.8	−.8	.64
4	−.4	−.4	.16
5	.0	.0	.00
6	.4	.4	.16
7	.8	.8	.64
8	1.2	1.2	1.44
9	1.5	1.5	2.25
			$\Sigma =$ 8.89

Definition 16.2 The correlation coefficient is the mean of the cross products of the z scores of two variables.

$$r = \frac{\Sigma z_1 z_2}{N}$$

We shall see that one need not go through computing z scores to obtain the value of the correlation coefficient that is a measure of the degree to which two variables correlate. The sign of the correlation coefficient indicates whether the correlation is negative or positive. The correlation coefficient always takes some value between -1 (when there is a perfect negative correlation) and $+1$ (when there is a perfect positive correlation). It is zero when there is absolutely no correlation between two variables; and it departs from zero in one direction or the other depending upon the strength of the correlation between the two variables.

The vertical line in Fig. 16.10 indicates the range of possible values of the correlation coefficient. The different segments of the line are labeled as they were in Fig. 16.9. Table 16.10 summarizes the important points discussed regarding correlation.

Fig. 16.10

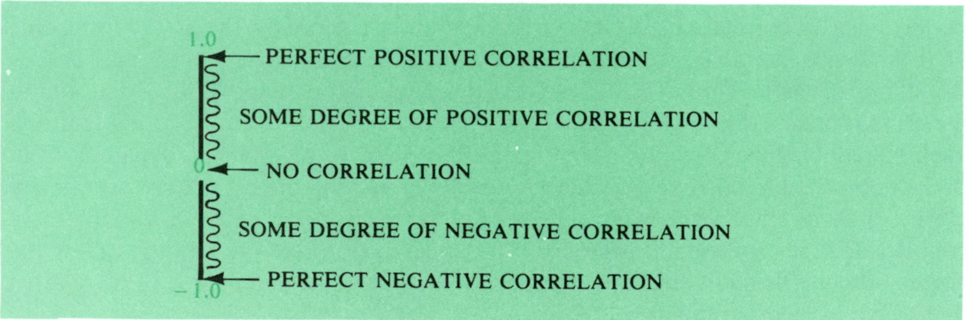

Table 16.10 Correlation

Type	Scattergram	r Value
Perfect positive	Rising straight line	$r = 1$
Positive, but not perfect	Rising, not straight line	$0 < r < 1$
No correlation	No tendency to rise or fall	$r = 0$
Negative, but not perfect	Falling, not straight line	$-1 < r < 0$
Perfect negative	Falling straight line	$r = -1$

16.4 Computing the Correlation Coefficient

Up to now we have substituted for the terms on each variable the appropriate z scores in order to get the "feel" of how each pair of variables went together. By translating the original terms in each case to z scores, we were able to communicate the meaning of the correlation coefficient most easily. However, as mentioned, it is not necessary in practice to make the translation in each case to z scores. One can operate upon the original scores in such a way as to compute from them the correlation coefficient; and quite naturally, since the substitution of z score values for the original terms is laborious, it is not usually done in practice.

We are going to illustrate now the usual method for computing the correlation coefficient. This method entails working with the original data and applying a formula that gives exactly the same result as if the data had first been put into z score form and then the mean z score cross product had been found. In other words, just as we computed the value of r before, we are going to do so once more. Only this time we are going to work from original data, or what is sometimes called "raw data," and we shall not trouble ourselves to translate the data to z score form.

Let us say that we wish to compute the correlation coefficient and that our two variables are IQ scores and scores on a musical aptitude test. Specifically, 20 students have each been given both tests and their scores are shown in Table 16.11. From now on we shall not bother to number the students. Nor shall we arrange them in any particular order since it is not necessary to do so.

Table 16.11 IQ and Musical Aptitude Scores

IQ Score	Musical Aptitude Test Score	IQ Score	Musical Aptitude Test Score
95	30	102	30
106	37	105	30
110	35	111	32
104	28	106	33
98	17	110	38
114	45	143	42
123	33	106	40
100	18	98	20
98	24	104	23
88	18	97	25

In order to use the formula for the correlation coefficient, which will be introduced soon, we must make various separate computations. Our first step is to choose one of the sets of scores arbitrarily and call it the X variable. We shall call IQ scores values of the X variable and musical aptitude scores values of the Y variable.

To compute the correlation coefficient, we make a table of preparatory computations (see Table 16.12). In the first column under the heading X go the values of the X variable. In the second column under the heading X^2 appear the squares of each of these scores. For instance, the first subject's X score is 95, which appears in the first column, and the square of this X score is 9,025, which appears in the second column. ($95^2 = 9{,}025$). Similarly, each subject's score on the X variable appears in the first column and the square of this score appears in the second one.

Each subject's score on the Y variable—the musical aptitude test—appears in Column (3). The square of each subject's score on the Y variable appears in the fourth column under the heading Y^2. For instance, the first subject's score on the musical aptitude test is 30, which appears in the third column, and the square of this score is 900, which appears in the fourth column.

Table 16.12 Computations for the Correlation Coefficient Where IQ Scores are the X Variable and Musical Aptitude Scores are the Y Variable

(1) X	(2) X^2	(3) Y	(4) Y^2	(5) XY
95	9,025	30	900	2,850
106	11,236	37	1,369	3,922
110	12,100	35	1,225	3,850
104	10,816	28	784	2,912
98	9,604	17	289	1,666
114	12,996	45	2,025	5,130
123	15,129	33	1,089	4,059
100	10,000	18	324	1,800
98	9,604	24	576	2,352
88	7,744	18	324	1,584
102	10,404	30	900	3,060
105	11,025	30	900	3,150
111	12,321	32	1,024	3,552
106	11,236	33	1,089	3,498
110	12,100	38	1,444	4,180
143	20,449	42	1,764	6,006
106	11,236	40	1,600	4,240
98	9,604	20	400	1,960
104	10,816	23	529	2,392
97	9,409	25	625	2,425
$\Sigma X = 2{,}118$	$\Sigma X^2 = 226{,}854$	$\Sigma Y = 598$	$\Sigma Y^2 = 19{,}180$	$\Sigma XY = 64{,}588$

Each subject's score on the X variable has been multiplied by his or her score on the Y variable to yield the value in the fifth column. For instance, the first subject got an X value of 95 and a Y value of 30; the product of these two numbers, which is 2,850, appears in the fifth column. The second subject's X score of 106 has been multiplied by the Y score of 37 to yield the value 3,922, which appears in the fifth column. Note that the heading of the fifth column is XY, indicating that each value in this column is a cross product for a single subject.

Beneath each column at the bottom of Table 16.12 appears the total for that column. Remember that the symbol Σ stands for "the sum of." The sum of the 20 X values, indicated as ΣX, is 2,118. The sum of the 20 X^2 values, indicated as ΣX^2, is

226,854. The sum of the 20 Y values, indicated as ΣY, is 598. The sum of the 20 Y^2 values, indicated as ΣY^2, is 19,180. Finally, the sum of the 20 XY values, indicated as ΣXY, is 64,588. The value of N, the number of pairs of scores, is of course 20.

Actually, it was to obtain the totals that we went through the entire procedure, for these totals are the important "pieces" that we use when we compute the correlation coefficient.

Now we are ready to put down the formula for working from original data and computing the correlation coefficient. In Formula (16.1) the correlation coefficient is indicated by the letter r. This is sometimes called the Pearson product-moment correlation coefficient.

$$r = \frac{N\Sigma XY - (\Sigma X)(\Sigma Y)}{\sqrt{N\Sigma X^2 - (\Sigma X)^2}\sqrt{N\Sigma Y^2 - (\Sigma Y)^2}} \tag{16.1}$$

To solve Formula (16.1) we must insert six "pieces of information" in their appropriate places. The "pieces of information" are the following facts:

$$\Sigma X = 2{,}118 \qquad \Sigma Y = 598 \qquad \Sigma XY = 64{,}588$$
$$\Sigma X^2 = 226{,}854 \qquad \Sigma Y^2 = 19{,}180 \qquad N = 20$$

Now we insert each numerical value in place of its alegebraic equivalent in Formula (16.1), and we get:

$$r = \frac{(20)(64{,}588) - (2{,}118)(598)}{\sqrt{(20)(226{,}854) - (2{,}118)^2}\sqrt{(20)(19{,}180) - (598)^2}}$$

$$r = \frac{25{,}196}{\sqrt{51{,}156}\sqrt{25{,}996}}$$

$$r = .69$$

Our ultimate finding is that the correlation coefficient is .69. Thus there is some degree of positive correlation between scores on the musical aptitude test and IQ scores for the students tested. We have gone through a rather laborious procedure, including the computations that we had to make in Table 16.12. However, this procedure is a standard one and it must be thoroughly understood from beginning to end. The reader should be clear exactly how, from the data of Table 16.12, the various totals were found. These totals were the crucial "pieces" substituted in Formula (16.1). It should also be clear how the substitutions were made and the computations were done in solving Formula (16.1). The reader will very likely have occasion to repeat the entire procedure, and will have to set up a table like Table 16.12 and to use Formula (16.1) to solve for the correlation coefficient.

Sometimes an experimenter already knows the population means and the standard deviations of both variables, in which case a different computing formula (also equivalent) is helpful.

$$r = \frac{\dfrac{\Sigma XY}{N} - \mu_X \mu_Y}{\sigma_X \sigma_Y} \qquad (16.2)$$

For the problem discussed, suppose we knew the following.

$$\Sigma XY = 64{,}588 \qquad \sigma_X = 11.31$$
$$\mu_X = 105.9 \qquad \sigma_Y = 8.06$$
$$\mu_Y = 29.9$$

Using Formula (16.2), the reader should verify that .69 is again obtained for the correlation coefficient.

Note that had we converted the original scores to z scores, we might simply have found the mean of the z score cross products. This mean, which is the correlation coefficient, would also have turned out to be .69. However, the task of making the conversions to z scores would have been much more laborious than our task was here. We might best describe the z score procedure as one that is important to consider in order to gain insight into the meaning of the correlation coefficient. On the other hand, the procedure just described is one that can be followed blindly for computing the value of r. The fact that one may operate blindly is a disadvantage in the sense that one does not gain insight, but it is an advantage for the worker who understands the meaning of the correlation coefficient and wishes to compute its value in a particular case as quickly as possible. Readers who are interested may demonstrate to themselves that taking the mean z score cross product is algebraically equivalent to solving Formula (16.1) for the value of the correlation coefficient.

The correlation coefficient is an enormously important measure. Nearly every meaningful kind of relationship between two variables involves at least some degree of correlation between them. The first step for researchers when looking for a relationship is usually to sketch a scattergram and then compute the value of the correlation coefficient. The correlation coefficient is by far the simplest measure of relationship between two continuous variables, so far as computing is concerned. For data in grouped frequency distributions the same basic computational approach as in Table 16.12 may be used. However, as in Chapter 6, the midpoint of each interval on each variable is used to represent the value of each score that has been grouped in that interval. In practice one almost always uses commercially prepared charts for computing r from grouped frequency distributions. These charts come complete with a detailed list of steps and with a formula essentially the same as Formula (16.1), perhaps in some mathematically equivalent form.

The use of the correlation coefficient is evidenced in virtually every field where data are gathered. The reader should review the meaning of the correlation coefficient and understand it well, for it is hard to find a text relating to education, psychology, or the social sciences that is not replete with correlation coefficients.

16.5 The Test of Significance for a Correlation Coefficient

Up to this point we have considered only the descriptive value of the correlation between two variables in a finite population. More often than not, we are trying to determine if there is any correlation between two variables in some vast population. Inferential methods come to our aid as we draw a sample from the infinite population and consider the *null hypothesis that the correlation between the two variables in question is zero*. For instance, one might wish to determine whether scores on an anxiety scale are correlated with IQ scores among 10-year-olds. In this case, one has in mind the vast population of 10-year-olds to whom the anxiety scale and the IQ test might be given. There exist methods for using a correlation coefficient obtained from a sample in order to test the hypothesis that the correlation is zero in the population from which the sample is drawn. We shall describe one such procedure by illustrating it.

Suppose that a researcher has given both the anxiety scale and the IQ test to 30 10-year-olds, and the obtained value of r is .40. This time it is not the value $r = .40$ that is of ultimate concern; rather, the researcher wishes to use it to make a decision about whether the correlation is zero between anxiety and IQ in the population that yielded the 30 cases. She knows that, even were this correlation exactly zero, there would likely be a correlation of some value other than zero in a sample such as hers. The question is whether her obtained value is sufficiently far from zero to indicate that she should reject the null hypothesis that the correlation is zero in the population.

As usual, our researcher must pick a significance level, and we shall say it is the .05 level. Next she asks: Under the hypothesis that the correlation is zero in the population, what would be the probability of obtaining a sample value of r as far from zero as mine? In other words, our experimenter conceives of her sample as one of a vast number of random samples from the same population. She conceives of her obtained r value as one of a collection of r values, each of which would have been yielded by a different sample.

Under the hypothesis that the correlation is zero in the population, the obtained values of r in different samples would vary around zero. There would be r values below zero and others above it. The question is whether the value of .40 is far enough away from zero to indicate rejecting the hypothesis.

We shall consider here a procedure applicable when the assumption can be made that the variables of interest are roughly *normally distributed*. In Chapter 19 we shall consider a different approach that is applicable when the assumption cannot be made that the variables are normally distributed.

When the variables are roughly normally distributed the following procedure may be followed. The researcher who has computed an r value should compute t from Formula (16.3).

$$t = r\,\frac{\sqrt{N-2}}{\sqrt{1-r^2}} \qquad (16.3)$$

This t value has $N - 2$ d.f.

To illustrate, in the case described, $r = .40$ and $N = 30$. Therefore,

$$t = \frac{.40\sqrt{30-2}}{\sqrt{1-(.40)^2}} = \frac{.40\sqrt{28}}{\sqrt{.84}} = 2.31$$

The obtained t value has exactly the same meaning as the t values that we computed in the past.

Formula (16.3) is sometimes called a *conversion*. Earlier our researcher had to think of each of the other researchers as obtaining a value of r from a sample. Now they must be considered as having gone through the next step, which means having computed a value of t. Under the null hypothesis, the distribution of the obtained t values would have 28 degrees of freedom (see Fig. 16.11).

Fig. 16.11

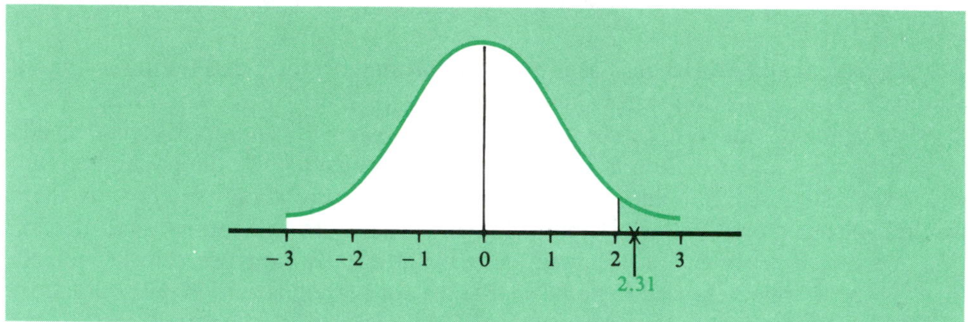

As for our researcher, using her knowledge of the t distribution for 28 d.f., she must locate her obtained t value in this t distribution. (This t value is a representative of the r value that she computed.) At the .05 significance level, her obtained t value of 2.31 is significant. Therefore she is able to reject the null hypothesis, which was that the correlation in the population is zero. She may conclude that anxiety and IQ are at least somewhat correlated in the population from which the sample was drawn.

Had the obtained value of r been .18, for instance, then the value of t would have been .97, as the reader may verify. The null hypothesis could not be rejected for 28 d.f. Note that a negative correlation would lead to a negative value of t but the procedure for testing the hypothesis would be the same. Also, the desire to prove that the correlation in the population differs from zero in a particular direction would lead to a one-tail test, exactly as in the situations described earlier. For instance, the desire to prove that the population correlation is less than zero would lead to a one-tail test where only an extreme negative value would lead a researcher to reject the null hypothesis.

The technique described involves what statisticians call the use of a *transformation*. Formula (16.3) was used to transform the statistic r into the statistic t since the latter has a distribution that is familiar. Thus the test of the hypothesis was directly involving t values, but the implication of the finding was immediate with regard to the original issue, which was whether or not the population correlation was zero.

Problem Set A

16.1 For 10 pairs of values from variables X and Y, $\Sigma X = 348$, $\Sigma Y = 356$, $\Sigma X^2 = 12{,}475$, $\Sigma Y^2 = 14{,}270$, and $\Sigma XY = 12{,}784$. Compute the correlation coefficient using Formula (16.1).

16.2 Consider the following sets of measures:

X	14	27	39	67
Y	43	15	9	2
Z	3.1	8.6	8.5	12

Compute the correlation coefficient between
a. variables X and Y
b. variables X and Z
c. variables Y and Z

16.3 In the following pairs of measures X represents the number of sunny days in a given month and Y represents the number of tourists who registered for lessons during the month at Jacques' Ski School.

X	14	18	22	20	26
Y	140	95	155	125	185

a. Plot a scattergram.
b. Compute r using Formula (16.1).
c. Compute r using z score products.
d. Plot the z scores of each pair on a scattergram.
e. Sketch in the best-fitting lines on the scattergrams in parts a and d (as best you can).

16.4 A psychologist looking for correlation between percent of minimum daily requirement of vitamin E ingested (X) and amount of aggressiveness as indicated on an oral-written test (Y) obtains the following scores from five 15-year-old boys.

X	72	67	65	60	56
Y	75	80	85	84	86

a. Plot the scattergram.
b. Compute r using Formula (16.1).
c. Compute r using Formula (16.2).

16.5 The table below shows the weight, W, in grams of a certain chemical that will dissolve in 1,000 grams of water at temperature $T°$ centigrade. Compute the correlation coefficient.

T	0	10	20	40	60	80	100	110
W	480	552	650	725	840	970	1,034	1,100

16.6 In a nationwide controlled study of 83 hospital units, the correlation coefficient between the nurse/patient ratio and the mean length of stay per patient was $r = .55$. Do these facts support the hypothesis that the two variables are positively correlated at the .01 significance level?

16.7 Compute the correlation coefficient between the amount of consumer credit (in billions of dollars) and the number of commercial failures (in hundreds) between 1970 and 1979.

	Year									
	1970	*1971*	*1972*	*1973*	*1974*	*1975*	*1976*	*1977*	*1978*	*1979*
Credit	31	33	38	41	42	49	53	55	56	62
Failures	92	80	76	89	111	110	127	137	150	141

16.8 Thirty-five large (over 500 enrollment) high schools are studied as to the number of incidents of vandalism and attendance at home basketball games. The correlation coefficient is found to be $r = -.33$. On the basis of these findings can the null hypothesis that vandalism and attendance are not correlated be rejected at the .01 level? Use a one-tail test.

16.9 Presented below are pairs of measures on two variables, W and Z, for six children. W is the age in months at which the child took his or her first step and Z is the score on an examination designed to measure how well the child relates to adults of the opposite sex.

	Child					
	1	*2*	*3*	*4*	*5*	*6*
W	8.9	9.7	10.5	11.3	12.4	12.9
Z	43	50.	37	52	49	60

Test the hypothesis that the two variables are related at the .05 level using a two-tail test.

16.10 Jose's, Inc., plans to locate a food processing plant in Centerville. One of the board of directors' major concerns in selecting the exact site is the availability and type of labor. The first vice-president recommends building as closely as possible to the five largest housing developments, and so the board selects 10 already established industrial plants in the city to see if there is any correlation between worker turnover and the mean distance the plant is located from the five largest housing developments in the area. X = percent of turnover per year; Y = mean distance in miles.

X	Y	X	Y
21.9	2.18	37.9	2.79
26.5	2.34	39.1	2.90
29.5	2.23	41.6	2.65
32.2	2.47	44.3	3.38
33.0	2.51	45.9	2.97
33.2	2.14	48.7	3.32
35.4	2.42	50.4	3.46
36.9	3.04		

Do the sample results support the first vice-president's hypothesis that the two variables are positively correlated at the .05 significance level?

16.11 The table below gives the chest girth in inches and lung capacity in cubic inches of college freshmen.

Chest Girth	*Lung Capacity*	*Chest Girth*	*Lung Capacity*
38.9	311	37.6	226
35.0	305	34.5	278
31.3	330	32.6	310
30.3	210	37.5	275
38.0	269	34.3	220
31.5	238	34.4	219
30.8	305	37.2	265
33.7	219		

a. Compute r and test it for significance. Use a two-tail test at the .05 level of significance.
b. What assumption must be made in part a?

16.12 Consider the following pairs of measures.

X	17	29	53	81	90
Y	23	17	38	49	35

a. Compute r.
b. Add 10 to each X value and then compute r.
c. Add 5 to each Y value and then compute r.
d. Add 10 to each X value and 5 to each Y value and compute r.
e. Generalize the conclusion you draw from parts a–d.
f. Redo b–e, replacing the words "add to" with "multiply by."

16.13 Show that taking the mean of the z score cross products is equivalent to evaluating Formula (16.1).

Problem Set B

16.14 Below are the values assumed by two variables in a six-member population.

X	12	14	18	21	35	42
Y	11	12	11	10	5	8

a. Find the correlation coefficient r.
b. How does Y tend to react as X increases?

16.15 Given the following information, compute the correlation coefficient r using Formula (16.1): $N = 8$, $\Sigma X = 26$, $\Sigma X^2 = 139$, $\Sigma Y = 28$, $\Sigma Y^2 = 198$, $\Sigma XY = 160$.

16.16 Which of the following pairs of variables would you expect to be (a) positively correlated? (b) negatively correlated? (c) not correlated?

1. hardness and strength of a metal
2. yearly income and number of overdue accounts
3. inflation rate and food costs
4. miles per hour and miles per gallon
5. distance of a planet from the earth and the amount of xenon gas in its atmosphere
6. age and maintenance cost of a computer
7. alcohol consumption and reaction time
8. number of acres planted and yield per acre
9. advertising budget and sales
10. age of child and ability to read

16.17 Recent accidents have aroused concern over nuclear safety. Six sites for waste disposal are studied. The number of tons of radioactive waste stored at the site (T) and the incidence (number per thousand) of cancer in a 100-mile radius of the site (C) are recorded below.

T	120	138.6	145.9	167.7	169.4	173
C	37	35	43	48	49	47

a. Calculate the coefficient that reflects the correlation between the variables at the six sites using Formula (16.1).

b. Calculate the correlation coefficient using z score cross products.
c. Plot the z scores on a scattergram and draw in the best-fitting straight line.

16.18 The annual inflation rate (X) and the mean GPA of college students (Y) for the past five years are given below.

X	7.3	7.1	8.5	9.4	11.7
Y	2.3	2.4	2.4	2.5	2.6

a. Plot a scattergram.
b. Draw in the best-fitting line.
c. Calculate r using Formula (16.1).
d. Calculate r using Formula (16.2).

16.19 A researcher is studying the influence a woman's father exerts on her selection of a mate. In the process, he collects the following data from 10 married women.

X *(height of father)*	Y *(height of husband)*	X	Y
65″	68″	70″	65″
66.5″	75″	71.5″	70″
67″	72.5″	72.5″	68.5″
68.5″	74″	74″	71″
68.5″	66.5″	75″	72″

Compute the correlation coefficient for these 10 cases.

16.20 In assessing the success of its campaign to educate teenagers in proper nutrition, the Department of Agriculture collects weekly data for two consecutive years (104 weeks). For the two-year period, the campaign's advertising-education budget and weekly consumption of "junk food" are found to have a correlation coefficient $r = -.37$. Does this support the hypothesis that as the size of the budget increases, consumption of junk food decreases ($r < 0$)? Use the .01 level of significance.

16.21 The number of highway deaths continues to climb year after year despite new efforts to enforce speed limits and improve road surfaces. Nevertheless, Safety Director Carver believes that speed and number of accidents are related and collects the following data from a dozen states.

X *(percent of increase in highway deaths)*	T *(mean speed on highways)*
5.0	53.2
5.3	54.7
7.0	55.0
8.0	55.8
9.4	56.8
11.0	52.0
11.3	57.3
12.0	59.0
15.0	55.5
17.1	63.0
23.0	69.4
25.7	68.3

Compute the correlation coefficient.

16.22 Is there any significant correlation between the success (measured in games won) of I.O. University's male and female soccer teams? Results for the past 10 seasons may be found below. Use .01 significance level.

Male	*Female*	*Male*	*Female*
18	29	23	20
15	28	29	17
17	20	25	23
15	24	17	23
12	23	24	15

16.23 In the belief that the number of American poor receiving food stamps (X) and the percentage of school children diagnosed as malnourished (Y) are inversely related, Public Health Nurse Williams compiles the following data for 10 consecutive years.

X (in millions)	2.8	3.7	4.9	5.1	5.2
Y (in percent)	15.3	15.6	14.3	12.2	11.7
X (in millions)	10.8	12.3	14.9	18.5	8.4
Y (in percent)	8.0	6.9	6.8	5.9	12.3

a. Calculate the correlation coefficient r and test it for significance at the .05 significance level using a one-tail test.

b. Does the result allow you to conclude that the increase in the number of recipients is responsible for the decrease in the percentage of malnourished children?

16.24 Fifteen residents of a large metropolitan area were selected at random and surveyed. Each was asked to state his or her age and the mean number of hours of television watched per day. The results are given below.

Age	*Hours*	*Age*	*Hours*	*Age*	*Hours*
5	2.4	13	5.1	26	3.6
7	3.6	15	4.0	28	4.1
8	4.7	17	1.0	35	2.5
12	6.0	22	2.3	40	5.0
12	5.8	26	1.4	63	3.7

a. Compute the correlation coefficient between these measures.

b. Is the statistic found in part a significant at the .05 level using a one-tail test?

16.25 For two variables, X and Y, in a certain population it is known that $\mu_X = 15$, $\mu_Y = 73.4$, and $r = 0$. Draw a scattergram that might represent pairs of values for X and Y. Sketch in the best-fitting straight line.

16.26 If variables X and Y are perfectly and positively correlated, and variables Y and Z are perfectly and negatively correlated, what can be said about variables X and Z?

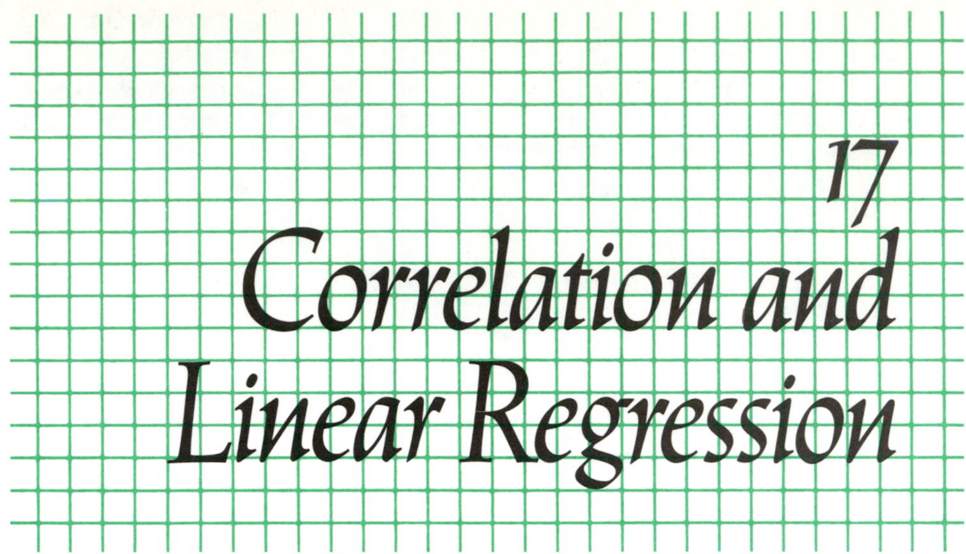

17 Correlation and Linear Regression

17.1 Introduction

We are now going to consider what is by far the most important case in which we make predictions—namely, that which arises when we use a straight line as our regression curve. The correlation coefficient has crucial meaning in this case, as we shall see. We shall proceed as before, by a close analysis of the scattergram. The purpose is to use it here to demonstrate some basic relationships. Later we shall see how to make use of these relationships without even drawing the scattergram for the population of interest. Keep in mind throughout this entire chapter that our primary goal is to find some expedient way to make predictions. This goal will be realized toward the end of the chapter when we arrive at an easy-to-use prediction formula. To understand the formula fully, however, we first need to gain a few insights studying scattergrams.

To use an illustrative situation, let us say that a vast number of students whose high school averages range from 65 to 95 have been going to the state university. Once again we shall say that there exists a relationship between their high school averages and their college freshman averages, and it is such that the higher a student's high school average, the better he or she tends to do as a college freshman; that is, positive correlation exists between the variables. We shall again assume that a vast number of students' high school and college freshman averages have been recorded. The relationship has been consistent and there is every reason to believe that it will continue.

Fig. 17.1 is a scattergram showing the performance of a number of students. High school averages are recorded along the horizontal axis and freshman averages at the state university are recorded along the vertical axis. Note that each dot in Fig.

17.1 tells us by its location both a single student's high school average and freshman average. For instance, the dot that has been encircled represents the performances of a student whose high school average was 70 and whose average at the state university was 74.

Fig. 17.1

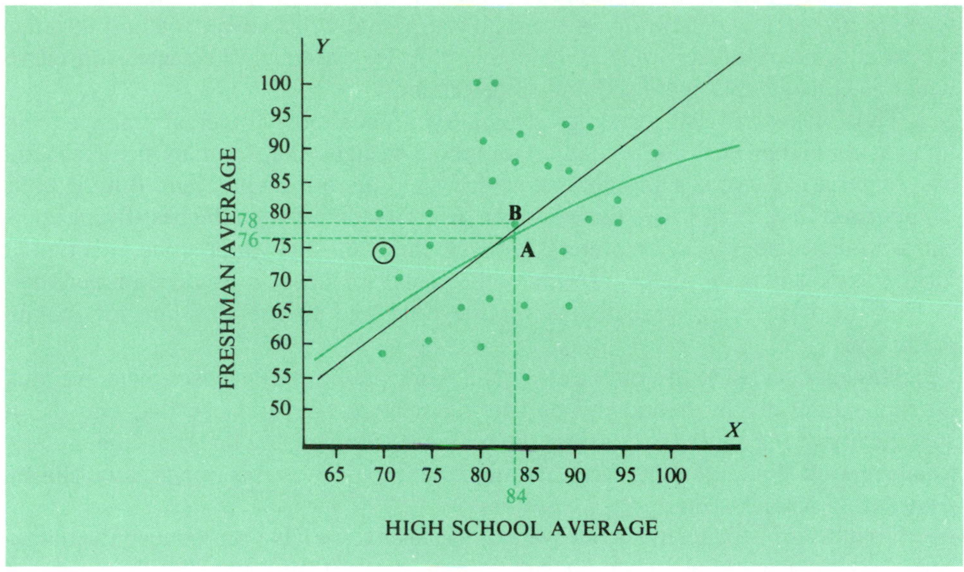

The curve drawn in Fig. 17.1 is the regression curve as defined in Chapter 15. In other words, one should refer to the height of the curve (based on past performances) to make the best possible prediction of what a student's freshman average at the state university will be. One finds the point on the horizontal axis corresponding to the high school average attained by the particular student and locates the height of the curve as it passes over that point. It is this height that becomes the predicted freshman average for the student. For instance, suppose a student comes along whose high school average was 84. The regression curve is at Point *A* in Fig. 17.1 as it passes over the value 84 on the horizontal axis. The height at Point A is 76, as we may see by referring to the vertical axis. Therefore, the best prediction is that the student's freshman average will be 76.

Next let us look at the straight line drawn in Fig. 17.1. Though the curve might be somewhat more accurate, we might also make it a practice to use the straight line for predictive purposes. For instance, in the case of the student whose high school average was 84, we would follow the straight line to Point *B* and predict a freshman average at the state university of 78.

We have defined the regression curve as the single continuous path that comes closest to all the dots. Thus the errors made when predicting from the curve would, by definition, lead to an error variance smaller than that made when predicting from the line. However, in practice there are enormous advantages to making predictions from a straight line. For instance, there is a direct way to "fit" a straight line to data and to use it for predictive purposes. One need not have a vast number of cases to "fit" a relatively accurate straight line, as we shall see. On the other hand, a large number of cases are needed to fit an accurate curve that can be used for future cases. It is easy to extend a straight line in either direction, but unless a curve is accurately drawn, it is not so easy to determine its path and so to extend it. On balance, then, *the advantage of using a best-fitting straight line for predictive purposes is that the line and its formula are easy to determine*, as we shall see. Usually this advantage far outweighs the small amount of accuracy we are giving up by not using the regression curve, which is considerably more difficult to determine.

Thus, instead of using the regression curve, it is almost universal practice to set up as a stipulation that our prediction path be a straight line. In many instances, the best regression curve is a straight line anyhow and there is no problem. But the point is important that, even where there is some discrepancy between the best-fitting curve and a straight line, it may be preferable to use the line. Thus, in practice, it is typical for the worker in many fields, especially the social sciences, to proceed immediately to find the best-fitting straight line for the data and to use this line for making predictions.

Having decided to use only a straight line for predictive purposes, what we must do quite naturally is to find the best-fitting straight line and use that one. More specifically, we are going to consider the problem of finding the line that comes closer to all the dots than any other line that might be drawn. It is this line that we call the **best-fitting straight line.**

To further distinguish this line we are discussing, we have reproduced the data of Fig. 17.1 in Fig. 17.2, where once again this best-fitting line has been drawn. Consider what happens when we use this line to make our prediction for each individual student. That is, given each student's high school average, we would refer to the appropriate point on the line to predict what the student's college freshman average would be. In each case, our error would be the distance between our chosen

Fig. 17.2

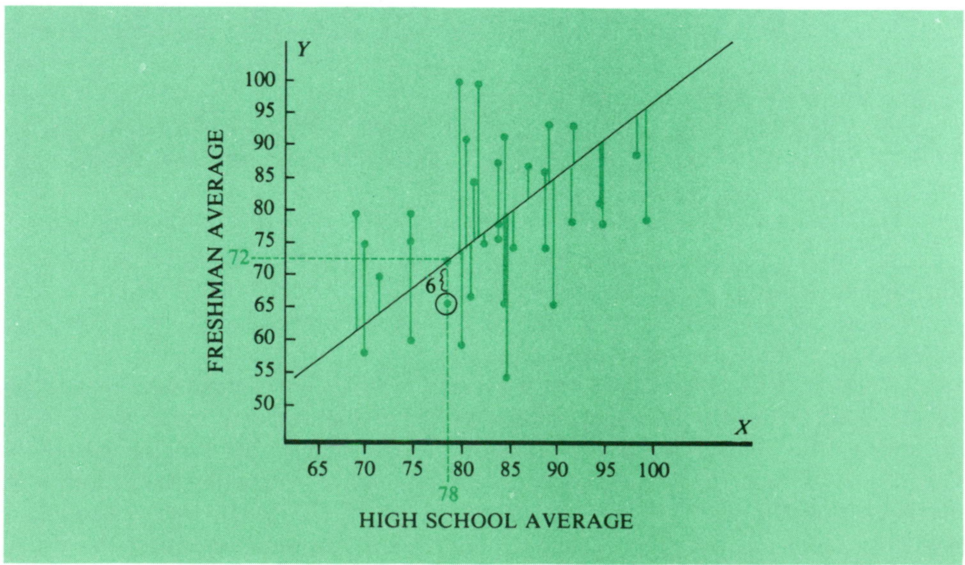

point on the line and the dot representing the actual performance of the student of interest.

For instance, the high school average of the student whose performance is encircled in Fig. 17.2 was 78. Suppose this student had come along and we knew only his high school average. Given only this fact, we would have gone ahead to find the height of the line over the point 78, and our prediction would have been that the student's freshman average was to be 72. Since the student's actual freshman average is 66, our error would have been six points, indicated by the length of the dotted line in Fig. 17.2. This dotted line connects the point on the regression line (which represented our prediction) to the dot indicating the student's actual performance. The length of the dotted line represents our error in the given case. The various vertical lines in Fig. 17.2 similarly stand for the errors that we would have made in individual cases.

Now suppose that we were to square each of these errors and then to find the mean of all the squares thus obtained. The best-fitting line may be defined as the one that would make this mean of the squared errors less than if any other line had been used. We said that in Fig. 17.2 we have drawn the best-fitting straight line. This means that the mean of the squared errors in Fig. 17.2 is less than it would have been had we drawn any other line in the same scattergram and computed the error variance from it. Thus intuitively we call the best-fitting straight line in a scattergram the one that comes closest to all the dots. In more precise language, we may describe it as the one that makes the mean of the squared errors less than if any other line had been drawn. The two statements are equivalent. (This line is often called the "least-squares line" and in many cases the method we are discussing here is called the "least-squares method.") As we shall see, there is a computational way to determine exactly where this best-fitting line should go, and we will make use of it.

Much of what has been said can be summed up in the following definition.

Definition 17.1 The best-fitting straight line to a scattergram is the line that gives the smallest possible mean of the squared errors. That is, when the vertical distances from the points to the line are used, the best-fitting straight line comes closer to all the points than any other straight line that might be drawn.

17.2 The Geometric Meaning of the Correlation Coefficient

Suppose now that we were to take our data and transform the scores on each variable into z scores. For the variable, high school averages, we would have to find its mean and standard deviation. Then for each student we would have to determine how many standard deviations below or above the mean his or her high school average was, and the value obtained for each particular student would be his or her z score. For instance, if a student's high school average was one standard deviation above the mean, then the z score on this variable would be 1.0. Similarly, the student would have a z score on the variable, freshman average.

In essence, we would be able to represent the performance of each student by that student's pair of z scores. One would be the z score of his or her high school average and the other would be the z score of his or her freshman average at the state university. Suppose that, having computed for each student a pair of z scores, we were now to draw a scattergram. This time the units of measurement on each variable would be z scores. This scattergram is shown in Fig. 17.3 as it would appear. The z scores of the student's high school averages are indicated along the horizontal axis, and those of their freshman averages are indicated along the vertical axis. Once more the best-fitting straight line has been drawn in Fig. 17.3.

Fig. 17.3

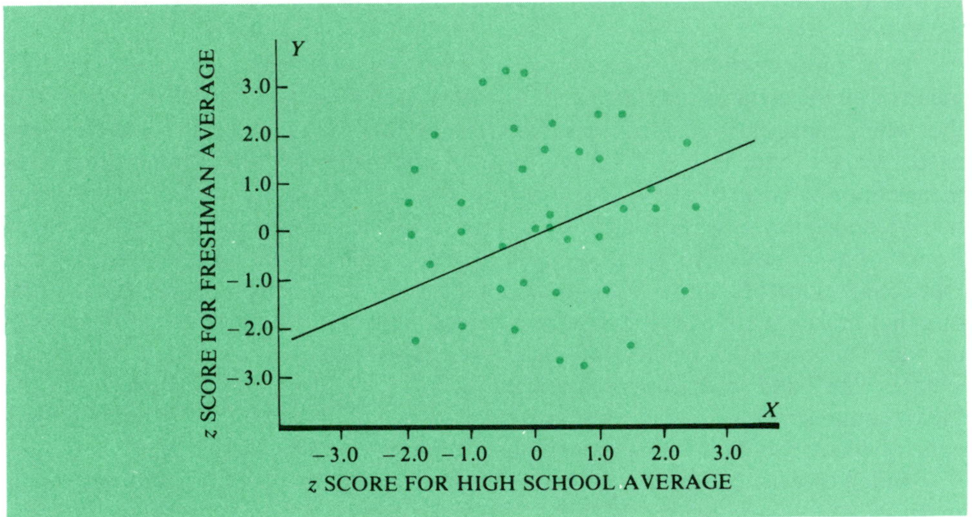

As the reader may know, a line drawn on a graph generally has some slope. That is, a point moving on the line would always be rising (or dropping) at some speed in proportion to how fast it is moving horizontally. To determine the slope of a line, we may pick any point on the line at random. From that point we measure off some distance to the right. For instance, for the line in Fig. 17.4 we have randomly picked Point A. We then draw a horizontal line any number of units that we decide upon. Let us say that in Fig. 17.4 we have chosen to measure off five units to the right, bringing us to Point B. Then from this point we draw a line vertically to the original line in question. In Fig. 17.4 our vertical dotted line meets the original line at Point C, creating a triangle. We shall say that there the length of the line BC turns out to be two units.

We may conclude from our computations that for every five units that a point moves from left to right on our line, it ascends two units. The line AB is sometimes called the *run* of our original line, and the line BC is sometimes called its *rise*.

The **slope** of a line is its rise divided by its run, which in this case is 2/5. The slope of a line is the fraction that tells us how fast a point moving along the line is

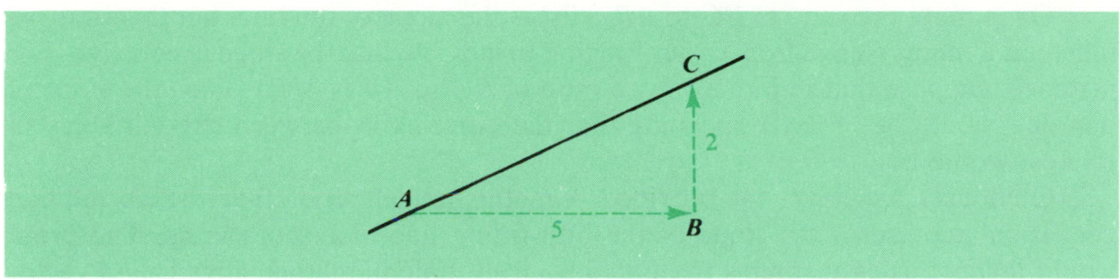

Fig. 17.4

rising compared with the speed that it is moving horizontally. We may pick any point on the line as Point *A* and draw our horizontal line any length to Point B. Then we draw our vertical line *BC*. The ratio of the segment *BC* to the segment *AB* will always be the same. Thus if we had made *AB* 10 units, line *BC* would have been four units, so that once again the ratio of *BC* to *AB* would be 2/5.

Now let us go back to the situation in which we drew our *z* score scattergram for the variables, high school average and freshman average. This scattergram was Fig. 17.3 and on it was drawn the best-fitting straight line. This best-fitting straight line has a slope; or, to put it another way, its rise divided by its run is some fraction. **The correlation coefficient for two variables is the same as the slope of the best-fitting line in the *z* score scattergram.** Thus suppose the correlation between high school averages and freshman averages is .40. Then the slope of the line in Fig. 17.3 would turn out to be .40 or 2/5.

Once again the translation into *z* scores reveals a meaning of the correlation coefficient. Naturally, one does not go to the trouble of translating the score on both variables into *z* scores, then making up a *z* score scattergram, and finally drawing in the best-fitting straight line. The point is merely that if one went to this trouble, it would turn out that the exact correlation between the two variables would reveal itself as the slope of that line. As in Chapter 16, we have conceived of translating the scores into *z* score units simply for the purpose of making clear an insight.

There occurs a variation in terms of slope when we consider a negative correlation. Then, as scores increase on the *X* variable, they decrease on the *Y* variable. This means that even in *z* score terms, the best-fitting straight line moves downward instead of upward. As we move horizontally from our chosen Point *A*, we find that the best-fitting straight line for the data is dropping below us. As in Fig. 17.5, we must draw our vertical line *BC* downward in order to return to the best-fitting straight line. In this case we say that our best-fitting straight line has a negative slope.

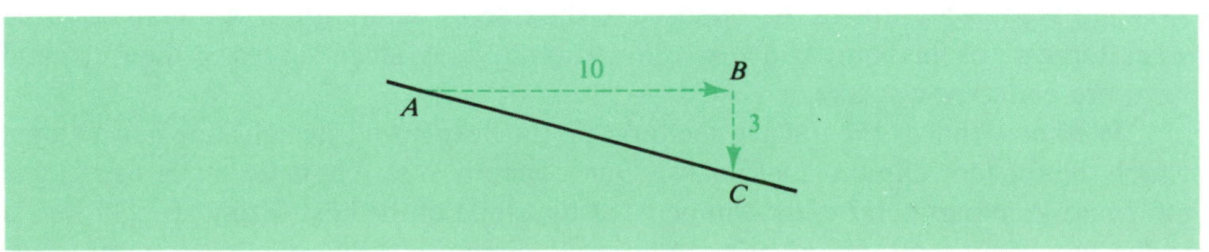

Fig. 17.5

Once more the ratio of BC to AB tells us the slope of the line, but this time we must put a minus sign in front of our fraction to indicate that the slope is negative. For instance, suppose that in Fig. 17.5 the ratio of BC to AB is 3/10. Then the slope of our line would be − 3/10, meaning that the correlation between the variables of interest would be − .30.

Obviously we have not provided a mathematical proof that the correlation coefficient represents the slope of the best-fitting line in z score units. The proof would involve using theorems from an advanced topic in mathematics called vector theory. But, as we shall see, the fact not only provides insight into the meaning of the correlation coefficient but also becomes useful to us in deriving more practical facts.

17.3 The Best-Fitting Line in the Scattergram of Original Data

We shall begin this time by supposing that we have in our possession the z score scattergram for the data described. That is, we have taken the trouble to convert the students' high school grades and their university freshman grades into z scores and plotted them as in Fig. 17.3. We shall say further that the correlation between the two variables is .40, which means that the slope of the best-fitting line is .40, or 2/5. This best-fitting straight line is also indicated in Fig. 17.6.

Fig. 17.6

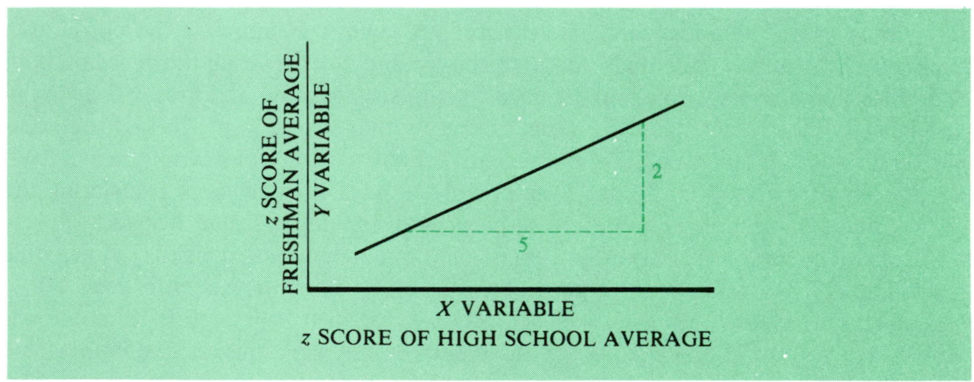

As mentioned, a scattergram for the two variables in their original forms might also be constructed. We have presented this scattergram for our data in Fig. 17.1. In that illustration, high school averages in their original form are plotted along the X axis and are the X variable. The university freshman averages are plotted along the Y axis and are the Y variable. As opposed to the z score scattergram, we shall describe a scattergram of this kind as a **raw score scattergram,** since scores in their original form are called raw scores.

We know that in the z score scattergram the best-fitting straight line has a slope, which is the correlation coefficient. Our concern now shall be with using this information in particular to determine what the slope of the best-fitting straight line is in the raw score scattergram. We are about to compare the z score scattergram with the raw score scattergram to solve our problem.

In Fig. 17.3 the *z* scores of the *X* variable are plotted along the horizontal axis. Each *z* score unit represents a certain number of points. In particular, each *z* score unit represents exactly the number of points contained in one standard deviation of this *X* variable. We shall say that the standard deviation of high school averages, which is the *X* variable, is 9 points. Thus the high school average with a *z* score of 1.0, which is exactly one standard deviation above the mean, is 9 points above the mean. The high school average with a *z* score of 2.0 would be 18 points above the mean, and so on. Similarly, we shall say that the standard deviation of the *Y* variable is 10 points, which would mean that each *z* score unit along the vertical axis represents 10 points.

Now look again at the scattergram and the best-fitting line in Fig. 17.3. Remember that in *z* score units the slope of this best-fitting line for the two variables was 2/5. But we have seen that a shift of five *z* score units to the right means a shift of 45 points. Fig. 17.7 shows how five *z* score units on the *X* variable represent 45 points of original *X* scores.

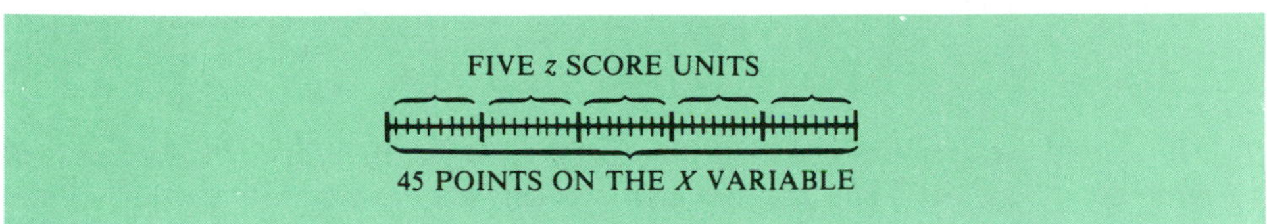

Fig. 17.7

Similarly each change of one *z* score unit on the *Y* variable represents a change of 10 points. Thus an increase of two *z* score units on the *Y* variable means an increase of 20 points in terms of original scores (see Fig. 17.8).

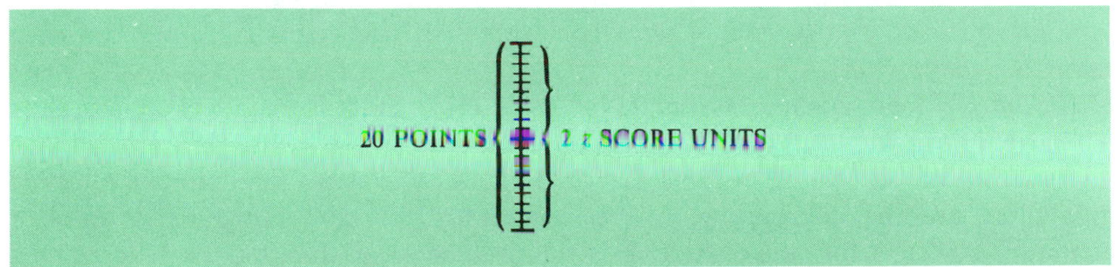

Fig. 17.8

We have supposed that for high school averages and college freshman averages the correlation was .40. Thus the slope of the best-fitting line in terms of *z* scores is 2/5. Now we can interpret our "slope triangle" in terms of original units on each variable. We see that five *z* score units on the *X* variable stand for 45 original units, and that two *z* score units on the *Y* variable stand for 20. Thus, as we see in Fig. 17.9, the slope of the best-fitting line in the raw score scattergram for the two variables described is 20/45.

We have seen how to go from the *z* score slope to the raw score slope. The *z* score slope was the fraction 2/5. The numerator represented *z* score change on the *Y* variable, so we multiplied it by σ_Y. Since $\sigma_Y = 10$, in effect we multiplied 2 by 10 to get our new numerator of 20. The denominator represented *z* score change on the *X* variable, so we multiplied it by σ_X. Since $\sigma_X = 9$, in effect we multiplied 5 by 9 to get our new denominator. To put what we did in its simplest form, we need only to take

Fig. 17.9

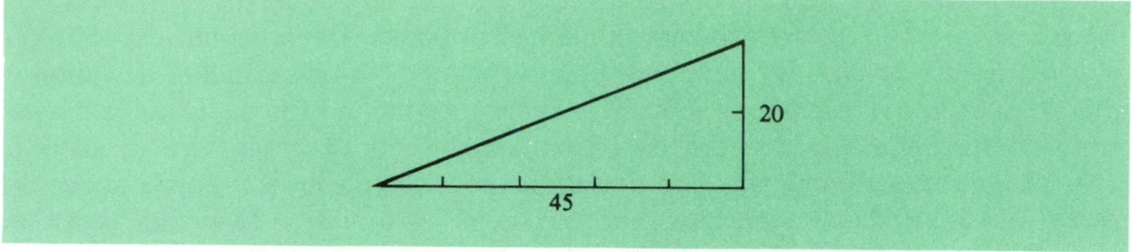

the z score slope and multiply it by σ_Y/σ_X in order to get the raw score slope.

The symbol b shall be used to stand for the slope of the best-fitting line in the raw score scattergram. This symbol will be satisfactory so long as we consistently use the letter X to refer to the variable measured along the horizontal axis and the letter Y to refer to the other variable. To summarize what we have said, the formula for b is as follows:

$$b = r \frac{\sigma_Y}{\sigma_X} \tag{17.1}$$

It should be emphasized that though *r is a straightforward measure of correlation, b is not.* Thus the research worker may indicate the value of the correlation coefficient he has obtained to throw light on how his variables are related, but he may not use his obtained value of b to do the same thing. It should be noted that though r had a lower limit of -1 and an upper limit of $+1$, the same is not true of b. In cases where σ_Y is much larger than σ_X, for instance, even where r is small the value of b may turn out to be considerably larger than 1.0. For instance, suppose $r = .10$, $\sigma_Y = 40$ and $\sigma_X = 2$. Then when we solve for b using Formula (17.1), it turns out that $b = 2.0$.

Actually there is another formula for b that looks more forbidding than Formula (17.1) but does not necessitate the computing of r, σ_Y, or σ_X. To use this formula, one must set up the same kind of table as Table 16.12. The totals that one derives from this table are the appropriate "pieces" not only for computing r but also for computing b. The formula (where X is the predictor variable and Y the predicted variable) is:

$$b = \frac{N\Sigma XY - (\Sigma X)(\Sigma Y)}{N\Sigma X^2 - (\Sigma X)^2} \tag{17.2}$$

To illustrate the use of the formulas just given, we shall look back at the data in Table 16.12 (IQ scores and scores on a musical aptitude test for 20 subjects), which has been reproduced in Table 17.1. In order to solve for b using Formula (17.1) we would first have to compute the values of σ_X and σ_Y. Using Formula (3.2), we find that $\sigma_X = 226.2$ and $\sigma_Y = 161.2$. Thus using Formula (17.1) we get

$$b = (.69)\frac{161.2}{226.2} = .49$$

Table 17.1 IQ Scores (X) and Musical Aptitude Scores (Y)

(1) X	(2) X^2	(3) Y	(4) Y^2	(5) XY
95	9,025	30	900	2,850
106	11,236	37	1,369	3,922
110	12,100	35	1,225	3,850
104	10,816	28	784	2,912
98	9,604	17	289	1,666
114	12,996	45	2,025	5,130
123	15,129	33	1,089	4,059
100	10,000	18	324	1,800
98	9,604	24	576	2,352
88	7,744	18	324	1,584
102	10,404	30	900	3,060
105	11,025	30	900	3,150
111	12,321	32	1,024	3,552
106	11,236	33	1,089	3,498
110	12,100	38	1,444	4,180
143	20,449	42	1,764	6,006
106	11,236	40	1,600	4,240
98	9,604	20	400	1,960
104	10,816	23	529	2,392
97	9,409	25	625	2,425
$\Sigma X = 2{,}118$	$\Sigma X^2 = 226{,}854$	$\Sigma Y = 598$	$\Sigma Y^2 = 19{,}180$	$\Sigma XY = 64{,}588$

Table 17.1 yields the totals that are the various "pieces" that go into Formula (17.2). Using this formula we get:

$$b = \frac{(20)(64{,}588) - (2{,}118)(598)}{(20)(226{,}854) - (2{,}118)^2}$$

$$b = .49$$

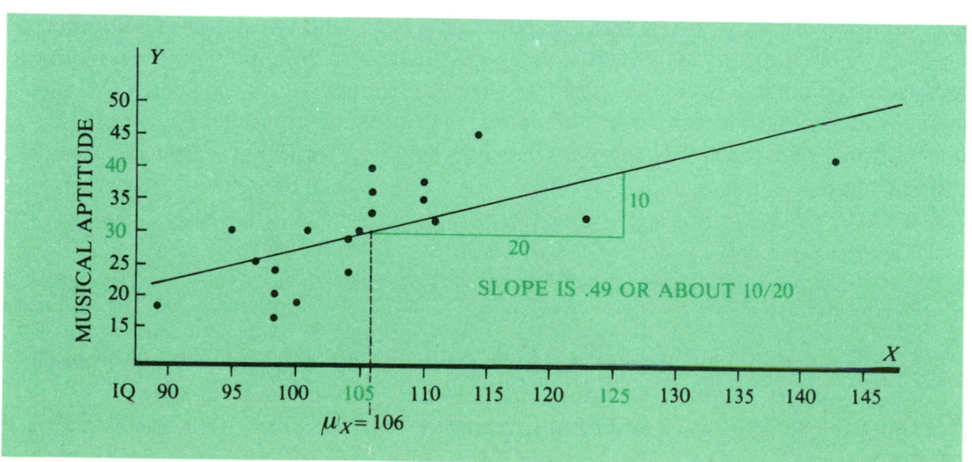

Fig. 17.10

Incidentally, the reader who likes to work with symbols will find it rather easy to show that Formula (17.1) is equivalent to Formula (17.2). Finally, having computed the value of b for our situation, the scattergram of Fig. 17.10 is presented to illustrate its meaning.

17.4 The Best-Fitting Line and Prediction

Note that one needs only to possess raw data, and has no need of a scattergram, to compute the value of b. On the other hand, without having the scattergram in mind, it is unimaginable that one could have proper insight into the meaning of b. In the same way, one need not have drawn a scattergram in order to make predictions in the manner to be described. However, it is virtually necessary for us to go back to the scattergram once more in order to make clear the standard method for making predictions using the best-fitting straight line. Thus we shall go back to the scattergram to develop this method, though we need not actually draw a scattergram if and when we wish to make use of the method.

Let us now consider a new situation and say that we have been giving an engineering aptitude test to a large population of students at the outset of their college careers as engineering majors. We have been recording these students' test scores and also their final college averages. In other words, each student's aptitude-test score and final college average as an engineering major have been recorded. Our purpose is ultimately to use the aptitude-test scores to predict how students will fare so that we can make certain decisions regarding them as they enter.

In this situation the test scores obviously compose the predictor variable. In other words, we shall refer to the test scores as values of the X variable and record them along the horizontal axis. The students' engineering averages are values of the predicted variable. These values shall constitute our Y variable and shall be recorded along the vertical axis.

Suppose that for the population of students that we have tested the following facts are true. On the aptitude test the mean score is 80 and the standard deviation is 8 points. The students' engineering averages have a mean of 75 and a standard deviation of 6 points. The correlation between the students' aptitude scores and their engineering averages is .80. From the data given we can compute by Formula (17.1) that the value of b is .60. We shall put these findings in a form that enables us to refer to them readily.

$$\mu_X = 80 \qquad \mu_Y = 75 \qquad r = .80$$
$$\sigma_X = 8 \qquad \sigma_Y = 6 \qquad b = .60$$

Now we are going to look at the scattergram for the data described in terms of the original scores and the best-fitting straight line. This scattergram appears in Fig. 17.11 and the slope of the best-fitting straight line is .60, as indicated. Note that the values $X = 80$ and $Y = 75$ have been labeled. A horizontal line passing through the point on the vertical axis where $Y = \mu_Y$ has been drawn covering all points in the scattergram at which $Y = 75$. Similarly, a vertical line passes through all points at which $X = \mu_X$. Thus the horizontal line passes through all points at which the Y variable is at its mean, and the vertical line passes through all points at which the X variable is at its mean.

Fig. 17.11

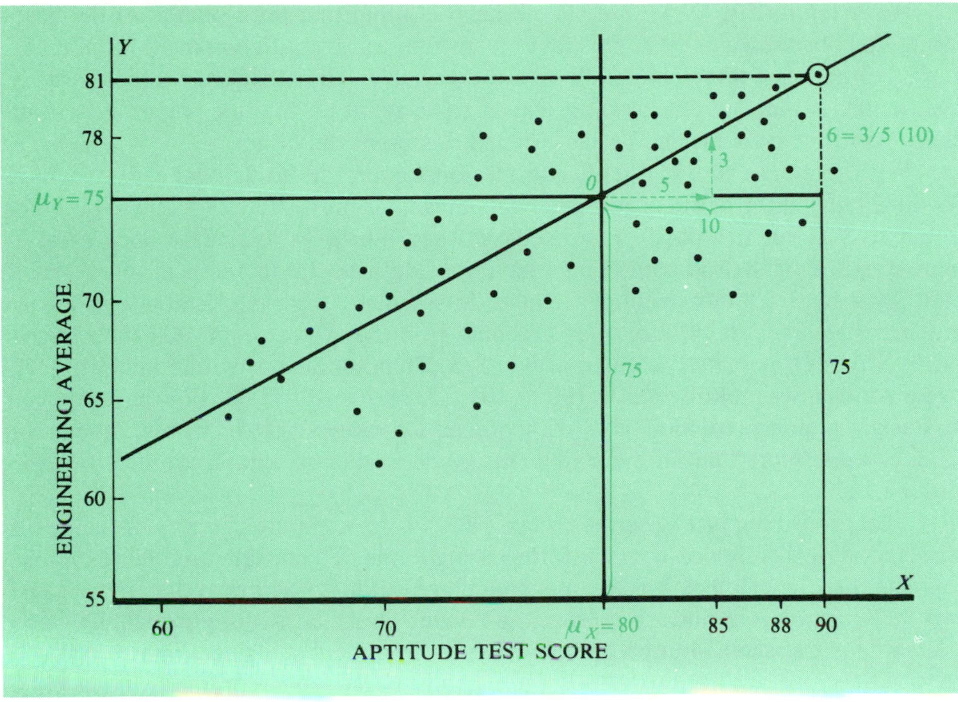

Most important is the fact that *these two lines representing the means of the two variables and the best-fitting straight line intersect at a single point*. We shall call this point the origin. It is marked O in Fig. 17.11. It is always the case that all three lines—the regression line and the lines indicating the means of the two variables—coincide at a single point.

We shall assume that the data recorded in Fig. 17.11 are typical of what we may expect for students in the future. As each new entering freshman receives a score on our aptitude test, our task will be to predict what this student's ultimate engineering average will be.

So far we are merely referring to the scattergram. As explained, we make our prediction in each case by locating the student's test score on the X axis and then finding the point on the best-fitting line directly above it. For instance, suppose a student gets a 90 on our test. To make a prediction of what the student's engineering average will be, we look directly upward to the point on the best-fitting line over the value of 90 on the horizontal axis. This point is encircled. Having found this point, we now look over at the Y axis where we see that it has a Y value of 81. Thus our prediction is that this student's college engineering average will be 81.

Our procedure is identical in each case. Given a student's score on the X variable, we simply refer to the point on the best-fitting line directly over it. The height of this point becomes our prediction of the student's score on the Y variable. Now we are going to substitute for this procedure a numerical way of doing the same thing. In other words, given any X value (for example, 90), there is a numerical way of determining the corresponding Y value on the best-fitting line. We shall make use of the scattergram to explain this procedure.

Referring to Fig. 17.11, we should keep in mind that the slope, b, of the best-fitting line has the value 3/5. This fact means that for every five-point shift on the X variable, there is a three-point shift on the Y variable. The other crucial fact, already mentioned, is that there necessarily exists some point on the line where X is at its mean and Y is at its mean. We have called this point the origin.

With reference to Fig. 17.11, consider once more the student whose X score is 90. Beginning at the origin, where X is at its mean, we must shift 10 points to the right to get to 90. But the slope of 3/5 tells us that, whatever amount X increases, Y increases 3/5 of that amount. Thus, having calculated the increase in X, we must multiply it by 3/5 to get the increase in Y. In essence, as far as X is above its mean, the increase in Y is 3/5 of that amount. For the X value of 90, the shift is $(90 - 80) = 10$ points. The value of b is 3/5. Thus, to determine the increase in Y over its mean, we multiply 3/5 by $(90 - 80)$. $(3/5)(90 - 80) = 6$. For the individual in question, our prediction is a Y score at $\mu_Y + 6$. In other words, it will be $75 + 6 = 81$. Note that all the statements made in this paragraph are illustrated in Fig. 17.11.

The procedure just described may be used to compute for any X value its corresponding Y value on the best-fitting straight line. To consider another example, suppose that the value of X is 88, meaning that a student's score on the aptitude test was 88. Our task is to find the appropriate Y value on the best-fitting straight line, and this value is to become our prediction of what the student's engineering average will be.

This time it is from the X score of 88 that we subtract 80 (see Fig. 17.12). The difference, 8, tells us that the X score represents a shift of 8 points. Next we multiply the slope, b, by this shift to determine how much Y has ascended above its mean. Since $b = 3/5$ we multiply 3/5 by $(88 - 80)$. Our product, 4.8, represents the shift of Y above its mean. To the mean of Y, which is 75, we add this increase of 4.8 points and get 79.8.

Thus, where X is 88, the corresponding Y value on the best-fitting straight line is 79.8. In other words, for the student whose test score is 88, we predict that his or her engineering average will be 79.8, as we would have had we been able to make a perfect visual reading of the appropriate point on the best-fitting line. Note that in Fig. 17.12, where $X = 88$, the height of the line at Point A is 79.8.

In each instance thus far we have started out with a supposedly obtained value of X and have computed the Y value, or height, of the best-fitting line at the appropriate point. This Y value, which is always a point on the best-fitting line, is in each case our prediction. The usual symbol for a prediction of this kind is $\tilde{Y}$, which we read "predicted Y."[1]

[1]The mark above the Y is called a "tilde." $\tilde{Y}$ is sometimes called "Y tilde" as well as "predicted Y."

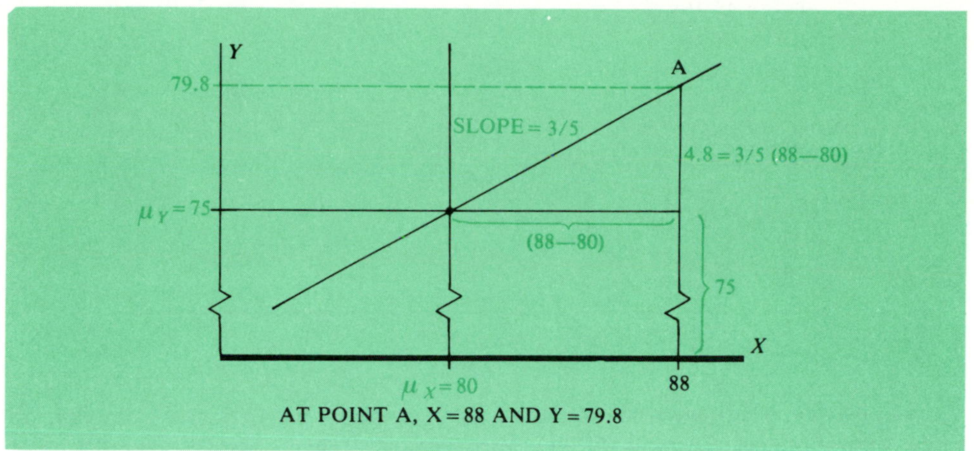

Fig. 17.12

We have pointed out that, given any value of X, there is the implication that the appropriate $\widetilde{Y}$ value is a distance of $b\ (X - \mu_X)$ from the mean of the Y values. Thus at any value of X the height of the line is $\mu_Y + b\ (X - \mu_X)$ (see Fig. 17.13). We have seen that for any value of X we make the prediction $\widetilde{Y}$. Now we see that given any value of X, the appropriate value of $\widetilde{Y}$ is given by Formula (17.3).

$$\widetilde{Y} = \mu_Y + b(X - \mu_X) \tag{17.3}$$

Formula (17.3) holds even if the given value of X is less than μ_X. In this case, the "shift" on the X variable is negative and reflects in the prediction. The value of $\widetilde{Y}$ yielded by Formula (17.3) is still the appropriate value on the line. Look back at Fig. 17.11. Suppose that a student comes along whose test score was 60. Where X is 60, our prediction from the scattergram is that the student's $\widetilde{Y}$ score will be about 63. Applying Formula (17.3) we get

$$\begin{aligned}\widetilde{Y} &= 75 + \tfrac{3}{5}(60 - 80)\\ &= 75 - 12\\ &= 63\end{aligned}$$

This time the value $(X - \mu_X)$ turned out to be negative, indicating that the shift in X from its mean was negative or to the left. As a consequence the formula led us to take the value μ_Y and diminish it rather than add to it.

Here is a good place to stop and note that the major goal we set forth at the beginning of this chapter has been realized. We now have a formula, (17.3), which allows us to make predictions without reference to a scattergram. Given paired data in a population we can easily determine μ_Y, μ_X, and b. Once these values have replaced the proper symbols in Formula (17.3), the prediction formula is complete and ready to use for making predictions.

Fig. 17.13

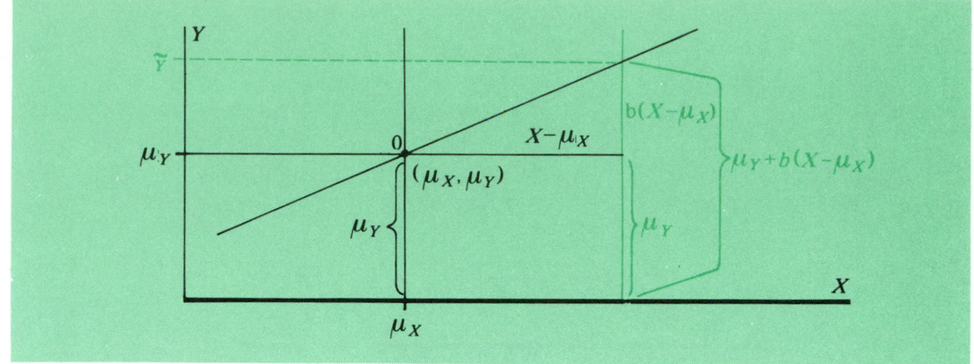

In practice, where we wish to make a vast number of predictions we can first do some simplifying of Formula (17.3). First we note that $\mu_Y = 75$, $b = 3/5$, and $\mu_X = 80$. It is only the value of X that will be new each time when we come to make predictions. Therefore we may write Formula (17.3) once more and then put in the values that we already have:

$$\begin{aligned} \widetilde{Y} &= \mu_Y + b(X - \mu_X) \\ &= 75 + \tfrac{3}{5}(X - 80) = 75 + \tfrac{3}{5}X - 48 \\ &= 27 + \tfrac{3}{5}X \end{aligned}$$

The formula above for $\widetilde{Y}$ in its final form would undoubtedly be simpler to work with than the original formula, especially if we had many successive predictions to make. Therefore, a simplification of the kind just done is often carried out, especially where many predictions are to be made. However, Formula (17.3) remains the crucial one so far as understanding is concerned, and this last simplification is not absolutely necessary.

Formula (17.3) is sometimes called a **prediction equation** and sometimes a **regression equation**. It is extremely important. It enables one to do arithmetically with perfect precision what one might try to do visually with a scattergram. That is, it is aimed at enabling a worker to take scores on one variable and to make rapid-fire predictions of scores on another. One assumption of course is that a definite linear relationship has been discovered. Another is that the facts, such as the correlation coefficient, derived from the previous collection of cases shall be largely the same for the new cases about which predictions are to be made.

One more illustration is in order to make clear how Formula (17.3) is used. A large number of U.S. Army lieutenants were given a leadership ability scale on which they received scores. The top score possible was 100. They were also given scores on their leadership ability after having served one year. Here the top score possible was 25. The test scores of course shall be said to constitute the X variable and the leadership scores the Y variable. The findings for this group were as follows:

$$\mu_X = 74 \qquad \mu_Y = 14 \qquad r = .48$$
$$\sigma_X = 8 \qquad \sigma_Y = 3 \qquad b = .18$$

Now assuming that what was true for this collection of subjects will be largely true for others like them, we may wish to make some predictions for future cases. Suppose in particular that a subject gets 70 on the leadership ability scale. We want to predict what his or her leadership ability score will be. To proceed, we solve Formula (17.3).

$$\begin{aligned} \widetilde{Y} &= \mu_Y + b(X - \mu_X) \\ &= 14 + .18(70 - 74) \\ &= 13.28 \end{aligned}$$

The reader should verify that, for the individual whose test score is 90, our prediction is that his or her leadership ability score will be 16.88. For the individual whose test score is exactly 74, our prediction is a leadership ability score of exactly 14. It is always true, when we are making linear predictions, that *for subjects whose scores were exactly the mean on one variable, we end by predicting that their scores will be exactly the mean on the other.*

17.5 Correlation and the Accuracy of Linear Prediction

An obvious question is how much advantage we gain in the long run by making predictions from the best-fitting straight line. The answer is that the advantage that we gain in accuracy depends completely on the strength of the correlation. Where the correlation is trivial, then it hardly pays to use the best-fitting straight line to make predictions. We might as well use a blind guess (μ_Y) and save the trouble of determining the prediction formula. In fact, when the correlation coefficient is zero, the best-fitting line is the line $\widetilde{Y} = \mu_Y$. But the stronger the correlation, the more precise the best-fitting line is in coming close to all the dots. Thus the stronger the correlation, the more it pays to use the best-fitting line for predictive purposes. As we shall see, when the correlation coefficient indicates perfect correlation ($r = 1$ or $r = -1$) the regression curve and the best-fitting line are one and the same.

There is a specific relationship between the magnitude of the correlation coefficient (and its sign is not relevant here) and the degree to which our enlightened predictions are improvements in accuracy over using the best blind prediction, μ_Y. Remember that with no information the best predictive device is to ascertain the mean of the Y variable and to use this value as the prediction for each individual case. Let us consider, for instance, the situation where the X variable and the Y variable have a correlation of .60. The only facts we need are that the mean of the Y variable is 44 and the standard deviation of the Y variable is 10. In sum, $r = .60$, $\mu_Y = 44$, and $\sigma_Y = 10$.

Fig. 17.14A

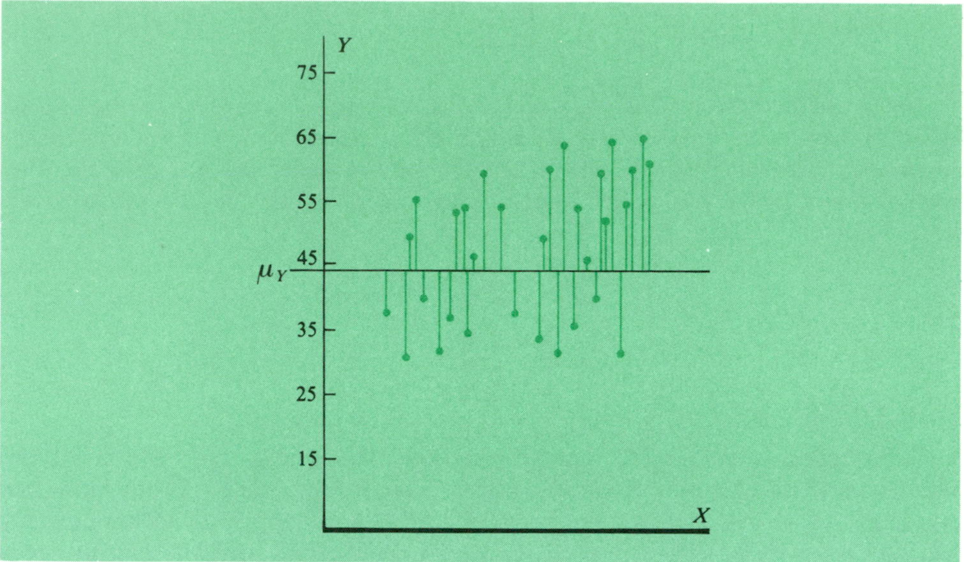

The scattergram for this situation has been drawn in Fig. 17.14A, where μ_Y is represented by the horizontal line at the height of 44. With no information concerning a relationship, it is this value 44 that would be the best one to use as the prediction for each new case. As mentioned, it is most efficient to use μ_Y as one's guess concerning each new case when one has no outside information. This guess would of course lead to errors, indicated in Fig. 17.14A. As mentioned, we are saying that the variance of the population of Y terms is 100 ($\sigma_Y^2 = 100$). When we guess the mean each time, this value becomes the variance of our errors.

In Fig. 17.14B the scattergram has been reproduced with the best-fitting line drawn in. It is based on the correlation of .60, which we supposed. Once again using this line for predictions, there would be errors, which would have an error variance. The question is: Based on the correlation of .60, how much did we reduce the error variance by shifting from the use of μ_Y as a predictor to the use of values on the best-fitting line? Specifically, we said that the error variance in Fig. 17.14A is 100. By how much do we reduce this error variance when we make predictions using the best-fitting straight line, as in Fig. 17.14B?

Remember that $r = .60$. It may be shown that what one must do is to square this value. $r^2 = (.60)^2 = .36$. The value of $r^2 = .36$ tells us that when we use the best-fitting straight line, the variance of our errors will be less than the original variance by .36 of that original variance. Here the error variance in Fig. 17.14A was 100. *The fact that $r^2 = .36$ tells us that 36 percent of this variance no longer exists* when we predict as in Fig. 17.14B. Of the original 100 units, there remain only 64. Using the best-fitting line as in Fig. 17.14B, the predictions have an error variance of 64.

The value of r^2 is sometimes described as indicating the **proportion of variance accounted for by a correlation.** For the example involving the engineering students whose final grade averages and aptitude test scores were correlated at .8, this means that 64% ($r^2 = .8^2 = .64$) of the variance in the final grades can be explained by

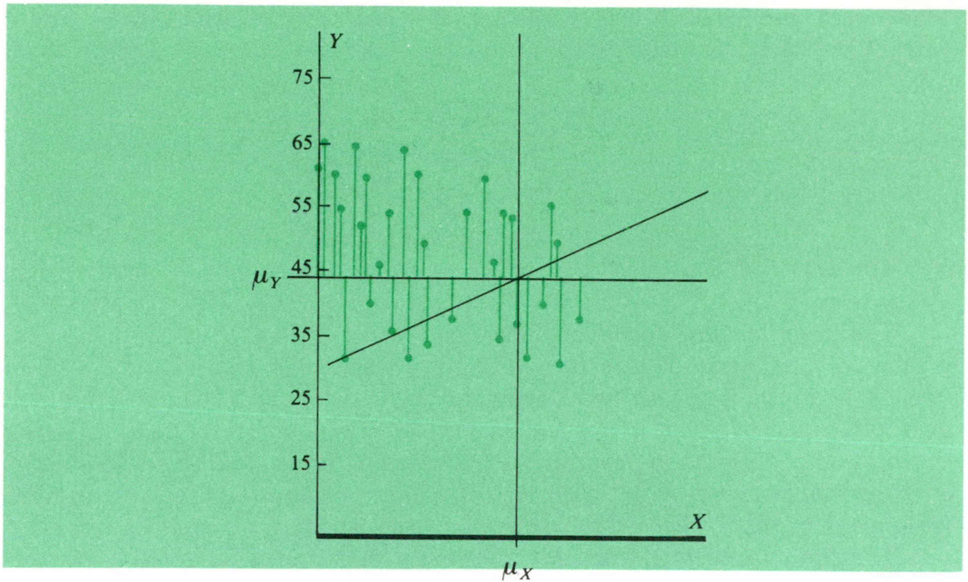

Fig. 17.14B

differences in aptitude. As mentioned, the error variance when we use the mean as a repeated prediction (as in Fig. 17.14A) is σ_Y^2. What remains of the error variance when we use knowledge of an X variable to predict is designated by $\sigma_{Y \cdot X}^2$. The subscript is to indicate that the variance is of the Y variable, but that predictions are based on knowledge of the X variable. The value of $\sigma_{Y \cdot X}^2$ in 17.14B is 64.

Thus the original error variance (as in Fig. 17.14A) is σ_Y^2. The proportion of this variance eliminated by using the best-fitting line is r^2 and the actual amount of original variance eliminated is $r^2\sigma_Y^2$. Thus, using the best-fitting straight line, we find that the remaining error variance is given by

$\sigma_{Y \cdot X}{}^2$	$=$	σ_Y^2	$-$	$r^2\sigma_Y^2$	
error variance using the best-fitting line		original error variance		the amount eliminated	(17.4)

Suppose that in some community the mean high school graduating mark is 82 and the variance of marks is 80. If we make a prediction for each entering student, knowing nothing about the person, our best approach is to predict that the student will get an 82. By this procedure, our errors will have a variance equal to the variance of the marks themselves. The error variance is 80.

$$\sigma_Y^2 = 80$$

Now suppose we know each student's IQ score upon entering. IQ scores on our test correlate .6 with graduating marks. This time, for each student we make a special prediction based on the student's IQ score. Our predictions are better in the long run.

How much better? $r^2\sigma_Y^2$ is the amount of variance eliminated. Here, $r^2\sigma_Y^2 = (.6)^2(80) = 28.80$. Our new error variance $\sigma_{Y \cdot X}{}^2$ is less than the old one by 28.80.

$$\sigma_{Y \cdot X}{}^2 = \sigma_Y^2 - r^2\sigma_Y^2$$

$$\sigma_{Y \cdot X}{}^2 = 80.00 - 28.80$$

$$\sigma_{Y \cdot X}{}^2 = 51.20$$

The name for the square root of the error variance is the **standard error of estimate** $\sigma_{Y \cdot X}$.

The major fact that should be understood is that the higher the correlation, the greater the proportion of reduction in error variance when we use the best-fitting straight line. The following two statements are easily verified by substituting the appropriate values for r into Formula (17.4). *When the correlation is perfect (meaning that r is either + 1.0 or − 1.0), then using the best-fitting line leads to an error variance of zero. Also, when the correlation is zero, then the original error variance is not at all reduced by using knowledge of the X variable.*

The first statement means that knowledge of scores of a variable perfectly correlated to the Y variable leads to predictions that are without error. The second statement means that knowledge of scores of an uncorrelated variable does not improve predictions made by use of the method described. In fact, when the X variable is uncorrelated with the Y variable, then, even when X scores are known, the best-fitting straight line turns out to be the horizontal line as drawn in Fig. 17.14A.

Prediction	Error Variance
μ_Y (blind guess)	σ_Y^2
Best-fitting line	$\sigma_Y^2 - r^2\sigma_Y^2$
Regression curve	Less than or equal to $\sigma_Y^2 - r^2\sigma_Y^2$

Problem Set A

17.1 Suppose you have a scattergram in which the variable measured on the horizontal axis is gallons of gas consumed and the variable measured on the vertical axis is miles traveled. Suppose also that all points on the diagram lie on a straight line. There are 20 dots and each dot represents the values the variables assumed for a specific automobile.

a. Two of the cars represented used 3 and 4.5 gallons of gas and traveled 60 and 90 miles, respectively. Draw the scattergram and represent these two cars.
b. Draw in the line on which all the points lie and find its slope.
c. What does the slope mean in terms of gallons and miles?
d. If the 20 values for each variable are converted to z scores and plotted on a z score scattergram, what would the slope of the line on which the dots lie be? Why?

e. Assuming the values assumed by the X variable have a standard deviation of .5, find σ_Y.

17.2 If the correlation coefficient between two variables is .58, the variance for the X values is 25, and the variance for the Y values is 16, what is the slope of the best-fitting straight line to the raw score scattergram?

17.3 The relationship in a certain North American city between the number of homes built in a year (X) and the mean percent appreciation per home for that year (Y) has remained the same for some time. The figures for the past five years are available as follows:

X	114	87	93	74	50
Y	14	12	10	9	7

a. Obtain the prediction formula for $\widetilde{Y}$

b. This year 100 new homes were built. Predict the mean appreciation rate per home.

c. If the number of new homes built is unknown, what prediction should be made as to mean appreciation?

17.4 The number of cigarettes a pregnant woman smokes per day (X) and the weight of her baby (Y) have been found to be correlated at $r = -.83$. This coefficient was arrived at by studying 1,000 births by mothers who smoked in which $\mu_X = 20$, $\mu_Y = 6.8$, $\sigma_X = 5$, and $\sigma_Y = .7$.

a. Use Formulas (17.1) and (17.3) to obtain the prediction formula for $\widetilde{Y}$.

b. Predict the weights of babies born to mothers who smoke 11, 18, and 35 cigarettes per day, respectively.

c. What percent of the variance in birth weights can be accounted for by correlation (explained by differences in the numbers of cigarettes smoked)?

17.5 a. What is the significance of the slopes of the best-fitting straight lines to the z score and raw score scattergrams? Are the slopes of the two lines related?

b. What distinguishes the best-fitting straight line to a scattergram from other lines?

17.6 A correlation of $-.32$ has been found to exist between the proportion of farmland (X) sown with wheat in five selected wheat-producing areas in North America during June and the price of bread (Y) in December. Furthermore, $\mu_X = .62$, $\mu_Y = .74$, $\sigma_X = .13$, and $\sigma_Y = .15$. The proportion of farmland devoted to wheat in the areas has been declining while the price of bread has continued to climb. This June only half of the available crop land was devoted to wheat.

a. Predict the price of bread next December.

b. Use Formula (17.4) to calculate the error variance associated with numerous predictions.

17.7 Underground miners of a certain mineral have long thought their occupation to be a health hazard. Health insurance officials gather the following data in hopes of predicting the number of days a given miner will be sick each year.

X (*number of years worked in the mines*)	Y (*number of sick days last year*)
14.0	11.5
9.5	11.5
15.5	12.0
12.5	10.0
18.5	15.0

a. Obtain the prediction formula for $\tilde{Y}$.
b. Predict the number of days a miner with the following number of years of experience in the mines will miss in a year due to illness: 10, 8.25, 15, 12.5.
c. Suppose a miner with 10 years of experience misses 13 days due to illness. What error is made if the prediction equation is used?
d. Calculate the error variance for prediction using the best-fitting straight line.

17.8 In the belief that extreme temperature changes are related to emotional well-being, a leading psychologist, Dr. I. M. Wells, obtains the following data from 10 doctors across the country who have well-established general practices. X represents the standard deviation of the temperature of the city in which the doctor is located. Y represents the average number of patients treated per year by the doctor for nervous or emotional disorders.

X	6.0	7.8	6.5	8.7	7.4	7.0	7.8	9.5	8.8	9.0
Y	66	70	75	80	82	85	90	92	95	98

a. Obtain the prediction formula for $\tilde{Y}$.
b. Predict the number of patients who would seek help for nervous or emotional problems from a well-established physician in a city whose temperature varies with a standard deviation of 10 degrees.
c. State the proportion of variance in Y that is accounted for by the correlation between X and Y.
d. Suggest other factors that might account for the variance in Y.

17.9 The lengths of sentences (Y) given out by Judge Just in 15 cases of a particular crime are presented below along with the ages of the offenders (X). All values are given in months.

X	Y	X	Y
311	57	275	65
223	59	237	74
282	57	242	58
362	79	245	59
357	67	331	64
233	60	326	62
284	70	247	66
265	64		

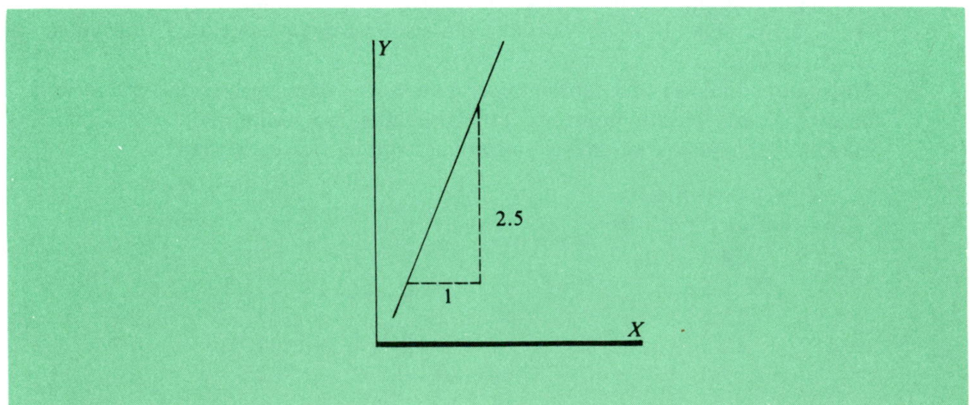

a. What sentence should a 28-year-old expect if convicted of the crime in question in Judge Just's court?
b. Compare the accuracy of the prediction formula in part a with the accuracy of the prediction formula in Problem 17.4.

17.10 Is the line in the scattergram at the bottom of page 384 the best-fitting line for a z score or raw score scattergram?

Problem Set B

17.11 Find the prediction equation for $\widetilde{Y}$ given the following: $\Sigma X = 47$, $\Sigma Y = 17$, $\Sigma X^2 = 473$, $\Sigma Y^2 = 52$, $\Sigma XY = 155$, $N = 8$.

17.12 Assuming the correlation between two variables is $-.43$, $\sigma_Y = .04$, and $\sigma_X = .13$, find the slope of the best-fitting straight line to (a) the z score scattergram, and (b) the raw score scattergram.

17.13 Presented below are pairs of measures obtained from five private universities. X indicates the mean annual cost to the school (in hundreds of dollars) of educating each student and Y indicates the percent of the educational costs that must be raised through methods other than tuition.

X	86	84	85	80	75
Y	72	67	65	60	56

a. Obtain the prediction formula for $\widetilde{Y}$.
b. Draw a scattergram and its best-fitting straight line. What is the slope of the line?
c. If the cost of educating a student rises to $9,000 per year, what part of the educational budget must be raised aside from tuition?
d. What proportion of the variance in Y (the percent of costs not borne by students) is accounted for by the variance in the cost to educate each student.

17.14 The correlation between the number of 16–18-year olds who drop out of school and the number of unemployed in a certain locality is .79. The number of dropouts at any time has a mean of 150,000 with a standard deviation of 10,000 while the number of unemployed has a mean of 900,000 and a standard deviation of 25,000.
a. Find the prediction formula for the number of unemployed, given the number of dropouts.
b. Predict the number of unemployed if there are 165,000 dropouts.

17.15 If the slope of the best-fitting straight line to the raw score scattergram is .60, $\sigma_Y^2 = 49$, and $\sigma_X^2 = 36$, what is the correlation coefficient r that describes the linear relationship between X and Y?

17.16 The rating given a teacher by a student is related to many things, among them the student's ability to perform in class. The evaluations given Professor Rerun by 10 students (Y) are presented below along with the class average for each student (X).

X	35	63	69	70	74	70	62	84	97	89
Y	0	49	59	63	75	75	77	79	85	97

Predict how students with the following averages would rate Professor Rerun.

(a) 65 (b) 75 (c) 85

17.17 The mean malpractice insurance premiums (X) in hundreds of dollars paid by physicians at 12 hospitals are given below along with the mean daily charges for a certain type of room (Y).

X	Y	X	Y
32.00	87.50	38.00	107.00
32.50	87.50	40.70	98.50
35.40	115.00	41.80	99.00
35.50	93.50	43.00	103.00
36.70	95.00	43.80	123.00
38.00	92.00	45.00	115.00

a. Obtain the prediction formula for $\widetilde{Y}$.
b. If the mean premium paid at Mercy Hospital is \$58.43, what would you predict the cost of a room to be?
c. What proportion of the variance in room rates can be explained by the scatter among malpractice insurance rates?
d. Calculate the error variance when using the prediction formula repeatedly.

17.18 Mr. Peters, president of Backgrounds, Inc., a large manufacturer of wall coverings, uses the regional unemployment rate (X) to predict the number of jobs that will go unfilled at his factory (Y). Records of the 12 previous quarters yield the following values.

X	Y	X	Y
5.7	107	6.3	96
5.9	123	7.0	77
6.5	84	6.9	79
6.3	89	7.0	73
7.3	61	7.1	50
8.2	60	10.0	57

a. Obtain the prediction formula for $\widetilde{Y}$.
b. Predict the number of unfilled jobs when the unemployment rate is 6%, 11%, and 7.5%.
c. Predict the number of unfilled jobs if the unemployment rate is unavailable.
d. Compare the accuracy of the predictions in parts b and c.

17.19 a. When $r = 0$, what is the best-fitting straight line?
b. When $r = -1$ or $r = +1$, what is the error variance for predicting Y based on a knowledge of X?
c. What can be said about the signs of r and b?

17.20 Show that Formulas (17.1) and (17.2) are equivalent.

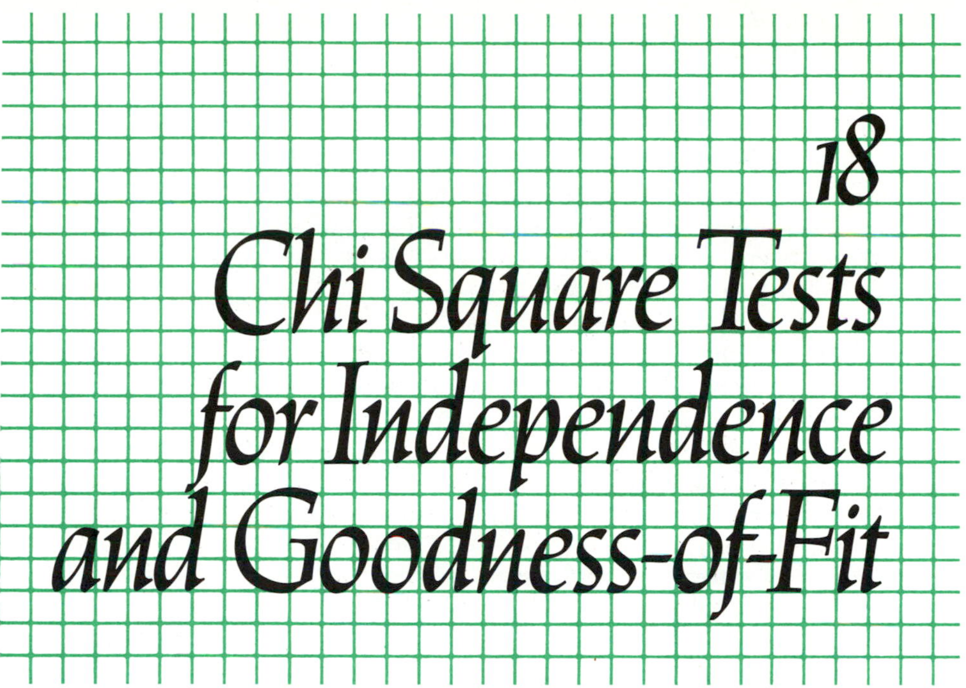

18
Chi Square Tests
for Independence
and Goodness-of-Fit

18.1 Introduction

The last few chapters have dealt with a very special kind of relationship that may or may not exist between two variables. That relationship, correlation, indicates a tendency for one variable either to increase or to decrease depending on the behavior of the other. The methods considered in dealing with correlation are most appropriate for variables that are *continuous* or that are viewed as continuous.

The first sections of this chapter also address the concept of relationship. Therefore, it will usually be necessary to identify two variables of interest in each population or problem situation. However, we now will deal primarily with approaches to problems involving *discrete* variables in which the values of the variables are given as *categories*. We shall see where, as in the past, our work will involve prediction. That is, knowing the value that one variable assumes will often be of great assistance in predicting the value that the second variable will assume. Our primary goal, however, will be the detection of relationships between discrete variables. As in our previous work, we shall consider both descriptive and inferential situations.

The latter discussions in this chapter again focus on the values assumed by a single variable. The variables, nonetheless, must remain categorical in nature. The methods discussed in this and the final chapter will greatly increase the number of hypotheses we will be able to consider statistically.

18.2 Independence

We shall begin by illustrating and defining the situation of unrelatedness or **independence** between two discrete variables. When we have understood the meaning of independence, the meaning of a relationship and some of its implications will become evident. Note that we have already considered one kind of independence; namely, that between two or more observations. Now we are considering independence in another form—that is, the *independence of variables*.

As usual our starting step is to specify a population of interest. For simplicity, say that Population A consists of 12 people and that we have scored each person on two variables, gender and political-party preference as indicated in the last election. The variable, gender, may assume either of two values, male and female. We shall say that the variable, party preference, must assume either the value Republican or the value Democrat. Table 18.1 summarizes the findings for the 12 people.

Table 18.1 Gender and Party Preference—Population A

	Republican	Democrat
Male	6	2
Female	3	1

We now pick one of the two variables at will and use its different values to break down the population into subpopulations. Specifically, we shall call the variable, gender, our primary variable. Each value of this variable characterizes a subpopulation and we are about to look at these subpopulations one by one.

First, we shall consider only the cases in which our primary variable assumes the value *male*. This means that we shall look at only the first row of Table 18.1. This row shows the eight cases having the value male. We see that when our primary variable assumes this value, the other variable, party preference, assumes the value Republican 75% of the time and the value Democrat 25% of the time. In other words, when we are considering only males, we find that 75% are Republicans and that 25% are Democrats. *The distribution of voting preferences among the males is such that 75% are Republicans and 25% are Democrats*. (We are going to use the word distribution repeatedly to refer to the way that terms distribute themselves without regard to how many there are in the subpopulation.)[1]

Now we shall let our primary variable assume the value *female*. This means that we shall look at only the second row of Table 18.1, which shows the four cases having the value female. We see that when our primary variable assumes this value, the other variable, party preference, assumes the value Republican 75% of the time and the value Democrat 25% of the time. But this is *exactly the same* situation that we had when the primary variable assumed the value male. Thus at each value of the variable, gender, the distribution of political preferences is the same. We therefore

[1]Thus a population of eight Republicans and two Democrats has the same distribution as a population of 80 Republicans and 20 Democrats. The percentages at the different values are the same.

say that gender and political-party preference are unrelated (or *independent*) in Population A. This leads us to make the following definition:

> **Definition 18.1 A. Two variables are called unrelated or independent in a population if at every value of one of them the distribution of the other is the same as at any other value. B. Two variables are said to be related when it is not true that at every value of one of them the distribution of the other is the same as at any other value.**

18.3 Illustrations of Relatedness

We shall now look at two illustrations of relatedness involving the same variables as before. Suppose that once again we record the gender and political-party preference of each of 12 people and they are as indicated in Table 18.2. Remember that this is a *population* that we are using for illustrative purposes.

Table 18.2 Gender and Party Preference—Population B

	Republican	Democrat	Totals
Male	7	1	8
Female	1	3	4
Totals	8	4	12

As before we arbitrarily pick the variable, gender, as the primary variable. We focus on those cases for which that variable assumes the value male. Among the males, seven out of eight are Republicans. In other words, when the variable, gender, assumes the value male, the distribution is such that 87.5% of the subpopulation are Republicans and 12.5% are Democrats.

However, when the primary variable assumes the value female, the distribution is such that 75% (three out of four) of the subpopulation are Democrats. Thus, at different values of the primary variable, the distribution of voting preferences *differs*. In Population B, the variables, gender and political preference are related. Note that among the males the Republicans are the majority, whereas among the females the Democrats are the majority.

Table 18.3 Gender and Party Preference—Population C

	Republican	Democrat	Totals
Male	1	7	8
Female	1	3	4
Totals	2	10	12

Finally, consider Population C (Table 18.3). In this population, when the variable, gender, assumes the value male, 87.5% of the members are Democrats and 12.5% are Republicans. When it assumes the value female, 75% of the members are Democrats and 25% are Republicans. Once again the variables, gender and party preference, are related. When we go from one value of the primary variable to another, the distribution of the other variable changes. Note that in Population C both subgroups are predominantly Democrats. Yet the majority is greater among the males than among the females.

The implication of relatedness for making predictions is important. Once again we shall start by considering the situations of unrelatedness. Suppose we are told that a person has been chosen at random from Population A. We are now asked to make a predictive statement concerning his or her party preference. Since three-quarters of the people in Population A are Republicans, we say that the probability is 3/4 that the person in question is Republican.

Next we are given the added information that the person is male. However, since three-quarters of the males are Republicans, we say again that the probability is 3/4 that the individual is a Republican. The knowledge that the person was female would have given rise to *exactly the same* probability statement. We have been given information concerning a variable unrelated to the variable being predicted. This information has not altered our predictions.

We shall now consider the effect of the same kind of information when the variables are related. This time suppose we are making a prediction about a person chosen from Population B and we know nothing about the person. Since 8/12 of the members of Population B are Republicans (Table 18.2), the statement we make is that the probability is 8/12 that this individual is Republican. As before, suppose we are told in addition that the person in question is a male. Since, among the males in Population B, 7/8 are Republicans, we now say that the probability is 7/8 that the person is a Republican. The information that the person is a female would have given rise to the statement that the probability is 3/4 that the person is a Democrat. The reason is that 3/4 of the females in this population are Democrats. The fact that gender is related to party preference in Population B implies that the gender of an individual, when it is known, influences our probability statement concerning the person's party preference. (Note the assumption that we are making predictions only from Population B.)

Our "blind" prediction for an individual from Population C would be that the probability is 10/12 that the person is a Democrat (Table 18.3). The information that the person is a male gives rise to the statement that the probability is 7/8 that the person is a Democrat. The information that the person is female gives rise to the statement that the probability is 3/4 that the person is a Democrat. In Population C, as in Population B, the gender of an individual when it is known influences our probability statement concerning his or her party preference.

More generally, we have seen that when two variables are independent, knowledge of the obtained value for one variable does not affect our prediction about the value for the other. When two variables are related and the relationship between them is known, knowledge of the obtained value on one variable does affect our prediction concerning the other.[2]

[2]More specifically, relatedness implies that there is at least one value of the unknown variable, which, when it turns up, is capable of altering the prediction we would have made in complete ignorance. In most practical applications, every value of the obtained variable gives rise to a unique set of predictive statements about the other variable.

The important problem is nearly always to determine the existence and nature of the relationships between variables in some vast population, not in some small collection of cases. As usual, we consider our observations as comprising a sample and we end by making inferences about some unseen population. Remember that in the foregoing we have been discussing populations. (Small populations were used for simplicity of discussion.)

We are now ready to consider *inferences* from a sample. The existence of some relationship in an obtained sample in itself tells us little. The reason is that even if two variables, like gender and political-party preference, are independent in some vast population, any particular sample drawn from this population is apt to show *some* relationship between the two variables merely by chance. That is, in any particular sample, it is likely that the proportion of Republicans among the males will not be exactly the same as the proportion of Republicans among the females. Therefore *the important issue is usually whether or not the relationship in the sample is marked enough to suggest that a relationship is actually present in the vast population that yielded the sample*. When the relationship in the sample is marked enough, we shall call it significant and infer that a relationship is also found in the population.

In experimental work, the purpose is usually to make a general statement about what is called an *infinite* population. An infinite population is one that contains so many terms that we can use our knowledge of theoretical distributions (such as the normal distribution) to make inferences about it.

For instance, the purpose of an experiment might be to determine whether or not the variable, gender, is related to the variable "reactivity to a particular drug." Almost certainly some kind of relationship between these two variables would appear in a sample. That is, it would be surprising if in a small sample the males and females reacted in exactly the same way to the drug. The issue, however, is whether or not the two variables are related in the vast population of people from which the sample was derived. In theory, we would even like to make a generalization to people not yet born. Thus we view our observations as comprising a random sample drawn from an infinite population.

18.4 Presentation of Findings in a Contingency Table

We are now ready to consider a method of inferring from a sample whether or not two variables are independent in the population that yielded the sample. The technique is best applied to categorical variables, although it is often applied to discrete variables that assume numerical values and to continuous variables.

To gain focus, we shall consider a specific problem. Suppose that we are again investigating voting preference (Republican or Democrat) among men and women. Accordingly, we obtain a sample of 150 people, of whom 100 are men and 50 are women. Analysis of the data reveals that among the men there are 66 Democrats and 34 Republicans and that among the women there are 27 Democrats and 23 Republicans. The findings appear in Table 18.4, which is called a **contingency table.** A contingency table is defined as a table in which the categories are discrete.

Table 18.4 Contingency Table For Voting Preference

	Democrat		Republican		Totals
Male	66	(1)	34	(2)	100
Female	27	(3)	23	(4)	50
Totals	93		57		150

Note that the two cells in the first row pertain to men and that the two in the second row pertain to women. The two cells in the first column pertain to Democrats and the two in the second column pertain to Republicans.

Each individual cell is in a particular row and a particular column. The large number in each cell is the *frequency* for that cell. Each cell has also been assigned a number in parentheses for purposes of identification. To determine the meaning of the frequency of any given cell, we must look at the title of its row and that of its column. For instance, Cell (1) (in the first row and first column) pertains to men and to Democrats. The interpretation of the number 66 is that there are 66 men who are Democrats in the sample. Similarly, Cell (4) is in the second row and second column. Thus its frequency of 23 means that there are 23 women who are Republicans in the sample.

The numbers in the margins of Table 18.4 are called *marginal totals*. The marginal total of the first column is 93, telling us there are 93 Democrats in the whole sample. The marginal total of the second column tells us that there are 57 Republicans in all. The marginal total of the first row tells us that there are 100 men in the sample, and the marginal total of the second row tells us that there are 50 women. The grand total of members of the sample, 150, appears in the lower right corner.

Let us digress to see whether or not the variables are related in the sample. We shall arbitrarily pick one of them, gender, and see what happens at each of its values. Among the men the proportion of Democrats is 66/100 (which is 33/50). Among the women the proportion of Democrats is 27/50. Thus the distribution of party preference is not identical at each value of the variable, gender. *The variables, gender and party preference, are related in our sample.*

18.5 Expected Frequencies

We shall soon decide by means of what is called the "chi square test" either to reject or not to reject the null hypothesis that the variables are independent *in the population*. This hypothesis of independence in the population gives rise to the expectancy that the variables will be independent in the sample also. The theoretical values that would constitute independence in the sample are called **expected frequencies**.

We must compute the expected frequency for each cell, because it is the number that would have appeared were the two variables absolutely independent in the sample. Once we have computed the expected frequencies for the cells, we may determine how far our obtained frequencies differ from them. The data provide us with marginal totals that in and of themselves tell us nothing concerning whether or not the variables are related. The row totals tell us merely that we have 100 men and 50 women. The column totals tell us merely that the sample contains 93 Democrats and 57 Republicans. Table 18.5 shows these marginal totals in *italics*. The marginal totals are now going to be used to determine the expected frequencies for each of the four cells.

Table 18.5 Expected Frequencies

	Democrat	Republican	Totals
Male	62 (1)	38 (2)	*100*
Female	31 (3)	19 (4)	*50*
Totals	*93*	*57*	150

The proportion of Democrats in the sample is 93/150. *Independence would mean that the proportion of Democrats among the men is identical to the proportion of Democrats among the women*. That is, independence would imply that 93/150 of the women are Democrats and 93/150 of the men are Democrats. There are 100 men in the sample and 93/150 of 100 is 62 $\left(\frac{93}{150}(100) = 62\right)$. Similarly, the expected frequency of Democrats among the 50 women is 31 $\left(\frac{93}{150}(50) = 31\right)$.

By analogous reasoning, 57/150 of the members of the sample are Republicans. Independence implies that 57/150 of the 100 men are Republicans and that the same proportion of the 50 women are Republicans. Thus the expected frequency of the Republicans among the men is 38 and among the women is 19. The expected frequencies just computed have been entered in the cells of Table 18.5. Note that the expected frequencies in any particular row or column add up to the appropriate marginal total. (For instance, in the first row $62 + 38 = 100$.)

In practice, the procedure for finding the expected frequency of a cell is mechanical. Each cell has a corresponding row total (the total of its row) and a corresponding column total (the total of its column). For instance, Cell (3) has a corresponding column total of 93 (since it is in the first column) and a corresponding row total of 50 (since it is in the second row). In effect, **the expected frequency of any cell has been found by multiplying its corresponding row total by its corresponding column total and dividing the product by the grand total for the**

sample. The grand total for our sample is 150. Thus in effect the expected frequency for Cell (3) has been found by multiplying 93 by 50 and dividing the product by 150. Note that we have computed each expected frequency by this procedure.

18.6 The Chi Square Test

We are now ready to test the null hypothesis that the two discrete variables are independent *in the population that yielded our sample.* If we are successful in rejecting this hypothesis, we will be able to accept the alternative that the two variables are related.

Specifically, we shall choose the .05 significance level. We are going to compute the *chi square statistic*, which indicates the discrepancy between our obtained frequencies (Table 18.4) and the expected frequencies (Table 18.5). The size of the chi square statistic will tell us whether to reject or not reject the null hypothesis.

Our procedure will be to compare our obtained chi square value with a value given in a chi square table. We shall reject or fail to reject the null hypothesis depending upon whether the value we obtain is smaller or larger than the appropriate value in the chi square table.

Table 18.6 shows the obtained frequencies and the *expected frequencies in italics* for each of the cells. The chi square statistic, to be computed, will be a measure of the discrepancies between these frequencies.

Table 18.6 Obtained and Expected Frequencies

	Democrat	Republican	Totals
Male	66 *62* (1)	34 *38* (2)	100
Female	27 *31* (3)	23 *19* (4)	50
Totals	93	57	150

Our first step is to list the obtained and the expected frequencies and to compute the difference between them for each cell. The magnitudes of these differences give us an impression of how far the variables depart from independence in the sample. The measure that we compute will be based on the squares of these differences. (Were we simply to add the differences, we would get a total of zero.) The differences and squared differences are shown in Columns (4) and (5) of Table 18.7.

In our example, each of the squared differences happened to have the value of 16. A squared difference, such as 16, in and of itself tells us little. We must consider the squared difference of each cell in proportion to the expected frequency of the same cell. Consequently our next step is to divide the squared difference for each cell by the expected frequency for that cell. The quotients for the different cells are listed in Column (6) of the table (for example, 16/62 = .258; 16/31 = .516; and so on).

The *sum* of the quotients in Column (6) is called the *chi square* statistic, and it is our final measure of departure from independence in the sample. *The larger the chi square statistic, the stronger is the relatedness in the sample.* We see in Table 18.7 that the value of the chi square statistic (denoted by χ^2) yielded by our sample is 2.037. Note that the procedure for computing the chi square value of 2.037 has been described, but it has not been justified here. The proof that it is a meaningful procedure is beyond the scope of this book. Stated symbolically we have

$$\chi^2 = \sum \frac{(O - E)^2}{E} \qquad (18.1)$$

where O is the obtained frequency and E is the expected frequency for the same cell.

Table 18.7 Computation of Chi Square

(1) Cell	(2) Observed Frequency O	(3) Expected Frequency E	(4) Difference $O - E$	(5) Square $(O - E)^2$	(6) $\frac{(O - E)^2}{E}$
1	66	62	4	16	.258
2	34	38	−4	16	.421
3	27	31	−4	16	.516
4	23	19	4	16	.842
					$\Sigma = 2.037 = \chi^2$

Appendix VI gives the values of chi square that would be needed for significance at various levels. However, the reader will note that this table is set up for various numbers of degrees of freedom. Before we can interpret our value of 2.037 correctly, we must determine the number of degrees of freedom that went into our particular chi square statistic. To this end, let us suppose once more that we have computed only the marginal totals from the data. These marginal totals, in *italics* in Table 18.8, provide no information concerning the possible relatedness of the variables.

Table 18.8 Degrees of Freedom in Obtained Frequencies

	Democrat	Republican	Totals
Male	*66* (1)	34 (2)	*100*
Female	27 (3)	23 (4)	*50*
Totals	*93*	*57*	150

Suppose we give ourselves *one* piece of information concerning the relationship of the variables; that is, suppose we allow ourselves knowledge of the obtained frequency in one cell. Specifically, we shall put the value *66* in the first cell. This has been done in *italics* in the table. Now the determination of the obtained frequency of one cell has simultaneously *fixed* the frequencies of each of the other cells. The frequencies of the two cells in the first column must total 93, so the frequency of Cell (3) must be 27 (*93* − *66* = 27). By further subtracting, it follows that the frequency of Cell (2) must be 34 and the frequency of Cell (4) must be 23. The frequency of one cell and the marginal totals (the numbers in *italics*) determine the frequencies for the other three cells.

Thus the totality of information concerning the relationship in the sample is contained in a single cell frequency. The frequency of only one cell can be assigned "freely." That is, given one frequency we are able to determine all the others by making proper subtractions from the marginal totals. It follows that the chi square measure of relatedness that we just computed is based upon one degree of freedom. A general method of enumerating degrees of freedom in chi square problems is discussed in the next section.

Our next step is to compare the chi square value derived from our sample with the appropriate value in the table in Appendix VI. The result of this comparison will determine whether or not we reject the null hypothesis that the variables are independent *in the population.* If the chi square value that we have computed from our sample turns out to be smaller than the appropriate value in Appendix VI, then our decision must be to fail to reject the null hypothesis. If the chi square value that we have computed turns out to be larger than the appropriate value in the table, then our decision will be to reject the null hypothesis and to accept its alternative.

Remember that we are testing our hypothesis at the .05 significance level. We go back to Appendix VI and use the row for one degree of freedom (which is the top row). The appropriate value given in this row for the .05 significance level is 3.84, and we must compare our finding with this value. Since our chi square value was 2.04 and it is smaller than the value in the table, the conclusion is that we cannot reject the null hypothesis. In essence, our obtained value of 2.04 was a measure of the discrepancy between the expected frequencies under the null hypothesis and those frequencies that were obtained. The fact that our obtained chi square value was smaller than the necessary value for significance means that we did not find enough of a discrepancy from independence to lead us to reject the null hypothesis.

The chi square distribution is discussed in various elementary texts but we shall not examine its properties here. Actually, there exists a different chi square distribution for each number of degrees of freedom, as is shown by the different rows in Appendix VI. We computed a chi square value based upon one degree of freedom, so we referred to a value given in the first row of the table. We also used the column of the table that gives values for the .05 significance level. The other columns give cut-off points for other levels of significance and give the experimenter more freedom in choosing a significance level.

We can observe in Appendix VI that the value 3.84 is a cut-off point in the distribution of chi square with one degree of freedom. To see the implications of this cut-off point, consider a population in which two discrete variables are actually independent. Suppose that a vast number of random samples like ours are drawn from this population and that a chi square value is computed from each sample in the manner described. We are assuming that each chi square value is based on one degree of freedom. Under these conditions only five out of 100 samples would yield values of chi square as large as or larger than 3.84.

We hypothesized that our data comprised a random sample drawn from a population characterized by independence. Our sample yielded a chi square value (based on one degree of freedom) smaller than 3.84. Thus the finding was not sufficiently improbable to lead us to reject the null hypothesis that it came from a population characterized by independence. We were prepared to reject the hypothesis only upon obtaining a sample with a chi square value at least as large as 3.84. For instance, if our sample had yielded a chi square of 6 we would have deemed it so improbable under the null hypothesis that we would have rejected the null hypothesis.

As we shall see, the procedure and the logic are the same when the chi square test is based on any number of degrees of freedom. For each case the appropriate cut-off point is indicated in the chi square table for each of the various significance levels. Thus when we consider a chi square test using four degrees of freedom at the .05 level, we shall use 9.49 as the cut-off point. That is, we will fail to reject the null hypothesis that the variables are independent in the population if the obtained chi square value is less than 9.49. We will reject the null hypothesis upon obtaining a chi square value as great as or greater than 9.49. Such a rejection will, of course, enable us to conclude that the variables in question are related in the population from which the sample came.

18.7 Chi Square for More Complex Problems

The chi square test is the same when the variables of interest take on more than two possible values. For instance, suppose that we are concerned with the reaction of Blacks, Whites, and Puerto Ricans to a proposed bill in New York State. We wish to test the null hypothesis that each of these groups holds the same attitude toward the bill. Our theory or alternative hypothesis is that a difference in attitude exists from group to group. Members of the three groups are asked to indicate whether their reactions are favorable, indifferent, or unfavorable. (Incidentally, the variable, reaction to the bill, might be viewed by some as continuous; but so long as we are assigning its values to distinct categories, we may use the chi square test.)

Suppose that data are collected for a sample of 100 Blacks, 200 Whites, and 100 Puerto Ricans, and that the obtained values are as given in Table 18.9 without italics.

The *expected frequency* for any cell is calculated exactly in the manner described in the previous sections although the variables are not dichotomous. For each cell, we multiply its corresponding row total by its corresponding column total and divide the product by the grand total for the sample. For example, the expected frequency for Cell (5) in Table 18.9 is (90)(200)/400 = 45. The logic is that the proportion of indifferents in the whole sample is 90/400. Independence implies that the same proportion of indifferents is found among the 200 Whites; therefore, the

Table 18.9 A Three-By-Three Contingency Table

	Favorable			Indifferent			Unfavorable			Totals
Black	80	*52.5*	(1)	10	*22.5*	(2)	10	*25*	(3)	100
White	55	*105*	(4)	70	*45*	(5)	75	*50*	(6)	200
Puerto Rican	75	*52.5*	(7)	10	*22.5*	(8)	15	*25*	(9)	100
Totals	210			90			100			400

expectancy is that 45 Whites will be indifferent to the proposed bill. The expected frequency for each of the cells, found in the same way, appears in *italics* in the table.

The computation of the chi square statistic for the values in Table 18.9 appears in Table 18.10. The procedure is exactly the same as that used in the previous section except that now we have to work with nine cells instead of four. The sum of Column (6) is the obtained value of chi square and it is 101.12. It is the value given by Formula (18.1).

Table 18.10 Computation of Chi Square for a Three-By-Three Table

(1) Cell	(2) O	(3) E	(4) $O - E$	(5) $(O - E)^2$	(6) $\frac{(O - E)^2}{E}$
1	80	52.5	27.5	756.25	14.40
2	10	22.5	−12.5	156.25	6.94
3	10	25	−15	225	9.00
4	55	105	−50	2500	23.81
5	70	45	25	625	13.89
6	75	50	25	625	12.50
7	75	52.5	22.5	506.25	9.64
8	10	22.5	−12.5	156.25	6.94
9	15	25	−10	100	4.00
					$\Sigma = 101.12 = \chi^2$

Before we can use our obtained value of chi square and Appendix VI to test our hypothesis, we must determine the number of degrees of freedom involved. We will also state a general rule concerning the number of degrees of freedom to be used in similar problems of any degree of complexity. Once again we begin with just the marginal totals that in and of themselves tell us nothing concerning the possible relatedness of the variables. Again we shall allow ourselves knowledge of the frequency in one cell, say Cell (1). This obtained frequency and the marginal totals have been written in *italics* in Table 18.11. But one piece of information is now insufficient for us to determine the frequencies in the other cells. If we give ourselves a second obtained cell frequency and then a third we still cannot fill out the chi square table. Suppose, for example, that we are given the frequencies for Cells (1), (2), and (4). These are italicized in Table 18.11. It is seen that these are sufficient to fix the frequencies in some of the other cells, namely Cells (3) and (7). However, the frequencies of the remaining cells, (5), (6), (8), and (9), are still *free* to vary.

Table 18.11 Degrees of Freedom for a Three-By-Three Table

	Favorable	Indifferent	Unfavorable	Totals
Black	*80* (1)	*10* (2)	10 (3)	*100*
White	*55* (4)	(70) (5)	75 (6)	*200*
Puerto Rican	75 (7)	10 (8)	15 (9)	*100*
Totals	*210*	*90*	*100*	400

One more independent piece of information is all that we need, however. Once we know the frequency of Cell (5), we can determine all of the rest by subtraction from the marginal totals. This frequency has been encircled in Table 18.11. It is thus seen that four independent pieces of information have proved sufficient. The frequencies of the four cells in the upper left corner together with the marginal totals are sufficient to determine the frequencies of all of the cells in the third row and the third column.

In general, we can always infer the frequencies in the last row and in the last column by making subtractions from the marginal totals. **The number of degrees of freedom is equal to the number of cells remaining in the chi square table after the last row and the last column are eliminated. Symbolically we have**

$$\text{d.f.} = (R - 1)(C - 1) \qquad (18.2)$$

where there are *R* rows and *C* columns.

We now know that our obtained value of chi square, 101.12, is based upon four degrees of freedom $[(3-1)(3-1) = 2 \cdot 2 = 4]$. The cut-off point given in Appendix VI is 9.49 for the .05 level. Since our obtained chi square value is much larger than the appropriate value in the table, we reject the null hypothesis that the sample came from a population in which the variables are independent. We conclude that attitude toward the proposed bill is related to ethnic group.

Incidentally, after a significant chi square value has been obtained, one may attempt to make more specific statements by considering the data in some of the cells and not others. In our own example, we might, for instance, compare the Blacks and Whites. To do so, we necessarily disregard the final row of Table 18.9 and revise the marginal totals as in Table 18.12. As usual, the *expected frequencies* have been added in *italics*. The reader should verify that these new expected frequencies are correct. They have been necessitated by the change in the marginal totals.

Table 18.12 A Two-By-Three Contingency Table

	Favorable	Indifferent	Unfavorable	Totals
Black	80 *45* (1)	10 *27* (2)	10 *28* (3)	100
White	55 *90* (4)	70 *53* (5)	75 *57* (6)	200
Totals.	135	80	85	300

The reader should also verify that the chi square value for Table 18.12 is 74.2 as found by Formula (18.1). It is based upon two degrees of freedom as verified by Formula (18.2) and is significant at the .05 level. Thus we reject the hypothesis that the Whites and Blacks are identical in attitude toward the proposed bill.

The chi square test is often used in connection with discrete variables that take on numerical values and with continuous variables. Successive numerical intervals are set up before the data are gathered and these intervals serve as categories. For instance, we might wish to see whether IQ is related to scores on a music appreciation test, in which case the table for recording the data might look like Table 18.13.

When we put the values of a continuous variable into categories, we are admittedly losing information. For instance, by treating the data as in Table 18.13 we are not differentiating between individuals whose IQs are between 101 and 115. However, the possible relationships become so numerous when one or both variables are continuous that we must either categorize the variables, as we have done here, or make certain assumptions about them in order to test for independence. *One need*

Table 18.13 Pattern for a Five-By-Four Contingency Table

	Music Appreciation Score				
IQ	Below 26	26–50	51–75	76–100	Totals
71–85					
86–100					
101–115					
116–130					
131–145					
Totals					

make no assumptions about the distribution of continuous variables in order to use the chi square test. An alternative method that also requires no assumptions is discussed in Chapter 19.

18.8 Chi Square and "Goodness-of-Fit"

We have just seen how the chi square statistic is useful in testing for relatedness between two variables in a population. There are, however, numerous other situations in which chi square is helpful. We shall now consider one such group of problems, which involve only one variable. They collectively have been classified as **"goodness-of-fit."** We actually will be *testing whether or not an obtained sample could have come from a predetermined population.* The hypotheses, however, can often be stated so they will be more useful to a sociologist, economist, researcher, and the like.

Let us consider, for instance, a study being made by a newspaper editor to assess the effects of a proposed change in the subscription rates. At present 55% of the readers receive both daily and Sunday papers, 30% subscribe to Sunday papers only, and 15% have only daily papers delivered. Two hundred customers are surveyed and asked what they would subscribe to if the changes go into effect. Of these, 99 respond that they would receive both daily and Sunday papers, while 64 and 37 would opt for Sunday only and for daily deliveries, respectively.

Now the editor must ask if the sample results indicate that the subscribers would distribute themselves differently under a rate adjustment than they do now. Or is it possible that there would be no change, and that the sample proportions differ from the present population proportions only because of sampling procedures? The null hypothesis that is being considered is that the sample comes from the predetermined population. The predetermined population is, of course, one in which 55% of the members are Sunday-daily subscribers, 30% are Sunday-only subscribers, and 15% are daily subscribers. *The term "goodness-of-fit" comes from our asking how well the sample "fits" the population.* From the editor's point of view a more useful statement of the null hypothesis might be "After the rate adjustment, the proportion of subscribers for each subscription plan will not change." The alternative hypothesis is, naturally, that the proportions would change if a rate adjustment were instituted. The acceptance of the alternative theory (rejection of the null hypothesis) would no doubt call for more studies as to the nature of the change.

As has been done in the past, the sample results must be examined under the assumption that the null hypothesis is true. A vast number of random samples of size 200 could conceivably be selected. Since each would come from a population with a 55–30–15 percentage breakdown, the expectation would be that each sample would not deviate drastically from this breakdown. In fact, the expectation for any sample of size 200 would be 110 [$.55 \times 200$] daily-Sunday subscribers, 60 [$.30 \times 200$] Sunday-only subscribers, and 30 [$.15 \times 200$] daily subscribers. But the observed values were not 110, 60, and 30. They were 99, 64, and 37, as summarized in Table 18.14.

Table 18.14 Obtained and Expected Frequencies for "Goodness-of-Fit"

	Observed	Expected
Daily–Sunday	99	110
Sunday	64	60
Daily	37	30
Totals	200	200

If one is willing to accept a .05 chance of error if the null hypothesis is true, *are the observed and expected frequencies different enough from one another to conclude that the sample comes from a population other than the one in question?*

The appropriate procedure for the editor to follow is the utilization of the chi square statistic, which is the application of Formula (18.1). The calculations are carried out below in Table 18.15.

Table 18.15 Computation of Chi Square for Goodness of Fit

(1)	(2)	(3)	(4)	(5)
O	E	$O - E$	$(O - E)^2$	$\frac{(O - E)^2}{E}$
99	110	−11	121	1.10
64	60	4	16	.27
37	30	7	49	1.63
				$\Sigma = 3.00 = \chi^2$

Once χ^2 has been calculated, all that is left is to compare it with some critical value from Appendix VI. To use the Appendix table, however, one must determine the proper number of degrees of freedom. **In all chi square "goodness-of-fit" problems involving one variable the degrees of freedom will be $K - 1$, where K is the number of categories used.**

$$\text{d.f.} = K - 1 \qquad (18.3)$$

In the newspaper example three categories were used. Hence there are two degrees of freedom and Appendix VI yields a critical value of 5.99. Since our calculated value of 3.00 is less than the table value, the editor is unable to conclude that the subscription distribution will change significantly. Unless other factors or results change his mind, he would do well to proceed as though no change would occur.

As a second illustration let us turn our attention to a buyer of fashion merchandise who feels that women have no preference for hem length this season and plans to order an equal number of long, medium, short, and mini lengths. To be sure, she asks 80 randomly selected customers to state their preferences. *The null hypothesis being tested is that there is no difference in the number of women who would select each length.* Since there are four choices we would expect 20 women to opt for each length. The sample observed values, along with the expected values and the calculation of the chi square statistic, are found in Table 18.16.

Table 18.16 Chi Square—Goodness of Fit

(1)	(2)	(3)	(4)	(5)
O	E	$O - E$	$(O - E)^2$	$\frac{(O - E)^2}{E}$
14	20	−6	36	1.8
29	20	9	81	4.05
25	20	5	25	1.25
12	20	−8	64	3.2
				$\Sigma = 10.3 = \chi^2$

The number of degrees of freedom in this situation is 3 since we are dealing with four categories. Our calculated value, 10.3, is larger than the 7.82 needed to reject the null hypothesis at the .05 significance level so it merits rejection. In light of the evidence, the buyer should accept the result that women do prefer one length to another and should plan purchasing accordingly.

The method just described can be used to determine if any sample departs from any population as long as the population is well defined. Often it would be helpful for a researcher to know whether or not a certain population is normal. As long as the mean and standard deviation are known[3] we are able to use χ^2 to test the null hypothesis that it is normal. The procedure is similar to the one used in the previous examples. It uses the information found in Appendix III. We know that in a normal distribution with a mean of 100 and standard deviation of 10, for example, that 19% of the terms are expected between 100 and 105 (z scores of 0 and .5), 15% are expected between 105 and 110 (z scores of .5 and 1), and 16% are expected above 110 (z score of 1). Using these expected percentages we can determine how much a given sample deviates from what we would expect in a sample from a normal population.

As we have seen, the chi square statistic is a versatile tool available to the statistical researcher. It needs to be reemphasized that its use depends on our being able to divide the data into categories. Also note that unlike much of our previous work, *no assumptions were made regarding the normality of underlying distributions*. This fact inherently links the chi square tests for independence and "goodness-of-fit" with the broad group of tests we shall consider in Chapter 19, namely, nonparametric tests.

[3]More sophisticated methods allow them to be estimated.

Problem Set A

18.1 A 10-year-old survey indicates that 45% of the farmers in the Morie River Valley support a federally funded irrigation plan, 35% oppose it, and 20% are undecided. Since the question is coming up again in Congress this session, Senator Snook surveys 500 of his rural constituents. He finds that 250 favor the proposal's passage, 167 oppose its passage, and 83 are undecided. Do these results indicate any change in the proposal's support in the last 10 years at the .05 significance level?

18.2 Do the following sample results dealing with 200 cases of blown-out tires allow for rejection of the null hypothesis that a blowout is equally likely to occur on any of the four tires of an automobile?

Survey conducted by Brown County Safety Council citing location of blowouts on 200 cars:

43 left front	59 right front
48 left rear	50 right rear

Use the .01 level of significance.

18.3 With a certain type of back injury full recovery is almost certain. However, it sometimes takes up to two full years. An Italian physician, Dr. Vertebroni, suspects that the time to recovery is linked with the type of therapy administered. Of 50 patients he has treated using physical means, only 38 recovered during the first year and 12 during the second. Of 75 patients he has treated using a combination of drugs and bed rest, 43 recovered during the first year and 32 recovered during the second. Do these results support Dr. Vertebroni's theory at the .01 significance level?

18.4 A new bank card company, in the process of revising its guidelines for extending credit, is interested in detecting a relationship between a person's payment record and whether the person owns or rents a dwelling. The previous year's records of 90 people were selected at random with the following results.

Number of Missed Payments	*Own*	*Rent*
0–3	19	15
4–6	13	23
7–12	8	12

Should the company differentiate in its credit policy on the basis of ownership of dwelling? Test at the .05 level of significance.

18.5 The governing assembly of a large religious denomination must vote on its policy statement regarding the use and allocation of the nation's energy resources. A preliminary survey of 175 delegates gives the following results.

	Rural	*Urban*	*Suburban*
Favor	18	45	33
Oppose	10	25	18
No opinion	9	7	10

Do the different delegate groups (rural, urban, and suburban) have different opinions regarding the policy statement? Test at the .05 significance level.

18.6 High School Dean Dismal is concerned about the unusually large number of students who have been referred to his office for disrupting classroom procedures. In the hope of getting to the root of the problem, he tests his theory that the number of times a student is sent to him for that reason is related to the junior high school the student attended. He organizes his records for the past five years into the contingency table found below. Do they support the dean's belief at the .05 significance level?

	Number of Referrals		
School	*One or Two*	*Three or Four*	*Five or More*
Washington	18	23	7
Lincoln	5	6	7
Adams	20	5	19
Roosevelt	11	6	12

18.7 The director of a social agency that deals with family planning and adoption contends that the decision on the part of an unwed mother to put a child up for adoption is related to a great extent to the level of education of the mother. Do the following statistics support her contention at the .01 level of significance?

	Highest Level Achieved				
Mother's Decision	*Some High School*	*High School Diploma*	*Some College*	*College Degree*	*Graduate Degree*
Adoption	85	90	93	41	11
Keep child	38	59	102	100	15

18.8 The dramatic increase in automobile repair costs has spawned a number of studies as to the necessity of the work being done. Among them is a study aimed at trying to determine if the type of work performed is related to how necessary it was in the first place. Cars were taken to 50 randomly selected garages. The following table summarizes the work that was done as to the type of repair (brake, transmission, suspension, engine) and the need for repair (necessary, marginal, unnecessary).

	Necessary	*Marginal*	*Unnecessary*
Brake	33	17	12
Transmission	14	20	11
Suspension	25	5	5
Engine	40	10	31

Apply a chi square test at the .05 level of significance to see if the relationship is significant.

18.9 An attorney for a human rights group is trying to establish prejudice on the part of a large corporation in its assignment of employee working shifts. The company contends that there is no relationship between an employee's race and the shift to which the employee is assigned. The attorney randomly selects 125 workers.

a. Do her findings below demand rejection of the company's claim at the .05 level of significance?

	Shift	
	Day	*Night*
White	15	8
Black	12	12
Oriental	18	9
Spanish surname	15	10
Other	14	12

b. Is there any relatedness in the sample?

18.10 To plan its political comeback, the Pay-As-You-Go Party must ascertain the makeup of its support base. At its political zenith 50 years ago, 19% of its support came from each of the following groups: union labor, nonunion labor, retired, and white collar. The other 25% came from the ranks of the unemployed. Has there been any significant change in the party's support if a random sample of 200 contributors contains 48 union laborers, 67 nonunion laborers, 32 retired people, 23 white collar workers, and 30 unemployed people? Test at the .05 level of significance.

18.11 Can it be concluded that a sample of 150 terms did not come from a normal distribution with a mean of 50 and a standard deviation of 10 if 20 terms are less than 40, 27 terms are between 40 and 45, 32 terms are between 45 and 50, 27 terms are between 50 and 55, 21 terms are between 55 and 60, and 23 terms are greater than 60? Use the .05 significance level.

Problem Set B

18.12 A vending machine company wishes to determine how effective putting a telephone number on its machines will be in helping to keep the machines in working order. It selects 100 machines at random. On 50 of them it puts a telephone number. After six months it is found that 18 of the machines without numbers are in need of repair whereas only 12 of those with numbers need repair. Does this support the effectiveness of using numbers at the .05 significance level?

18.13 Because he is setting up next year's schedule, Coach Hatchett wants to know if there really is a home field advantage for his team. During the past three seasons they have won 50 games at home and 30 away, 20 games were lost at home and 40 away, and five were tied both at home and away. Is the home field advantage significant at the .05 level?

18.14 The number of motorcycles on the streets and highways has increased noticeably in the last few years. A five-year-old survey indicates that automobile drivers responded to the question "What is the major hazard presented by motorcycles?" in the following manner.

lack of visibility	30%
riding in large groups	15%
drivers not safety conscious	45%
motorcycle inherently unsafe	10%

A more recent survey of 75 automobile drivers shows 20, 15, 32, and 8 selecting the above responses. Do today's drivers differ from drivers of five years ago? Use the .05 significance level.

18.15 The U.S. Association of Pediatrics (USAP) recommends the breast feeding of infants for a variety of reasons. It feels, however, that mothers need to be educated as to the benefits. To determine where the major thrust of its educational campaign should be, the association is attempting to determine if the decision to breast feed is related to a mother's economic status. The records from University Hospital of 200 mothers are studied. The results are as follows:

	Economic Level			
	Poverty	*Low*	*Middle*	*Upper*
Bottle	30	15	11	12
Breast	7	18	19	29
Bottle and breast	5	23	7	19

Do the results support relatedness of the variables at the .05 significance level?

18.16 A special interest group is interested in determining if a person's parental status is related to his or her opinion on the upcoming school bond issue, and 140 voters are randomly selected, categorized, and asked for their opinions. The contingency table below summarizes the findings. Are the variables in question related at the .01 significance level?

	Favor	*Oppose*
Parents	34	4
Future parents	30	16
No plans to be parents	16	26
Retired or single	7	7

18.17 Dr. Abner Ormal, a leading psychologist, is conducting a study designed to detect a relationship between the number of children a couple has and how protective they are of those children. A test is given to each couple that rates their "protectiveness" as high, average, or low. All of the sample information appears in the following table.

Number of Children	*High*	*Average*	*Low*
1	12	15	7
2	13	14	8
3	20	21	13
4 or more	9	12	5

Do the results support Dr. Ormal's belief at the .05 level of significance?

18.18 Statistics has traditionally been considered a "high anxiety" course. Dr. Howard suspects, however, that performance in statistics is related to performance in general scholastically. To test her theory, Dr. Howard compares the grades of her last semester's statistics classes against the overall grades of the students.

		Overall Grades	
Statistics	*A or B*	*C*	*D or F*
A or B	17	15	5
C	8	30	9
D or F	4	23	25

Do the above results support Dr. Howard's theory at the .05 level of significance?

18.19 For her psychology lab assignment, Susie Kew is attempting to determine if birth order of white mice is related to ability to learn a specific behavior. Each of 15 first-, second-, third-, fourth-, and fifth-born mice is given 45 minutes to learn, then its birth order and success or failure are recorded. Here are Susie's results. The first number in parentheses indicates the number of successes, the second indicates the number of failures.

first-born	(9, 6)	fourth-born	(7, 8)
second-born	(10, 5)	fifth-born	(4, 11)
third-born	(8, 7)		

What conclusion can Susie draw at the .05 significance level?

18.20 A family planning counselor believes on the basis of her many interviews that the numbers of children people want differ according to the sizes of the families from which they come. She randomly surveys newly married people, and obtains the following results.

Size of Family	*Number Desired* 0	*1*	*2*	*3*	*4 or more*
1	8	14	13	20	15
2	8	10	9	5	8
3	5	8	8	11	5
4 or more	7	9	10	12	5

a. Do the sample results support the counselor's theory at the .05 significance level?
b. Are the variables related in the sample?

18.21 Ten years ago 23% of dog owners owned poodles. Other leading breeds and their percentages were: German shepherds 22%, collies 12%, spaniels 8%, all others 35%. A leading dog food manufacturer surveys 300 dog owners to determine, if possible, if these percentages have changed. Of the sample, 20% own poodles, 15% German shepherds, 14% collies, 12% spaniels, and 39% all others. Can the company conclude at the .01 level of significance that the percentages have changed during the last 10 years?

18.22 A sample of 1,000 terms comes from a distribution whose mean is 100 and standard deviation is 10. Of these, 35 terms are less than 80, 187 terms are between 80 and 90, 295 terms are between 90 and 100, 350 terms are between 100 and 110, 100 terms are between 110 and 120, and 33 terms are greater than 120. Can one conclude at the .01 significance level that the population sampled from is not normal?

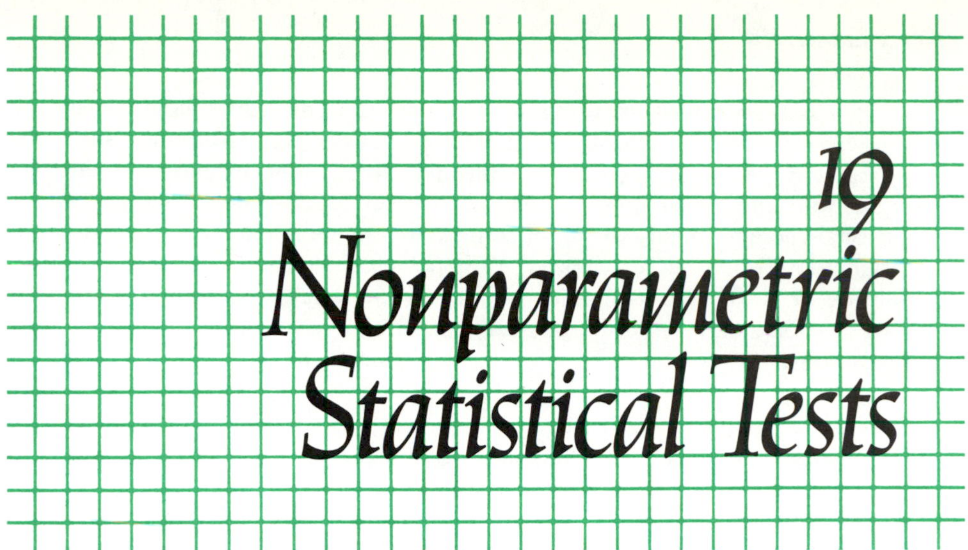

19 Nonparametric Statistical Tests

19.1 Introduction

We have discussed procedures for testing various hypotheses. Some of these hypotheses concerned population means, some concerned population variances, and the one in Chapter 16 was the hypothesis that the correlation coefficient in a population is zero. Remember that when testing these hypotheses, we have in each case assumed an approximately normal distribution in the population from which our samples were drawn.

In contrast, when we worked with contingency tables and did the chi square test in the previous chapter, we made no assumption of this kind. That is, we did not stipulate that the distributions of the variables involved had to be normal, or had to have any other specified shape, for that matter.

There is an important distinction between the procedures that necessitate making some assumption regarding the shape of a distribution and those for which we need not make any such assumption to proceed. In contrast with the first type of procedure, the latter type is called a **nonparametric** or **distribution-free** method of testing a hypothesis. Although the only instance of a nonparametric test discussed so far has been the chi square test, there exist many other such tests used for different purposes. The inability to assume normality in various situations is becoming increasingly of concern to researchers. As substitutes for the older parametric tests, which had been used where they were not indicated, the distribution-free hypothesis tests are becoming more and more popular.

To use a nonparametric test is a necessity when the distribution of interest is known not to be normal and when one can infer absolutely nothing regarding its shape. Where an assumption is not warranted, it is obviously inappropriate to make it implicitly by using a test that follows from it. On the other hand, where the assumption of normality is warranted, to renounce making this assumption and to use a nonparametric test would be a serious mistake. As in other instances, to renounce making a warranted assumption is to hold oneself back from drawing a proper conclusion. Where normality exists, the refusal to assume it and the decision to use a nonparametric test is the surrendering of a powerful tool for a weaker and more general one. It is true that nonparametric tests require fewer assumptions than the methods described earlier. But for this very reason they give less precise results than parametric tests when they are used in situations where parametric tests are applicable.

Nonparametric tests tend as a group to be easy to apply. Where such tests are applicable they are a great convenience. But especially where the actual data gathering has been painstaking, the fact that a nonparametric test is easy to carry out should not commend it to use. The careful research worker always questions first whether a parametric test can be used, since it would give the most accuracy, and if it appears reasonable to do so does use one. It is only after carefully ruling out the use of a parametric test that the worker uses a nonparametric one. Having issued these words of caution, we shall move on to discuss several nonparametric methods. Those presented here are among the more important and widely used of the nonparametric tests.

19.2 The Mann-Whitney Test[1]

In Sections 13.3 and 13.4 we discussed the t test for the difference between means of two distributions. We saw that this test has many important applications. However, its use was predicated on the assumption of normality and the independence of samples. This means the random selection of one sample does not influence the selection of the second sample. Another way of stating this is that we have independence between samples as well as within samples.

The situation often arises, however, in which we have independent samples and want to test for difference between the central tendencies of two populations, but are *unable to assume normality* in the underlying distributions. As long as the populations have approximately the same variability and shape, we may employ a nonparametric test known as the Mann-Whitney U test.

Suppose, for example, that we wish to help a sociologist determine if the mean ages of first offenders for drug-related crimes differ significantly between the East and West coasts. Our first step, of course, is to obtain the ages of first offenders in two samples, one from each coast as indicated in Table 19.1.

We might proceed by finding the mean of each sample and testing for significant difference between the two. In the case in which we have reason to believe the difference scores are approximately normally distributed we would be justified in doing so. However, often when two samples are drawn from the same population but are subject to different influences this is not possible. In the case at hand first offenders living on one coast might be subject to different social, economic, or

[1] This test is sometimes called the Wilcoxon Two Sample or Rank Sum Test.

Table 19.1 Mann-Whitney Test for Difference Between Ages of First Offenders

(A) East Coast	(B) West Coast
16	15
26	25
23	17
19	40
45	22
30	27
23	21
23	20
29	18
24	16
35	14
32	31

perhaps ethnic influences than those on the other. Our theory is that this situation results in a mean age difference. The null hypothesis, therefore, is that no meaningful difference exists.

The appropriate procedure, the Mann-Whitney U test, begins by combining all the data from both samples into one sample. The terms are then ranked from smallest to largest, and the original sample that each term came from is recorded. Let us agree to call the East coast sample A and the West coast sample B, although this assignment is strictly arbitrary.

Table 19.2 Mann-Whitney Test Ranked Scores

Ranked Data	Rank	Sample	Ranked Data	Rank	Sample
14	1	B	23	12	A
15	2	B	24	14	A
16	3.5	A	25	15	B
16	3.5	B	26	16	A
17	5	B	27	17	B
18	6	B	29	18	A
19	7	A	30	19	A
20	8	B	31	20	B
21	9	B	32	21	A
22	10	B	35	22	A
23	12	A	40	23	B
23	12	A	45	24	A

Note that the two scores of 16 that are ranked 3 and 4 divide those ranks between them, each receiving 3.5 [(3 + 4) / 2 = 3.5]. Similarly, the rank 12 appears three times in the table since the score 23 occupies ranks 11, 12, and 13, which have been divided among them [(11 + 12 + 13) / 3 = 12].

As we have observed in previous hypothesis-testing situations, our sample must yield some statistic or value of interest. In the Mann-Whitney test this value is represented by U and is defined to be the smaller of U_A and U_B where

$$U_A = N_A N_B + \frac{N_B(N_B + 1)}{2} - R_B \tag{19.1}$$

$$U_B = N_A N_B + \frac{N_A(N_A + 1)}{2} - R_A \tag{19.2}$$

In Formulas (19.1) and (19.2) N_A and N_B represent the sizes of samples A and B respectively ($N_A = 12$, $N_B = 12$ in our example). Note that it is not necessary for the sample sizes to be the same. R_A and R_B are the sums of the ranks of samples A and B, respectively. For our illustration

$$R_A = 3.5 + 7 + 12 + 12 + 12 + 14 + 16 + 18 + 19 + 21 + 22 + 24$$
$$= 180.5$$

$$R_B = 1 + 2 + 3.5 + 5 + 6 + 8 + 9 + 10 + 15 + 17 + 20 + 23 = 119.5$$

Now we can calculate U_A and U_B.

$$U_A = 12(12) + \frac{12(12 + 1)}{2} - 119.5 = 102.5$$

$$U_B = 12(12) + \frac{12(12 + 1)}{2} - 180.5 = 41.5$$

Since we defined our value of interest, U, to be the smaller of U_A and U_B, we have $U = 41.5$. As in previous testing situations we must ask ourselves if this U is significant. We reason in the usual way by viewing our U value as one of many U values, each obtained in a similar fashion. Although the verification of such is beyond the scope of this text, as long as each sample has more than 10 terms, the distribution of all U scores will be approximately normal and

$$\mu_U = \frac{N_A N_B}{2} \tag{19.3}$$

$$\sigma_U = \sqrt{\frac{N_A N_B(N_A + N_B + 1)}{12}} \tag{19.4}$$

By using Formulas (19.3) and (19.4) we can test the significance of our $U = 41.5$.

Since $\mu_U = \dfrac{12 \cdot 12}{2} = 72$ and $\sigma_U = \sqrt{\dfrac{144(12 + 12 + 1)}{12}} = 17.32$ we can convert U to a z score by the usual method.

$$z_U = \frac{41.5 - 72}{17.32} = -1.76$$

For a two-tail test at the .05 significance level we fail to reject the null hypothesis. We therefore are unable to conclude from our evidence that there is any significant difference between the ages of first offenders in drug-related crimes between the East and West coasts; see Fig. 19.1.

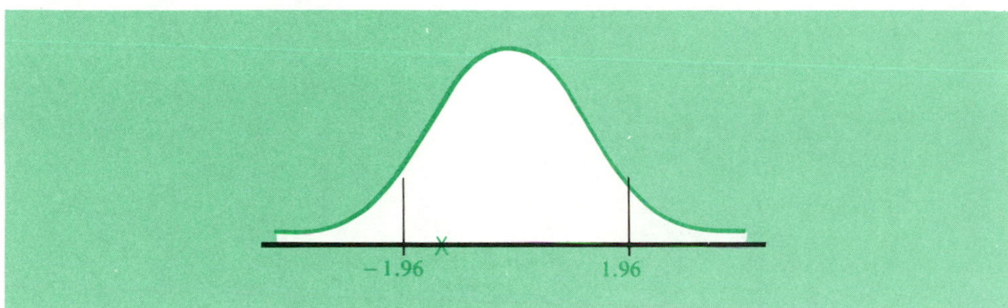

Fig. 19.1

19.3 The Sign Test—"Before" and "After" Data

In the previous section we discussed an alternative to the t test for the difference between the means of two distributions. Recall that we assumed that each sample was selected independently. In other words, we assumed that each sample was chosen in such a way as to have no effect on the choice of the other sample. This assumption is of course implicit in the definition of "random sample" and it must be kept in mind whenever data obtained from random samples are used.

The assumption of independence is not satisfied in some very important types of experiments. One of these is the "before" and "after" type of study similar to the situation discussed in Section 13.5. What we are about to examine is an alternative to the t test for matched groups, an alternative that can be employed when we are *unable or unwilling to assume normality* in the underlying distribution of difference scores.

Suppose, for example, that a researcher has only one class and wishes to determine whether a particular lecture that he gives the students in this class actually increases their incentive to perform well in arithmetic. He gives them a pre-test of their performances on handling arithmetic concepts; then he gives his lecture, after which he gives the class a post-test. He would like to use the results of the two tests to determine whether his lecture actually improved their scores.

In essence, each student gets a pre-test score and a post-test score, as indicated in the first two columns of Table 19.3 for the fifteen students whom we shall say compose his class. Note that here the scores in Column (1) are not independent of those in Column (2). Rather, each score in one group is coupled with a score in the other—namely, that produced by the same person. Because of this natural pairing, we cannot do the *t* test described in Section 13.3; but we can use the one in Section 13.5.

Table 19.3 Sign Test for Pre-Test and Post-Test Scores

(1) Pre-Test	(2) Post-Test	(3) Sign of Difference
65	75	+
80	79	−
50	90	+
92	90	−
85	80	−
75	85	+
45	68	+
89	89	0
60	82	+
65	75	+
64	75	+
63	60	−
83	87	+
75	76	+
69	60	−

We might take each student's post-test score and subtract from it his or her pre-test score to obtain what is sometimes called a "change" score. For the case in which we have reason to believe that these "change" scores are normally distributed, there does exist a variation of the *t* test. This test has frequent application and it is described in Section 13.5. More often it occurs that because of individual characteristics, some persons change drastically whereas others change hardly at all. In such a case, one drastic change in a negative direction, for instance, might be enough to balance out many changes in the positive direction. Where such a result is anticipated, there may be reason to believe that the obtained differences will not be normally distributed. In fact, an experimenter might properly wish to avoid using a measure that equates the units of change for a relatively few people with those of others whose scores do not seem to be so amenable to change. Specifically, the experimenter is interested only in the proportion of individuals who change in one direction as opposed to the other, and would like to give each individual equal weight regardless of how much he or she changed.

In such a case, the appropriate procedure is for the researcher to record for each individual only the sign, plus or minus, that indicates the direction in which that person changed. Thus, where a subject increases his or her score in going from pre-test to post-test, this fact is recorded by a plus sign. The fact of a decrease is recorded by a minus sign. The next step for the experimenter is to test whether the discrepancy between plus and minus signs is extreme enough to reach the decision that the discrepancy was not the result of chance.

The appropriate test to be used is known as the sign test, since it deals with only the signs of the individual changes and not with their magnitudes. Column (3) of Table 19.3 shows only the sign of the difference in performance of each student on the two tests. This has been recorded as + if the student did better on the post-test than on the pre-test and as − if his or her score went down from the first test to the second. In one case where there was no change, a 0 has been recorded in Column (3). This case will not be included in our treatment of the data. Thus we find that there are nine plus signs and five minus signs.

The theory to be proved is that the lecture given is effective in increasing students' performances on the arithmetic test. The null hypothesis to be tested therefore is that the lecture produces no change in students' achievement. It follows that if the method produces no meaningful change in students, then in any given sample we should expect no more students to receive better scores than receive poorer scores on the post-test as compared with the pre-test. In other words, our null hypothesis is the same as the hypothesis that we are as likely to obtain a + sign as a − sign. This is equivalent to the statement that we are as likely to obtain a head as a tail when tossing a fair coin. The situation in our example is somewhat equivalent to testing the hypothesis that a coin is fair if we have obtained nine heads and five tails in a sample consisting of 14 tosses.

However, our situation differs in one important respect from that involved in testing the fairness of a coin. We shall use a *one-tail test*, because we are interested in the specific alternative that the lecture is effective—that is, that it aids student achievement. Let us also decide in advance to use the .05 significance level.

Since we are dealing with a dichotomous variable, we now use the procedure described in Section 11.3. We suppose that our sample is one of a vast number of equal-sized samples obtained in the same way. Let us designate the plus sign as the one of interest, so that the outcome of our sample is nine. Under the null hypothesis the expectancy is for an equal number of pluses and minuses. Thus, in this vast number of samples of 14 cases each, some would contain no pluses, others would contain one plus, others two pluses, and so on. In other words, the number of pluses in each sample is the outcome of interest, and in different samples this number would range from zero to 14.

Our concern is with locating our outcome of nine in the distribution of outcomes that would be found, assuming that the null hypothesis is true. Applying Theorem 9.1, we calculate the mean and standard deviation of this distribution of outcomes. Under the null hypothesis, the probability, P, of a plus is $P = 1/2$. $N = 14$. The mean,

$$\mu = NP = 14\,(½) = 7;\ \sigma = \sqrt{NP\,(1 - P)} = \sqrt{14\,(½)\,(½)} = 1.87$$

Since $NP = 7 > 5$ and $N(1 - P) = 7 > 5$, we are justified in applying Theorem 9.1.

Recalling our discussions of tests of hypotheses concerning proportions, we now convert our outcome of 9 to a z score. Remember that we are approximating a discrete variable with a continuous one and consequently must use the .5 correction factor. If the outcome is above the mean we subtract .5. If it is below the mean we add .5. Using Formula (11.1) we have

$$z = \frac{(O \pm .5) - NP}{\sqrt{NP - (1 - P)}} = \frac{(9 - .5) - 7}{1.87} = \frac{1.5}{1.87} = .80$$

Since this z value is smaller than the 1.65 required for rejection of our null hypothesis in a one-tail test at the .05 level, we fail to reject the hypothesis. We cannot then accept the alternative that the lecture effectively improved student performance.

As another example, suppose that the experiment had been carried out with 30 students and had resulted in 22 students receiving better scores, three showing no change, and five receiving poorer scores on the post-test as compared with the pre-test. We would then have 22 plus signs and five minus signs for the 27 students for which there was an observable change. If we again test the hypothesis of no change at the .05 level, following the same procedure as before, we have

$$z = \frac{(O \pm .5) - NP}{\sqrt{NP(1 - P)}} = \frac{21.5 - 13.5}{\sqrt{27(1/2)(1/2)}} = \frac{8}{2.60} = 3.08$$

Since this exceeds the value of 1.65 needed for rejection of the null hypothesis, we accordingly do reject the hypothesis. This, then, proves the theory that the teaching method is effective, using the .05 level.

19.4 The Wilcoxon Signed-Ranks Test—Naturally Paired Samples

We have just finished discussing an alternative to the t test for matched groups, the sign test. A very useful and easy-to-apply test, it does have one drawback that is important in certain cases. It fails to use all the information available; namely, *how much change* is registered. In contrast, the Wilcoxon signed-ranks test, another alternative for the matched groups t test, takes into account not only the direction of change but also the size of the change relative to the sizes of other changes. It accomplishes this by ranking each change, hence the name signed-ranks test. We shall explore exactly how this is done shortly.

The test in question, which we shall refer to as simply the "Wilcoxon" test, is most often applied when some natural pairing exists between the members of two samples. This pairing puts us again in the realm of *dependent* samples. Each member of the first sample must have a partner in the second sample. This, of course, rules out the Mann-Whitney test even though we are testing for difference between two groups. We want to take advantage of the matching of members between samples whenever possible as it reduces the probability of chance differences.

Bear in mind we are unable to identify a normal distribution of difference scores and are therefore turning to a nonparametric test. If an assumption of normality is justified, we would be remiss in not using the t test for matched groups.

To illustrate the procedure followed in the Wilcoxon test, let us consider a biological research scientist who is studying the reaction of cancer cells to two types of treatment. Her theory is that there is a difference in the numbers of cells that survive the two treatments. To control as many variables as possible the scientist selects two each of a number of different types of cancer cell specimens. She selects two specimens of malignant cells taken from the same lung, two from the same liver, two from the same throat, and so on. Each specimen must be kept under exactly the same conditions as its counterpart except for the type of cancer treatment to which it is subject. After each specimen is carefully treated, the data are recorded as in Table 19.4. The numbers represent the percentage of cells in each specimen that respond favorably to the treatment being used.

Table 19.4 Wilcoxon Signed-Ranks Data for Treated Cancer Specimens

Type of Cell	(1) Sample A	(2) Sample B
Lung	90	85
Liver	94	90
Throat	78	80
Bowel	79	70
Brain	84	84
Skin	75	65
Lymph	74	63
Pancreas	99	89
Bone	72	73

Since she is looking for differences between the two treatments, the scientist figures the change in each couple by subtracting Column (2) figures from Column (1) figures and records the results in Column (3) in Table 19.5. The differences are then ranked from smallest to largest in Column (4). In ranking the differences attention is given only to the absolute value of each term. In other words, negative scores are treated as though they were positive for ranking purposes. Differences of zero are not

Table 19.5 Wilcoxon Signed-Ranks

(1) Sample A	(2) Sample B	(3) Differences	(4) Ranks (R)
90	85	5	4
94	90	4	3
78	80	−2	2
79	70	9	5
84	84	0	—
75	65	10	6.5
74	63	11	8
99	89	10	6.5
72	73	−1	1

ranked and are ignored for the remainder of the test. Tied scores divide their ranks equally as in the Mann-Whitney test. That is why the tied scores of 10 that occupy ranks 6 and 7 each receive a rank of 6.5 [(6 + 7) / 2 = 6.5].

Once the rankings and differences have been determined for each pair, two sums are calculated. The first, ΣR^+, is the sum of the ranks associated with positive difference scores, and the second, ΣR^-, is the sum of the ranks associated with negative difference scores. In the researcher's case

$$\Sigma R^+ = 4 + 3 + 5 + 6.5 + 8 + 6.5 = 33$$

$$\Sigma R^- = 2 + 1 = 3$$

If the null hypothesis is true there is no difference in the specimen reactions to the two cancer treatments. This would make one expect the positive differences to be balanced by the negative differences and the sum of the ranks assigned to the positive differences to equal the sum of the ranks assigned to the negative differences. Since the sum of all the ranks is 36 (1 + 2 + 3 + 4 + 5 + 6 + 7 + 8 = 36), the truth of the null hypothesis would imply $\Sigma R^+ = \Sigma R^- = 18$. Obviously, our sample results deviate from that expectation. The question that must certainly be asked is whether or not the obtained results differ enough from the expected results to call for rejection of the null hypothesis. To answer this question one of the two results, ΣR^+ or ΣR^-, must be selected for study. The smaller of the two is the one selected as the value of interest, and it is designated by V. So in the example $V = 3$. This V must be compared with a vast number of other values selected in the same manner to determine if it has a low enough probability of occurring to allow rejection of the null hypothesis.

The distribution of all possible V values is approximately normal[2] with the following mean and standard deviation.

$$\mu_V = \frac{N(N+1)}{4} \tag{19.5}$$

$$\sigma_V = \sqrt{\frac{N(N+1)(2N+1)}{24}} \tag{19.6}$$

where N indicates the number of pairs tested. In the cell-testing situation, $N = 8$ (the zero difference is not counted).

The value of interest, *3*, can now be transformed to a z score in the usual way using $\mu_V = \dfrac{8(8+1)}{4} = 18$ and $\sigma_V = \sqrt{\dfrac{8(8+1)(2 \cdot 8 + 1)}{24}} = 7.14$. Hence $z_V = \dfrac{V - \mu_V}{\sigma_V}$

[2] The approximation is better with larger sample sizes. If more exactness is required tables are available with critical V values for small samples.

$$= \frac{3 - 18}{7.14} = -2.10$$, which is smaller than the -1.96 necessary for rejection at the .05 level for a two-tail test.

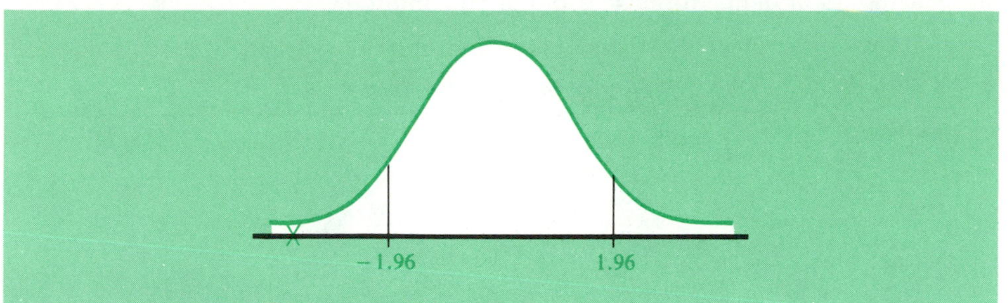

Fig. 19.2

On the basis of these statistical findings, our researcher is justified in concluding that a significant difference exists in the ways that cancer cells react to the two treatments in question.

19.5 The Rank Correlation Test—Alternative to Significance of Pearson r

One of the oldest and for many years the most widely used of all nonparametric methods is that known as *rank correlation*. It was originally devised as a shortcut method for estimating the value of a correlation coefficient. As the reader will soon see, it has the advantage of circumventing much of the computation involved in the calculation of the Pearson product-moment correlation coefficient. But in view of what has been said, when it is appropriate to compute a product-moment correlation coefficient and to test it for significance, it is often not satisfactory merely to compute a rank correlation coefficient instead.

In the application of this method we make no assumptions whatsoever about the distributions of the underlying populations. We do not assume normality either to compute the rank correlation coefficient or to test the hypothesis that there is zero correlation in the population. It is only necessary for us to be able to arrange the sample observations in rank order. Sometimes the ranking is actually done by judges, and at other times we must do the actual ranking of numerical scores we have obtained—when we are unable to assume that the scores are normally distributed. Each situation will be illustrated to show how we proceed to compute the rank correlation.

First we shall consider the situation in which the actual ranks are given to us, and later we shall consider the one where we rank the scores ourselves. Table 19.6 shows the ratings given to 10 contestants by two judges in an oratory contest. The 10 contestants are listed in Column (1). The judges are designated A and B. Each contestant's rating received from Judge A appears in Column (2) and from Judge B in Column (3). For instance, the first contestant received a rating of 1 from Judge A and a rating of 3 from Judge B. Similarly, each contestant's pair of ratings appears in Columns (2) and (3).

Our purpose now is to measure the correlation between Columns (2) and (3). In other words, our purpose is to measure the degree to which Judges A and B correlate in their ratings of the performances of the 10 contestants.

Table 19.6 Rank Correlation for Oratory Contest Ratings

(1) Contestant	(2) Judge A X	(3) Judge B Y	(4) $X - Y$	(5) $(X - Y)^2$
M	1	3	−2	4
N	2	2	0	0
O	3.5	1	2.5	6.25
P	3.5	8	−4.5	20.25
Q	5	7	−2	4
R	6	9	−3	9
S	7	4	3	9
T	8	6	2	4
U	9	5	4	16
V	10	10	0	0
				$\Sigma(X - Y)^2 = 72.50$

Let us look briefly at the ratings by Judge A, which appear in Column (2). The two ratings of 3.5 appearing in Column (2) indicate that Judge A was unable to decide between two of the contestants and was willing to divide third and fourth ranks between them $\left(\frac{3 + 4}{2} = 3.5\right)$. The ranks given by Judges A and B have been designated as variables X and Y, respectively. These are our variables of interest in this example.

The formula for the rank correlation coefficient requires the value of the square of the difference of the ranks given to each contestant. Accordingly, the differences, $X - Y$, have been found and are entered in Column (4) of Table 19.6. These differences have been squared and the squared values are shown in Column (5) of the table.

The formula that has been devised mathematically for the rank correlation coefficient is

$$R = 1 - \frac{6\Sigma(X - Y)^2}{N(N^2 - 1)} \qquad (19.7)$$

In Formula (19.7), R represents the coefficient of rank correlation that we are to find. $\Sigma(X - Y)^2$ represents the sum of the squares of the differences in ranks. That is, $\Sigma(X - Y)^2$ represents the sum of the terms in Column (5) of Table 19.6. Finally, N

stands for the number of individuals who have been ranked. We see that in our problem $\Sigma(X - Y)^2$ turns out to be 72.50, and $N = 10$. Thus, we have

$$R = 1 - \frac{6(72.50)}{10(10^2 - 1)}$$
$$= 1 - .44$$
$$= .56$$

We have found the value of R, the rank correlation coefficient for the ratings given by Judge A and Judge B, to be .56. Note that in this particular problem our original data were ranks. When this is the case—that is, when we do not have scores but only ranks to begin with—then the formula for r given in Chapter 16 would yield the same value as Formula (19.7),[3] except that Formula (19.7) is easier to apply. However, we still cannot consider our finding of .56 as a value of r, but we must think of it as a value of R. The distinction becomes meaningful when we come to test whether our finding is significantly different from zero.

Where our original data are ranks, we must of course compute a rank coefficient. Where our data are scores, and we suspect that the scores on even one variable are far from normal, we must hesitate before interpreting the correlation coefficient. A chance pairing of extreme scores might have accounted for much of the correlation (positive or negative). In such a case we may proceed by converting our scores to ranks, calculating R, and proceeding to test it for significance—a procedure we will shortly examine.

Without assuming normality, we could not ultimately take our obtained correlation coefficient and use it to test the hypothesis that the correlation in the population from which the sample came is zero. On the other hand, the rank coefficient has the advantage of leading to a test of this hypothesis that does not necessitate the assumption of normality. Therefore, when normality seems unlikely and we wish to correlate two variables, what we do is rank the scores on each variable ourselves in order to compute the rank coefficient.

As an illustration, consider the scores made by a class of 15 students on each of two tests, A and B. Let us say that the scores on Test A tended to be either very low or very high, since students were either well or poorly prepared for it. Clearly the scores on this test were far from normally distributed. Therefore, in order to measure the correlation between scores on the two tests, we are led to compute a rank correlation coefficient.

The scores of the 15 students on Tests A and B are given in Columns (1) and (3) of Table 19.7. (This time we have omitted a left-hand column listing the students themselves.) Our first step is to assign a rank to each score. The ranks for the scores on Test A are given in Column (2). These have been obtained by giving the highest score the rank 1, the next highest score the rank 2, and so on. Similarly, the ranks for the scores on Test B have been found and are given in Column (4). The reason for the

[3] The values are identical only provided that there are no ties in rank.

appearance of rank 7 three times in Column (4) is that three students had scores of 81 on Test B. These three students occupy ranks 6, 7, and 8, which have been divided among them $\left(\frac{6+7+8}{3} = 7\right)$. The values in Columns (2) and (4) have been labeled X and Y. They are the two variables of interest in our problem.

Next the values of $X - Y$, the differences in rank, have been found and entered in Column (5) and the squares of these values are in Column (6) of Table 19.7. The sum of the numbers in Column (6) is $\Sigma(X - Y)^2 = 164$. Applying Formula (19.7) we have

$$R = 1 - \frac{6\Sigma(X - Y)^2}{N(N^2 - 1)} = 1 - \frac{6(164)}{15(225 - 1)} = 1 - \frac{984}{15(224)}$$

$$= 1 - \frac{984}{3360} = 1 - .29 = .71$$

This is the value of the rank correlation coefficient between scores on Tests A and B as found for the 15 students.

Table 19.7 Rank Correlation for Test Scores

(1) Test A	(2) Rank	(3) Test B	(4) Rank	(5)	(6)
	X		Y	$X - Y$	$(X - Y)^2$
99	1	96	2	−1	1
98	2	95	3	−1	1
96	3	100	1	2	4
94	4	81	7	−3	9
88	5	88	5	0	0
70	6	65	12	−6	36
62	7	81	7	0	0
54	8	58	14	−6	36
47	9	90	4	5	25
41	10	81	7	3	9
35	11	72	10	1	1
33	12	50	15	−3	9
28	13	80	9	4	16
25	14	60	13	1	1
22	15	70	11	4	16
					$\Sigma(X - Y)^2 = 164$

In Section 16.5 we discussed the test for significance of a value of the Pearson product-moment correlation coefficient, r. The test presented there was subject to the assumption that the variables of interest could be taken as samples derived from normal distributions. However, the rank correlation coefficient, R, is calculated directly from ranks, and hence there is no assumption about the distributions of the variables of interest. Remember that this is true of all the nonparametric methods that we have discussed or shall discuss in this chapter.

It may be shown that, even when the variables are not normal, we may test R for significance as we have tested r. That is, we may use the transformation:

$$t = R\,\frac{\sqrt{N-2}}{\sqrt{1-R^2}} \tag{19.8}$$

where d.f. $= N - 2$.

Note that had we computed r instead of R in our second problem, we would not have been correct in using this transformation. In other words, when one or both variables are not normal, it is not correct to compute r and then to use Formula (16.3). One must first rank the variables and then compute R, which is apt to have quite a different value. Then one can use Formula (19.8), which transforms the obtained value of R into a t value, to test the null hypothesis that the correlation in the population is zero.

For example, suppose we consider the test-score data of the previous example as constituting a sample. Our theory is that there is a positive correlation between the two tests. Therefore we test the null hypothesis of no correlation between scores on Tests A and B. We shall use the .05 level of significance. We have $R = .71$ and $N = 15$. Using Formula (19.8) we have $t = .71\sqrt{15-2}/\sqrt{1-(.71)^2} = 3.64$. Remember, we are predicting that the correlation in the population departs from zero in the positive direction. Therefore we are doing a one-tail test for 13 d.f. Since the necessary value of t for significance at the .05 level for a one-tail test is 1.77, we are able to reject our null hypothesis and consider our theory as proved. The reader should verify that the value of R that we obtained is large enough so that it would have led to a t value that is significantly different from zero at the .01 level for 13 d.f. also.

The reader should also test the null hypothesis for the value $R = .56$ obtained for the oratory contest ranks in the first example of this section, and should verify that this hypothesis cannot be rejected at the .01 level for 8 d.f. The obtained value of $R = .56$ is one that would occur too often by chance to be considered significant at that level.

19.6 The Runs Test for Randomness

In all of our previous discussions of sampling and tests based on samples, it was assumed that the samples were random. This assumption may not always be justified. For instance, samples of the prices of stocks on the New York Stock Exchange taken at different times over a period of several months would likely be influenced by any general trends of the market over that period of time. As another example, samples of the speeds of automobiles passing a given point that are taken by recording the speeds of several cars in succession may very well be suspected of lacking randomness. Drivers who are caught behind a "slow poke" will also be driving slowly, while drivers following a speeder may also be exceeding the speed limit, perhaps unconsciously. In each of these examples, we see that there would be a tendency for successive items included in the samples to have similar values: high stock prices or slow speeds. Such a succession of similar values is known as a "run." The examples just cited would be suspected of producing long runs and, hence, relatively few runs in a given sample.

Statistical methods of the nonparametric type have been devised for testing data for the presence of an abnormal number of runs. A sample with too many runs is just as suspect as one with too few runs. Too many runs tend to indicate that the data occur in definite cycles. The tests based on runs, then, provide us with information for drawing conclusions about the randomness of a sample. Such information is particularly important in instances where it is necessary to utilize data recorded over a period of time in order to make predictions about future events. Such would be the case, for example, in certain aspects of meteorology, in studying the incidence of infectious diseases, and in studying fatalities resulting from drowning or other types of accidents.

To start with a simple illustration of a test based on runs, let us suppose that the speeds of cars on a certain residential street have been checked and compared with the speed limit, which is 25 miles per hour. For each car an "a" has been recorded if it was traveling at a speed above 25 mph, and a "b" if it was traveling at a speed below 25 mph. Cars traveling at exactly 25 mph have been omitted from the list. Let us suppose that 18 successive entries have been obtained as follows: a, b, b, b, a, a, a, a, a, b, b, a, a, b, b, b, b, b. We now state:

Definition 19.1 A run is a sequence of identical symbols that is preceded and followed by a different symbol (or no symbol at all).

Thus the first run in the sequence above consists of the single letter "a"; the next run consists of three "b's"; the next of five "a's"; and so on. In all there are six runs in the sequence of 18 letters.

Now, we may suspect that our sample has too few runs for it to be a random sample. The influence of "slow pokes" and speeders may have caused long runs so that the total number of runs in the sample is relatively small. We need some way to test the relative size of the number of runs in comparison to the sample. The mathematicians have come to our aid again, and have devised such a test. The application of this test does not depend on the lengths of the individual runs; but it does depend on the total number of times that each symbol appears in the sequence. We observe, in our example, that there are 8 "a's" and 10 "b's." Our theory is that the sample is not random so *the null hypothesis to be tested is that the sample is random.*

The method of approach to our test based on runs should be familiar by now. We assume that each of a vast number of investigators obtains a random sample consisting of 8 "a's" and 10 "b's." Under the null hypothesis, there is no tendency either toward or away from runs, but of course the obtained number of runs would vary from one sample to the next. This number of runs per sample is commonly designated by the letter u. Remember that we are assuming the hypothesis is true and conceiving of a collection of random samples, each consisting of 8 symbols of one kind and 10 of another, and it is under this condition that we are considering the distribution of u values that would be obtained.

Had there been different numbers of symbols involved in our sample, we would be concerned with a different distribution of u values, or what we call a different sampling distribution of u. That is, we do not think merely of a sample size, N, but rather of a sample of N_1 symbols of one kind and N_2 symbols of another kind. Here $N_1 = 8$ and $N_2 = 10$ (though we might have made it the reverse). In other words, in our case we conceive of our sample as one of a vast collection of similar random samples, in each of which there are 8 symbols of one kind and 10 symbols of another. This fact is important, since for each pair of values N_1 and N_2 we are led to consider a unique distribution of u values that would be found in different samples.

Our next step is to determine whether our obtained u value is sufficiently below or above the expectancy to indicate rejecting the null hypothesis. In our particular problem where $N_1 = 8$ and $N_2 = 10$, our obtained u value turned out to be 6. Specifically, we ask whether this obtained value is sufficiently far from the expectancy to indicate rejecting the null hypothesis. Appendix VII is set up for testing our hypothesis at the .05 level of significance. For each particular pair of values of N_1 and N_2 it give the critical values of u. Note that each cell entry in Appendix VII contains two numbers. The smaller number is called the left-tail critical value of u, and the larger number is the right-tail critical value of u, for a two-tail test at the .05 level. This means that the rejection region consists of values at or below the smaller number and the values at or above the larger number. If a one-tail test is to be performed at the .025 level the same critical values may be used as those found in Appendix VII. Of course only one critical score will be used.

Specifically, for any pair of values of N_1 and N_2, we find the appropriate cell in Appendix VII, which means the cell whose row is given by the smaller of the values of N_1 and N_2 and whose column is given by the larger of these values. Once we have found that cell, we reject the hypothesis only if our obtained value of u is either equal to or less than the smaller number in the cell or equal to or greater than the larger number in the cell. In our example, $N_1 = 8$ and $N_2 = 10$, and the observed value of u is 6. In using the table, since 8 is the smaller of N_1 and N_2, we find the row with 8 in the left margin; and since 10 is the larger we find the column with 10 at the top. We thus find the cell that is in this row and column. The two numbers in this cell are 5 and 15. This means that we would reject as not random a sample having 5 or fewer runs or a sample having 15 or more runs. Since in our sample $u = 6$, we fail to reject at the .05 level the hypothesis that our sample is random, and thus we have not proved our theory.

As a variation on our example, suppose that the first "a" had not been present in the sequence based on speed data. The sequence would then contain 7 "a's" and 10 "b's" and would have five runs. We now have $N_1 = 7$, $N_2 = 10$, and $u = 5$. In the table in Appendix VII we look in the cell that appears in the column with 10 at the top and in the row with 7 in the left-hand margin. The critical values given for u in this cell are 5 and 14. Since our observed value is 5, we would now reject the hypothesis that the sample is random and thus consider proved the theory that it is not a random sample. In this new example there are too few runs and the reason may involve the "slow pokes" and speeders previously cited.

The test using runs as we have presented it may be used in any situation where each item of the data is given as one or the other of two symbols. Another example would be the situation in which there are data concerning defective and satisfactory articles produced by a manufacturing process. Suppose we believe some piece of machinery is malfunctioning because too few runs are observed. This theory calls for a one-tail test. A sample of 20 nondefective and three defective articles would have to have three or fewer runs to prove our theory at the .025 level of significance. Note how we use Appendix VII to do the runs test as a one-tail test at the .025 level of significance. In such a case we would pick one of the critical values in the appropriate cell as our theory dictated, and we would then reject the null hypothesis only upon obtaining a u value at or beyond that critical value. To prove that there are too few runs for randomness, we would reject the null hypothesis only upon obtaining a u value at or smaller than the lesser critical value given. To prove that there are too many runs for randomness, it would be correct to reject the null hypothesis only upon obtaining a u value at or larger than the greater critical value given. The runs test may be done as a one-tail test and also at other levels of significance, though, of course, in each case appropriate tables would be needed.

In the first example of this section, there was a "built-in" reference point. The speed limit was 25 miles per hour and so it was natural to classify the speeds as above and below this value. In some situations where it is desirable to apply the runs test, there is no such "built-in" reference point. In such cases we will have to supply a reference point before we can classify the data as "above" and "below" and check on runs.

Some examples of the type of data just mentioned are weather observations and stock market prices. These do not appear in a form containing only two symbols. Before we can apply the runs test to them, we will have to convert the data into such a form. Then we will be able to proceed in the same manner as in the previous section and we can obtain a check on the randomness of the data.

We shall use an example from weather observations. Let us suppose that the number of inches of snowfall in 25 successive winters in our city has been recorded. The data appear as follows: 32, 45, 49, 42, 30, 67, 72, 73, 79, 25, 28, 36, 49, 52, 21, 42, 49, 67, 43, 75, 21, 17, 44, 42, 96. We are interested in whether the data appear in essentially random fashion or in patterns of some sort. Too few runs would indicate that the data tend to fall in rather long cycles of heavy and light snowfall, thus resulting in long runs. Too many runs would indicate that winters of heavy and light snowfall tended to alternate, thus resulting in short runs.

Before proceeding we must decide on a reference point. The median suggests itself as a good choice, owing to its definition as a value that is larger than or equal to half of the other values in the distribution and equal to or smaller than half of them. In the set of data above there are 12 terms that are greater than 44 and 12 terms that are less than 44, so the median is 44.

We now replace each actual term by one of the symbols "a" and "b," depending on whether the term is above or below the median. However, we omit values at the median—in this case the one term that is 44. After each value above the median has been replaced by an "a" and each value below the median by a "b," the resulting series is: b, a, a, b, b, a, a, a, a, b, b, b, a, a, b, b, a, a, b, a, b, b, b, a. We have now reduced the data to the proper form and thus are ready to apply the runs test. Our theory is that the data do not fall in a random pattern. The null hypothesis to be tested is that the data do constitute a random sample.

We have 12 "a's" and 12 "b's" and there are 12 runs in the sequence. Hence we are ready to use the table in Appendix VII with $N_1 = 12$, $N_2 = 12$, and our observed value of u is 12. We look in the cell that appears in the column with 12 at the top and in the row with 12 in the left margin. The critical values given for u in this cell are 7 and 19. Hence we cannot reject our hypothesis of randomness at the .05 level for a two-tail test. Note that were we using a one-tail test, whether right-hand or left-hand, we still could not reject the hypothesis of randomness, this time at the .025 level.

Problem Set A

For all problems in this set of exercises no assumptions may be made regarding underlying distributions. Apply only nonparametric tests.

19.1 I. M. Rock & Co. and U. R. Stone, Ltd., two independent jewelers, claim to sell the most brilliant type available of a certain kind of gem. To answer whether or not there really is a difference between the brilliance levels of the stones sold by the two firms, an expert is brought in to examine their merchandise. Fifteen gems are randomly selected from each jeweler and rated for brilliance on a scale from 1 to 15. Use the following results to test for significant difference at the .01 significance level.

Rock Co.	14.1	7.0	9.8	10.7	13.0	12.0	12.4	11.0
	11.9	11.8	13.5	11.7	12.0	9.4	8.0	
Stone, Ltd.	9.3	8.1	12.9	14.9	13.8	12.1	7.1	5.9
	9.8	10.5	12.3	6.1	8.0	11.4	7.5	

19.2 In an attempt to justify a new training program for its sales people, Sanders and Cohorts randomly select 10 sales representatives who have completed the course and compare their records before and after participation in the program.

Salesperson	*Before*	*After*
A	93	91
B	81	84
C	74	75
D	50	52
E	103	100
F	24	25
G	95	98
H	80	79
I	75	77
J	40	45

Use the sign test at the .05 level of significance to see if the new program is effective.

19.3 Two golf pros, brothers Stan and Ed Olson, have a continuing feud as to which has the more effective teaching technique. They each agree to give lessons to eight golfers for a month and then hold a tournament. The student golfers are picked in pairs so that each pair is made up of two students whose abilities and potential are as closely matched as possible before the lessons begin. The tournament results are below.

Couple	*Stan*	*Ed*
1	103	100
2	95	89
3	90	95
4	87	88
5	80	95
6	120	130
7	75	84
8	100	93

Does the Wilcoxon test detect any significant difference at the .01 level of significance?

19.4 Could a medical researcher who has obtained the following results conclude at the .01 level of significance that there is correlation between a person's weight and the number of infectious diseases he or she contracts during a year?

Weight	169	147	105	250	175	160	118	123
Number of diseases	0	1	0	6	4	3	6	2

19.5 Some of the trees in Farmer Rich's apple orchard have been infected by an unknown parasite. To determine the best method of treatment, Farmer Rich must first determine if the trees are infected at random or in groups. A row of 20 trees are found to be arranged as follows (I stands for infected; N stands for not infected): I, N, I, N, I, I, I, N, N, I, I, I, I, N, I, N, N, N, I, I. What results will the runs test for randomness yield at the .05 level of significance?

19.6 Test the theory that athletes from Communist nations receive higher ranks in Olympic competition than those from non-Communist nations. Use the following random samples and the .05 level of significance.

Communist (ranks)					*Non-Communist (ranks)*				
1	9	4	5	9	1	8	9	10	8
6	8	1	2	3	7	3	3	14	11
4	3	15	2	9	2	1			

19.7 Consider the following 15 pairs of college entrance examination scores and college grade-point averages for four years as coming from populations not necessarily normal. Compute the rank correlation coefficient and test it for significance using a two-tail test at the .01 significance level.

Entrance Examination	*Grade-Point Average*	*Entrance Examination*	*Grade-Point Average*
419	2.18	579	2.79
465	2.34	591	2.90
495	2.23	616	2.65
522	2.47	643	3.38
530	2.51	659	2.97
532	2.14	687	3.32
554	2.42	704	3.46
569	3.04		

19.8 U-Bet Insurance Associates are trying to convince executive women that their lives are underinsured when compared with men in similar situations. It presents the following 12 examples. In each couple the male and female executives are comparable when income, financial obligations, age, and earning potential are considered.

Present Life Insurance (thousands of dollars)

Male	*Female*	*Male*	*Female*
210	180	250	280
150	125	190	150
500	480	450	480
85	90	180	120
325	315	100	50
450	425	135	120

Use the Wilcoxon test at the .05 level of significance to see if the company's claim is substantiated by the data.

19.9 Successive weights (in milligrams) of a certain chemical precipitate were recorded as follows: 17, 15, 18, 15, 18, 15, 15, 17, 17, 14, 19, 15, 18, 17, 15, 14, 17, 14, 18, 14. Could a researcher conclude at a .05 level of significance that "heavy" and "light" precipitates do not occur randomly?

19.10 A new petroleum additive is advertised as being able to increase gas mileage. To test this claim consumer advocate James R. Dodd tests the additive against regular gasoline in pairs of identical automobiles under identical driving conditions. Should the results below (given in miles per gallon) compel Mr. Dodd to speak out against the manufacturer? Use the Wilcoxon test at the .05 significance level.

Type of Car	*Car A (with additive)*	*Car B (without additive)*
Ford	15.7	15.0
Chevrolet	18.2	18.2
Oldsmobile	20.0	19.0
Mazda	25.3	25.0
Mercedes	23.1	20.0
Dodge	19.8	19.1
Cadillac	13.9	12.3
A.M.C.	17.4	17.5
Lincoln	15.3	16.7
Volkswagen	28.0	25.0

19.11 Two years ago Foster, Inc., a large manufacturer of depth finders, instituted an employee incentive program to reduce absenteeism. The records of 15 employees were randomly selected from the files and the average number of absences per quarter before and after the program was started are noted as below.

Employee	1	2	3	4	5	6	7	8
Before	1	0	3	4.5	7.5	1	0	3
After	3	0	2.5	0	7	0.5	4	2

Employee	9	10	11	12	13	14	15
Before	1.5	2	2	4	3	4	8
After	1	1	0	3.5	2	3	6

Do these data show any significant decline in absenteeism? Use the sign test at the .05 significance level.

19.12 In deciding which of two contractors to hire to install a new reactor system, the city manager calls for a study to determine if there is any difference between the numbers of accidents that have occurred in plants constructed by the two contractors. A random sample of already constructed plants yields the following information concerning number of accidents in the last 10 years. Conduct the appropriate test at the .05 level.

Contractor A					*Contractor B*			
30	28	33	26	28	40	45	33	27
29	24	32	38	28	30	29	35	30
23	25	29	32	35	39	32	41	34

19.13 A local theater operator has noticed that more people seem to attend motion picture showings at the beginning of each month than at the end. In fact, he maintains that attendance declines throughout each month. To test his theory he randomly selects 20 days during the past six months and records the daily attendance.

Date	1	31	9	16	15	17	2	29	25	20
Attendance	297	94	153	208	234	273	350	101	84	122

Date	3	5	18	30	15	8	8	11	13	14
Attendance	300	250	180	100	203	215	310	235	207	207

a. Calculate the rank correlation coefficient R for the above data.
b. Test R for significance using a one-tail test at the .05 level of significance.

Problem Set B

For all problems in this set of exercises no assumptions may be made regarding underlying distributions. Apply only nonparametric tests.

19.14 A city health officer believes that the swine flu does not occur at random times in her city. The numbers of cases reported in successive years are as follows: 5, 8, 6, 3, 1, 1, 5, 7, 9, 10, 8, 7, 4, 2, 7, 1, 5, 6, 8, 1, 7. Test the officer's belief using a two-tail test for randomness at the .05 significance level.

19.15 The weights of 12 people before they stopped smoking and six weeks after they stopped smoking are as follows:

Before	127	157	145	162	181	172	169	156	187	115	118	173
After	136	161	137	164	190	180	162	164	165	118	117	180

Do the above results support the theory that to quit smoking increases a person's weight? Apply the sign test at the .01 significance level.

19.16 To test for difference of effectiveness between two acne creams, 10 teenagers who had this problem were asked to use preparation A on one side of their face and preparation B on the other side for a month. A dermatologist then judged the improvement in each case. The results are given below. Higher scores show more improvement. Use the Wilcoxon test at the .01 level of significance to test for difference in the effectiveness of the two preparations.

Patient	*A*	*B*	*Patient*	*A*	*B*
1	85	80	6	45	70
2	15	10	7	73	79
3	70	77	8	64	60
4	65	62	9	47	65
5	50	49	10	95	97

19.17 B. K. Oil and Gas is considering the installation of a new type of pumping mechanism for its newly found oil fields in the Gulf of Mexico. It surveys 50 pumping stations where the new mechanism has replaced traditional techniques. Thirty of them report an increase in the volume of oil pumped. Fifteen show a decrease, and the remaining five report no change. If B. K. is willing to accept a 5% chance of making a type-one error (changing to the new system and realizing no gain), what decision should it make?

19.18 A college dean suspects that graduating classes with high grade-point averages appear in spurts (too few runs). She notes that in the last 30 years 18 classes have been high and 12 have been low. How many runs would have to appear to support her theory at the .025 significance level?

19.19 Dr. Swift, an avid jogger, maintains that aerobic exercise significantly reduces one's resting respiratory rate. His good friend, Dr. Sid N. Tary, takes issue with his claim and sets out to show there is no statistical basis for it at the .01 significance level. He randomly selects a number of healthy 25-year-old males, joggers and nonjoggers, and measures their respiratory rates as below. Whose claim is supported?

Nonjoggers 18, 16.1, 17.3, 15.1, 17, 17.3, 16.4, 16.5, 17.3, 15.9, 14, 14.9, 17.4

Joggers 17.3, 18.1, 14.9, 15.7, 15.8, 16.3, 16.4, 17.9, 17.8, 14.8, 17.2, 17.5

19.20 Believing that the national ranking of a major university's football team is negatively correlated with the national ranking of its basketball team, sports information director Leon Kat gathers the following pairs of rankings from 15 major colleges.

School	*Football Rank*	*Basketball Rank*
A	2	25
B	3	18
C	8	2
D	11	5
E	14	9
F	15	11
G	18	15
H	23	27
I	25	3
J	28	26
K	30	1
L	32	29
M	38	16
N	40	13

Compute the rank correlation coefficient, and test Leon Kat's claim at the .01 level of significance.

19.21 It is hoped that a new fertilizer will increase the heights of blooming plants. Nine pairs of matched plants are grown for six months. One of the plants in each pair is grown using a conventional fertilizer. To the other is applied the new compound. The height in inches of each plant is found below. Use a one-tail Wilcoxon test at the .05 significance level to test the null hypothesis that no difference exists between the heights of the two groups.

Pair	1	2	3	4	5	6	7	8	9
Conventional	14.3	9.0	8.0	7.3	5.0	10.2	7.5	4.0	3.0
New fertilizer	12.0	10.3	9.1	7.4	6.3	9.3	7.6	4.9	2.6

19.22 In a quality-control process, manufactured items are removed intermittently from a production line and scored by inspection. The scores for a particular time interval are 7.2, 7.3, 7.0, 6.9, 7.1, 7.2, 7.2, 7.3, 7.2, 7.0, 6.9, 6.8, 6.9, 7.0, 7.1, 7.2, 7.2, 7.3, 7.0, 6.9, in consecutive order. Test for randomness using a two-tail test at the .05 significance level.

19.23 Because of a large number of statement errors, Trust City Bank has computerized its statement department. Was the changeover successful in cutting the number of errors according to the following sample showing the number of errors reported per day for the first 20 days of the month before computerization and the first 20 days of the month after computerization?

Day	*Before*	*After*	*Day*	*Before*	*After*
1	15	13	11	10	9
2	17	18	12	9	10
3	25	20	13	9	5
4	30	35	14	8	3
5	30	15	15	12	9
6	22	20	16	8	4
7	18	15	17	7	7
8	17	10	18	6	4
9	15	10	19	7	1
10	12	12	20	9	0

Use the sign test at the .05 level of significance.

19.24 To substantiate its claims concerning the value of its product, a manufacturer of a treatment for vinyl car roofs treats 14 car tops with the product and sends each car through a car wash numerous times. Fourteen cars with untreated vinyl roofs are also subjected to numerous washings. The number of trips through the washer required before the roofs show a certain level of wear (as judged by a well-known impartial consumer advocate) are shown below. Test the manufacturer's claim of effectiveness at the .05 level of significance.

Treated				*Untreated*			
98	64	82	83	81	76	50	72
82	93	74	55	70	72	89	81
73	64	64	89	62	66	49	53
94	69			53	55		

19.25 The owners of "The Browsers," a chain of gift shops, wish to determine if an increase in advertising brings about a corresponding increase in sales. Twelve months are randomly selected for the study. The advertising budget and total sales are presented below (both in thousands of dollars).

Advertising	1.3	4.0	2.7	.75	4	2.8
Sales	80	115	95	95	120	125

Advertising	3.9	5.8	.9	1.8	3.1	2.3
Sales	130	183	77	125	130	142

Test the rank correlation coefficient R at the .01 level of significance.

Review IV

The final five chapters, Chapters 15 through 19, have equipped us to apply statistical methods to many more practical situations. They have done so by providing specific procedures to address problems in which (1) two variables are involved or (2) no assumptions of normality can be made for the distributions involved.

The identification of two variables in a population often greatly enhances our ability to predict what value one of them may assume. Of course, predictions can be made without reference to a second variable. In such cases the mean of the population is the best guess that can be made. Naturally, an error is involved with each guess or prediction; that error is the difference between the guess and the true value assumed by the variable. The error of a group of such guesses has been defined in the classical way as the mean of the squares of the individual errors. In the long run, for repeated guesses, this becomes the variance of the distribution. Using knowledge of a second variable to predict the value a first variable will assume (a method known as regression) clearly increases our accuracy in making predictions, since in most cases the mean of the squared errors is less than the variance. However, before regression can be of any value, the relationship between the two variables must be determined. Graphs of paired values, called scattergrams, often help suggest what the relationship and the regression (or prediction) curve might be. The precise regression curve locates the mean value obtained for the variable being predicted for a multitude of instances in which a specific value of the predictor variable is assumed. Procedures for constructing scattergrams and regression curves are discussed in Chapter 15.

Chapter 16 examines one specific kind of relationship—correlation. It often greatly simplifies the making of predictions. When two variables are correlated, one tends either to increase or to decrease uniformly as the other increases, depending on whether the correlation is positive or negative. The strength of the tendency is indicated by the correlation coefficient r. This coefficient ranges from $+1$ to -1; a stronger correlation is indicated by values far from zero and close to $+1$ or -1. r values of $+1$ and -1 are unusual, but they indicate perfect uniformity in the change of one variable as the other changes. Two procedures are commonly employed for determining the value of the correlation coefficient. Either the mean of the z score cross products is computed, or Formula (16.1), which requires only the original data, may be used.

Often it is desirable or necessary to detect correlation between two variables in a large or infinite population. In such cases, a random sample is taken from the population. Almost always some correlation, as measured by the correlation coefficient r, will be detected in the sample. Its significance can be determined by using the t test outlined in Chapter 16.

The concepts of regression and correlation are brought together in Chapter 17. Linear regression, a much used tool for predicting values, involves the use of a straight line as the regression curve. The most suitable line is the best-fitting straight line to the scattergram of paired values of the predictor and predicted variables. The equation of this line serves as the prediction formula for the value the variable will assume ($\widetilde{Y}$). The formula $\widetilde{Y} = \mu_Y + b(X - \mu_X)$ depends upon finding the slope b of the line, which in turn is known to depend upon r, the correlation between the variables. Either the relationship between b and the correlation coefficient, $b = r\,\dfrac{\sigma_Y}{\sigma_X}$, or Formula (17.2) may be used to calculate b. Once b, μ_X, and μ_Y have been found, the prediction formula may be written down and used. As Formula (17.4) reveals, the accuracy of predictions made using the prediction formula increases as the correlation coefficient moves away from zero in either the positive or negative direction. In fact, when $r = \pm 1$, predictions can be made with perfect accuracy, as the error variance will be zero.

The final two chapters of this text address nonparametric statistics. As more and more situations arise in which no assumptions of normality can be made in underlying distributions, nonparametric methods increase in importance. They call for no such assumptions.

The chi square test for independence described in Chapter 18 is used to detect relationships between two variables. Unlike the t test for significance of correlation, it is used for variables whose values can be placed in categories. The test compares observed frequencies with expected frequencies. If they are significantly different from one another, the null hypothesis stating independence may be rejected, thus allowing acceptance of the alternative that states that the variables are related. Once two variables are found to be related we know that knowledge of one will help determine the probability that the other will assume specified values.

The chi square statistic $\left(\Sigma \dfrac{(O - E)^2}{E}\right)$, which is used in the test for independence, is also used in tests for "goodness-of-fit." These tests determine if an observed sample is enough like a large population to be a random sample from it.

In addition to chi square tests, numerous other nonparametric tests have been developed for use in situations in which no assumptions regarding normality in underlying distributions are justified. Each nonparametric test replaces a parametric test that would be used if necessary assumptions could be made. Because they are easier to apply than parametric tests, the temptation is to use them as substitutes, even when the parametric test is clearly called for. This temptation must be resisted. The five tests described and illustrated in Chapter 19 (the Mann-Whitney test, the sign test, the Wilcoxon signed-ranks test, the rank correlation test, and the runs test) are among the most commonly employed. The reader should not only know how to apply each test, but should be able to identify situations in which they are called for.

Problem Set A

IV.1 A small business has gradually been converting its weekly billings to a computerized system. During this period, the mean number of reported errors per week has been 35 with a standard deviation of 10. The percent of bills handled by computer varies from week to week. The mean numbers of errors reported are given here for five percentages, each representing the percent of billings handled by computer.

Percent	*Mean Number of Errors*
.35	53
.55	42
.60	25
.75	16

a. Draw a scattergram and sketch the regression curve.
b. Predict the number of errors reported during weeks in which 40%, 50%, and 65% of the billings are handled by computer.
c. Below are the numbers of errors reported during 15 billing periods. The percent of billings handled by computer for the week is given at the top of each column.

40	*50*	*65*
48	41	25
47	39	23
53	42	22
45	44	21
45	43	22

Compute the error variance for each group, assuming the predictions in part b had been used.

d. Predict the number of errors reported in a week selected at random. If this prediction is made repeatedly, what will the error variance be in the long run?

IV.2 A leading sociologist has noted a change in society's attitude toward divorce and remarriage. In his book, *The Fork in the Road*, he discusses in detail 10 case histories in which divorced women remarried. The ages of the women and the time spans between divorce and remarriage are given here.

Age at Divorce	*Time to Remarriage (in years)*
23	1.0
25	2.5
29	7.0
32	3.0
35	5.5

Age at Divorce	Time to Remarriage (in years)
40	4.0
43	5.0
48	7.5
50	6.0
56	6.0

a. Plot a scattergram.
b. Calculate the correlation coefficient between the two variables using the raw data.
c. Calculate the correlation coefficient by finding the mean of the z score cross products.

IV.3 The school health nurse believes that children from larger families are absent more often than those from smaller families. School attendance records for 51 children show a correlation of $+.52$ between family size and absences. Is the nurse's belief supported at the .05 level of significance?

IV.4 The amount of a certain substance present in harvested oats varies with the amount of precipitation during the spring months. Ten samples of the crop taken from one agricultural region were tested for the presence of the substance. The results from each sample (presence of the substance in parts per million) follow.

Sample	(X) Rainfall	(Y) Toxic Substance
1	10.7	38
2	10.0	40
3	9.8	49
4	6.5	90
5	9.1	63
6	8.4	79
7	6.9	86
8	7.2	83
9	8.5	78
10	7.0	91

a. Plot a scattergram of the paired values.
b. Find the prediction formula for $\tilde{Y}$.
c. This year's rainfall was 9.5 inches during the spring months. Predict how much of the substance will be found in the region's oat crop this year.

IV.5 Successful treatment of patients confined for severe emotional and mental problems has long been known to depend partly upon patient contact with the outside world. The mean number of weekly visits for each of 16 patients at one facility was determined. Those means, along with the numbers of weeks of required hospitalization, are provided here.

Patient	Mean Number of Weekly Visitors	Length of Stay (in weeks)
A	0	29
B	2	14
C	3	17
D	2.5	14
E	3.5	9
F	6	5
G	2.7	13
H	5	7
I	4.8	6
J	7	6
K	3.4	12
L	1.2	17
M	1	23
N	0	22
O	1	25
P	3.2	11

a. Find the prediction formula for the length of stay, given the mean number of weekly visitors.

b. Assuming each of the three members of a new patient's family visits the patient twice a week, and there are no other visitors, predict the length of the patient's stay.

c. What proportion of the variance in lengths of stay can be explained by the variance in the mean number of weekly visits?

IV.6 When determining rates, an insurance company considers a home's distance from the nearest fire station. In one large metropolitan area, the mean rate charged is $215 per single-family dwelling. The mean distance from a fire station is 1.75 miles. If the rate for a home that is 2.5 miles from the nearest station is set at $295, what is the rate for a home that is only .5 mile from the nearest station? Assume a linear relationship.

IV.7 Truancy among the city's public school teachers has become an almost epidemic problem that is costing the district a considerable sum. The superintendent has ordered studies in an attempt to deal with the problem better. Among the data collected are the results of a random sample of 250 of the district's educators. Each is classified by the mean number of days missed per year: habitually absent (12 or more), frequently absent (9–11), and seldom absent (8 or fewer). Each teacher in the sample is also identified as having attended either a private or public college or university. The results are summarized here.

	Habitual	*Frequent*	*Seldom*
Private	38	35	25
Public	59	62	31

a. Are the variables related in the sample?
b. Do the results indicate that the variables involved are related in the population at the .05 significance level?

IV.8 In an attempt to find ways to appeal to the college-age market, fashion designers are constantly studying the traits of college students. Three years ago, 30% of all students on the nation's college campuses were 18–21 years of age, 25% were between 21 and 25, 15% were between 25 and 30, and 30% were older than 30. A recent sample of 200 students contained 43 between the ages of 18 and 21, 32 between 21 and 25, 68 between 25 and 30, and 57 over the age of 30. Does the age distribution of today's college students differ at the .10 significance level from the distribution three years ago?

IV.9 Children who learn to swim before they are a year old generally develop muscle coordination earlier than children who do not. Does the experience also have an effect upon their performance on intelligence tests? Fifteen pairs of identical twins were identified at birth as test subjects in an experiment aimed at helping to answer this question. One twin in each set was taught to swim. At age 3, the 30 subjects were verbally tested by a panel of three experts who were not told the reason for the test. The following intelligence scores were arrived at.

Twin Set	*Group A (Swimmers)*	*Group B (Nonswimmers)*
1	86	85
2	49	53
3	77	70
4	63	60
5	59	53
6	99	84
7	77	79
8	83	85
9	67	65
10	73	70
11	78	72
12	74	70
13	86	88
14	73	71
15	69	60

Can a significant difference be detected between the two groups? Use the Wilcoxon test at the .05 significance level.

IV.10 Two test groups of 20 randomly selected high school students are given identical material to study for the same time period. Then identical reading-comprehension tests are administered. The only difference in the treatment of the two groups is that Group A studied from manuscripts written in blue ink and Group B's material was printed in black. Do the test results presented here support a difference at the .01 level of significance? Make no assumptions regarding the underlying distributions.

Group A:	79	83	52	107	73	64	87	93	84
	77	103	107	42	57	66	71	79	83
	93	87							
Group B:	104	73	74	86	92	101	82	50	53
	67	84	97	73	72	68	99	49	67
	83	87							

IV.11 A manufacturer of electric toothbrushes employs six inspectors to check finished products for defects. After a product passes inspection, the inspector's number is stamped on it. Of 348 sets returned because of customer dissatisfaction, 67 were inspected by inspector A, 53 by B, 74 by C, 41 by D, 60 by E, and 53 by F. At the .05 significance level, do these statistics indicate that the number of complaints received is not the same for each inspector?

IV.12 In response to growing concern over teenage alcoholism, one school system initiated a study in an attempt to isolate factors that might contribute to the problem. A random sample of 435 high school students was asked to fill out questionnaires anonymously. Answers to the questions "How often do you consume alcohol?" and "How many hours per week are you employed outside of school?" are summarized below.

Hours	*Alcohol*			
Employed	*Never*	*Seldom*	*Occasionally*	*Frequently*
0–5	20	32	83	29
5–10	15	41	75	30
10–20	11	22	40	14
20–40	5	9	3	6

Can it be concluded that the number of hours a student is employed per week is related to the frequency with which the student consumes alcohol? Use the .05 level of significance.

IV.13 An industrial plant was granted a temporary permit to dump treated wastes into the East River. Environmentalists objected to the permit on the grounds that the waste contains a chemical that has been proved harmless to humans, animals, and fish, but that is suspected of being detrimental to the microscopic organisms that help complete the life cycle of the river. Each side has had one year to prove its case before a permanent permit is granted or denied. Because the level of organisms naturally varies during the year, the environmentalists took a different sample of river water on the 15th of each month during the test period. The results are now compared with the results obtained from samples taken on the 15th of each month prior to the granting of the temporary permit. Use the sign test and the .01 significance level to determine if the results indicate that the organism level has dropped during the one-year period that wastes have been dumped into the river.

Month	*Before*	*After*
January	86	89
February	93	93
March	105	100
April	115	107
May	123	120
June	119	129
July	114	100
August	111	82
September	108	97
October	77	63
November	85	72
December	80	70

IV.14 The editor of a consumer magazine is planning a feature article on paint jobs for used cars. In preparation, he takes similar used cars to 10 establishments whose refinishing services vary greatly in price. He has the finish replaced on each automobile. The cars are then subjected to identical environmental conditions for 11 months. At the end, the finishes are ranked according to durability. Larger ranks indicate greater durability. Do the results indicate that the durability of the finishes is positively correlated with the price at the .05 significance level? Make no assumptions of normality.

Car	*Durability*	*Price*
1	3	149.50
2	4	150.00
3	10	400.00
4	5	225.00
5	9	189.00
6	6	389.00
7	7	229.50
8	1	110.00
9	8	250.00
10	2	99.50

IV.15 Do winners at a carnival booth where darts are thrown at balloons appear at random? An observer from the city police department notes a string of 30 players in which winners (W) and losers (L) appear in the following order.

L L L W W L L W L L W W L L L L L L L L W L L W L W L L W W

Apply the appropriate test at the .05 significance level.

Problem Set B

IV.16 When a new manufacturing process is used, the cost per article declines as the number of articles produced per day increases. The means of the mean costs per article for six production levels are given below. (Note, we are taking the mean of many means.)

Production Level (number per day)	*Means of Mean Cost per Article (in dollars)*
121	95
103	103
96	108
85	115
80	122
76	129

a. Draw a regression curve and predict the mean cost per article for days on which 100, 115, and 90 articles are produced.

b. Suppose that for one week (five working days), production levels are held to 80 articles per day because of a shortage of raw materials. Suppose also that the mean costs for the five days prove to be \$125, \$120, \$123, \$123, and \$127. What would the prediction be for each of the days (based on the regression curve)? What would the error variance be?

IV.17 Compute the correlation coefficient between (a) weight and mean daily calorie intake and (b) weight and mean time spent exercising per day for the eight high school girls listed here.

Student	*Weight*	*Mean Calorie Intake*	*Time Spent Exercising (in minutes)*
Martha	108	2,475	25
Mary	97	2,350	15
Anne	115	2,500	10
Shelly	95	2,200	0
Sherry	103	2,580	5
Cindy	118	2,650	20
Susan	123	3,000	60
Nancy	85	2,000	15

IV.18 New secretaries hired by Alco International are generally assigned to the typing pool. They are told that the length of time they remain in the pool depends upon their performance. Among the criteria used for promotion are typing speed and accuracy. Sam Steno has been in the pool for nine months, and although he types 80 words per minute without error, he has yet to be promoted. The typing speeds (with no errors) and promotion times of the last 11 secretaries promoted are given here.

Words per Minute	*Months to Promotion*
65	5.8
73	4.3
62	6.4
59	8.9
73	7.3
81	2.0
77	3.6
62	4.0
51	5.9
49	5.0
61	8.5

Compute the correlation coefficient r and test it for significance at the .05 significance level.

IV.19 A prospective buyer has asked the owner of a home for copies of the electric bills for the past year. Using the bills and mean monthly temperatures provided by the weather bureau, the buyer constructs the following table.

Month	*(X)* *Mean Temperature*	*(Y)* *Kilowatt-Hours Used*
March	40	1,137
April	42	883
May	58	1,025
June	73	1,149
July	84	2,272
August	85	1,959
September	71	1,534
October	53	1,105
November	40	500
December	31	909
January	28	1,287
February	25	1,420

a. Plot a scattergram.
b. Draw in the best-fitting straight line. What is its slope?
c. Predict the number of kilowatt-hours that will be used during a month whose mean temperature is 50 degrees, 73 degrees, and 29 degrees.

IV.20 The amount of time that a mosquito repellent (KO-7) is effective is linearly related to the amount of repellent used. Twenty-five test patches sprayed with varying amounts of KO-7 were timed for effectiveness. The mean time for all 25 patches was 2.8 hours ($\sigma = .5$ hour). The mean amount of repellent per square foot tested was .45 ounce ($\sigma = .1$ ounce). The resulting formula predicts that a patch with .5 ounce per square foot will repel for three hours. Use the prediction formula and Formula (17.1) to determine r, the correlation coefficient.

IV.21 Seventy-five men are randomly approached in a shopping center and asked to select their favorite from among four samples of perfume. Do the results below indicate any difference in preference at the .05 level of significance?

Night Air	*Spring Splash*	*Rose Petals*	*Honey, Do!*
21	16	13	25

IV.22 The agent in charge of inventory and purchasing for Audio Corporation, which owns a nationwide chain of music stores, has been noting a gradual, but steady, decline in record sales during the past few years. The beginning of the decline corresponds with the time that inexpensive recording and tape-copying devices were introduced into the market. The agent collects weekly figures for his chain that indicate the mean number of LPs sold per store and the mean number of tape-copying devices sold per store. Below are the figures for 10 randomly selected weeks.

(X) *LPs*	*(Y)* *Tape Devices*
437	30
845	18
673	25
380	32
417	32
391	41
633	27
784	17
521	28
888	13

a. What is the prediction formula for $\widetilde{Y}$?

b. Predict the mean number of tape-copying devices that would be sold per store during a week in which the mean number of LP record sales is 550.

c. Calculate the error variance for prediction using the formula found in part a.

IV.23 Stress factors have long been associated with heart disease. In an attempt to determine if cheating on one's income tax is accompanied by the kind of stress related to heart ailments, Dr. A. Orta randomly selects 500 taxpayers. Seventy-five are found to be suffering from some type of heart problem. Of those, 22 have knowingly cheated on tax forms. Of the 425 individuals with healthy hearts, 80 have misrepresented the facts at income tax time. At the .01 level of significance, is heart disease related to one's willingness to file inaccurate tax information?

IV.24 The general education requirement in the English Department can be satisfied by completing one of five courses. Figures for the past 10 years indicate that of the total enrollment in these five courses, the percentage breakdown by course is

Humanities I	35
Composition I	22
English Literature	10
American Literature	15
Speech I	18

Preenrollment figures show 100 students signed up for Humanities I, 85 for Composition I, 30 for English Literature, 35 for American Literature, and 93 for Speech I. If the preenrolled students represent a random sample of students, does it appear that the proportion of students selecting each option may change? Use the .01 significance level.

IV.25 Seventh National Bank and Trust established 24-hour computerized banking centers at five strategic locations a year ago. The services provided are to be reevaluated annually. Each site offers services in four areas: savings, checking, loan repayment, and bill paying. The following contingency table gives the number of transactions that fits into each category from random samples of size 100 from each site.

	Savings	*Checking*	*Loan Repayment*	*Bill Paying*
Residential	34	30	8	28
Industrial	22	39	23	16
Commercial	30	36	20	14
Shopping Center	25	40	15	20
Apartment Complexes	29	36	21	14

At the .05 level of significance, is the service utilized related to the location of the facility?

IV.26 A job recruiter visits two different universities each fall. Job applicants are asked to fill out standard applications that call for references from three professors. The reference form asks the professor to score the student on a scale from 1 to 25. The interviewer has noted that faculty reference scores seem to be higher as a general rule at one of the schools (School B) even though the caliber of the applicants seems no higher. To test his theory, he finds the means of the three faculty scores for 15 applicants from each school. Use the results below to test the recruiter's belief at the .05 significance level.

School A (mean scores): 15.7, 23.3, 20, 18, 16.7, 16.3, 16.3, 17, 18.3, 21, 15, 21.3, 19, 18.7, 16

School B (mean scores): 24.7, 17, 23.7, 24.3, 21, 15, 16.7, 24.3, 16.3, 17.7, 18, 19, 21, 24, 22

IV.27 The statistics class is performing an experiment to determine if one of two popular calculators is faster than the other for most students. The 50 members of the class are divided into two groups of 25. Members of Group I work with calculator A first, while members of Group II work with calculator

B first. Each student must practice on one calculator for 20 minutes and then perform a number of computations while being timed. Then the student must practice for 20 minutes using calculator B and perform similar computations, again while being timed. Only the completion times (in minutes) of the 28 students who obtained correct results for all computations on both calculators are used. They appear below.

Student	*A (time)*	*B (time)*	*Student*	*A (time)*	*B (time)*
1	12	16	15	16	12
2	11	17	16	15	13
3	10	15	17	14	18
4	15	19	18	12	19
5	9	12	19	14	20
6	13	11	20	13	19
7	14	16	21	11	18
8	16	18	22	8	16
9	15	15	23	9	19
10	18	14	24	11	16
11	10	12	25	12	18
12	15	11	26	13	18
13	21	14	27	17	18
14	17	11	28	17	19

What conclusion should the class arrive at if a sign test is performed at the .05 significance level?

IV.28 A small-town grocer believes that ice cream sales are positively correlated with the number of flu cases in the town's elementary school. For 10 weeks he records his ice cream sales (in gallons), and his wife, who is the principal of the elementary school, reports the number of absences attributed to the flu. The results follow.

Week	*Ice Cream Sales*	*Number of Flu Cases*
1	35	15
2	29	12
3	41	21
4	28	8
5	27	6
6	36	5
7	18	14
8	32	11
9	23	6
10	25	5

Calculate the rank correlation coefficient and test it for significance at the .05 significance level.

IV.29 The feature editor of a college newspaper has been assigned to write a story dealing with student satisfaction or dissatisfaction with the campus cafeteria.

He interviews a string of 25 students as they leave the dining hall one Thursday evening. The feelings of the 25 students, satisfied (S) or dissatisfied (D), are: S S S D D D D S S D D S S S S D D D D S S S D D D. Do the responses fail to appear randomly at the .05 significance level?

IV.30 In an attempt to cut the number of industrial accidents that occur each year, the state Health and Safety Commission has prepared a two-day seminar, which it presents without charge to groups of persons employed in hazardous industries (those in which 50 or more accidents requiring medical attention occurred during the previous year). Fifteen companies requested the presentation for their employees last year. The numbers of accidents for the 12-month period preceding the seminar and the 12-month period following the seminar are given below. Use the Wilcoxon test at the .01 level of significance to determine if the program has been effective.

	Number of Accidents	
Company	*Before*	*After*
A	68	52
B	75	73
C	83	85
D	107	100
E	52	57
F	77	78
G	89	82
H	97	90
I	92	71
J	84	83
K	118	89
L	135	115
M	56	59
N	158	120
O	75	72

Appendixes

I	Tables of Square Roots	A2
II	Normal Curve z Scores and Percent of Scores Exceeded	A3
III	Normal Curve Areas	A4
IV	The t Distribution	A5
V	Values of F	A6
VI	Values of χ^2	A10
VII	Critical Values of u (Total Runs)	A11
VIII	Math Review	A12
IX	Answers to Odd-Numbered Problems	A15

I Tables of Square Roots

n	$\sqrt{n}$	$\sqrt{10n}$	n	$\sqrt{n}$	$\sqrt{10n}$	n	$\sqrt{n}$	$\sqrt{10n}$
1.00	1.00000	3.16228	4.10	2.02485	6.40312	7.10	2.66458	8.42615
1.10	1.04881	3.31662	4.20	2.04939	6.48074	7.20	2.68328	8.48528
1.20	1.09545	3.46410	4.30	2.07364	6.55744	7.30	2.70185	8.54400
1.30	1.14018	3.60555	4.40	2.09762	6.63325	7.40	2.72029	8.60233
1.40	1.18322	3.74166	4.50	2.12132	6.70820	7.50	2.73861	8.66025
1.50	1.22474	3.87298	4.60	2.14476	6.78233	7.60	2.75681	8.71780
1.60	1.26491	4.00000	4.70	2.16795	6.85565	7.70	2.77489	8.77496
1.70	1.30384	4.12311	4.80	2.19089	6.92820	7.80	2.79285	8.83176
1.80	1.34164	4.24264	4.90	2.21359	7.00000	7.90	2.81069	8.88819
1.90	1.37840	4.35890	5.00	2.23607	7.07107	8.00	2.82843	8.94427
2.00	1.41421	4.47214	5.10	2.25832	7.14143	8.10	2.84605	9.00000
2.10	1.44914	4.58258	5.20	2.28035	7.21110	8.20	2.86356	9.05539
2.20	1.48324	4.69042	5.30	2.30217	7.28011	8.30	2.88097	9.11043
2.30	1.51658	4.79583	5.40	2.32379	7.34847	8.40	2.89828	9.16515
2.40	1.54919	4.89898	5.50	2.34521	7.41620	8.50	2.91548	9.21954
2.50	1.58114	5.00000	5.60	2.36643	7.48331	8.60	2.93258	9.27362
2.60	1.61245	5.09902	5.70	2.38747	7.54983	8.70	2.94958	9.32738
2.70	1.64317	5.19615	5.80	2.40832	7.61577	8.80	2.96648	9.38083
2.80	1.67332	5.29150	5.90	2.42899	7.68115	8.90	2.98329	9.43398
2.90	1.70294	5.38516	6.00	2.44949	7.74597	9.00	3.00000	9.48683
3.00	1.73205	5.47723	6.10	2.46982	7.81025	9.10	3.01662	9.53939
3.10	1.76068	5.56776	6.20	2.48998	7.87401	9.20	3.03315	9.59166
3.20	1.78885	5.65685	6.30	2.50998	7.93725	9.30	3.04959	9.64365
3.30	1.81659	5.74456	6.40	2.52982	8.00000	9.40	3.06594	9.69536
3.40	1.84391	5.83095	6.50	2.54951	8.06226	9.50	3.08221	9.74679
3.50	1.87083	5.91608	6.60	2.56905	8.12404	9.60	3.09839	9.79796
3.60	1.89737	6.00000	6.70	2.58844	8.18535	9.70	3.11448	9.84886
3.70	1.92354	6.08276	6.80	2.60768	8.24621	9.80	3.13050	9.89949
3.80	1.94936	6.16441	6.90	2.62679	8.30662	9.90	3.14643	9.94987
3.90	1.97484	6.24500	7.00	2.64575	8.36660	10.00	3.16228	10.00000
4.00	2.00000	6.32456						

II *Normal Curve z Scores and Percent of Scores Exceeded*

(The proportion of area given is that to the left of the given z score. It should be noted that all area values have been rounded to three decimal places.)

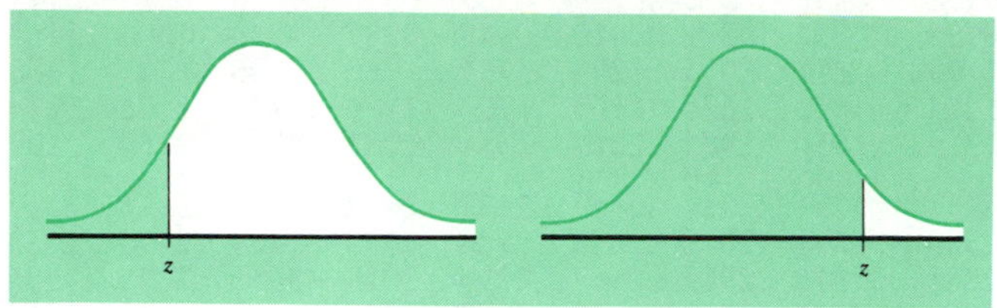

z	AREA	z	AREA	z	AREA
−4.0	.000	−1.3	.097	1.4	.919
−3.9	.000	−1.2	.115	1.5	.933
−3.8	.000	−1.1	.136	1.6	.945
−3.7	.000	−1.0	.159	1.7	.955
−3.6	.000	−0.9	.184	1.8	.964
−3.5	.000	−0.8	.212	1.9	.971
−3.4	.000	−0.7	.242	2.0	.977
−3.3	.001	−0.6	.274	2.1	982
−3.2	.001	−0.5	.308	2.2	.986
−3.1	.001	−0.4	.345	2.3	.989
−3.0	.001	−0.3	.382	2.4	.992
−2.9	.002	−0.2	.421	2.5	.994
−2.8	.003	−0.1	.460	2.6	.995
−2.7	.004	0.0	.500	2.7	.996
−2.6	.005	0.1	.540	2.8	.997
−2.5	.006	0.2	.579	2.9	.998
−2.4	.008	0.3	.618	3.0	.999
−2.3	.011	0.4	.655	3.1	.999
−2.2	.014	0.5	.692	3.2	.999
−2.1	.018	0.6	.726	3.3	.999
−2.0	.023	0.7	.758	3.4	1.000
−1.9	.029	0.8	.788	3.5	1.000
−1.8	.036	0.9	.816	3.6	1.000
−1.7	.045	1.0	.841	3.7	1.000
−1.6	.055	1.1	.864	3.8	1.000
−1.5	.067	1.2	.885	3.9	1.000
−1.4	.081	1.3	.903	4.0	1.000

III Normal Curve Areas

z	.00	.01	.02	.03	.04	.05	.06	.07	.08	.09
0.0	.0000	.0040	.0080	.0120	.0160	.0199	.0239	.0279	.0319	.0359
0.1	.0398	.0438	.0478	.0517	.0557	.0596	.0636	.0675	.0714	.0753
0.2	.0793	.0832	.0871	.0910	.0948	.0987	.1026	.1064	.1103	.1141
0.3	.1179	.1217	.1255	.1293	.1331	.1368	.1406	.1443	.1480	.1517
0.4	.1554	.1591	.1628	.1664	.1700	.1736	.1772	.1808	.1844	.1879
0.5	.1915	.1950	.1985	.2019	.2054	.2088	.2123	.2157	.2190	.2224
0.6	.2257	.2291	.2324	.2357	.2389	.2422	.2454	.2486	.2517	.2549
0.7	.2580	.2611	.2642	.2673	.2704	.2734	.2764	.2794	.2823	.2852
0.8	.2881	.2910	.2939	.2967	.2995	.3023	.3051	.3078	.3106	.3133
0.9	.3159	.3186	.3212	.3238	.3264	.3289	.3315	.3340	.3365	.3389
1.0	.3413	.3438	.3461	.3485	.3508	.3531	.3554	.3577	.3599	.3621
1.1	.3643	.3665	.3686	.3708	.3729	.3749	.3770	.3790	.3810	.3830
1.2	.3849	.3869	.3888	.3907	.3925	.3944	.3962	.3980	.3997	.4015
1.3	.4032	.4049	.4066	.4082	.4099	.4115	.4131	.4147	.4162	.4177
1.4	.4192	.4207	.4222	.4236	.4251	.4265	.4279	.4292	.4306	.4319
1.5	.4332	.4345	.4357	.4370	.4382	.4394	.4406	.4418	.4429	.4441
1.6	.4452	.4463	.4474	.4484	.4495	.4505	.4515	.4525	.4535	.4545
1.7	.4554	.4564	.4573	.4582	.4591	.4599	.4608	.4616	.4625	.4633
1.8	.4641	.4649	.4656	.4664	.4671	.4678	.4686	.4693	.4699	.4706
1.9	.4713	.4719	.4726	.4732	.4738	.4744	.4750	.4756	.4761	.4767
2.0	.4772	.4778	.4783	.4788	.4793	.4798	.4803	.4808	.4812	.4817
2.1	.4821	.4826	.4830	.4834	.4838	.4842	.4846	.4850	.4854	.4857
2.2	.4861	.4864	.4868	.4871	.4875	.4878	.4881	.4884	.4887	.4890
2.3	.4893	.4896	.4898	.4901	.4904	.4906	.4909	.4911	.4913	.4916
2.4	.4918	.4920	.4922	.4925	.4927	.4929	.4931	.4932	.4934	.4936
2.5	.4938	.4940	.4941	.4943	.4945	.4946	.4948	.4949	.4951	.4952
2.6	.4953	.4955	.4956	.4957	.4959	.4960	.4961	.4962	.4963	.4964
2.7	.4965	.4966	.4967	.4968	.4969	.4970	.4971	.4972	.4973	.4974
2.8	.4974	.4975	.4976	.4977	.4977	.4978	.4979	.4979	.4980	.4981
2.9	.4981	.4982	.4982	.4983	.4984	.4984	.4985	.4985	.4986	.4986
3.0	.4987	.4987	.4987	.4988	.4988	.4989	.4989	.4989	.4990	.4990
3.1	.49903									
3.2	.49931									
3.3	.49952									
3.4	.49966									
3.5	.49977									
3.6	.49984									
3.7	.49989									
3.8	.49993									
3.9	.49995									
4.0	.50000									

IV The *t* Distribution

	Proportion of Area to Left of *t* Value			
	.95	.975	.99	.995
	Level of Significance for One-Tail Test			
	.05	.025	.01	.005
	Level of Significance for Two-Tail Test			
d.f.	.10	.05	.02	.01
1	6.31	12.71	31.82	63.66
2	2.92	4.30	6.96	9.92
3	2.35	3.18	4.54	5.84
4	2.13	2.78	3.75	4.60
5	2.02	2.57	3.36	4.03
6	1.94	2.45	3.14	3.71
7	1.90	2.36	3.00	3.50
8	1.86	2.31	2.90	3.36
9	1.83	2.26	2.82	3.25
10	1.81	2.23	2.76	3.17
11	1.80	2.20	2.72	3.11
12	1.78	2.18	2.68	3.06
13	1.77	2.16	2.65	3.01
14	1.76	2.14	2.62	2.98
15	1.75	2.13	2.60	2.95
16	1.75	2.12	2.58	2.92
17	1.74	2.11	2.57	2.90
18	1.73	2.10	2.55	2.88
19	1.73	2.09	2.54	2.86
20	1.72	2.09	2.53	2.84
21	1.72	2.08	2.52	2.83
22	1.72	2.07	2.51	2.82
23	1.71	2.07	2.50	2.81
24	1.71	2.06	2.49	2.80
25	1.71	2.06	2.48	2.79
26	1.71	2.06	2.48	2.78
27	1.70	2.05	2.47	2.77
28	1.70	2.05	2.47	2.76
29	1.70	2.04	2.46	2.76
30	1.70	2.04	2.46	2.75
31	1.70	2.04	2.45	2.75
32	1.69	2.03	2.45	2.74
33	1.69	2.03	2.45	2.74
40	1.68	2.02	2.42	2.70
60	1.67	2.00	2.39	2.66
120	1.66	1.98	2.36	2.62
∞	1.65	1.96	2.33	2.58

V Values of F

The 5% (Roman Type) and 1% (Boldface Type) Points for the Distribution of *F*

n_2	n_1 degrees of freedom (for greater estimate of variance)																								n_2
	1	2	3	4	5	6	7	8	9	10	11	12	14	16	20	24	30	40	50	75	100	200	500	∞	
1	161 **4,052**	200 **4,999**	216 **5,403**	225 **5,625**	230 **5,764**	234 **5,859**	237 **5,928**	239 **5,981**	241 **6,022**	242 **6,056**	243 **6,082**	244 **6,106**	245 **6,142**	246 **6,169**	248 **6,208**	249 **6,234**	250 **6,258**	251 **6,286**	252 **6,302**	253 **6,323**	253 **6,334**	254 **6,352**	254 **6,361**	254 **6,366**	1
2	18.51 **98.49**	19.00 **99.00**	19.16 **99.17**	19.25 **99.25**	19.30 **99.30**	19.33 **99.33**	19.36 **99.34**	19.37 **99.36**	19.38 **99.38**	19.39 **99.40**	19.40 **99.41**	19.41 **99.42**	19.42 **99.43**	19.43 **99.44**	19.44 **99.45**	19.45 **99.46**	19.46 **99.47**	19.47 **99.48**	19.47 **99.48**	19.48 **99.49**	19.49 **99.49**	19.49 **99.49**	19.50 **99.50**	19.50 **99.50**	2
3	10.13 **34.12**	9.55 **30.82**	9.28 **29.46**	9.12 **28.71**	9.01 **28.24**	8.94 **27.91**	8.88 **27.67**	8.84 **27.49**	8.81 **27.34**	8.78 **27.23**	8.76 **27.13**	8.74 **27.05**	8.71 **26.92**	8.69 **26.83**	8.66 **26.69**	8.64 **26.60**	8.62 **26.50**	8.60 **26.41**	8.58 **26.35**	8.57 **26.27**	8.56 **26.23**	8.54 **26.18**	8.54 **26.14**	8.53 **26.12**	3
4	7.71 **21.20**	6.94 **18.00**	6.59 **16.69**	6.39 **15.98**	6.26 **15.52**	6.16 **15.21**	6.09 **14.98**	6.04 **14.80**	6.00 **14.66**	5.96 **14.54**	5.93 **14.45**	5.91 **14.37**	5.87 **14.24**	5.84 **14.15**	5.80 **14.02**	5.77 **13.93**	5.74 **13.83**	5.71 **13.74**	5.70 **13.69**	5.68 **13.61**	5.66 **13.57**	5.65 **13.52**	5.64 **13.48**	5.63 **13.46**	4
5	6.61 **16.26**	5.79 **13.27**	5.41 **12.06**	5.19 **11.39**	5.05 **10.97**	4.95 **10.67**	4.88 **10.45**	4.82 **10.27**	4.78 **10.15**	4.74 **10.05**	4.70 **9.96**	4.68 **9.89**	4.64 **9.77**	4.60 **9.68**	4.56 **9.55**	4.53 **9.47**	4.50 **9.38**	4.46 **9.29**	4.44 **9.24**	4.42 **9.17**	4.40 **9.13**	4.38 **9.07**	4.37 **9.04**	4.36 **9.02**	5
6	5.99 **13.74**	5.14 **10.92**	4.76 **9.78**	4.53 **9.15**	4.39 **8.75**	4.28 **8.47**	4.21 **8.26**	4.15 **8.10**	4.10 **7.98**	4.06 **7.87**	4.03 **7.79**	4.00 **7.72**	3.96 **7.60**	3.92 **7.52**	3.87 **7.39**	3.84 **7.31**	3.81 **7.23**	3.77 **7.14**	3.75 **7.09**	3.72 **7.02**	3.71 **6.99**	3.69 **6.94**	3.68 **6.90**	3.67 **6.88**	6
7	5.59 **12.25**	4.74 **9.55**	4.35 **8.45**	4.12 **7.85**	3.97 **7.46**	3.87 **7.19**	3.79 **7.00**	3.73 **6.84**	3.68 **6.71**	3.63 **6.62**	3.60 **6.54**	3.57 **6.47**	3.52 **6.35**	3.49 **6.27**	3.44 **6.15**	3.41 **6.07**	3.38 **5.98**	3.34 **5.90**	3.32 **5.85**	3.29 **5.78**	3.28 **5.75**	3.25 **5.70**	3.24 **5.67**	3.23 **5.65**	7
8	5.32 **11.26**	4.46 **8.65**	4.07 **7.59**	3.84 **7.01**	3.69 **6.63**	3.58 **6.37**	3.50 **6.19**	3.44 **6.03**	3.39 **5.91**	3.34 **5.82**	3.31 **5.74**	3.28 **5.67**	3.23 **5.56**	3.20 **5.48**	3.15 **5.36**	3.12 **5.28**	3.08 **5.20**	3.05 **5.11**	3.03 **5.06**	3.00 **5.00**	2.98 **4.96**	2.96 **4.91**	2.94 **4.88**	2.93 **4.86**	8
9	5.12 **10.56**	4.26 **8.02**	3.86 **6.99**	3.63 **6.42**	3.48 **6.06**	3.37 **5.80**	3.29 **5.62**	3.23 **5.47**	3.18 **5.35**	3.13 **5.26**	3.10 **5.18**	3.07 **5.11**	3.02 **5.00**	2.98 **4.92**	2.93 **4.80**	2.90 **4.73**	2.86 **4.64**	2.82 **4.56**	2.80 **4.51**	2.77 **4.45**	2.76 **4.41**	2.73 **4.36**	2.72 **4.33**	2.71 **4.31**	9
10	4.96 **10.04**	4.10 **7.56**	3.71 **6.55**	3.48 **5.99**	3.33 **5.64**	3.22 **5.39**	3.14 **5.21**	3.07 **5.06**	3.02 **4.95**	2.97 **4.85**	2.94 **4.78**	2.91 **4.71**	2.86 **4.60**	2.82 **4.52**	2.77 **4.41**	2.74 **4.33**	2.70 **4.25**	2.67 **4.17**	2.64 **4.12**	2.61 **4.05**	2.59 **4.01**	2.56 **3.96**	2.55 **3.93**	2.54 **3.91**	10
11	4.84 **9.65**	3.98 **7.20**	3.59 **6.22**	3.36 **5.67**	3.20 **5.32**	3.09 **5.07**	3.01 **4.88**	2.95 **4.74**	2.90 **4.63**	2.86 **4.54**	2.82 **4.46**	2.79 **4.40**	2.74 **4.29**	2.70 **4.21**	2.65 **4.10**	2.61 **4.02**	2.57 **3.94**	2.53 **3.86**	2.50 **3.80**	2.47 **3.74**	2.45 **3.70**	2.42 **3.66**	2.41 **3.62**	2.40 **3.60**	11
12	4.75 **9.33**	3.88 **6.93**	3.49 **5.95**	3.26 **5.41**	3.11 **5.06**	3.00 **4.82**	2.92 **4.65**	2.85 **4.50**	2.80 **4.39**	2.76 **4.30**	2.72 **4.22**	2.69 **4.16**	2.64 **4.05**	2.60 **3.98**	2.54 **3.86**	2.50 **3.78**	2.46 **3.70**	2.42 **3.61**	2.40 **3.56**	2.36 **3.49**	2.35 **3.46**	2.32 **3.41**	2.31 **3.38**	2.30 **3.36**	12
13	4.67 **9.07**	3.80 **6.70**	3.41 **5.74**	3.18 **5.20**	3.02 **4.86**	2.92 **4.62**	2.84 **4.44**	2.77 **4.30**	2.72 **4.19**	2.67 **4.10**	2.63 **4.02**	2.60 **3.96**	2.55 **3.85**	2.51 **3.78**	2.46 **3.67**	2.42 **3.59**	2.38 **3.51**	2.34 **3.42**	2.32 **3.37**	2.28 **3.30**	2.26 **3.27**	2.24 **3.21**	2.22 **3.18**	2.21 **3.16**	13

n_2	n_1 degrees of freedom (for greater estimate of variance)																								n_2
	1	2	3	4	5	6	7	8	9	10	11	12	14	16	20	24	30	40	50	75	100	200	500	∞	
14	4.60	3.74	3.34	3.11	2.96	2.85	2.77	2.70	2.65	2.60	2.56	2.53	2.48	2.44	2.39	2.35	2.31	2.27	2.24	2.21	2.19	2.16	2.14	2.13	14
	8.86	6.51	5.56	5.03	4.69	4.46	4.28	4.14	4.03	3.94	3.86	3.80	3.70	3.62	3.51	3.43	3.34	3.26	3.21	3.14	3.11	3.06	3.02	3.00	
15	4.54	3.68	3.29	3.06	2.90	2.79	2.70	2.64	2.59	2.55	2.51	2.48	2.43	2.39	2.33	2.29	2.25	2.21	2.18	2.15	2.12	2.10	2.08	2.07	15
	8.68	6.36	5.42	4.89	4.56	4.32	4.14	4.00	3.89	3.80	3.73	3.67	3.56	3.48	3.36	3.29	3.20	3.12	3.07	3.00	2.97	2.92	2.89	2.87	
16	4.49	3.63	3.24	3.01	2.85	2.74	2.66	2.59	2.54	2.49	2.45	2.42	2.37	2.33	2.28	2.24	2.20	2.16	2.13	2.09	2.07	2.04	2.02	2.01	16
	8.53	6.23	5.29	4.77	4.44	4.20	4.03	3.89	3.78	3.69	3.61	3.55	3.45	3.37	3.25	3.18	3.10	3.01	2.96	2.89	2.86	2.80	2.77	2.75	
17	4.45	3.59	3.20	2.96	2.81	2.70	2.62	2.55	2.50	2.45	2.41	2.38	2.33	2.29	2.23	2.19	2.15	2.11	2.08	2.04	2.02	1.99	1.97	1.96	17
	8.40	6.11	5.18	4.67	4.34	4.10	3.93	3.79	3.68	3.59	3.52	3.45	3.35	3.27	3.16	3.08	3.00	2.92	2.86	2.79	2.76	2.70	2.67	2.65	
18	4.41	3.55	3.16	2.93	2.77	2.66	2.58	2.51	2.46	2.41	2.37	2.34	2.29	2.25	2.19	2.15	2.11	2.07	2.04	2.00	1.98	1.95	1.93	1.92	18
	8.28	6.01	5.09	4.58	4.25	4.01	3.85	3.71	3.60	3.51	3.44	3.37	3.27	3.19	3.07	3.00	2.91	2.83	2.78	2.71	2.68	2.62	2.59	2.57	
19	4.38	3.52	3.13	2.90	2.74	2.63	2.55	2.48	2.43	2.38	2.34	2.31	2.26	2.21	2.15	2.11	2.07	2.02	2.00	1.96	1.94	1.91	1.90	1.88	19
	8.18	5.93	5.01	4.50	4.17	3.94	3.77	3.63	3.52	3.43	3.36	3.30	3.19	3.12	3.00	2.92	2.84	2.76	2.70	2.63	2.60	2.54	2.51	2.49	
20	4.35	3.49	3.10	2.87	2.71	2.60	2.52	2.45	2.40	2.35	2.31	2.28	2.23	2.18	2.12	2.08	2.04	1.99	1.96	1.92	1.90	1.87	1.85	1.84	20
	8.10	5.85	4.94	4.43	4.10	3.87	3.71	3.56	3.45	3.37	3.30	3.23	3.13	3.05	2.94	2.86	2.77	2.69	2.63	2.56	2.53	2.47	2.44	2.42	
21	4.32	3.47	3.07	2.84	2.68	2.57	2.49	2.42	2.37	2.32	2.28	2.25	2.20	2.15	2.09	2.05	2.00	1.96	1.93	1.89	1.87	1.84	1.82	1.81	21
	8.02	5.78	4.87	4.37	4.04	3.81	3.65	3.51	3.40	3.31	3.24	3.17	3.07	2.99	2.88	2.80	2.72	2.63	2.58	2.51	2.47	2.42	2.38	2.36	
22	4.30	3.44	3.05	2.82	2.66	2.55	2.47	2.40	2.35	2.30	2.26	2.23	2.18	2.13	2.07	2.03	1.98	1.93	1.91	1.87	1.84	1.81	1.80	1.78	22
	7.94	5.72	4.82	4.31	3.99	3.76	3.59	3.45	3.35	3.26	3.18	3.12	3.02	2.94	2.83	2.75	2.67	2.58	2.53	2.46	2.42	2.37	2.33	2.31	
23	4.28	3.42	3.03	2.80	2.64	2.53	2.45	2.38	2.32	2.28	2.24	2.20	2.14	2.10	2.04	2.00	1.96	1.91	1.88	1.84	1.82	1.79	1.77	1.76	23
	7.88	5.66	4.76	4.26	3.94	3.71	3.54	3.41	3.30	3.21	3.14	3.07	2.97	2.89	2.78	2.70	2.62	2.53	2.48	2.41	2,37	2.32	2.28	2.26	
24	4.26	3.40	3.01	2.78	2.62	2.51	2.43	2.36	2.30	2.26	2.22	2.18	2.13	2.09	2.03	1.98	1.94	1.89	1.86	1.82	1.80	1.76	1.74	1.73	24
	7.82	5.61	4.72	4.22	3.90	3.67	3.50	3.36	3.25	3.17	3.09	3.03	2.93	2.85	2.74	2.66	2.58	2.49	2.44	2.36	2.33	2.27	2.23	2.21	
25	4.24	3.38	2.99	2.76	2.60	2.49	2.41	2.34	2.28	2.24	2.20	2.16	2.11	2.06	2.00	1.96	1.92	1.87	1.84	1.80	1.77	1.74	1.72	1.71	25
	7.77	5.57	4.68	4.18	3.86	3.63	3.46	3.32	3.21	3.13	3.05	2.99	2.89	2.81	2.70	2.62	2.54	2.45	2.40	2.32	2.29	2.23	2.19	2.17	
26	4.22	3.37	2.98	2.74	2.59	2.47	2.39	2.32	2.27	2.22	2.18	2.15	2.10	2.05	1.99	1.95	1.90	1.85	1.82	1.78	1.76	1.72	1.70	1.69	26
	7.72	5.53	4.64	4.14	3.82	3.59	3.42	3.29	3.17	3.09	3.02	2.96	2.86	2.77	2.66	2.58	2.50	2.41	2.36	2.28	2.25	2.19	2.15	2.13	

n_2	n_1 degrees of freedom (for greater estimate of variance)																								n_2
	1	2	3	4	5	6	7	8	9	10	11	12	14	16	20	24	30	40	50	75	100	200	500	∞	
27	4.21 **7.68**	3.35 **5.49**	2.96 **4.60**	2.73 **4.11**	2.57 **3.79**	2.46 **3.56**	2.37 **3.39**	2.30 **3.26**	2.25 **3.14**	2.20 **3.06**	2.16 **2.98**	2.13 **2.93**	2.08 **2.83**	2.03 **2.74**	1.97 **2.63**	1.93 **2.55**	1.88 **2.47**	1.84 **2.38**	1.80 **2.33**	1.76 **2.25**	1.74 **2.21**	1.71 **2.16**	1.68 **2.12**	1.67 **2.10**	27
28	4.20 **7.64**	3.34 **5.45**	2.95 **4.57**	2.71 **4.07**	2.56 **3.76**	2.44 **3.53**	2.36 **3.36**	2.29 **3.23**	2.24 **3.11**	2.19 **3.03**	2.15 **2.95**	2.12 **2.90**	2.06 **2.80**	2.02 **2.71**	1.96 **2.60**	1.91 **2.52**	1.87 **2.44**	1.81 **2.35**	1.78 **2.30**	1.75 **2.22**	1.72 **2.18**	1.69 **2.13**	1.67 **2.09**	1.65 **2.06**	28
29	4.18 **7.60**	3.33 **5.42**	2.93 **4.54**	2.70 **4.04**	2.54 **3.73**	2.43 **3.50**	2.35 **3.33**	2.28 **3.20**	2.22 **3.08**	2.18 **3.00**	2.14 **2.92**	2.10 **2.87**	2.05 **2.77**	2.00 **2.68**	1.94 **2.57**	1.90 **2.49**	1.85 **2.41**	1.80 **2.32**	1.77 **2.27**	1.73 **2.19**	1.71 **2.15**	1.68 **2.10**	1.65 **2.06**	1.64 **2.03**	29
30	4.17 **7.56**	3.32 **5.39**	2.92 **4.51**	2.69 **4.02**	2.53 **3.70**	2.42 **3.47**	2.34 **3.30**	2.27 **3.17**	2.21 **3.06**	2.16 **2.98**	2.12 **2.90**	2.09 **2.84**	2.04 **2.74**	1.99 **2.66**	1.93 **2.55**	1.89 **2.47**	1.84 **2.38**	1.79 **2.29**	1.76 **2.24**	1.72 **2.16**	1.69 **2.13**	1.66 **2.07**	1.64 **2.03**	1.62 **2.01**	30
32	4.15 **7.50**	3.30 **5.34**	2.90 **4.46**	2.67 **3.97**	2.51 **3.66**	2.40 **3.42**	2.32 **3.25**	2.25 **3.12**	2.19 **3.01**	2.14 **2.94**	2.10 **2.86**	2.07 **2.80**	2.02 **2.70**	1.97 **2.62**	1.91 **2.51**	1.86 **2.42**	1.82 **2.34**	1.76 **2.25**	1.74 **2.20**	1.69 **2.12**	1.67 **2.08**	1.64 **2.02**	1.61 **1.98**	1.59 **1.96**	32
34	4.13 **7.44**	3.28 **5.29**	2.88 **4.42**	2.65 **3.93**	2.49 **3.61**	2.38 **3.38**	2.30 **3.21**	2.23 **3.08**	2.17 **2.97**	2.12 **2.89**	2.08 **2.82**	2.05 **2.76**	2.00 **2.66**	1.95 **2.58**	1.89 **2.47**	1.84 **2.38**	1.80 **2.30**	1.74 **2.21**	1.71 **2.15**	1.67 **2.08**	1.64 **2.04**	1.61 **1.98**	1.59 **1.94**	1.57 **1.91**	34
36	4.11 **7.39**	3.26 **5.25**	2.86 **4.38**	2.63 **3.89**	2.48 **3.58**	2.36 **3.35**	2.28 **3.18**	2.21 **3.04**	2.15 **2.94**	2.10 **2.86**	2.06 **2.78**	2.03 **2.72**	1.98 **2.62**	1.93 **2.54**	1.87 **2.43**	1.82 **2.35**	1.78 **2.26**	1.72 **2.17**	1.69 **2.12**	1.65 **2.04**	1.62 **2.00**	1.59 **1.94**	1.56 **1.90**	1.55 **1.87**	36
38	4.10 **7.35**	3.25 **5.21**	2.85 **4.34**	2.62 **3.86**	2.46 **3.54**	2.35 **3.32**	2.26 **3.15**	2.19 **3.02**	2.14 **2.91**	2.09 **2.82**	2.05 **2.75**	2.02 **2.69**	1.96 **2.59**	1.92 **2.51**	1.85 **2.40**	1.80 **2.32**	1.76 **2.22**	1.71 **2.14**	1.67 **2.08**	1.63 **2.00**	1.60 **1.97**	1.57 **1.90**	1.54 **1.86**	1.53 **1.84**	38
40	4.08 **7.31**	3.23 **5.18**	2.84 **4.31**	2.61 **3.83**	2.45 **3.51**	2.34 **3.29**	2.25 **3.12**	2.18 **2.99**	2.12 **2.88**	2.07 **2.80**	2.04 **2.73**	2.00 **2.66**	1.95 **2.56**	1.90 **2.49**	1.84 **2.37**	1.79 **2.29**	1.74 **2.20**	1.69 **2.11**	1.66 **2.05**	1.61 **1.97**	1.59 **1.94**	1.55 **1.88**	1.53 **1.84**	1.51 **1.81**	40
42	4.07 **7.27**	3.22 **5.15**	2.83 **4.29**	2.59 **3.80**	2.44 **3.49**	3.32 **3.26**	2.24 **3.10**	2.17 **2.96**	2.11 **2.86**	2.06 **2.77**	2.02 **2.70**	1.99 **2.64**	1.94 **2.54**	1.89 **2.46**	1.82 **2.35**	1.78 **2.26**	1.73 **2.17**	1.68 **2.08**	1.64 **2.02**	1.60 **1.94**	1.57 **1.91**	1.54 **1.85**	1.51 **1.80**	1.49 **1.78**	42
44	4.06 **7.24**	3.21 **5.12**	2.82 **4.26**	2.58 **3.78**	2.43 **3.46**	2.31 **3.24**	2.23 **3.07**	2.16 **2.94**	2.10 **2.84**	2.05 **2.75**	2.01 **2.68**	1.98 **2.62**	1.92 **2.52**	1.88 **2.44**	1.81 **2.32**	1.76 **2.24**	1.72 **2.15**	1.66 **2.06**	1.63 **2.00**	1.58 **1.92**	1.56 **1.88**	1.52 **1.82**	1.50 **1.78**	1.48 **1.75**	44
46	4.05 **7.21**	3.20 **5.10**	2.81 **4.24**	2.57 **3.76**	2.42 **3.44**	2.30 **3.22**	2.22 **3.05**	2.14 **2.92**	2.09 **2.82**	2.04 **2.73**	2.00 **2.66**	1.97 **2.60**	1.91 **2.50**	1.87 **2.42**	1.80 **2.30**	1.75 **2.22**	1.71 **2.13**	1.65 **2.04**	1.62 **1.98**	1.57 **1.90**	1.54 **1.86**	1.51 **1.80**	1.48 **1.76**	1.46 **1.72**	46
48	4.04 **7.19**	3.19 **5.08**	2.80 **4.22**	2.56 **3.74**	2.41 **3.42**	2.30 **3.20**	2.21 **3.04**	2.14 **2.90**	2.08 **2.80**	2.03 **2.71**	1.99 **2.64**	1.96 **2.58**	1.90 **2.48**	1.86 **2.40**	1.79 **2.28**	1.74 **2.20**	1.70 **2.11**	1.64 **2.02**	1.61 **1.96**	1.56 **1.88**	1.53 **1.84**	1.50 **1.78**	1.47 **1.73**	1.45 **1.70**	48

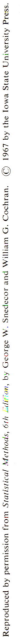

n_2	n_1 degrees of freedom (for greater estimate of variance)																								n_2
	1	2	3	4	5	6	7	8	9	10	11	12	14	16	20	24	30	40	50	75	100	200	500	∞	
50	4.03 **7.17**	3.18 **5.06**	2.79 **4.20**	2.56 **3.72**	2.40 **3.41**	2.29 **3.18**	2.20 **3.02**	2.13 **2.88**	2.07 **2.78**	2.02 **2.70**	1.98 **2.62**	1.95 **2.56**	1.90 **2.46**	1.85 **2.39**	1.78 **2.26**	1.74 **2.18**	1.69 **2.10**	1.63 **2.00**	1.60 **1.94**	1.55 **1.86**	1.52 **1.82**	1.48 **1.76**	1.46 **1.71**	1.44 **1.68**	50
55	4.02 **7.12**	3.17 **5.01**	2.78 **4.16**	2.54 **3.68**	2.38 **3.37**	2.27 **3.15**	2.18 **2.98**	2.11 **2.85**	2.05 **2.75**	2.00 **2.66**	1.97 **2.59**	1.93 **2.53**	1.88 **2.43**	1.83 **2.35**	1.76 **2.23**	1.72 **2.15**	1.67 **2.06**	1.61 **1.96**	1.58 **1.90**	1.52 **1.82**	1.50 **1.78**	1.46 **1.71**	1.43 **1.66**	1.41 **1.64**	55
60	4.00 **7.08**	3.15 **4.98**	2.76 **4.13**	2.52 **3.65**	2.37 **3.34**	2.25 **3.12**	2.17 **2.95**	2.10 **2.82**	2.04 **2.72**	1.99 **2.63**	1.95 **2.56**	1.92 **2.50**	1.86 **2.40**	1.81 **2.32**	1.75 **2.20**	1.70 **2.12**	1.65 **2.03**	1.59 **1.93**	1.56 **1.87**	1.50 **1.79**	1.48 **1.74**	1.44 **1.68**	1.41 **1.63**	1.39 **1.60**	60
65	3.99 **7.04**	3.14 **4.95**	2.75 **4.10**	2.51 **3.62**	2.36 **3.31**	2.24 **3.09**	2.15 **2.93**	2.08 **2.79**	2.02 **2.70**	1.98 **2.61**	1.94 **2.54**	1.90 **2.47**	1.85 **2.37**	1.80 **2.30**	1.73 **2.18**	1.68 **2.09**	1.63 **2.00**	1.57 **1.90**	1.54 **1.84**	1.49 **1.76**	1.46 **1.71**	1.42 **1.64**	1.39 **1.60**	1.37 **1.56**	65
70	3.98 **7.01**	3.13 **4.92**	2.74 **4.08**	2.50 **3.60**	2.35 **3.29**	2.23 **3.07**	2.14 **2.91**	2.07 **2.77**	2.01 **2.67**	1.97 **2.59**	1.93 **2.51**	1.89 **2.45**	1.84 **2.35**	1.79 **2.28**	1.72 **2.15**	1.67 **2.07**	1.62 **1.98**	1.56 **1.88**	1.53 **1.82**	1.47 **1.74**	1.45 **1.69**	1.40 **1.62**	1.37 **1.56**	1.35 **1.53**	70
80	3.96 **6.96**	3.11 **4.88**	2.72 **4.04**	2.48 **3.56**	2.33 **3.25**	2.21 **3.04**	2.12 **2.87**	2.05 **2.74**	1.99 **2.64**	1.95 **2.55**	1.91 **2.48**	1.88 **2.41**	1.82 **2.32**	1.77 **2.24**	1.70 **2.11**	1.65 **2.03**	1.60 **1.94**	1.54 **1.84**	1.51 **1.78**	1.45 **1.70**	1.42 **1.65**	1.38 **1.57**	1.35 **1.52**	1.32 **1.49**	80
100	3.94 **6.90**	3.09 **4.82**	2.70 **3.98**	2.46 **3.51**	2.30 **3.20**	2.19 **2.99**	2.10 **2.82**	2.03 **2.69**	1.97 **2.59**	1.92 **2.51**	1.88 **2.43**	1.85 **2.36**	1.79 **2.26**	1.75 **2.19**	1.68 **2.06**	1.63 **1.98**	1.57 **1.89**	1.51 **1.79**	1.48 **1.73**	1.42 **1.64**	1.39 **1.59**	1.34 **1.51**	1.30 **1.46**	1.28 **1.43**	100
125	3.92 **6.84**	3.07 **4.78**	2.68 **3.94**	2.44 **3.47**	2.29 **3.17**	2.17 **2.95**	2.08 **2.79**	2.01 **2.65**	1.95 **2.56**	1.90 **2.47**	1.86 **2.40**	1.83 **2.33**	1.77 **2.23**	1.72 **2.15**	1.65 **2.03**	1.60 **1.94**	1.55 **1.85**	1.49 **1.75**	1.45 **1.68**	1.39 **1.59**	1.36 **1.54**	1.31 **1.46**	1.27 **1.40**	1.25 **1.37**	125
150	3.91 **6.81**	3.06 **4.75**	2.67 **3.91**	2.43 **3.44**	2.27 **3.14**	2.16 **2.92**	2.07 **2.76**	2.00 **2.62**	1.94 **2.53**	1.89 **2.44**	1.85 **2.37**	1.82 **2.30**	1.76 **2.20**	1.71 **2.12**	1.64 **2.00**	1.59 **1.91**	1.54 **1.83**	1.47 **1.72**	1.44 **1.66**	1.37 **1.56**	1.34 **1.51**	1.29 **1.43**	1.25 **1.37**	1.22 **1.33**	150
200	3.89 **6.76**	3.04 **4.71**	2.65 **3.88**	3.41 **3.41**	2.26 **3.11**	2.14 **2.90**	2.05 **2.73**	1.98 **2.60**	1.92 **2.50**	1.87 **2.41**	1.83 **2.34**	1.80 **2.28**	1.74 **2.17**	1.69 **2.09**	1.62 **1.97**	1.57 **1.88**	1.52 **1.79**	1.45 **1.69**	1.42 **1.62**	1.35 **1.53**	1.32 **1.48**	1.26 **1.39**	1.22 **1.33**	1.99 **1.28**	200
400	3.86 **6.70**	3.02 **4.66**	2.62 **3.83**	2.39 **3.36**	2.23 **3.06**	2.12 **2.85**	2.03 **2.69**	1.96 **2.55**	1.90 **2.46**	1.85 **2.37**	1.81 **2.29**	1.78 **2.23**	1.72 **2.12**	1.67 **2.04**	1.60 **1.92**	1.54 **1.84**	1.49 **1.74**	1.42 **1.64**	1.38 **1.57**	1.32 **1.47**	1.28 **1.42**	1.22 **1.32**	1.16 **1.24**	1.13 **1.19**	400
1000	3.85 **6.66**	3.00 **4.62**	2.61 **3.80**	2.38 **3.34**	2.22 **3.04**	2.10 **2.82**	2.02 **2.66**	1.95 **2.53**	1.89 **2.43**	1.84 **2.34**	1.80 **2.26**	1.76 **2.20**	1.70 **2.09**	1.65 **2.01**	1.58 **1.89**	1.53 **1.81**	1.47 **1.71**	1.41 **1.61**	1.36 **1.54**	1.30 **1.44**	1.26 **1.38**	1.19 **1.28**	1.13 **1.19**	1.08 **1.11**	1000
∞	3.84 **6.64**	2.99 **4.60**	2.60 **3.78**	2.37 **3.32**	2.21 **3.02**	2.09 **2.80**	2.01 **2.64**	1.94 **2.51**	1.88 **2.41**	1.83 **2.32**	1.79 **2.24**	1.75 **2.18**	1.69 **2.07**	1.64 **1.99**	1.57 **1.87**	1.52 **1.79**	1.46 **1.69**	1.40 **1.59**	1.35 **1.52**	1.28 **1.41**	1.24 **1.36**	1.17 **1.25**	1.11 **1.15**	1.00 **1.00**	∞

(The significance level for each value is given at the top of the column.)

d. f.	.01	.05	.10
1	6.64	3.84	2.71
2	9.21	5.99	4.60
3	11.34	7.82	6.25
4	13.28	9.49	7.78
5	15.09	11.07	9.24
6	16.81	12.59	10.64
7	18.48	14.07	12.02
8	20.09	15.51	13.36
9	21.67	16.92	14.68
10	23.21	18.31	15.99
11	24.72	19.68	17.28
12	26.22	21.03	18.55
13	27.69	22.36	19.81
14	29.14	23.68	21.06
15	30.58	25.00	22.31
16	32.00	26.97	23.54
17	33.41	27.59	24.77
18	34.80	28.87	25.99
19	36.19	30.14	27.20
20	37.57	31.41	28.41
21	38.93	32.67	29.62
22	40.29	33.92	30.81
23	41.64	35.17	32.01
24	42.98	36.42	33.20
25	44.31	37.65	34.38
26	45.64	38.88	35.56
27	46.96	40.11	36.74
28	48.28	41.34	37.92
29	49.59	42.56	39.09
30	50.89	43.77	40.26

Appendix VI is abridged from Table IV of Fisher & Yates, *Statistical Tables for Biological, Agricultural, and Medical Research,* published by Oliver & Boyd Ltd., Edinburgh, and by permission of the authors and publishers.

VII Critical Values of *u* (Total Runs)

(All values are at the .05 significance level. The larger of n_1 and n_2 is to be read at the top and the smaller is to be read in the left margin.)

	5	6	7	8	9	10	11	12	13	14	15	16	17	18	19	20
2								2 6	2 6	2 6	2 6	2 6	2 6	2 6	2 6	2 6
3		2 8	2 8	2 8	2 8	2 8	2 8	2 8	2 8	2 8	3 8	3 8	3 8	3 8	3 8	3 8
4	2 9	2 9	2 10	3 10	3 10	3 10	3 10	3 10	3 10	3 10	3 10	4 10	4 10	4 10	4 10	4 10
5	2 10	3 10	3 11	3 11	3 12	3 12	4 12	4 12	4 12	4 12	4 12	4 12	4 12	5 12	5 12	5 12
6		3 11	3 12	3 12	4 13	4 13	4 13	4 13	5 14	5 14	5 14	5 14	5 14	5 14	6 14	6 14
7			3 13	4 13	4 14	5 14	5 14	5 14	5 15	5 15	6 15	6 16	6 16	6 16	6 16	6 16
8				4 14	5 14	5 15	5 15	6 16	6 16	6 16	6 16	6 17	7 17	7 17	7 17	7 17
9					5 15	5 16	6 16	6 16	6 17	7 17	7 18	7 18	7 18	8 18	8 18	8 18
10						6 16	6 17	7 17	7 18	7 18	7 18	8 19	8 19	8 19	8 20	9 20
11							7 17	7 18	7 19	8 19	8 19	8 20	9 20	9 20	9 21	9 21
12								7 19	8 19	8 20	8 20	9 21	9 21	9 21	10 22	10 22
13									8 20	9 20	9 21	9 21	10 22	10 22	10 23	10 23
14										9 21	9 22	10 22	10 23	10 23	11 23	11 24
15											10 22	10 23	11 23	11 24	11 24	12 25
16												11 23	11 24	11 25	12 25	12 25
17													11 25	12 25	12 26	13 26
18														12 26	13 26	13 27
19															13 27	13 27
20																14 28

Taken with permission from "Tables for Testing Randomness of Grouping in a Sequence of Alternatives" by C. Eisenhart and F. Smed, *Annals of Mathematical Statistics,* XIV (1943), 66.

VIII Math Review

I The Calculator

The hand-held calculator is making life easier and easier for the student studying statistics. A calculator need not be expensive, but should include *at least* the following.

1. The four basic functions (addition, subtraction, multiplication, and division).
2. Square (X^2) and square root ($\sqrt{X}$) functions.
3. One or more storage units (memories) for intermediate results.

Additional features (built-in operations to sum lists of numbers or lists of squares of numbers or to calculate means, variances, correlation coefficients, etc.) are available, but the deciding factor in selecting a calculator should be ease of operation. If you can obtain the following results easily on a certain model, it should serve you well in an elementary statistics course.

1. $17 + 18 + 56 = 91$
2. $3/4(17 + 18 + 56) = 68.25$
3. $(49)^2 = 2401$
4. $\sqrt{15} = 3.872983$
5. $(\sqrt{15} + 7)^2 = 118.2218$
6. $\sqrt{99} + \sqrt{18} = 14.19251$
7. $(-4) + 15 - 77 + 3/4 = -65.25$
8. $\sqrt{99}\,\sqrt{18} = 42.21374$
9. $3.98 + \frac{4}{9}(77) = 38.20222$
10. $17 - 1.96\frac{.39}{\sqrt{18}} = 16.81983$

Even an expensive calculator is no substitute for a knowledge and understanding of the basic arithmetic and algebraic principles presented here.

II Signed Numbers

Rules

1. When adding numbers with like signs, add the absolute values and use the common sign.
2. When adding numbers with opposite signs, subtract the absolute value of the smaller from the absolute value of the larger and use the sign of the larger.
3. When subtracting, change the sign of the number being subtracted and add the results.
4. The product and quotient of two numbers with the same sign are positive.
5. The product and quotient of two numbers with different signs are negative.

Examples (Each example illustrates the rule with the same number.)

1. $+15 + 77 = 92 \qquad -44 + (-8) = -52$
2. $-15 + 77 = 62 \qquad 44 + (-77) = -33$
3. $-15 - (77) = -15 + (-77) = -92 \qquad 8 - (-4) = 8 + 4 = 12$
4. $8 \cdot 7 = 56 \qquad \frac{90}{18} = 5$

   $-8 \cdot -7 = 56 \qquad \frac{-90}{-18} = 5$

5. $-8 \cdot 7 = -56 \qquad \frac{-90}{18} = \frac{-18}{90} = -5$

III Exponents

3^7: 3 is called the "base"; 7 is called the "exponent" or "power."

Rules

1. b^n means to multiply the base b by itself n times.
2. $b^n \cdot b^m = b^{n+m}$ When multiplying two numbers *with the same base*, add exponents.
3. $\frac{b^n}{b^m} = b^{n-m} = \frac{1}{b^{m-n}}$ When dividing numbers with the same base, subtract exponents.
4. $(b^n)^m = b^{n \cdot m}$ When raising a number with a power to a power, multiply exponents.

Examples

1. $3^4 = 3 \cdot 3 \cdot 3 \cdot 3 = 81$
2. $4^2 \cdot 4^3 = 4^5 = 1{,}024$
3. $\frac{4^8}{4^4} = 4^4 = 256 \qquad \frac{4^5}{4^7} = \frac{1}{4^2} = \frac{1}{16}$
4. $(3^3)^2 = 3^6 = 729 \qquad (3^8)^{1/2} = 3^4 = 81$

IV Order of Operations

Rule

When simplifying an expression involving more than one operation, do the operations in the following order.

1. As indicated by pairs of grouping symbols (parentheses, brackets, etc.).
2. Raising to powers and taking square roots.
3. Multiplication and division.
4. Addition and subtraction.

Examples

$6 + 5 \cdot 3 = 6 + 15 = 21$

$2^3 - 4 = 8 - 4 = 4$

$3 \cdot 2^2 = 3 \cdot 4 = 12$

$4 + \frac{1}{2}\frac{3}{\sqrt{36}} = 4 + \frac{1}{2} \cdot \frac{3}{6} = 4 + \frac{1}{4} = 4.25$

$$\left[3 + \frac{1}{2}(4 - 3^2)\right] + 8 = \left[3 + \frac{1}{2}(4 - 9)\right] + 8$$

$$= \left[3 + \frac{1}{2}(-5)\right] + 8$$

$$= \left[3 - \frac{5}{2}\right] + 8 = \frac{1}{2} + 8 = 8.5$$

V Square Roots

Square roots are most easily found using calculators or tables (see Appendix I). Two rules may be used to simplify calculations.

Rules

1. $\sqrt{a}\sqrt{b} = \sqrt{ab}$ 2. $\dfrac{\sqrt{a}}{\sqrt{b}} = \sqrt{\dfrac{a}{b}}$

Examples

1. $\sqrt{32}\sqrt{18} = \sqrt{576} = 24$ 2. $\dfrac{\sqrt{38}}{\sqrt{5}} = \sqrt{7.6} = 2.75681$

VI Algebraic Operations

Rules

1. The same term may be added to or subtracted from both sides of an equation.
2. A term may be moved from one side of an equation to the other if its sign is changed.
3. Both sides of an equation may be multiplied or divided by the same term.

Examples

1. $X + 4 = 7$

   $X + 4 - 4 = 7 - 4$

   $X = 3$

2. $-X + 9 = 15$

   $9 - 15 = X$

   $-6 = X$

3. $3y = 11$

   $\frac{1}{3}\,3y = \frac{1}{3}\,11$

   $y = \frac{11}{3}$

Combining rules

$X = \frac{3}{7}Y + 14$

$X - 14 = \frac{3}{7}Y$

$\frac{7}{3}(X - 14) = \frac{7}{3}\frac{3}{7}Y$

$\frac{7}{3}(X - 14) = Y$

IX Answers to Odd-Numbered Problems

Note: Your answers may differ slightly from those given here, because rounding was used in computing these answers.

Chapter 2

2.1 mean: 6.4, median: 5, mode: 11

2.3 (a) .54 (b) .50 (c) .475

2.5 $\mu = 42.86$, median: 37

2.7 $48.47

2.9 $8.11

2.11 (a) 96 (c) 2,522 (d) 3,062

2.13 (a) health (b) business

2.15 The mode if there is one. If no mode, the median if one or more registers a very low or high temperature. The mean if the readings are all different but cluster. The mean or median may also be computed after extreme value(s) are dropped.

2.17 $\mu = 30.35$, median: 30.25

2.19 −69

2.21 (a) 74.5 (b) 107.5

2.23 (a) 410.33 (b) 512.91

2.25 (a) 15 (b) 240 (c) The neighboring state; $240 > 150$

2.27 The Olympic average will be greater than the actual average by 4.764 centimeters.

2.29 (a) 5,378.8 (b) 5,648 (c) 5,042.63

Chapter 3

3.1 (a) 24 (b) 156 (c) 576 (d) 0 (e) 20 (f) 60 (g) 0 (h) 60 (i) 159 (j) 10

3.3 $\sigma^2_W = 4.67$
$\sigma_W = 2.16$
$\sigma^2_Y = 996.5$
$\sigma_Y = 31.57$
$\sigma^2_Z = 25$
$\sigma_Z = 5$

3.5 (a) $\sigma^2 = 5.64$; $\sigma = 2.37$
(b) (days)^2 or days squared.

3.7 (a) $\sigma^2 = 4{,}496$
$\sigma = 67.05$
(b) 4,496 and 67.05

3.9 (a) $\mu = 843.75$
$\sigma^2 = 51{,}503.35$
$\sigma = 226.94$
(b) $\mu = 759.38$
$\sigma^2 = 41{,}717.71$
$\sigma = 204.25$

3.11 (a) $\mu = 32.59$
$\sigma^2 = 68.83$
$\sigma = 8.30$
(b) $\mu = 42.59$, $\sigma^2 = 68.83$, $\sigma = 8.30$
(c) $\mu = 6.52$, $\sigma^2 = 2.75$, $\sigma = 1.66$
(d) $\mu = 42.37$, $\sigma^2 = 116.32$
$\sigma = 10.79$

3.13 $\sigma = 8.57$, $\mu = 28.4$
(a) 14 (b) 25 (c) 25

3.15 (a) T (b) F (c) F (d) F (e) T (f) F (g) F (h) T (i) T (j) T (k) F (l) F

3.17 $\sigma^2_A = 8.33$, $\sigma_A = 2.89$
$\sigma^2_B = 24.4$, $\sigma_B = 4.94$
$\sigma^2_C = 50.25$, $\sigma_C = 7.09$

3.19 (a) 69 (b) 1,109 (c) 4,761 (d) 74 (e) 94 (f) 1,924 (g) 1,114 (h) 39.5 (i) 37 (j) 156.8

3.21 (a) $\sigma^2 = 88{,}708.45$, $\sigma = 297.84$
(b) $\sigma^2 = 88{,}708.45$, $\sigma = 297.84$

3.23 (a) $\sigma^2 = 4.16$, $\sigma = 2.04$
(b) Average deviation: 1.70

3.25 (a) $\sigma^2 = 115.1$, $\sigma = 10.73$
(b) $\sigma^2 = 14.10$, $\sigma = 3.76$

3.27 (a) $\sigma^2 = 86.55$, $\sigma = 9.30$
(b) $\sigma^2 = 21.64$, $\sigma = 4.65$
(c) $\sigma^2 = 86.55$, $\sigma = 9.30$

3.29 Yes. Given any standard deviation there is some distribution whose variability it describes. Shifting that distribution (adding or subtracting the same constant to or from every term) can produce a distribution with the same σ and any μ, including $\mu = \sigma$.

Chapter 4

4.1 10, 30, 50, 70, 90

4.3 (a) −2.2 (b) 1.4 (c) 3.2 (d) −1.6 (e) .4

4.5 (a) 48 (b) 70 (c) 3 (d) 93 (e) 28

4.7 (a) $\mu = 47.3$, $\sigma = 2.39$
(b) $z_{45} = -.96$, $z_{42.5} = -2.01$
$z_{48.5} = .50$, $z_{51} = 1.55$
(c) $T_{45} = 40$, $T_{42.5} = 30$
$T_{48.5} = 55$, $T_{51} = 66$

4.9 (a) 24 in. (b) 28.5 in. (c) 32.33 in. (d) 13.47 in. (e) 1.5 in.

4.11 34

4.13 (a) 17 (b) 33 (c) 44 (d) 57 (e) 76 (f) 89 (g) 99

4.15 z score distribution has $\mu = 0, \sigma = 1$. Multiplying by 10 (each term) yields a distribution with $\mu = 0$, $\sigma = 10$. Adding 50 to each term yields a T distribution with $\mu = 50$, $\sigma = 10$.

4.17 (a) 46 (b) fourth day

4.19 (a) $T_5 = 53$, $T_2 = 44$
$T_{-7} = 19$, $T_{-4.5} = 26$
(b) for $z = 1.55$, $X = 9.43$
for $z = -2.13$, $X = -3.46$

4.21 (a) $\mu = 45.67$, $\sigma = 19.15$
(b) $z_{\text{Feb}} = 1.43$, $z_{\text{Aug}} = .75$,
$z_{\text{May}} = -1.97$
(c) $T_{\text{Feb}} = 64$, $T_{\text{Aug}} = 58$,
$T_{\text{May}} = 30$

4.23 (a) $\mu = 546.46$, $\sigma = 168.81$
(b) $z_{373} = -1.03$, $z_{415} = -.78$
$z_{500} = -.28$, $z_{400} = -.87$

4.25 (a) $PR_{4.6} = 50$ (b) $PR_{50} = 56$
(c) $PR_{3.7} = 15$

4.27 (a) $PR_{3.8} = 6$
$PR_{14.4} = 86$
$PR_{5.5} = 24$
(b) 6 86 24
(c) 6 86 24
(d) No change in either case.

4.29 1 2 3 4 5 6 7 8 9 1,000
$PR_9 = 95$

Chapter 5

5.1 discrete: c, e, g, h, j
continuous: a, b, d, f, i

5.3 (a) 499 (b) 115 (c) 3 (d) 1,000 (e) 9

5.5 (a) discrete
(b)

Interval	*Boundaries*
70–84	69.5–84.5
55–69	54.5–69.5
40–54	39.5–54.5
25–39	24.5–39.5
10–24	9.5–24.5

Midpoints	*F*
77	7
62	2
47	7
32	5
17	15

Bar Graph

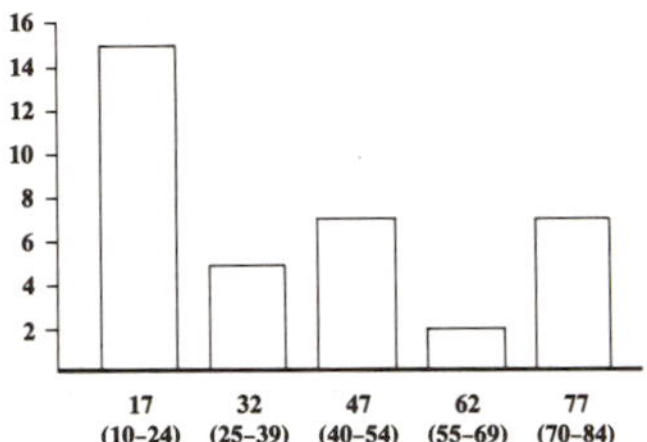

Histogram

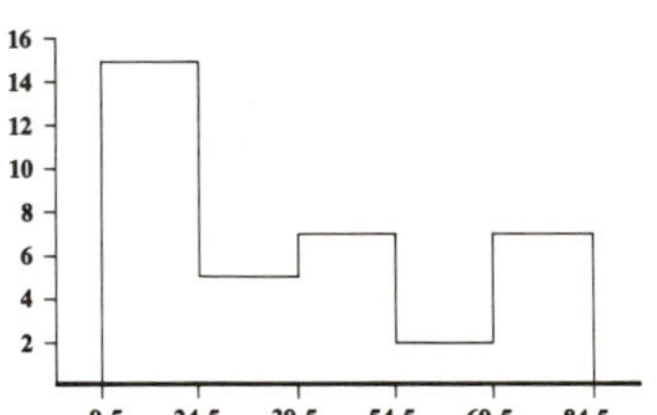

5.7 (a)

Boundaries	*Mid-points*	*F*
21.5–24.5	23	2
18.5–21.5	20	3
15.5–18.5	17	5
12.5–15.5	14	2
9.5–12.5	11	7
6.5– 9.5	8	13
3.5– 6.5	5	7

(b)

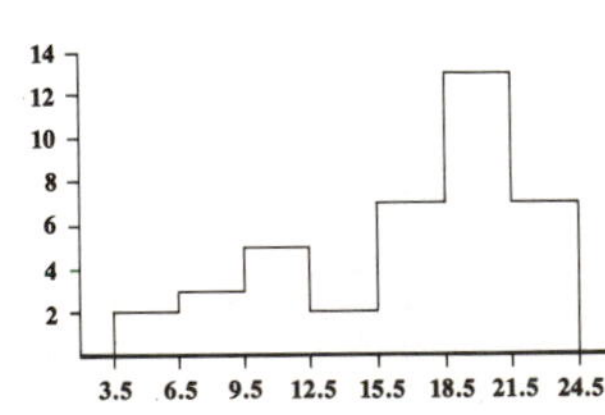

5.9

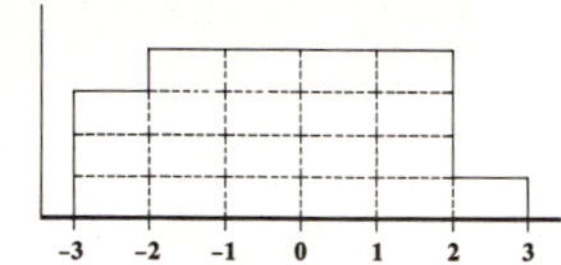

5.11 (a)

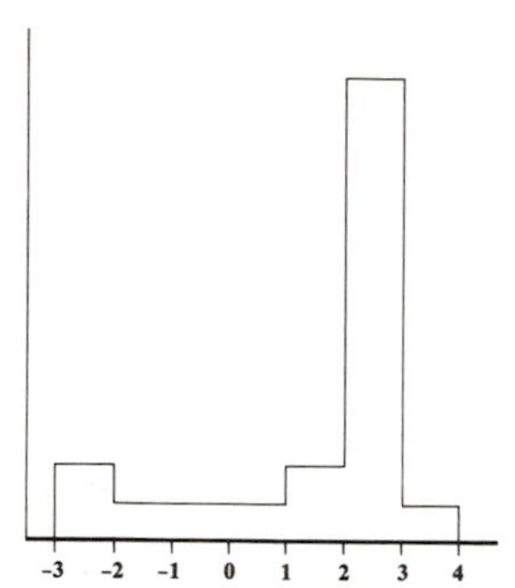

(b) Pairings for more terms. "Percent below" carried out to more decimal places.

5.13 200

5.15

Boundaries	Midpoints	F
0–1	.5	45
1–2	1.5	55
2–3	2.5	113
3–4	3.5	172
4–5	4.5	88
5–6	5.5	46
6–7	6.5	51
7–8	7.5	45
8–9	8.5	9
9–10	9.5	11

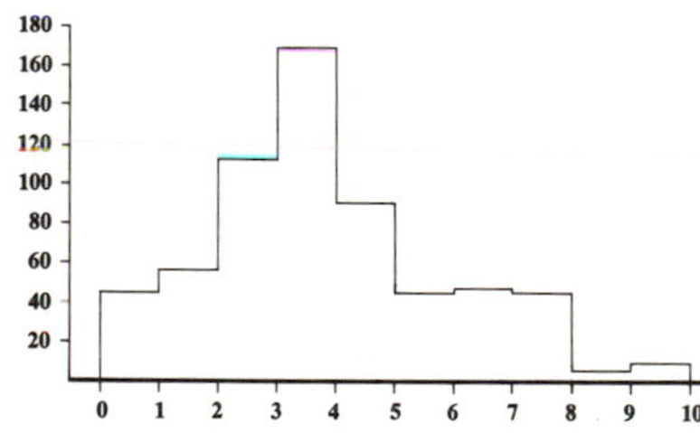

5.17 discrete: b, c, f, h;
continuous: a, d, e, g, i, j

5.19

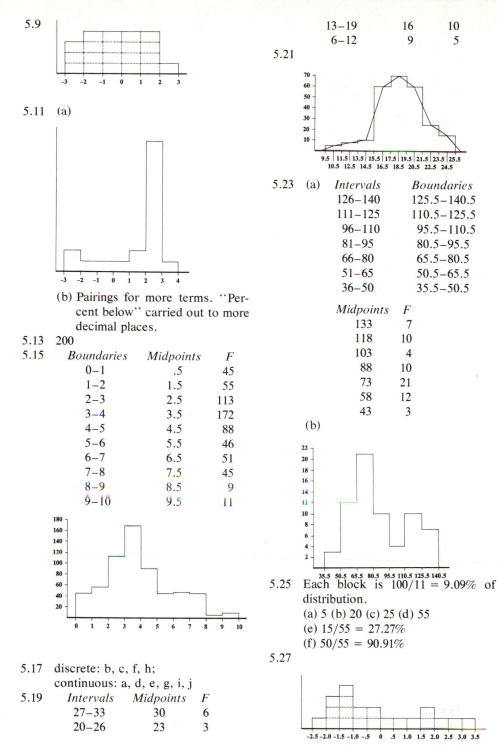

Intervals	Midpoints	F
27–33	30	6
20–26	23	3
13–19	16	10
6–12	9	5

5.21

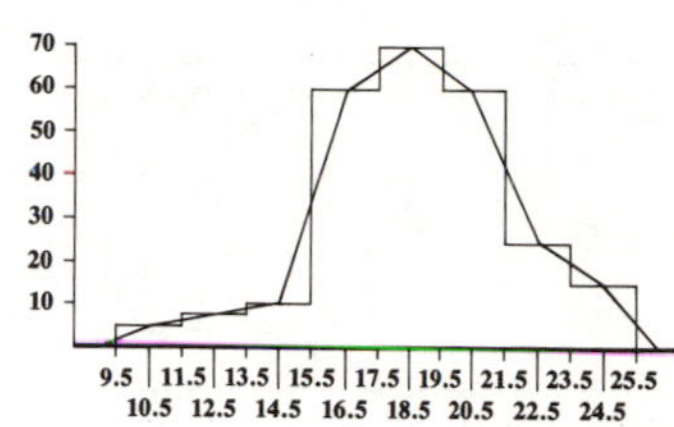

5.23 (a)

Intervals	Boundaries
126–140	125.5–140.5
111–125	110.5–125.5
96–110	95.5–110.5
81–95	80.5–95.5
66–80	65.5–80.5
51–65	50.5–65.5
36–50	35.5–50.5

Midpoints	F
133	7
118	10
103	4
88	10
73	21
58	12
43	3

(b)

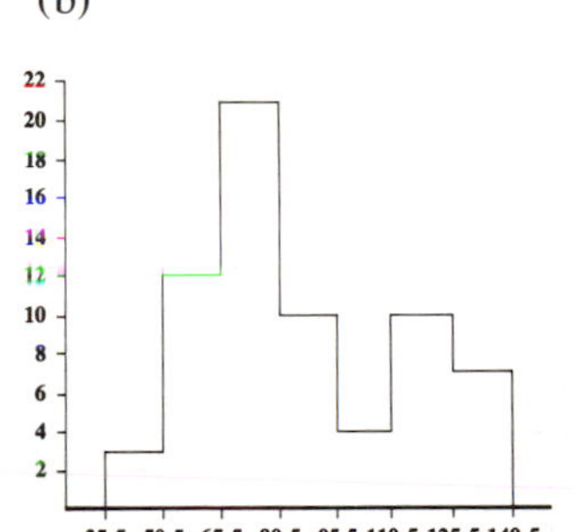

5.25 Each block is 100/11 = 9.09% of distribution.
(a) 5 (b) 20 (c) 25 (d) 55
(e) 15/55 = 27.27%
(f) 50/55 = 90.91%

5.27

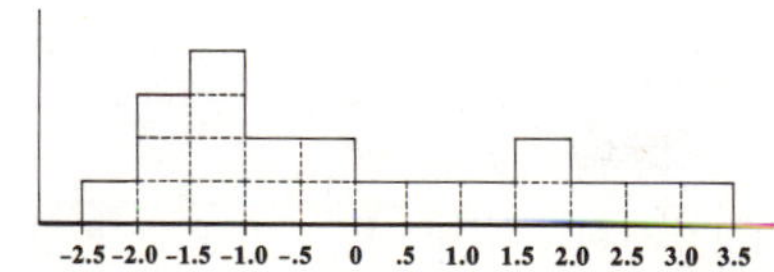

5.29

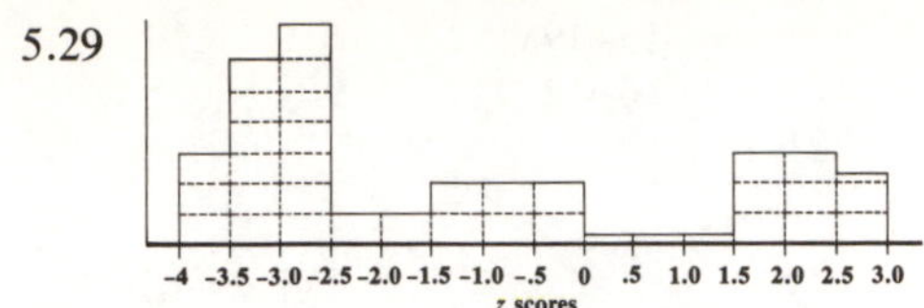

Chapter 6

6.1

Boundaries	Mid-points	F	CF
25.5–30.5	28	14	100
20.5–25.5	23	18	86
15.5–20.5	18	23	68
10.5–15.5	13	20	45
5.5–10.5	8	17	25
.5–5.5	3	8	8

6.3 (a) 25 (b) 6 (c) 73

6.5 (a) 500 (b) 87 (c) 925 (d) 1,283.33

6.7 (a) crude mode: 525, $\mu = 974.17$, $\sigma = 419.47$
(b) $\mu = 974.17$, $\sigma = 420$, discrepancy due to rounding

6.9 $Q_2 = 28.45$, $D_1 = 11.47$, $P_{45} = 27.03$

6.11

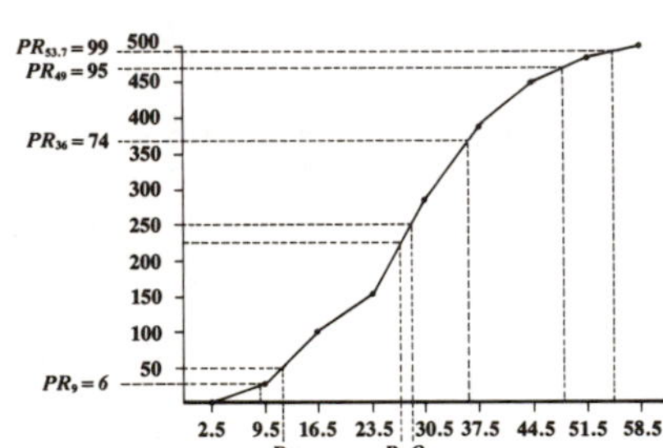

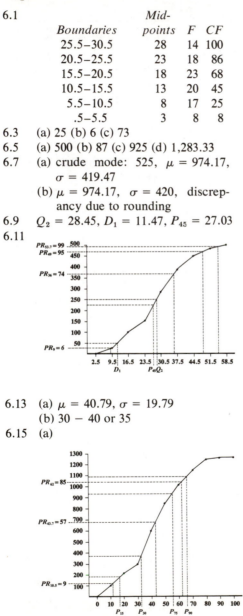

6.13 (a) $\mu = 40.79$, $\sigma = 19.79$
(b) 30 – 40 or 35

6.15 (a)

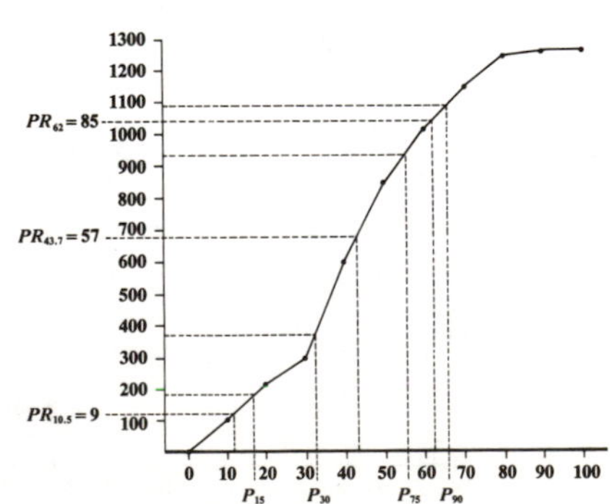

(b)

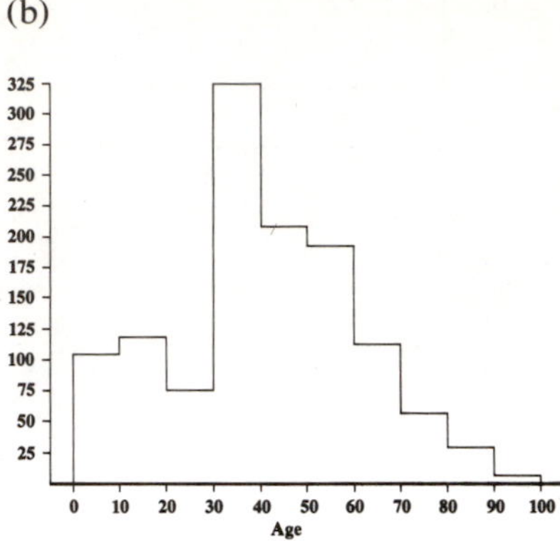

6.17 (a) $\sigma^2 = 18.92$, $\sigma = 4.35$
(b) $z_{6.5} = -.49$, $z_{8.75} = .03$, $z_{16} = 1.69$

6.19 20–30 years and 50–70 years

6.21 (a) 25 (b) 55 (c) 77 (d) 3

6.23 (a)

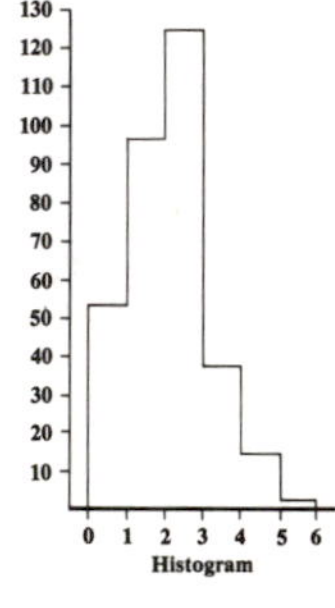

(b)

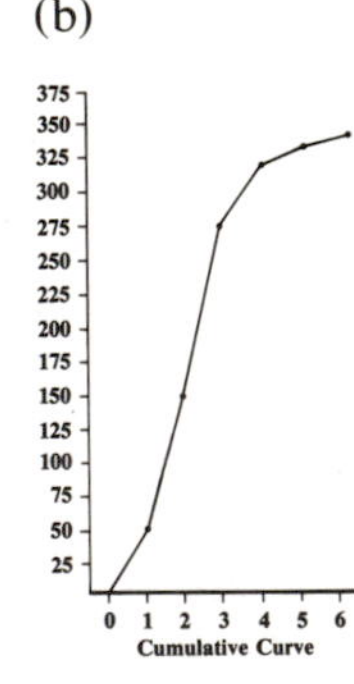

6.25 *New Boundaries*

2.5–3.0	
2.0–2.5	new mean: 1.06
1.5–2.0	
1.0–1.5	new standard
0.5–1.0	deviation: .53
0.0–0.5	

6.27 $\sigma^2 = 42.16$, $\sigma = 6.49$, $\sigma^2 = 42.165$, $\sigma = 6.49$

6.29

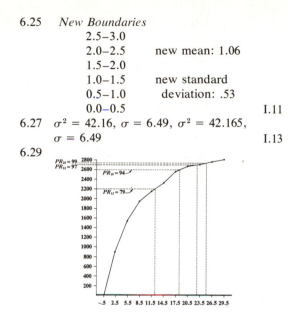

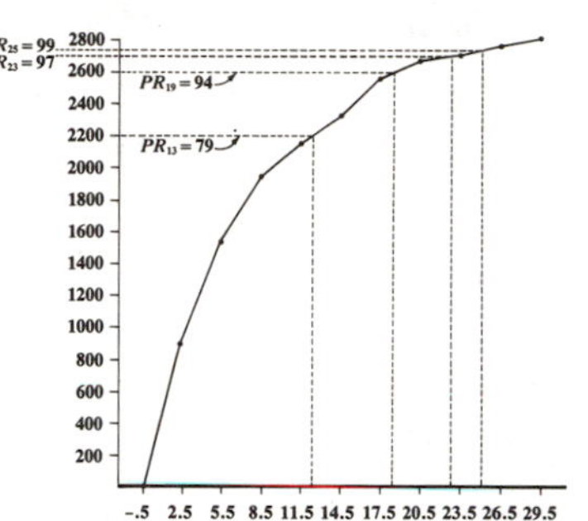

Review I

I.1 X: $\mu = 5.33$, median $= 5.5$, no mode, $\sigma^2 = 8.23$, $\sigma = 2.87$

Y: $\mu = 27.86$, median $= 22$, mode $= 22$, $\sigma^2 = 563.39$, $\sigma = 23.74$

Z: $\mu = 21.6$, median $= 19$, mode $= 19$, $\sigma^2 = 150.11$, $\sigma = 12.25$

I.3 (a) $\mu = 67.8$, median $= 70$, no mode

(b) $\sigma^2 = 50.96$, $\sigma = 7.14$

(c) $z = -1.09$, $PR = 30$

I.5 (a) $\mu = 543.75$, $\sigma^2 = 3{,}736.44$, $\sigma = 61.13$

(b) $\mu = 593.75$, $\sigma^2 = 3{,}736.44$, $\sigma = 61.13$

(c) 815.63, 8,406.99, 91.70

(d) 489.38, 3,026.52, 55.02

I.7 $(X - \mu)$

−36.75	−8.75	
58.25	−18.75	$\Sigma(X - \mu) = 0$
−53.75	131.25	
−3.75	−67.75	

I.9 (a) $z_{27} = -1.67$, $z_{42} = .71$
$z_{35} = -.40$, $z_{53.5} = 2.54$

(b) $T_{27} = 33$, $T_{42} = 57$, $T_{35} = 46$, $T_{53.5} = 75$

(c) -1.5: $X = 28.05$
-2.75: $X = 20.175$
1.2: $X = 45.06$

(d) 85

I.11 discrete: b, c, f, h, j
continuous: a, d, e, g, i

I.13 (a)

X	*CF*
23	500
20	489
17	419
14	367
11	283
8	266
5	222
2	137

(b)

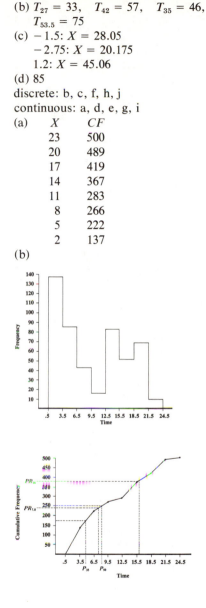

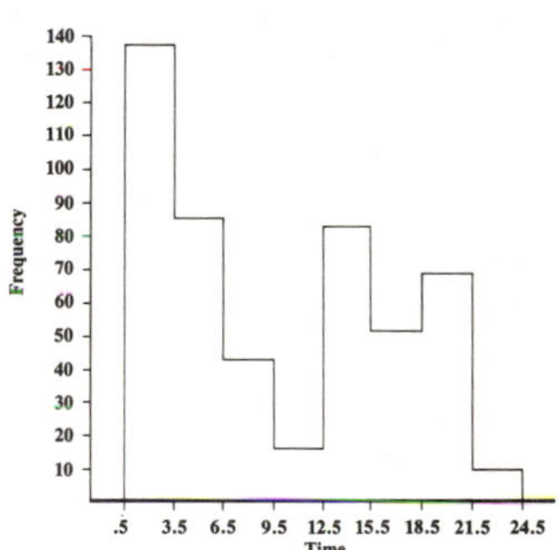

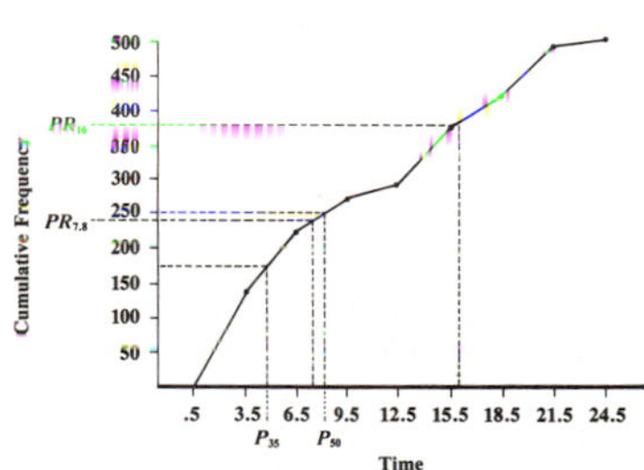

(c) $P_{50} = 8.41$, $P_{35} = 4.84$, $PR_{7.8} = 48$, $PR_{16} = 75$

I.15

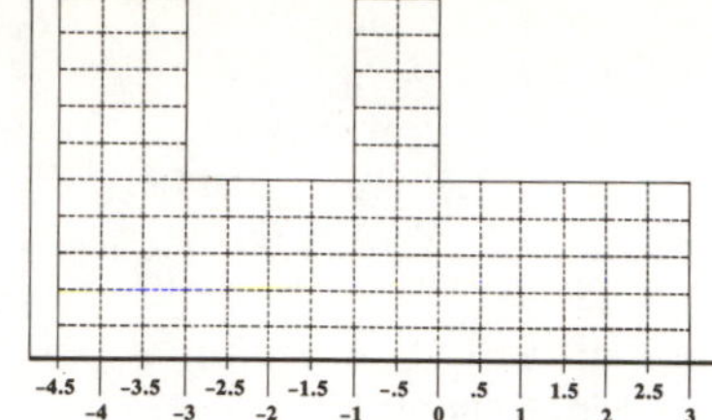

Area under the curve represents fraction of terms.

I.17 (a) 8 (b) 8 (c) 12 (d) 72 (e) 1,290 (f) 5,184 (g) 0 (h) 426 (i) 71 (j) 2,167

I.19 (a) $z_{21.7} = .98$, $z_{15.3} = -1.35$, $z_{12} = -2.55$
(b) $T_{21.7} = 60$, $T_{15.3} = 36$, $T_{12} = 24$
(c) 3: $X = 27.25$, $-.37$: $X = 17.98$
-1.8: $X = 14.05$, 1.75: $X = 23.81$

I.21 (a) 57,820 (b) 14,271.80 (c) Both are divided by 100.

I.23 (a) 10/20 = 50% (b) 2/20 = 10%
(c) a vertical frequency axis

I.25 Instructor: $z_9 = -.5$, $z_{22} = .25$, $z_{18} = 1$
Average: .25
Associate professor: $z_{13} = .29$, $z_{29} = .73$, $z_{17} = -.4$
Average: .21
Full professor: $z_{20} = 1$, $z_{25} = -.43$, $z_{17} = -.68$
Average: $-.04$
The full professor did best (fastest).

I.27 (a)

Time	*X*	*F*	*CF*
24.5–29.5	27	9	60
19.5–24.5	22	10	51
14.5–19.5	17	19	41
9.5–14.5	12	15	22
4.5–9.5	7	7	7

(b) *Histogram*

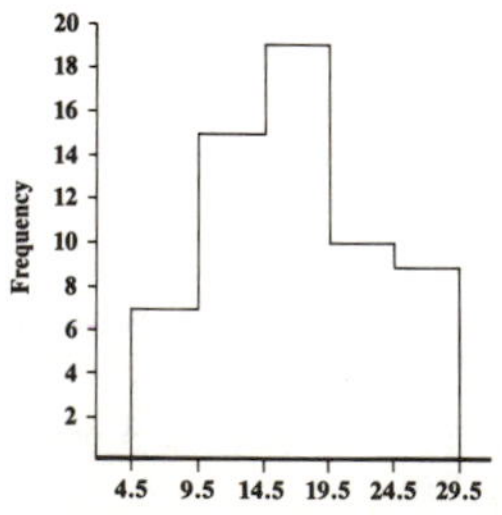

(c) *Cumulative Frequency Curve*

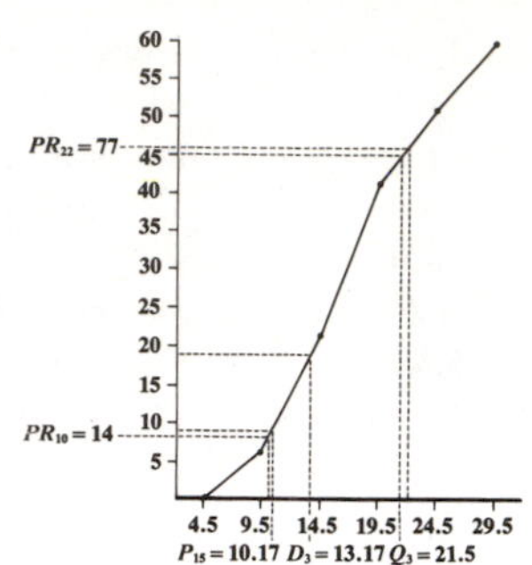

I.29 (a)

X	*CF*
74,000	1,000
67,000	981
60,000	970
53,000	923
46,000	880
39,000	880
32,000	825
25,000	698
18,000	533
11,000	244
4,000	51

(b) 20,700.69
(c) $P_9 = 8{,}914.51$, $Q_3 = 31{,}366.14$, $D_4 = 18{,}278.55$
(d) $PR_{30,000} = 73$, $PR_{8,700} = 8$

Chapter 7

7.1 (a) 41.92% (b) 49.38% (c) 5.29% (d) 3.42% (e) 73.51% (f) .95% (g) 2.51% (h) 89.2% (i) 50% (j) 96%

7.3 (a) 84 (b) 93 (c) 2 (d) 1 (e) 88 (f) 89 (g) 0 (h) 81 (i) 17 (j) 53

7.5 (a) 1.88 (b) .39 (c) $-.67$ (d) $-.52$ (e) .67 (f) $-.13$ (g) -1.04 (h) .99 (i) -1.28 (j) .08

7.7 (a) .3634
(b) .1076
(c) .0377
(d) .9034
(e) .2517
(f) .3241

7.9 (a) 12.10% (b) 74.20% (c) 80.78% (d) 11.12% (e) 68.26% (f) 13.68% (g) 12.12%

7.11 (a) .90
(b) -1.47
(c) $Z_a = -.61$, $Z_b = .09$
(d) $z_a = -1.34$, $z_b = .97$
(e) $z = .65$
(f) 1.90
(g) -1.19
(h) $-.60$

7.13 (a) .0475 (b) 3.7475–4.7525 hours (3 hours 45 minutes–4 hours 45 minutes)

7.15 7,592 subjects

7.17 (a) $-.97$ (b) .40 (c) .56 (d) $-.34$ (e) $-.82$ (f) 1.19 (g) 1.95 (h) $-.18$

7.19 (a) .6554 (b) .0694 (c) .1151 (d) .1151 (e) .901 (f) .9802 (g) .6709 (h) .8911

7.21 (a) .1950 (b) .1193 (c) .0564 (d) .5306 (e) .1257 (f) .3724

7.23 (a) 30.85%
(b) 37
(c) 24.56

7.25 (a) $z_{14} = -.25$ $z_{18} = .75$ 9.87% + 27.34% = 37.21%
(b) 21 (c) 66.6 mph

7.27 (a) 69.95–78.05 (b) Yes

7.29 .9616–1.0224

Chapter 8

8.1 (a) (Theorem 8.1) (c) (Theorem 8.2) (d) (Theorem 8.1) (f) (Theorem 8.2)

8.3 (a) $\mu_{\bar{X}} = 50$, $\sigma_{\bar{X}} = .83$ (b) .1151
(c) .99984

8.5 (a) 18.41% (b) 1.22%

8.7 (a) .0027 (b) .6726

8.9 2.28%

8.11 (a) $\mu_{\bar{X}} = 10.75$, $\sigma_{\bar{X}} = .15$ (b) 4.75%
(c) 10.876

8.13 (a) 511.03 (b) 495.86 (c) 492.89–507.11

8.15 (a) $\mu = 8.6$, $\sigma = 4.13$
(b) (3,3) (6,3) (8,3) (11,3) (15,3)
(3,6) (6,6) (8,6) (11,6) (15,6)
(3,8) (6,8) (8,8) (11,8) (15,8)
(3,11) (6,11) (8,11) (11,11) (15,11)
(3,15) (6,15) (8,15) (11,15) (15,15)
(c) 3 4.5 5.5 7 9
4.5 6 7 8.5 10.5
5.5 7 8 9.5 11.5
7 8.5 9.5 11 13
9 10.5 11.5 13 15
(d) 8.6
(e) 2.92

8.17 (a) $\mu_{\bar{X}} = 500$, $\sigma_{\bar{X}} = 6.25$ (b) 21.19%
(c) 42.07%

8.19 (a) 5.71% (b) 393.04

8.21 (a) 83.40% (b) 99.38% (c) 30.21%

8.23 65.91%

8.25 (a) 470 (b) 109 (c) 109 (d) 282

8.27 .8502

8.29 (a) 84.72%
(b) 8.0798–8.1358

Chapter 9

9.1 (a) 1/728 (b) 1/728 (c) 224/728 = 28/91 (d) 128/728 = 16/91 (e) 0 (f) 29/728 (g) 127/728 (h) 77/728 = 11/104

9.3 (a) .3446
(b) .0026
(c) .8621

9.5 (a) .1230
(b) The sample is not random or the mean and/or standard deviation (17.5 and 3) are incorrect.

9.7 (a) .1112
(b) .8065
(c) .0119
(d) .6032

9.9 (a) $S = \{R_1R_2, R_1R_3, R_1F_1, R_1F_2, R_2R_1, R_2R_3, R_2F_1, R_2F_2, R_3R_1, R_3R_2, R_3F_1, R_3F_2, F_1R_1, F_1R_2, F_1R_3, F_1F_2, F_2R_1, F_2R_2, F_2R_3, F_2F_1\}$
(b) 18/20 = .9 (c) 2/20 = .1
(d) Their sum is 1. One of the events (at least one rotten or no rotten) must occur.

9.11 (a) It is discrete; it is finite.
(b) .0062

9.13 (a) 1/4 (b) 6/12 = 1/2 (c) 2/12 = 1/6

9.15 .0606

9.17 .9147

9.19 (a) .8078 (b) .9772 (c) .9995

9.21 .7164

9.23 .5793, 0

9.25 20 plates; chicken entree, 1/25

9.27 (a) .0287 (b) 198,234.99 barrels/day
9.29 (a) 19/27 (b) 3/27 = 1/9 (c) 18/24 = 3/4

Chapter 10

In these solutions *P* (type-two error) represents the probability of a type-two error.

10.1 type one—a, type two—c
10.3 type one—a, type two—b
10.5 .2670

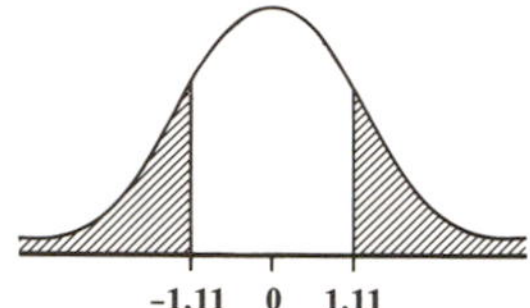

10.7 *P* (type-two error) = .0475
Power of the test = .9525

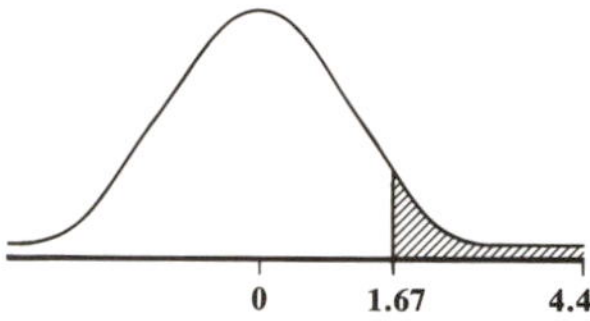

10.9 (a) 63.17–66.83 (b) .1336
10.11 .1336
10.13 (a) The number placed per year is distributed normally (since N = 5 < 30).
Only social work majors are placed in social work jobs.
All sample members respond.
(b) .1362
If a type-one error occurs, the hypothesis is true but rejected. The placement office is charged with making inaccurate claims.
(c) *P* (type-two error) = .3725
Power of the test = .6275
10.15 1.6128–1.8872
10.17 type one—a, type two—d
10.19 35.5–54.5
10.21 .3734
10.23 .1868
10.25 *P* (type-two error) = .1986
Power of the test = .8014
10.27 .1446
Risks involved:
If accepted hypothesis is too large, the manufacturer will overproduce or will accept the device for production when it should not be.
If accepted hypothesis is too small, the manufacturer will underproduce and fail to meet demand.
10.29 14.63–16.77

Review II

II.1 (a) .6517 (b) .2177 (c) .9646 (d) .2586 (e) .18 (f) .30 (g) .20
II.3 (a) 39.70% (b) 66.98 (c) .2420
II.5 (a) .7324 (b) .8365 (c) .1041 (d) .5205 (e) .4795
II.7 (a) .005 (b) .1314
II.9 (a) .00034 (b) .8023
II.11 .4753
II.13 (a) H: $\mu = 485$, ($\sigma = 25$) (b) .0456
(c) .2061, .7939
(d) Type-one: The machine is working properly, but is serviced anyway.
Type-two: The machine is not working properly, but the problem is not detected.
II.15 (a) Y, Z—distributions of sample means
W—distribution of sample sums
(b) all are discrete (c) $\mu_X = \mu_Y = \mu_Z$
(d) F (e) F (f) T (g) T
II.17 (a) $\mu = 28$, $\sigma = 2.37$ (b) 14.69%
(c) 26 (actually 26 exceeds 14.69%)
(d) 74
II.19 .8051, .0087
II.21 750
II.23 16
II.25 (a) 9,288 (b) 85% (c) 49.6%
II.27 .4472
Risk: True—rejected; 30% have heartworms but either the problem is underestimated (nothing done) or overestimated (overkill).
II.29 If $\mu = 200$, 0; If $\mu = 275$, .5.

Chapter 11

11.1 (a) H_0 (f) H_0

(b) H_1	(g) H_0
(c) H_0	(h) H_1
(d) H_1	(i) H_1
(e) H_1	(j) H_0

11.3 $H_1: \mu > 22$, $H_0: \mu \leq 22$, $z = 5.33$, Reject H_0

11.5 (a) mean
(b) $H_1: \mu < 5.5$, $H_0: \mu \geq 5.5$, $z = 6$, Reject H_0

11.7 $H_1: \mu \neq 7.5$, $H_0: \mu = 7.5$, $z = -5.5$, Reject H_0

11.9 (a) $\bar{X} \leq .72$, $\bar{X} \geq .78$ (b) No (c) $\bar{X} \leq .73$, $\bar{X} \geq .77$ (d) It increases.

11.11 $H_1: P > .45$, $H_0: P \leq .45$, $z = 1.10$, Do not reject H_0

11.13 (a) $H_1: \mu \neq 45$, $H_0: \mu = 45$, $z = -1.65$, Do not reject H_0
(b) 24

11.15 $H_1: P \neq .53$, $H_0: P = .53$, $z = -2.20$, Reject H_0

11.17 (a) mean (b) one-tail test (c) $H_1: \mu > 55{,}000$, $H_0: \mu = 55{,}000$, $z = 3.54$, Reject H_0

11.19 $H_1: \mu < 48$, $H_0: \mu \geq 48$, $z = -3.23$, Reject H_0

11.21 (a) $\pm$ 1.7 (b) .1003

11.23 H_1: The sample is not random. ($\mu \neq 8$), H_0: The sample is random. ($\mu = 8$), $z = 2.35$, Reject H_0

11.25 51

Chapter 12

12.1 (a) 868.83 (b) 89,412.16 (c) 299.02
(d) 5

12.3 (a) 84.1 (b) $s^2 = 127.52$, $s = 11.29$, d.f. = 9 (c) 117.33

12.5 (a) $\bar{X} = 46.42$, $s^2 = 1.61$, $s = 1.27$
(b) $\bar{X}$, s^2, and s are unbiased estimates for μ, σ^2, and σ, respectively.
(c) 63

12.7 $22.92 < \mu < 24.58$

12.9 $4.53 < \mu < 5.53$

12.11 (a) 95: $21.8825 < \mu < 22.6175$
99: $21.7662 < \mu < 22.7338$
(b) 98: $21.8131 < \mu < 22.6869$
(c) It gets bigger.

12.13 4

12.15 (a) $\bar{X} = 1{,}212.13$, $s^2 = 4{,}757$, $s = 68.97$
(b) 7
(c) Yes, because the number of degrees of freedom would be increased from 7 to 8.

12.17 (a) \$2,329
(b) s^2 = \$1,675,588
s = \$1,294.45
(c) 34

12.19 $7.8776 < \mu < 8.1224$

12.21 (a) $\$1{,}491.46 < \mu < \$1{,}708.54$
(b) $\$1{,}460.44 < \mu < \$1{,}739.56$
(c) larger $\sigma \rightarrow$ larger intervals

12.23 9

12.25 $4.825 < \mu < 5.1750$

Chapter 13

13.1 $H_1: \mu \neq 100$, $H_0: \mu = 100$, $t = -8.75$, Reject H_0

13.3 $H_1: \mu > 1.50$
$H_0: \mu = 1.50$, $t = 1.67$, Do not reject H_0

13.5 $H_1: \mu_G > \mu_L$
$H_0: \mu_G \leq \mu_L$, $t = 2.30$, Do not reject H_0

13.7 $H_1: \mu_{\bar{d}} > 0$
$H_0: \mu_{\bar{d}} = 0$, $t = 2.30$, Reject H_0

13.9 $H_1: \mu > 495$
$H_0: \mu = 495$, $t = 2.70$, Reject H_0

13.11 $H_1: \mu_M \neq \mu_W$
$H_0: \mu_M = \mu_W$, $t = -.60$, Do not reject H_0

13.13 $H_1: \mu_{\bar{d}} \neq 0$
$H_0: \mu_{\bar{d}} = 0$, $t = 3.50$, Reject H_0

13.15 $H_1: \mu_{\bar{d}} \neq 0$
$H_0: \mu_{\bar{d}} = 0$, $t = -.06$, Do not reject H_0

13.17 $H_1: \mu < 55$
$H_0: \mu = 55$, $t = -1.69$, Do not reject H_0

13.19 $H_1: \mu_p \neq \mu_c$
$H_0: \mu_p = \mu_c$, $t = 2.98$, Reject H_0

13.21 $H_1: \mu_E \neq \mu_N$
$H_0: \mu_E = \mu_N$, $t = 4.05$, Reject H_0

13.23 $H_1: \mu_{\bar{d}} > 0$
$H_0: \mu_{\bar{d}} = 0$, $t = 1.89$, Reject H_0

13.25 $H_1: \mu \neq 6$
$H_0: \mu = 6, \bar{X} = 6.825, t = 1.76$, Do not reject H_0

13.27 $H_1: \mu > 3$
$H_0: \mu \leq 3, t = 2.75$, Reject H_0

13.29 $H_1: \mu_1 < \mu_2$
$H_0: \mu_1 = \mu_2, t = -.67$, Do not reject H_0

Chapter 14

14.1 $H_1: \sigma_N > \sigma_A$
$H_0: \sigma_N = \sigma_A, F = 1.6$, Do not reject H_0

14.3 $H_1: \sigma_T > \sigma_W$
$H_0: \sigma_T = \sigma_W, F = 1.81$, Reject H_0

14.5 H_1: Means not equal, H_0: Means equal, $F = 242.7/42.3 = 5.74$, Reject H_0

14.7 H_1: Means not equal., H_0: Means equal., $F = 14.6/.93 = 15.70$, Reject H_0

14.9 H_1: The fertilizers differ in effectiveness.
$H_0: \mu_1 = \mu_2 = \mu_3 = \mu_4, F = 10.72/5.21 = 2.06$, Do not reject H_0

14.11 $H_1: \sigma_L > \sigma_M$
$H_0: \sigma_L = \sigma_M, F = 1.13$, Do not reject H_0

14.13 (a) H_1: The means of A, B, C, and D are not all equal
$H_0: \mu_A = \mu_B = \mu_C = \mu_D, F = 7.19/.87 = 8.26$, Reject H_0
(b) H_1: The means of A, B, and C are not all equal.
$H_0: \mu_A = \mu_B = \mu_C, F = .2/.54 = .37$, Do not reject H_0
(c) Space observation mean $\neq$ Earth observation mean

14.15 The test for significance tests whether or not s_B^2 is significantly larger than (not different from) s_W^2. This then translates into a statement about the difference of the K means.

14.17 $H_1: \sigma_{II} > \sigma_I$
$H_0: \sigma_{II} = \sigma_I, F = 2.12$, Do not reject H_0

14.19 (a) A distribution of all possible ratios s_1^2/s_2^2. s_1^2 is an unbiased estimate of σ^2, the variance of a known normal distribution. s_1^2 has 10 degrees of freedom since it is based on a sample of size 11. s_2^2 is an unbiased estimate of σ^2, the variance of the same known normal distribution. It has 20 degrees of freedom since it is based on a sample of size 21.
(b) less than one
(c) $N_1 < N_2 \rightarrow P_{50} < 1; N_1 > N_2 \rightarrow P_{50} > 1$.
(d) $F_{10,10}$ has a median of 1. It is not symmetric.

14.21 H_1: The yields differ.
H_0: The yields are the same., $F = 1.54$, Do not reject H_0

14.23 H_1: The mean lengths differ.
$H_0: \mu_1 = \mu_2 = \mu_3 = \mu_4 = \mu_5, F = 7.42$, Reject H_0

14.25 H_1: The stress quotient means are not the same for all groups.
$H_0: \mu_1 = \mu_2 = \mu_3 = \mu_4, F = 331.57/228.76 = 1.45$, Do not reject H_0

14.27 H_1: The three groups have different mean batting averages.
$H_0: \mu_1 = \mu_2 = \mu_3, F = 4116.62/2698.42 = 1.53$, Do not reject H_0

14.29 H_1: The mean salary for presidents differs among the three types of schools.
$H_0: \mu_1 = \mu_2 = \mu_3$
$SS_B = 2.60\overline{22} \times 10^8$
$SS_W = 3.53\overline{33} \times 10^8$
$s_B^2 = .8674 \times 10^8$
$s_W^2 = .5888 \times 10^8$
$F = 1.47, F_{3,6} = 4.76$
Do not reject H_0, Do not accept H_1

Review III

III.1 $H_1: P \neq .15$
$H_0: P = .15, z = -1.11$, Do not reject H_0

III.3 (a) $H_1: P > .15$
(b) $X = 169.13$
170 is best answer to keep the

probability of a type-one error under .05.
(c) 89 or more for $N = 500$, 327 or more for $N = 2{,}000$

III.5 $\bar{X} = 35.19$
$s^2 = 3{,}520.82$
$s = 59.34$

III.7 95: $594.83 < \mu < 709.17$
99. $576.75 < \mu < 727.25$

III.9 84.22; 82.90

III.11 $H_1: \mu < 20$
$H_0: \mu \geq 20$, $t = -1.18$, Do not reject H_0

III.13 $H_1: \mu_{\bar{d}} > 0$
$H_0: \mu_{\bar{d}} = 0$, $t = 3.13$, Reject H_0

III.15 .0869

III.17 $H_1: P \neq .08$
$H_0: P = .08$, $z = -2.71$, Reject H_0 at either .01 or .05

III.19 (a) 6,553 (b) $s^2 = 2{,}686{,}934$, $s = 1{,}639.19$ (c) $5{,}561 < \mu < 7{,}545$

III.21 $H_1: \mu < 25$
$H_0: \mu \geq 25$, $t = 1.23$, Do not reject H_0

III.23 $H_1: \sigma_B^2 > \sigma_A^2$
$H_0: \sigma_B^2 \leq \sigma_A^2$, $F = 2.57$, Reject H_0

III.25 $H_1: \mu > 10$
$H_0: \mu \leq 10$, $t = .65$, Do not reject H_0

III.27 H_0: The four pH mean levels differ.
H_0: The means are equal., $F = 4.08$, Reject H_0

III.29 $H_1: \mu_{\bar{d}} > 0$
$H_0: \mu_{\bar{d}} \leq 0$, $t = 3.23$, Reject H_0

Chapter 15

15.1

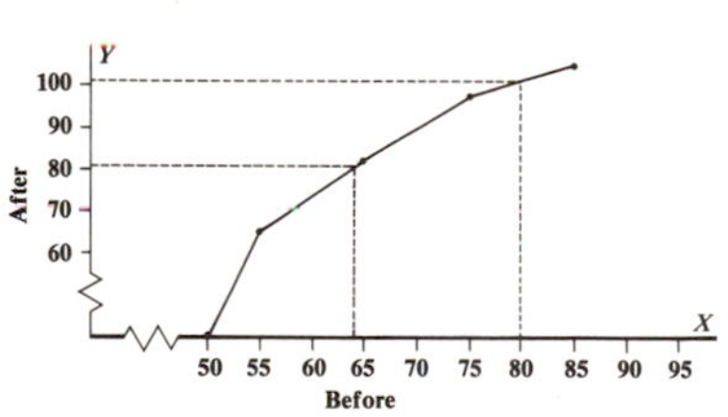

(a) \$75,000
(b) \$97,000
(c) \$82,000
(d) \$102,000
(e) \$81,000

15.3 (a) \$66,000, \$82,000, \$97,000, \$105,-000
(b) \$79,000 − \$66,000 = \$13,000
(c) \$6,000,000

15.5 (a) \$10,845
(b) \$10,845
(c) \$12,500
(d) $(1{,}200)^2 = 1{,}440{,}000$
(e) $(900)^2 = 810{,}000$

15.7 (a) \$6,800, \$8,300, \$12,500, \$13,400, \$15,100
(b) (1) 3.48/4 = .87, \$870,000
(2) 2.73/4 = .6825, \$682,500
(3) 4.37/4 = 1.0925, \$1,092,500
(4) 5.7/4 = 1.425, \$1,425,000
(5) 3.66/4 = .915, \$915,000

15.9 (a) when there is no relationship between the two variables
(b) no. 2
(c)

(d) $\tilde{Y} = \mu_Y$

15.11

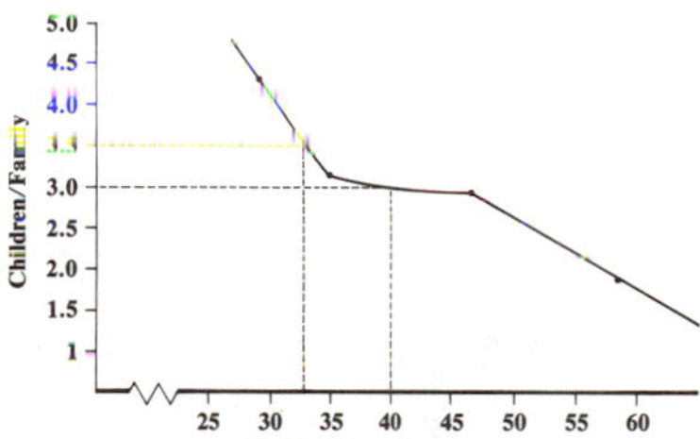

(a) about three
(b) No. Examples: weight at age 3 of a 45-year-old man; number of fish in lake 10 years ago when the dam was 15 feet below spill-

over; percent of MDR of vitamin C that children received in 1950
(c) about 3.5
(d) These data show no "cause–effect" relationship, nor do they deny one.
Other possible causes: decrease in family incomes; inflation; higher mean ages at marriage

15.13

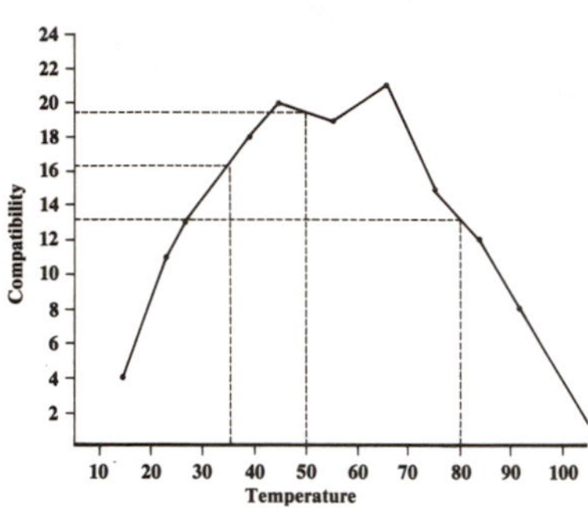

(a) 16.8
(b) 13
(c) 19.5
(d) 19

15.15 (a) −1 (one foot below ideal)
(b) 16
(c) −4
(d) 4
(e) As one increases, the other appears to increase.

15.17 The mean creek level for *all* years with 15 inches of snowfall is 2.9 feet below the ideal level.

15.19 1. something relating death ages of the 150 cases (other than the relationship found in formula), perhaps location since all are from same state
2. error in formula
3. change in relationship between time it was devised and time claims were made

Chapter 16

16.1 .52

16.3 (a)

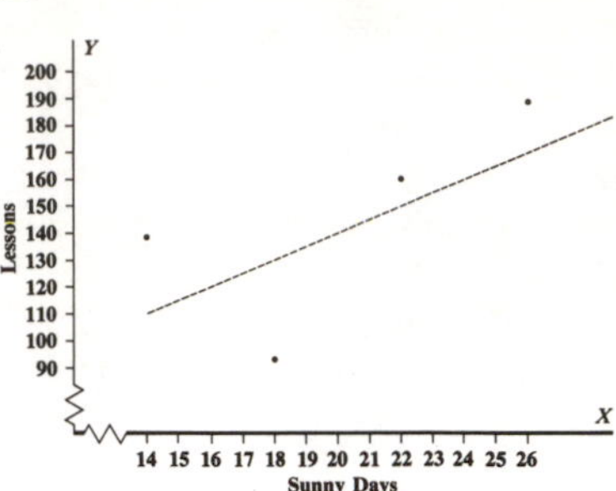

(b) .65
(c) .65
(d)

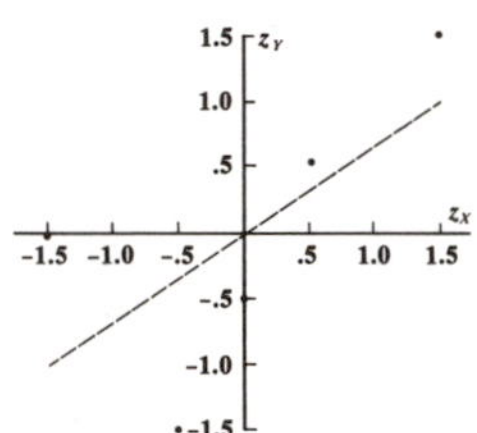

(e) See scattergrams.

16.5 $r = .996$ or 1

16.7 $r = .91$

16.9 $r = .64$, $H_1: r \neq 0$, $H_0: r = 0$, $t = 1.66$, Do not reject H_0

16.11 (a) $r = .04$, $H_1: r \neq 0$, $H_0: r = 0$, $r = .14$
Do not reject H_0
(b) The distributions of chest girth and lung capacity are normal or approximately so.

16.15 .93

16.17 (a) $r = .90$
(b) $r = 5.4258/6 = .90$, $\mu_T = 152.43$, $\sigma_T = 19.30$, $\mu_C = 43.17$, $\sigma_C = 5.40$
(c)

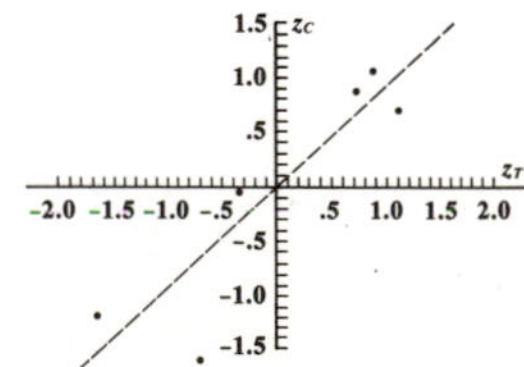

16.19 $r = -.05$

16.21 $r = .90$

16.23 (a) $r = -.94$, $H_1: r < 0$, $H_0: r = 0$, $r = -7.82$
Reject H_0
(b) Correlation implies no cause–effect relationship. Therefore answer is no.

16.25 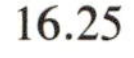

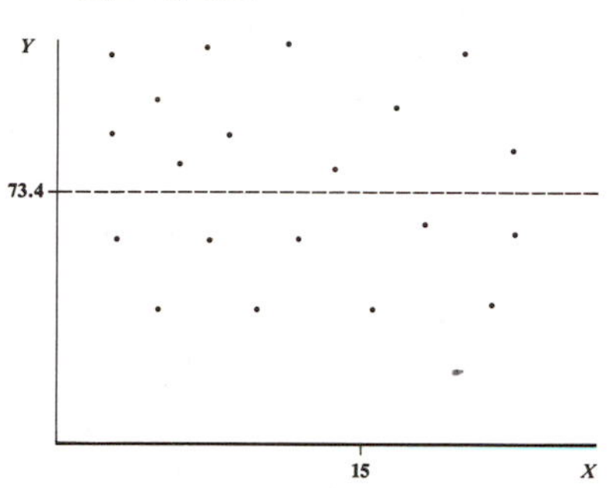

Chapter 17

17.1 (a)

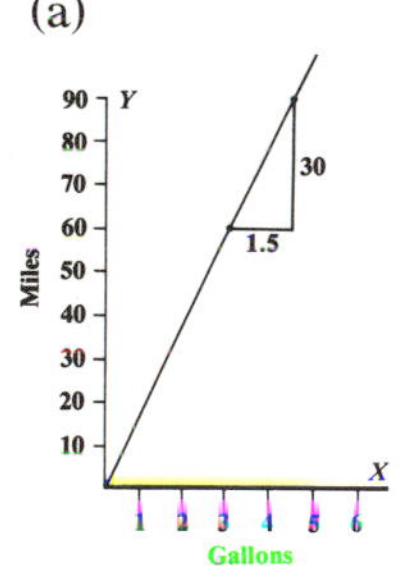

(b) 20
(c) 20 miles per gallon
(d) 1 because the slope is r and the points are colinear
(e) 10

17.3 (a) $\tilde{Y} = 10.4 + .11\,(X - 83.6)$
(b) 12.204%
(c) 10.4%

17.5 (a) z score: slope is r, correlation coefficient
raw score: slope is rate of change; how much Y changes for each unit change in X
The slopes are related by the formula $b = r\,\sigma_Y/\sigma_X$.
(b) It is closer to the dots than any other straight line (vertical distances used). It gives a smaller mean of squared errors than any other line.

17.7 (a) $\tilde{Y} = 12 + .42\,(X - 14)$
(b) $\tilde{Y} = 10.32$
$\tilde{Y} = 9.59$
$\tilde{Y} = 12.42$
$\tilde{Y} = 11.37$
(c) 2.68 days
(d) $r = .76$, $\sigma_{\tilde{Y}}^2 = 2.7$, $\sigma_{Y \cdot \tilde{X}}^2 = 1.14$

17.9 (a) $\tilde{Y} = 66.80$ months
(b) $r = .37$, $\sigma_{\tilde{Y}}^2 = 38.17$
$\sigma_{X \cdot \tilde{Y}}^2 = 32.83$
For Problem 17.4
$\sigma_{Y \cdot \tilde{X}}^2 = .26$
The error variance is less in Problem 17.4.

17.11 $Y = 2.125 + .28\,(X - 5.876)$

17.13 (a) $\tilde{Y} = 64 + 1.28\,(X - 82)$
(b) Slope $= b = 1.28$

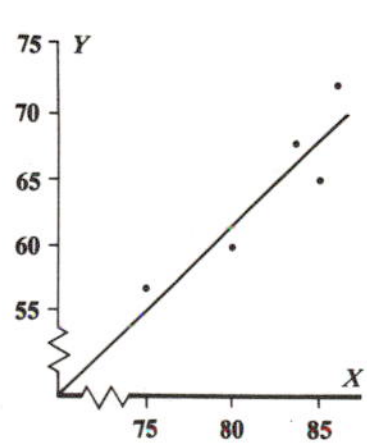

(c) 74.24%
(d) $r^2 = .86$

17.15 $r = .51$

17.17 (a) $\tilde{Y} = 101.33 + 1.76\,(X - 38.53)$
(b) $136.35
(c) $r^2 = .44$
(d) 68.64

17.19 (a) $\tilde{Y} = \mu_Y$
(b) 0
(c) They are the same.

Chapter 18

18.1 H_1: The proposal's support has changed.
H_0: $P_1 = .45$, $P_2 = .35$, $P_3 = .2$, $\chi^2 = 6.04 > 5.99$, Reject H_0

18.3 H_0: Time to recovery and type of treatment are independent. $\chi^2 = 4.59 > 3.84$, Reject H_0

18.5 H_0: Variables are independent., $\chi^2 = 4.74 < 9.49$, Do not reject H_0

18.7 H_0: Decision and education are independent., $\chi^2 = 50.09 > 13.28$, Reject H_0

18.9 (a) H_0: Shift and race are independent., $\chi^2 = 2.1 < 9.49$, Do not reject H_0.
(b) Yes, since $\chi^2 \neq 0$

18.11 H_0: Distribution is normal., $\chi^2 = 2.22 < 9.49$, Do not reject H_0.

18.13 H_0: Variables are independent., $\chi^2 = 11.66 > 5.99$, Reject H_0.

18.15 H_0: Decision and economic level are independent., $\chi^2 = 39.09 > 12.59$, Reject H_0.

18.17 H_0: Variables are independent., $\chi^2 = .61 < 12.59$, Do not reject H_0.

18.19 H_0: Success and birth order are independent., $\chi^2 = 5.66 < 9.49$, Do not reject H_0.

18.21 H_0: $P_1 = .23$, $P_2 = .22$, $P_3 = .12$, $P_4 = .08$, $P_5 = .35$; $\chi^2 = 16.22 > 13.28$, Reject H_0.

Chapter 19

19.1 H_1: There is a difference in brilliance., $U = 85$, $z = -1.14$, Do not reject H_0

19.3 H_1: Methods differ in effectiveness., $V = 11$, $z = -.98$, Do not reject H_0.

19.5 H_1: Not random., $u = 11$, Do not reject H_0.

19.7 H_1: Variables are correlated., $R = .86$, $t = 6.07$, Reject H_0

19.9 H_1: Not random, $u = 16$, Reject H_0.

19.11 H_1: Program is effective., $z = 2.41$, Reject H_0.

19.13 H_1: Variables are negatively correlated., $R = -.83$, $t = -6.31$, Reject H_0.

19.15 H_1: Weight increases., $z = .87$, Do not reject H_0.

19.17 H_1: New pump is more effective., $z = 2.09$, Reject H_0.

19.19 H_1: Rate lower for joggers., $U = 70.5$, $z = -.41$, Do not reject H_0.

19.21 H_1: Heights increase., $V = 16.5$, $z = -.71$, Do not reject H_0.

19.23 H_1: Changeover successful., $z = 2.59$, Reject H_0

19.25 H_1: Variables are positively correlated., $R = .6$, $t = 2.37$, Do not reject H_0.

Review IV

IV.1 (a)

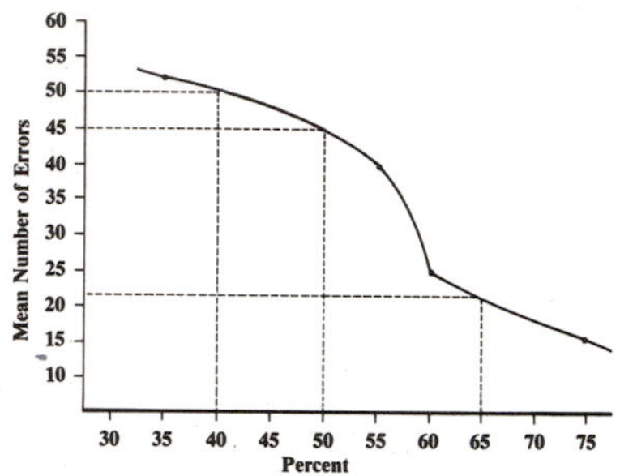

(b) 50, 45, 22
(c) $72/5 = 14.4$, $66/5 = 13.2$, $11/5 = 2.2$
(d) 35, 100

IV.3 H_1: $r > 0$
H_0: $r \leq 0$, $t = 4.28$, Reject H_0

IV.5 (a) $\tilde{Y} = 14.375 - 3.25\,(X - 2.894)$
(b) 4.27
(c) $r^2 = .83$

IV.7 H_0: The variables are independent.
$\chi^2 = 1.0915 < 5.99$
Do not reject H_0

IV.9 H_1: Swimmers perform better.
$V = 21.5$
$z_V = -2.17$
Reject H_0

IV.11 H_0: All inspectors receive the same number of complaints.
$\chi^2 = 11.7242 > 11.07$
Reject H_0

IV.13 H_1: The wastes are harmful to microorganisms.
$z = 1.81$
Do not reject H_0

IV.15 H_1: Winners do not appear at random.
$u = 14$
Do not reject H_0

IV.17 (a) .92 (b) .60

IV.19 (a)

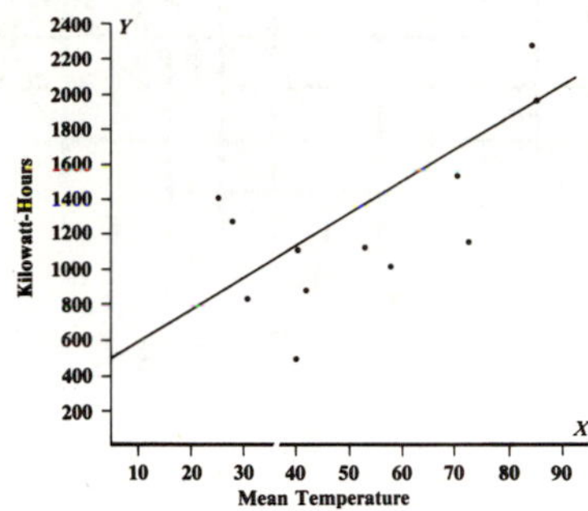

(b) $b = 14.46$

(c) $\tilde{Y} = 1{,}265 + 14.46\ (X - 52.5)$

1,228.85

1,561.43

925.19

IV.21 H_0: No perfume is preferred.
$\chi^2 = 4.5199 < 7.82$
Do not reject H_0

IV.23 H_0: Cheating and heart disease are independent.
$\chi^2 = 4.3364 > 3.84$
Reject H_0

IV.25 H_0: Service utilized is not related to location.
$\chi^2 = 20.661 < 21.03$
Do not reject H_0

IV.27 H_1: The calculators differ in speed.
$z = 2.31$
Reject H_0

IV.29 H_1: Responses fail to appear randomly.
$u = 8$
Reject H_0

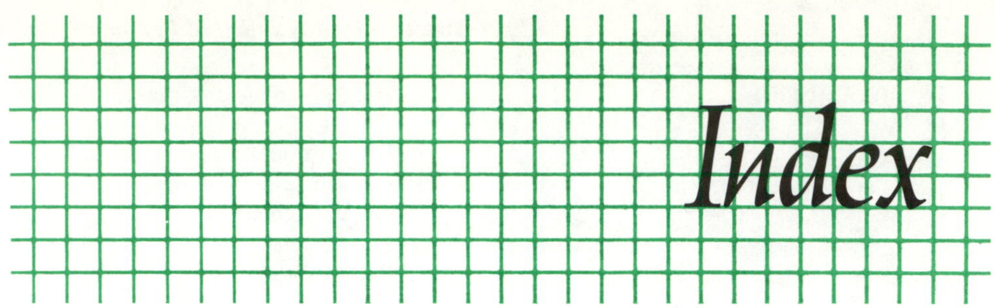

Index

Abuses of statistics, 6–7
Alternative hypothesis, 209, 222
Analysis of variance, 252–305
  comparing sample means, 286–287
  introduction to, 287–293
  null hypothesis and, 298
  t test for difference between means, 259–263
    application of, 263–267
  t test for matched groups, 267–270
  table of, 296
  technique of, 293–298
  theoretical model of F distribution, 279–286
  worksheet, 297
Anthropology, statistics and, 4
Area under a graph, 77
Average deviation, 31
Average IQ, 4
Averages, 6

Balance point
  differences from, 16
  the mean and, 14–15
Bar graphs, 66–67
  applications, 80–81
Batting averages, 1
Best-fitting straight line, 366–367
  prediction and, 374–379
  in scattergram of original data, 370–374
Biased coins, long-run outcomes with, 186–191
Bimodal distribution, 13–14
Binomial distribution, normal approximation to, 173–176
Blind prediction, 317–320

Calculus, 3
Causal relationships, 317
Central Commission for Statistics, 3
Central Limit Theorem, 144–146
  nature and, 146–147
  for sample means, 147–149
Central Limit Theorem for Means, 148
Chi square
  computation of, 395
    for contingency table, 398
    for three-by-three contingency table, 398
  goodness-of-fit and, 401–403
Chi square statistic, 394, 395
Chi square test, 394–397
  for complex problems, 397–401
Coded scores, 101–102
Computational formula for variance, 39–40
Computers, 4
Consequential relations, 317
Contingency table, 391–392
  computation of chi square for, 398
  five-by-four, 400
  three-by-three, 398, 399
  two-by-three, 400
Continuous variables, 65, 387
  rounding off, 67–68
Conversion formula, 358
Correlation, 337–363
  and accuracy of linear prediction, 379–382
  concept of, 337–346
  correlation coefficient, 350–356
    computing, 353–356
    defined, 352
    geometric meaning of, 367–370
    Pearson product-moment, 350–356, 419, 422
    test of significance for, 357–358

for two variables, 369
linear regression and, 364–386
accuracy of linear prediction, 379–382
best-fitting line and prediction, 374–379
best-fitting line in scattergram of original data, 370–374
no correlation between two variables, 343–346
perfect negative, 340–343, 347, 349
perfect positive, 338–340, 347, 349
and proportion of variance, 380–381
as the relationship between two sets of *z* scores, 346–350
Correlation coefficient, 350–356
computing, 353–356
defined, 352
geometric meaning of, 367–370
Pearson product-moment, 350–356, 419, 422
test of significance for, 357–358
for two variables, 369
Correlation theory, 3
Crude mode, 97
Cumulative curve, 95–96
Curve, as approximate graph, 75–76
Curvilinear relationship, 345

Darwin, Charles, 3
Deciles, 94
Decision making, risk and, 181–199
Definable outcomes, 162
Degrees of freedom, 235
concept, 235–237
dependability of estimate and, 236
t distribution and, 256–257
DeMoivre, Abraham, 3
Descriptive statistics, 4
Desk calculators, 4
Deviation
average, 31
mean, of *z* scores, 55–56
sample standard, 240
standard, 32–34
definition of, 32
for grouped data, coded method, 105–107
for grouped data, direct method, 103–104
how changing terms affect, 34–36
of *z* scores, 51–52, 55–56
Dichotomous variable, 173
Discrete distribution, 173
Discrete variables, 64–67, 387
tabulating and graphing values of, 65–67
Distribution
definition of, 4
family of, 281
position of term in, 46–63
applications, 57
notation, 57
percentile ranks, 46–49
percentiles, 46–49
standard scores, 56–57
T scores, 56–57
z scores, 49–53, 55–56
z scores and percentile ranks, 53–55
Distribution of sample means
computing errors using, 191–194
locating a sample mean in, 152–156
properties of, 149–152
Distribution of shots, 132–135
Distribution-free tests, 409

Economics, statistics and, 3
Education, statistics and, 4
Electronic computers, 4
Error variance, 320
Errors
computing, using distribution of sample means, 191–194
type-one, 185
type-two, 186
Estimate
defined, 230
dependability of degrees of freedom and, 236
parameter value and, 318
standard error of, 382
unbiasedness of, 230–235
Estimation, 230–251
degrees of freedom concept, 235–237
formulas for s^2, 241
interval, 241–243, 245
of population mean, 243–247
unbiasedness of, 230–235
of the variance, 237–241

Estimator, 231
  of parameter, 232
Events, 165–167
Expected frequencies, 392–394

F distribution, theoretical model of, 279–286
F test, 252–305
  comparing sample means, 286–287
  t tests for difference between means, 259–263
    application of, 263–267
  t test for matched groups, 267–270
  theoretical model of *F* distribution, 279–286
Fair coins, long-run outcomes with, 186–191
Family of distributions, 281
Fermat, Pierre de, 162
Fisher, Sir Ronald, 3
Five-by-four contingency table, 400
Fraction, 163
Freedom, degrees of, 235
  concept of, 235–237
  dependability of estimates and, 236
  t distribution and, 256–257
Frequency, 392
  definition of, 66
  expected, 392–394
Frequency polygon, 75
  applications, 81
Frequency table, 65–66

Galton, Sir Francis, 3
Gauss, Carl, 3
Goodness-of-fit
  chi square and, 401–403
  origin of term, 401
Graphs, 64–88
  area under, 77
  bar, 66–67
  curves as approximate, 75–76
  data in shaping, 78–80
  discrete variable values, 65–67
  z scores and, 79–80
Grouped data, computing, 89–112
  coded methods of, 99–102, 105–107
  computation of median and other percentiles, 92–93
  cumulative curve, 95–96
  deciles, 94
  direct methods of, 97–99, 103–104
  median from histogram, 89–91
  the mode for, 97
  percentile ranks from, 94–95
  percentiles from histogram, 89–91
  quartiles, 94
  variance for
    coded method, 105–107
    direct method, 103–104
Grouping data, 65–88
  applications, 80–81
  continuous case, 72–75
  continuous variables, 65, 387
  curves, 75–76
  discrete case, 69–71
  discrete variables, 64–67, 387
    tabulating and graphing values of, 65–67
  rounding off values, 67–68
Grouping methods, 68

Heterogeneity, variance and, 36
Histograms
  applications, 80–81
  the median from, 89–91
  percentiles from a, 89–91
History of statistics, 2–4
Homoscedasticity, 322
Horizontal relationship, 345
How to Lie with Statistics (Huff), 7
Huff, Darrell, 7
Hypothesis testing, 183–186, 208–229
  one-tail tests, 222–226, 282, 415
  rejection of null hypothesis, 210–217
  three testing problems, 217–222

Independence of variables, 388–389
Infinite population, 391
Infinitude of terms, 122
Insurance companies, 2
Interval estimation, 241–243, 245
  of population mean, 243–247

Inverse relationship, 317
IQ, average, 4

Landon, Alfred, 7
Least-squares line, 367
Least-squares method, 367
Life insurance rates, 2
Line, slope of, 368–369
Linear prediction, accuracy of, correlation and, 379–382
Linear regression, correlation and, 364–386
  accuracy of linear prediction, 379–382
  best-fitting line and prediction, 374–379
  best-fitting line in scattergram of original data, 370–374
Literary Digest, 7
Long-run outcomes, with fair and biased coins, 186–191

Mann-Whitney *U* test, 410–413, 416
Marginal totals, 392
Matched groups, *t* test for, 267–270
Mean deviation of *z* scores, 55–56
Means, 2, 6, 11–14
  applications of, 21–22
  balance point and, 14–15
  Central Limit Theorem for, 148
  definition of, 11
  difference between, *t* test for, 259–267
  differences from, 15, 16
  for grouped data
    coded method, 99–102
    direct method, 97–99
  how changing terms affect, 17–21
  of normal distribution, 126
  population, interval estimation and, 243–247
  sampling distribution of, 150
  squares of differences from, 16, 17
  two properties of, 14–17
  See also Sample means
Median, the, 6, 11–14
  applications of, 21–22
  computation of, 92–93
  definition of, 12
  from a histogram, 89–91
  of normal distribution, 126
Modal interval, 97
Mode, the, 6, 11–14
  applications of, 21–22
  definition of, 12
  for grouped data, 97
  of normal distribution, 126
Mortality tables, 2

Nature, Central Limit Theorem and, 146–147
Negative relationship, 317
Newton, Sir Isaac, 3
Nightingale, Florence, 3
Nonparametric statistical tests, 409–433
  Mann-Whitney *U* test, 410–413, 416
  rank correlation test, 419–423
  runs test for randomness, 423–427
  sign test, 413–416
  Wilcoxon signed-ranks test, 416–419
Normal curve, normal distribution and, 122–124
Normal distribution, 122–143
  applications, 136
  definition of, 124
  determining scores from percentile ranks, 135–136
  distribution of shot, 132–135
  four properties of, 124–126
  mean, median, and mode of, 126
  normal curve and, 122–124
  probability and, 168–170
  t distribution and, 257
  terms using, 126–132
Null hypothesis, 209–217, 264, 287
  analysis of variance and, 298
  rejection of, 210–217

Objections to statistics, 7–9
One-tail tests, 222–226, 282, 415
Outcomes
  definable, 162
  long-run, with fair and biased coins, 186–191
  probability of, definition, 163

Parameter value, estimate and, 318
Parameters, 222
  estimator of, 232
Pascal, Blaise, 162
Pearson, Karl, 3
Pearson product-moment correlation coefficient, 350–356, 419, 422
Percentile ranks, 46–49
  applications, 57
  definition of, 47
  determining scores from, 135–136
  from grouped data, 94–95
  position of terms in distribution, 46–49
  z scores and, 53–55
Percentiles, 3, 46–49
  applications, 57
  computation of, 92–93
  from a histogram, 89–91
Perfect negative correlation, 340–343, 347, 349
Perfect positive correlation, 338–340, 347, 349
Point estimate, 245
Population, 4–5
  infinite, 391
  interval estimation of, 243–247
Predicted variable, predictor variable and, 320–328
Predicted *Y*, 376 *n*
Prediction, 316–336
  best-fitting line and, 374–379
  blind, 317–320
  defined, 318
  predictor and predicted variables, 320–328
  regression and, 316–336
  regression curve and, 331–332
Prediction equation, 378
Predictor variable, predicted variable and, 320–328
Probability, 162–180, 242
  in everyday language, 164–165
  normal approximation to binomial distribution, 173–176
  normal distribution and, 168–170
  of an outcome, definition, 163
  sample means and, 171–173
  sample spaces and events, 165–167
  sampling experiment and, 167–168
Proportion, 163
Proportion of variance, 380–381
Psychology, statistics and, 3, 4
Pythagoras, 2

Quartiles, 94
Quetelet, Adolph, 3

Random sample, 5–6
  stratified, 6
Randomness, 5
  runs test for, 423–427
Rank correlation test, 419–423
Rank sum test, 410 *n*
Raw data, 4
Raw score scattergram, 370
Regression
  defined, 329
  prediction and, 316–336
Regression curve, prediction curve and, 331–332
Regression equation, 378
Related variables, 389–391
Relatedness, 389–391
Reviews
  I, 113–121
  II, 200–207
  III, 306–315
  IV, 434–445
Risk, decision making and, 181–199
Roosevelt, Franklin D., 7
Rounded-off value, 68
Run, defined, 424
Run of a line, 368
Runs test for randomness, 423–427

S^2, computational formulas for, 241
Sample means
  Central Limit Theorem for, 147–149
  comparing more than two, 286–287
  distribution of
    computing errors using, 191–194
    locating a sample mean in, 152–156
    properties of, 149–152
  probability and, 171–173
Sample spaces, 165–167

Sample standard deviation, 240
Sample variance, 240
Sampling distribution of means, 150
Sampling experiment, probability and, 167–168
Sampling statistics, 4, 5
Scattergrams, 325, 330, 338, 341, 342, 343, 352, 364, 365, 374, 375
  of original data, best-fitting line in, 370–374
  raw score, 370
Scores, from percentile ranks, 135–136
Shots, distribution of, 132–135
Sigma (Σ), use of, 22
Sign test, 413–416
Significance level, 210–217
Skepticism, 7–9
Slope of a line, 368–369
Sociology, statistics and, 4
Square of the sum, 36–39
Standard deviation, 32–24
  definition of, 32
  for grouped data
    coded method, 105–107
    direct method, 103–104
  how changing terms affect, 34–36
  z scores and, 51–52, 55–56
Standard error of estimate, 382
Standard scores, 56–57
  applications, 57
Statistical inference, 5
Statistical literacy, 4
Statistical methods, description of, 1–2
Statistical surveys, 2, 4
Statistical tests, nonparametric, 409–433
  Mann-Whitney *U* test, 410–413, 416
  rank correlation test, 419–423
  runs test for randomness, 423–427
  sign test, 413–416
  Wilcoxon signed-ranks test, 416–419
Stratified random sample, 6
Studies in the History of Statistical Method (Walker), 3 *n*
Sum of the squares, 36–39

T curve, 257
T distribution
  degrees of freedom and, 256–257
  normal distribution and, 257
T ratio, 259
T scores, 56–57
  applications, 57
T test
  for difference between means, 259–263
    application of, 263–267
  for matched groups, 267–270
Terms, infinitude of, 122
Theoretical distribution, 122
Three-by-three contingency table, 398
  computation of chi square for, 398
  degree of freedom for, 399
Three-hypothesis-testing problems, 217–222
Tilde, 376 *n*
Transformation, 358
Two variables
  correlation coefficient for, 369
  no correlation between, 343–346
Two-by-three contingency table, 400
Two-tail tests, 224
Type-one error, 185
Type-two error, 186

U scores, 101–102
Unbiased estimate of the variance, 233
Unbiasedness of estimates, 230–235
Unrelated variables, 389

Value, rounded-off, 68
Variables
  continuous, 65, 387
    rounding off, 67–68
  dichotomous, 173
  discrete, 64–67, 387
  independence of, 388
  independent, 388–389
  related, 389–391
  unrelated, 389
  X, 323
  Y, 323
Variability, 28–29
  notation, 36–39
Variance, 30–32
  analysis of, 252–305
    comparing sample means, 286–287
    introduction to, 287–293

null hypothesis and, 298
t test for difference between means, 259–267
t test for matched groups, 267–270
table of, 296
technique of, 293–298
theoretical model of *F* distribution, 279–286
worksheet, 297
applications of, 36
computational formula for, 39–40
error, 320
estimating, 237–241
for grouped data
coded method, 105–107
direct method, 103–104
heterogeneity and, 36
how changing terms affect, 34–36
proportion of, correlation and, 380–381
sample, 240
unbiased estimate of, 233

Walker, Helen M., 3 *n*
Wechsler Adult Intelligence Scales, 65, 130
Wilcoxon signed-ranks test, 416–419
Wilcoxon two sample test, 410 *n*

X variable, 323

4. t score for difference of a matched pair.

$$t = \frac{\Sigma\, d}{\sqrt{\dfrac{N\, \Sigma\, d^2 - (\Sigma\, d)^2}{N - 1}}} \tag{13.7}$$

I. Value of F for the F distribution

$$F = \frac{s_1^2}{s_2^2} \tag{14.1}$$

J. Analysis of Variance

1. Variance between groups.

$$s_B^2 = N \left[\frac{\Sigma\, \bar{X}^2 - \dfrac{(\Sigma\, \bar{X})^2}{K}}{K - 1} \right] \tag{14.5}$$

2. Variance within the groups.

$$s_W^2 = \frac{(N_1 - 1)s_1^2 + (N_2 - 1)s_2^2 + \cdots + (N_K - 1)s_K^2}{N_T - K} \tag{14.6}$$

3. Computations for analysis of variance.

$$G = \frac{(\Sigma\, X_1)^2}{N_1} + \frac{(\Sigma\, X_2)^2}{N_2} + \cdots + \frac{(\Sigma\, X_K)^2}{N_K} \tag{14.7}$$

$$SS_B = G - \frac{(\Sigma\, X_T)^2}{N_T} \tag{14.8}$$

$$SS_W = \Sigma\, X_T^2 - G \tag{14.9}$$

$$s_B^2 = \frac{SS_B}{K - 1} \tag{14.10}$$

$$s_W^2 = \frac{SS_W}{N_T - K} \tag{14.11}$$

K. Computation of the Correlation Coefficient

1. From z scores.

$$r = \frac{\Sigma\, z_X z_Y}{N}$$

2. From original data.

$$r = \frac{N\, \Sigma\, XY - (\Sigma\, X)(\Sigma\, Y)}{\sqrt{N\, \Sigma\, X^2 - (\Sigma\, X)^2}\sqrt{N\, \Sigma\, Y^2 - (\Sigma\, Y)^2}} \tag{16.1}$$

3. Significance of correlation coefficient.

$$t = \frac{r\sqrt{N - 2}}{\sqrt{1 - r^2}} \tag{16.3}$$

L. Prediction

1. Slope of raw score scattergram (Y vs. X) from r.

$$b = r\frac{\sigma_Y}{\sigma_X} \tag{17.1}$$